VOLUME ONE HUNDRED AND SEVENTY ONE

PROGRESS IN MOLECULAR BIOLOGY AND TRANSLATIONAL SCIENCE

The Microbiome in Health and Disease

VOLUME ONE HUNDRED AND SEVENTY ONE

PROGRESS IN MOLECULAR BIOLOGY AND TRANSLATIONAL SCIENCE

The Microbiome in Health and Disease

Edited by

JUN SUN

Department of Medicine, Department of Microbiology/Immunology, UIC Cancer Center, College of Medicine, University of Illinois at Chicago, Chicago, IL, United States

ELSEVIER

ACADEMIC PRESS

An imprint of Elsevier

Academic Press is an imprint of Elsevier
50 Hampshire Street, 5th Floor, Cambridge, MA 02139, United States
525 B Street, Suite 1650, San Diego, CA 92101, United States
The Boulevard, Langford Lane, Kidlington, Oxford OX5 1GB, United Kingdom
125 London Wall, London EC2Y 5AS, United Kingdom

First edition 2020

ISBN: 978-0-12-820000-1
ISSN: 1877-1173

For information on all Academic Press publications
visit our website at https://www.elsevier.com/books-and-journals

Publisher: Zoe Kruze
Acquisitions Editor: Ashlie M. Jackman
Editorial Project Manager: Leticia M. Lima
Production Project Manager: James Selvam
Cover Designer: Matthew Limbert

Typeset by SPi Global, India

Contents

Contributors

Rafael Ballan
Department of Pharmaceutical and Biochemical Technology, School of Pharmaceutical Sciences; Food Research Center, University of São Paulo, São Paulo, SP, Brazil

Carolina Battistini
Department of Pharmaceutical and Biochemical Technology, School of Pharmaceutical Sciences; Food Research Center, University of São Paulo, São Paulo, SP, Brazil

Stephen L. Chan
State Key Laboratory of Translational Oncology, Department of Clinical Oncology, Sir YK Pao Centre for Cancer, Institute of Digestive Disease, The Chinese University of Hong Kong; Hand in Hand Cancer Foundation Limited, Hong Kong Special Administrative Region, China

Katerina Claud
Department of Pediatrics/Neonatology, University of Chicago, Chicago, IL, United States

Wendy J. Dahl
Department of Food Science and Human Nutrition, University of Florida, Gainesville, FL, United States

Yukuang Guo
Department of Pharmaceutical Sciences, Center for Biomolecular Sciences, College of Pharmacy, University of Illinois at Chicago, Chicago, IL, United States

Emma E. Hamilton-Williams
The University of Queensland Diamantina Institute, The University of Queensland, Woolloongabba, QLD, Australia

Casey Hensley
Department of Biomedical Sciences and Pathobiology, Virginia-Maryland College of Veterinary Medicine, Virginia Polytechnic Institute and State University; Integrated Life Science Building, Blacksburg, VA, United States

Hyunyoung Jeong
Department of Pharmaceutical Sciences, Center for Biomolecular Sciences, College of Pharmacy, University of Illinois at Chicago, Chicago, IL, United States

Takahiro Kageyama
Department of Microbiology and Immunology, University of Illinois at Chicago, College of Medicine, Chicago, IL, United States

Ikuko Kato
Department of Oncology; Department of Pathology, Wayne State University School of Medicine, Detroit, MI, United States

Reina Kurakake
Department of Microbiology and Immunology, University of Illinois at Chicago, College of Medicine, Chicago, IL, United States

Jason M. Lambert
Department of Food Science and Human Nutrition, University of Florida, Gainesville, FL, United States

Hyunwoo Lee
Department of Pharmaceutical Sciences, Center for Biomolecular Sciences, College of Pharmacy, University of Illinois at Chicago, Chicago, IL, United States

Hassan M. Mahsoub
Department of Biomedical Sciences and Pathobiology, Virginia-Maryland College of Veterinary Medicine, Virginia Polytechnic Institute and State University; Integrated Life Science Building, Blacksburg, VA, United States

Takashi Matsuo
Department of Microbiology and Immunology, University of Illinois at Chicago, College of Medicine, Chicago, IL, United States

Carol J. Miller
Department of Civil and Environmental Engineering, Wayne State University, Healthy Urban Waters, Detroit, MI, United States

Emilee Parsons
Department of Pediatrics/Neonatology, University of Chicago, Chicago, IL, United States

Elaine O. Petrof
Department of Medicine, Queen's University, Kingston, ON, Canada

Ashwin K. Ramesh
Department of Biomedical Sciences and Pathobiology, Virginia-Maryland College of Veterinary Medicine, Virginia Polytechnic Institute and State University; Integrated Life Science Building, Blacksburg, VA, United States

Rongrong Ren
Department of Gastroenterology and Hepatology, Chinese PLA General Hospital, Beijing, China

Daniela Rivero Mendoza
Department of Food Science and Human Nutrition, University of Florida, Gainesville, FL, United States

Susana Marta Isay Saad
Department of Pharmaceutical and Biochemical Technology, School of Pharmaceutical Sciences; Food Research Center, University of São Paulo, São Paulo, SP, Brazil

Teruyuki Sano
Department of Microbiology and Immunology, University of Illinois at Chicago, College of Medicine, Chicago, IL, United States

Matthew Silbergleit
Wayne State University School of Medicine, Detroit, MI, United States

Jun Sun
Department of Medicine, College of Medicine, University of Illinois at Chicago, Chicago, IL, United States

Bree J. Tillett
The University of Queensland Diamantina Institute, The University of Queensland, Woolloongabba, QLD, Australia

Adrian A. Vasquez
Department of Civil and Environmental Engineering, Wayne State University, Healthy Urban Waters, Detroit, MI, United States

Zikai Wang
Department of Gastroenterology and Hepatology, Chinese PLA General Hospital, Beijing, China

Douglas Xavier-Santos
Department of Pharmaceutical and Biochemical Technology, School of Pharmaceutical Sciences; Food Research Center, University of São Paulo, São Paulo, SP, Brazil; Department of Civil, Chemical and Environmental Engineering, University of Genoa, Pole of Chemical Engineering, Genoa, Italy

Yinglin Xia
Department of Medicine, University of Illinois at Chicago, Chicago, IL, United States

Yunsheng Yang
Department of Gastroenterology and Hepatology, Chinese PLA General Hospital, Beijing, China

Lijuan Yuan
Department of Biomedical Sciences and Pathobiology, Virginia-Maryland College of Veterinary Medicine, Virginia Polytechnic Institute and State University; Integrated Life Science Building, Blacksburg, VA, United States

Peng Zhou
Department of Biomedical Sciences and Pathobiology, Virginia-Maryland College of Veterinary Medicine, Virginia Polytechnic Institute and State University; Integrated Life Science Building, Blacksburg, VA, United States

Preface

Microbiome studies have grown in sophistication over the past 20 years. It is now 2020. When we discuss the microbiome in health and disease in a book, our expectation on the topic is much higher than what we had years ago.

We now know that microbiome is not just bacteria; we now know that the role of microbiome is not just in intestine; and we now know that 16S rRNA sequencing is not sufficient to identify all the microbes. We have uncovered important mechanistic links between the microbiome and aspects of host health. We start to explore microbiome for the potential applications and therapeutic strategies in human diseases.

I started to work on the proposal of this Elsevier book invited by Dr. David Teplow in December 2018. I believe it is a timely topic to be included in the series *Progress in Molecular Biology and Translational Science.* I planned to bring in experts from various aspects of microbiome research filed to discuss: not just bacteria in microbiome, but also virome and fungus, not only intestinal microbiome, but also microbiome in other organs, and not just digestive diseases, but also gut-brain axis and other diseases, not just adult hood, but also infant microbiome and their development, not just mechanisms in diseases, but also analysis tools to help us work on the mechanisms in microbiome studies. In the final version, we are able to cover the research progress as follows:

Dr. Petrof and her team have reviewed the infant microbiome and implications for the development of central nervous system. Dr. Yuan et al. discuss microbiota in viral infection and disease in humans and farm animals. Dr. Jeong's chapter is about gut microbiota and drug metabolism. You will find the comprehensive review of relationship between T cells and microbiota from an immunologist Dr. Sano. Drs. Kato and Sun lead the topic on oral and intestinal bacterial exotoxins in carcinogenesis. Further, Dr. Yang reviews the fungi and bacteria interactions in mucosal microbiome in gastric lesions. An interesting piece from Dr. Hamilton-Williams focused on microbiota derived factors in type 1 diabetes. We will learn the current progress in nutrition and microbiota by Dr. Dahl and probiotics and prebiotics interacting with the microbiota by Dr. Saad's team. "Microbiome and cancer treatment: are we ready to apply in clinics?" Around this question, Dr. Chan's chapter is built around this interesting and practical perspective. Last, but not least, we need suitable bioinformatic and biostatistical tools to analyze microbiome data and identify critical factors. Dr. Xia discusses the current and future correlation and association analysis in microbiome study and how to integrate microbiome data with other multiomics.

You may notice that the microbiome research is getting global. I am so glad to invite the scientists around the world to contribute to the book. Our international experts are from North America, South America, Australia, and Asia. In this changing climate, we should be more collaborative, but not isolated, we should be more tolerant, but not biased. We need to work together to bring our intelligence and discovery to the table for sharing.

I am finalizing this book project during the time of COVID-19 outbreak. "The world is changing before our eyes. We scientists and supporters of science must rise together to meet this unprecedented moment." I would like to quote the statement from Sudip S. Parikh, Chief Executive Officer, AAAS. In the very trying time, we should ensure science is well integrated into decision making by giving leaders the scientific information. On a daily base, indispensable information through research is the key to move on in the modern life. I believe that microbiome research is not just a hot area, but will become an established steady field for us to explore innovative ideas for prevention, diagnosis, and therapy of human diseases. Our common goal is for the general health of human beings.

I would like to thank all authors for their contribution and editors for their hard work, especially during this pandemic. I would like to thank all reviewers who contributed to the peer-review of each chapter. Their comments and suggestions have helped us to improve our manuscripts. I also want to acknowledge the funds from the NIH, DOD, and VA that support my research in microbiome. Finally, I am deeply thankful for my parents, husband, and sons. Their love encourages me to explore the new world.

I hope you enjoy your reading on microbiome and entertain your own microbiome in the meantime.

Jun Sun, PhD, AGA Fellow for APS, Professor
Division of Gastroenterology and Hepatology
Department of Medicine, University of Illinois at Chicago
Chicago, IL, United States

CHAPTER ONE

The infant microbiome and implications for central nervous system development

Emilee Parsons[a], Katerina Claud[a], Elaine O. Petrof[b,*]
[a]Department of Pediatrics/Neonatology, University of Chicago, Chicago, IL, United States
[b]Department of Medicine, Queen's University, Kingston, ON, Canada
*Corresponding author: e-mail address: epetrof1@gmail.com

Contents

Abstract

Neurodevelopmental impairment remains a significant morbidity in former very low birth weight premature infants. There is increasing evidence the microbiome affects neurodevelopment but mechanistic causes are largely unknown. There are many factors which affect the developing microbiome in infants including mode of delivery, feeding, medications, and environmental exposures. The overall impact of these factors may differ between premature and term infants. The microbiome and brain have well recognized bidirectional communication pathways via neural, hormonal, and immunologic mechanisms. Understanding the interplay between these different pathways has been possible with the use of animal models, particularly germ-free mice. The intricate relationship between the microbiome and the brain remains a research priority not only to improve the care, but to also improve the long-term neurodevelopmental outcomes in this vulnerable population.

1. Introduction

Preterm birth rates and resultant morbidity and mortality from complications of prematurity continue to rise despite continued improvement of

Progress in Molecular Biology and Translational Science, Volume 171
ISSN 1877-1173
https://doi.org/10.1016/bs.pmbts.2020.04.007

maternal and neonatal care. Preterm neonates of less than 32 weeks gestation with birth weights under 1500 g are particularly vulnerable to medical complications such as necrotizing enterocolitis (NEC), interventricular hemorrhage, infection, and bronchopulmonary dysplasia. All these aforementioned complications increase the risk of poor neurodevelopmental outcomes in this population such as cerebral palsy, hearing and vision impairment, and learning disabilities. The overall incidence of neurodevelopment impairment in premature infants remains high, with increasing incidence in growth restricted infants and earlier gestational ages. While the causes of these morbidities are multifactorial, there is increasing interest in the role the microbiome may play in their development. In particular, several studies have indicated a role for the microbiome in NEC.[1–3] Furthermore, there is accumulating evidence of a connection between the microbiome and neurodevelopment. Understanding links among microbiome, morbidities, and neurodevelopment may enable interventions to improve preterm infant outcomes. It is well known the microbiome of a preterm infant differs from a term infant in terms of overall diversity and dominant species of bacteria. In addition, since brain and microbiome development occur in parallel in neonates, there are potential therapeutic target windows of time for possible protective strategies for these processes. The microbiome has been linked to other neurologic conditions such as autism, schizophrenia, and Alzheimer's disease.[4–6] However, the impact of the microbiome on the developing brain in preterm infants is largely unknown.

2. Factors influencing the developing microbiome

Colonization and maturation of the infant microbiome is a process which undergoes patterns of succession from birth through the preschool years when it starts to resemble the microbiome of an adult. Within the first year of life, the infant microbiome contains more oxygen-tolerant and flagellated taxa as well as a higher proportion of Gram-negative species, all of which exhibit rapid growth. Over time, the velocity of taxonomic growth and turnover slows down and becomes dominated by anaerobic (oxygen intolerant) and Gram-positive species.[7] The "mature" microbiome is much more resilient and resistant to changes in composition when exposed to various factors such as illnesses, environmental exposures, and diet.[8] In contrast, during the critical period of microbiome development in the first year of life, it is very susceptible to change due to environmental factors which potentially may alter its mature composition.[9]

Immediately after birth the infant undergoes rapid colonization of the microbiome following exposure to maternal vaginal, fecal, and skin microbiota.[10] In addition to maternal microbiome composition, one of the first major influences in the developing microbiome is mode of delivery. Numerous studies have shown a difference in bacterial composition of fecal samples in healthy term infants born vaginally versus via cesarean section delivery.[11] There is emerging evidence of long-term implications of mode of delivery, such as higher rates of asthma and allergic rhinitis in school age children born via cesarean delivery. This suggests that immunologic development may be affected by changes in bacterial composition of the gut.[12]

Another factor influencing microbiome development is the diet of the infant. There is mounting evidence that breastfed infants and formula fed infants have differences in intestinal microbes. Breastfed infants have higher amounts of *Lactobacillus* and *Bifidobacteria* species as well as overall higher diversity of bacteria, likely due to a combination of microbes within the breast milk and exposure to maternal skin microbes while nursing.[13] Human breastmilk also contains antimicrobial factors such as lactoferrin and secretory IgA,[14] which serve to further shape and select the microbial members of the developing infant microbiome. It is hypothesized that the maternal diet while breastfeeding can also impact the developing microbiome; however, there is limited data on this particular topic.[15] As the infant is introduced to solid foods and weaning from human milk or formula, there is a marked shift in microbiome composition and it develops the ability to digest and process complex carbohydrates, plant polysaccharides, as well as artificial substances found in processed foods. In addition, the gut microbiota begins to perform synthesis of vitamins such as vitamin K and B complex vitamins.[16]

Even after weaning, diet can strongly affect development of the microbial community in the infant gut.[17] Severe malnutrition can result in an "immature" microbiota, which in turn can exert effects on neurodevelopment. A prime example of this is environmental enteropathy (EE), a childhood condition of chronic undernutrition encountered in many developing countries which results in stunted growth and neurodevelopmental delay. A study comparing microbiota composition (fecal samples) of healthy children to children with severe malnutiriton from the same geographic area over the first 2 years of life found that the microbiota community development in the malnourished group was also stunted, i.e., was in a more immature state, compared to the chronological age of the infant and to the healthy control infants of the same age.[18] These findings were in keeping with previous murine studies, which used germ-free mice colonized with human fecal microbiota samples from twins discordant for Kwashiorkor.[19] In both the human and animal studies, supplementing the diet resulted in only a transient improvement in weight gain and microbiome maturity, both of which quickly reverted back, once feeding treatment was stopped. Collectively, these results indicate that nutrition in infancy also plays a critical role in proper development of the human microbiome.

Environmental exposures can also affect the development in the microbiome in neonates, and there is no better example of this than the medications used in the neonatal intensive care unit. Premature infants are likely more susceptible to these factors, simply due to their being inherently more ill (and thus more likely to be in the NICU) than healthy term babies. Antibiotics can alter microbial patterns in infants, particularly with exposure to broad spectrum antibiotics and prolonged courses.[20] One study involving preterm infants exposed to broad spectrum antibiotics showed decreased species richness with proliferation of pathogenic bacteria including *Klebsiella* spp. and *E. coli*.[21] The effects of the changes in microbial composition from antibiotic exposure may persist well past the neonatal period. A variety of studies related to antibiotic exposure in infancy have shown an association with childhood obesity as well as allergies and asthma, suggesting a change in metabolism as well as in the developing immune system.[22] While the effects of antimicrobials on perturbing the intestinal microbiota are well recognized, the effect of other commonly used drugs and their contribution to intestinal dysbiosis is less well described. H2 blockers have been associated with decreased diversity, increased Enterobacteriacea, decreased firmicutes, and increased incidence of NEC.[23,24] Additionally, proton pump inhibitors (PPIs) are some of the most widely prescribed drugs globally[25] and are also used in the NICU. Although the effects of PPIs on the microbiota of neonates have not been studied in detail, in adults approximately 20% of the PPI dose is secreted in stool[26,27] and PPI use in adults is associated with reduced GI microbial diversity.[28–30] Furthermore, when tested against a reference panel of 40 (mostly commensal) human gut bacteria, a large screen of over 1000 non-antibiotic pharmaceuticals found that 27% of these drugs impaired bacterial growth.[31] Therefore, it is likely that more data will emerge on previously unappreciated drug exposures that have an impact the developing infant microbiome.

3. Difference in composition and development of the preterm and term infant microbiome

Another key factor influencing the developing microbiome is the gestational age of the infant at birth.[32] Full term is defined as 37 weeks or greater gestational age at birth while a preterm infant is less than 37 weeks gestational age at birth. Studies comparing fecal samples of term infants to preterm infants have demonstrated that preterm infants have an overall lower diversity of bacteria and a difference in the dominant species of bacteria. Preterm infants have higher amounts of *Enterococcus, Escherichia,* and *Klebsiella* species which can potentially be pathogenic in these infants and lower amounts of *Lactobacillus, Bifidobacteria,* and *Bacteroides*.[33] These bacterial taxa play an important role, both in paving the way for new species to colonize and in training the immune system to appropriately recognize pathogens. *Bifidobacteria*, among the first colonizers of the term infant gut, seem to play a critical role as "pioneers" to promote the growth of succession species. Through a process known as metabolic cross-feeding, other species utilize metabolites produced by *Bifidobacteria* as food to establish themselves in the gut.[34] These newly-established anaerobes (e.g., *Roseburia* spp.) in turn produce metabolites important for gut health such as butyrate, a short-chain fatty acid with important gut trophic effects.[35] *Bifidobacteria* species also produce lactate and large quantities of acetate, which help to inhibit pathogen growth and shape bacterial community formation by decreasing the gut lumenal pH.[36] *Bifidobacteria* also enhance epithelial barrier function,[37] and likely promote training and early development of the infant immune system, as suggested by their ability to improve immunologic T cell responses to vaccines.[38] *Bacteroides* species are also early colonizers but, in contrast to *Bifidobacteria*, they are able to utilize a larger repertoire of substrates (in addition to human milk oligosaccharides) as nutrients,[39] thus allowing them to easily adapt and thrive in the changing lumenal conditions of the growing neonate. *Bacteroides* species likewise appear to play an important role in early development of the infant immune system, and play a pivotal role in T cell and Treg development.[40]

Additionally, the preterm microbiome colonizes slower and can take longer to reach a mature state. The earlier the gestational age, the longer the microbiome takes to reach this mature state.[41] There are many factors explaining the differences between term and preterm microbial patterns and speed of maturation, the majority of which are environmental. A higher proportion of preterm infants are born via cesarean delivery and many of these infants are kept in a relatively sterile environment.[11] Many preterm infants cannot be held by their mothers or have exposure to maternal skin microbes due to clinical factors requiring intervention, such as respiratory distress, hemodynamic instability, and the presence of central umbilical catheters. In addition, many preterm infants have a delay in enteral feeding and take a longer time to get to full feeds. Due to a high metabolic demand, preterm infants require fortification of breast milk and preterm formula to meet their nutritional needs and to promote healthy growth. Insufficient supply of maternal breast milk is also a common occurrence in the NICU which can necessitate donor breast milk or formula as the dominant food source for these infants.[42] Since preterm infants are at high risk for infections, many receive prophylactic antibiotic therapy, sometimes multiple times throughout their NICU admission. There is also evidence preterm neonates exhibit differences in microbial patterns depending on the specific NICU to which they are admitted.[43] In a multicenter study, preterm infants were noted to have significant differences in microbial patterns, plasmid content, and overall diversity between different centers.[44]

To summarize, preterm infants have differences in initial composition and subsequent development of the microbiome from term infants. Understanding these differences and the impact of environmental factors, medications, and other exposures can offer insight in neuroprotective measures for these preterm infants.

4. Microbiome-gut-brain axis

A preterm infant undergoes rapid and critical development of the microbiome and the brain in parallel in the postnatal period. There are well recognized and complex bidirectional communication systems between the gut and brain including neural, endocrine, and immune pathways.[45]

The neural pathway allows communication between the gut and brain through the vagus nerve. The vagus nerve, also known as the tenth cranial nerve, is a large component of the parasympathetic nervous system and exerts control on multiple organs, including the gut.[46] An important function of the vagus nerve in the gut is to maintain consistent motility by regulating peristalsis. However, it can also send signals from the brain to promote an anti-inflammatory environment via the cholinergic anti-inflammatory pathway.[47] This efferent pathway of the vagus nerve, in response to inflammatory cytokines, releases acetylcholine which binds to receptors on cytokine secreting cell membranes which prevents further release of pro-inflammatory cytokines.[48] In addition, the afferent fibers of the vagus nerve in the gut can transmit information in response to inflammation and changes in bacterial composition, which may manifest as symptoms in the brain like anxiety.[49]

The endocrine pathway primarily involves the hypothalamic-pituitary-adrenal (HPA) axis which is central to the body's stress response and production of the hormone, cortisol. Acute and chronic stress can influence the environment within the gut by decreasing the diversity of bacteria and creating a pro-inflammatory state.[50] A premature infant's gut is inherently more "leaky," due to a still-developing and immature intestinal epithelial barrier; a proinflammatory state can have dangerous consequences as the relative immaturity of the gut and increased permeability can cause leakage of bacteria or microbiome produced metabolites into the systemic circulation. This can directly affect the brain as the blood-brain-barrier in premature infants is also quite immature and much more susceptible to inflammation.[51]

5. Animal models

Utilizing animal models, specifically germ-free mice, has shed light into how microorganisms affect physiology and various diseases, and helped determine the mechanisms through which the microbiome affects a host. Germ free mice have been critical in exploring changes in the microbiome and its relation to neurodevelopment. These animals are raised in a pathogen free environment and have no microorganisms living in or on them. Early studies demonstrated that germ free mice have altered behavior such as an exaggerated hypothalamic-pituitary response to stress and impaired memory.[50] Interestingly, if the germ free mice were colonized with the microbiota from a control mouse early enough in life, their behavior was normalized. Mice that were colonized later in life, after the weaning period, did not have normalization of behavior. This suggests a critical window of the influence of the microbiome on long term neurodevelopment. A report modeling chorioamnionitis by injecting lipopolysaccharide (a cell wall component of Gram-negative bacteria) in a pregnant mouse resulted in a litter demonstrating abnormal neuronal development and neuroinflammation.[52]

Other germ-free mouse studies have shown that the microbiome is required for normal development of the blood-brain-barrier,[53] for behavior related to social cognition,[54,55] and the absence of a gut microbiota is accompanied by neurotransmitter and neurochemical changes in the brain.[56] Taken together, these studies suggest that early modification of the microbiome may alter the long term neurodevelopment of the animal and may provide a potential window for therapeutic intervention.

Another approach with animal studies is to "humanize" the microbiome of the germ free mouse by transfaunating a human fecal sample from a specific disease state into the mouse.[57] One animal model designed to specifically study the developing microbiome involves transfaunating a pregnant mouse to study the subsequent outcomes in the pups. In these experiments a germ free pregnant dam is colonized with a human microbiome community of interest. In one such study, germ free, pregnant mice were transfaunated with fecal samples from twins, one obese and one not. The resultant pups with the obese fecal sample had increased body mass and adipose tissue while the other pups with the non-obese fecal sample had normal body mass.[58] Another model transfaunated pregnant mice with fecal samples from a preterm infant with poor growth and those pups not only had poor growth, but evidence of altered intestinal barrier function. In addition, the pups had both systemic and neuroinflammation as well as decreased numbers of neurons and decreased myelination.[59,60]

Autism spectrum disorder (ASD) is a neurodevelopmental disorder characterized by the presence of impaired social interactions and stereotypic behaviors. While the disorder is diagnosed by these altered behaviors, immune and gastrointestinal dysfunction are common comorbidities. Preterm infants are at increased risk for ASD. Recent studies have established a link between the gut microbiome and autism. A 2019 study which investigated the stool transplantation of microbiota from typically developing individuals and ASD individuals into germ free mice determined that colonization with an ASD microbiota can induce autistic behaviors in the mice and alter the splicing of relevant ASD genes in the brain.[4] The ASD colonized mice were found to be lacking in the microbiome associated metabolites taurine and 5AV, both $GABA_A$ receptor agonists related to neural receptors. The display of autistic behaviors such as impaired social interaction in the mice suggests that the gut microbiome is a causal factor of ASD behaviors.

Altogether, these animal studies demonstrate that alterations in the microbiome can have a profound influence on the developing brain and impact cognitive function and social interactions. While these animal studies have limitations on our complete understanding of the developing human microbiome, they continue to shed light on mechanistic causes between dysbiosis of the microbiome and resultant neurologic outcomes.

6. Concluding remarks

With preterm births on the rise, there are increasing numbers of former preterm infants with abnormal neurodevelopment ranging in severity. These children may require much assistance and be a source of great stress for their families and community. It is imperative to understand the various factors affecting neurodevelopment of preterm infants to guide medical care.

The understanding of a preterm microbiome and its interactions with the developing brain remains a research priority. Better insight into mechanistic causes of dysbiosis of the microbiome and resultant neurologic outcomes will allow for potential therapeutic interventions and clinical practice alterations, in order to optimize microbiome succession and preterm outcomes (Fig. 1).

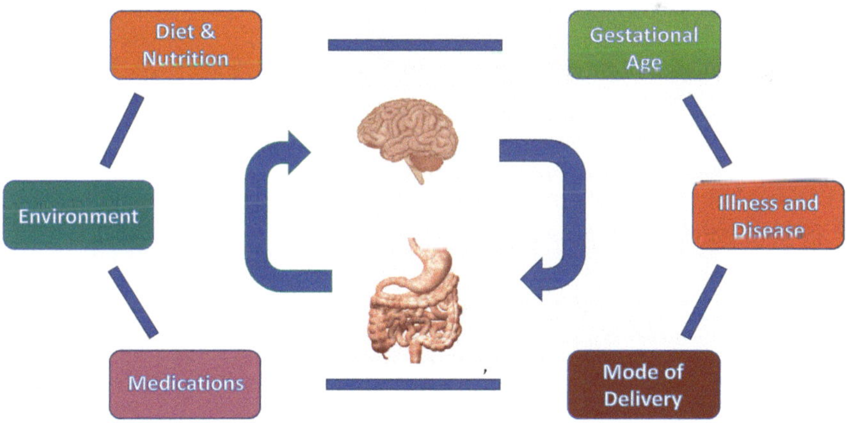

Fig. 1 The bidirectional communication pathways between the developing brain and gut microbiome in infants is affected by many external factors. These factors, including gestational age, mode of delivery, environment, feeding regime, medication exposure, and specific environment, can impact the composition and diversity of the developing microbiome. The long term complications from these effects on the microbiome and subsequent neurodevelopment of the infant remains an important research priority in order to optimize care of premature infants.

References

1. Mai V, Young CM, Ukhanova M, et al. Fecal microbiota in premature infants prior to necrotizing enterocolitis. *PLoS One*. 2011;6:e20647. https://doi.org/10.1371/journal.pone.0020647.
2. Olm MR, Bhattacharya N, Crits-Christoph A, et al. Necrotizing enterocolitis is preceded by increased gut bacterial replication, Klebsiella, and fimbriae-encoding bacteria. *Sci Adv*. 2019;5:eaax5727. https://doi.org/10.1126/sciadv.aax5727.
3. Wang Y, Hoenig JD, Malin KJ, et al. 16S rRNA gene-based analysis of fecal microbiota from preterm infants with and without necrotizing enterocolitis. *ISME J*. 2009;3:944–954. https://doi.org/10.1038/ismej.2009.37.
4. Sharon G, Cruz NJ, Kang D-W, et al. Human gut microbiota from autism spectrum disorder promote behavioral symptoms in mice. *Cell*. 2019;177:1600–1618.e17. https://doi.org/10.1016/j.cell.2019.05.004.
5. Wang X, Sun G, Feng T, et al. Sodium oligomannate therapeutically remodels gut microbiota and suppresses gut bacterial amino acids-shaped neuroinflammation to inhibit Alzheimer's disease progression. *Cell Res*. 2019;29:787–803. https://doi.org/10.1038/s41422-019-0216-x.
6. Zheng P, Zeng B, Liu M, et al. The gut microbiome from patients with schizophrenia modulates the glutamate-glutamine-GABA cycle and schizophrenia-relevant behaviors in mice. *Sci Adv*. 2019;5:eaau8317. https://doi.org/10.1126/sciadv.aau8317.
7. Guittar J, Shade A, Litchman E. Trait-based community assembly and succession of the infant gut microbiome. *Nat Commun*. 2019;10:1–11. https://doi.org/10.1038/s41467-019-08377-w.
8. Matamoros S, Gras-Leguen C, Le Vacon F, Potel G, de La Cochetiere M-F. Development of intestinal microbiota in infants and its impact on health. *Trends Microbiol*. 2013;21:167–173. https://doi.org/10.1016/j.tim.2012.12.001.
9. Arboleya S, Sánchez B, Milani C, et al. Intestinal microbiota development in preterm neonates and effect of perinatal antibiotics. *J Pediatr*. 2015;166:538–544. https://doi.org/10.1016/j.jpeds.2014.09.041.
10. Ferretti P, Pasolli E, Tett A, et al. Mother-to-infant microbial transmission from different body sites shapes the developing infant gut microbiome. *Cell Host Microbe*. 2018;24:133–145.e5. https://doi.org/10.1016/j.chom.2018.06.005.
11. Biasucci G, Rubini M, Riboni S, Morelli L, Bessi E, Retetangos C. Mode of delivery affects the bacterial community in the newborn gut. *Early Hum Dev*. 2010;86:13–15. https://doi.org/10.1016/j.earlhumdev.2010.01.004. Proceedings from the 2nd International Conference on: Nutrition of the Preterm Infant: Current Issues.
12. Neu J, Rushing J. Cesarean versus vaginal delivery: long-term infant outcomes and the hygiene hypothesis. *Clin Perinatol*. 2011;38:321–331. https://doi.org/10.1016/j.clp.2011.03.008.
13. Bezirtzoglou E, Tsiotsias A, Welling GW. Microbiota profile in feces of breast- and formula-fed newborns by using fluorescence in situ hybridization (FISH). Anaerobe. 2011;17:478–482. https://doi.org/10.1016/j.anaerobe.2011.03.009.
14. Prentice A, Ewing G, Roberts SB, et al. The nutritional role of breast-Milk IgA and lactoferrin. *Acta Paediatr*. 1987;76:592–598. https://doi.org/10.1111/j.1651-2227.1987.tb10526.x.
15. Gomez-Gallego C, Garcia-Mantrana I, Salminen S, Collado MC. The human milk microbiome and factors influencing its composition and activity. *Semin Fetal Neonatal Med*. 2016;21:400–405. https://doi.org/10.1016/j.siny.2016.05.003.
16. Barko PC, McMichael MA, Swanson KS, Williams DA. The gastrointestinal microbiome: a review. *J Vet Intern Med*. 2018;32:9–25. https://doi.org/10.1111/jvim.14875.

17. Blanton LV, Barratt MJ, Charbonneau MR, Ahmed T, Gordon JI. Childhood undernutrition, the gut microbiota, and microbiota-directed therapeutics. *Science*. 2016;352:1533. https://doi.org/10.1126/science.aad9359.
18. Subramanian S, Huq S, Yatsunenko T, et al. Persistent gut microbiota immaturity in malnourished Bangladeshi children. *Nature*. 2014;510:417–421. https://doi.org/10.1038/nature13421.
19. Smith MI, Yatsunenko T, Manary MJ, et al. Gut microbiomes of Malawian twin pairs discordant for kwashiorkor. *Science*. 2013;339:548–554. https://doi.org/10.1126/science.1229000.
20. Greenwood C, Morrow AL, Lagomarcino AJ, et al. Early empiric antibiotic use in preterm infants is associated with lower bacterial diversity and higher relative abundance of enterobacter. *J Pediatr*. 2014;165:23–29. https://doi.org/10.1016/j.jpeds.2014.01.010.
21. Gasparrini AJ, Crofts TS, Gibson MK, Tarr PI, Warner BB, Dantas G. Antibiotic perturbation of the preterm infant gut microbiome and resistome. *Gut Microbes*. 2016;7:443–449. https://doi.org/10.1080/19490976.2016.1218584.
22. Murphy R, Stewart AW, Braithwaite I, et al. Antibiotic treatment during infancy and increased body mass index in boys: an international cross-sectional study. *Int J Obes (Lond)*. 2014;38:1115–1119. https://doi.org/10.1038/ijo.2013.218.
23. Gupta R, Tran L, Norori J, et al. Histamine-2 receptor blockers Alter the fecal microbiota in premature infants. *J Pediatr Gastroenterol Nutr*. 2013;56:397–400. https://doi.org/10.1097/MPG.0b013e318282a8c2.
24. Scholz SM, Greiner W. An exclusive human milk diet for very low birth weight newborns—A cost-effectiveness and EVPI study for Germany. *PLoS One*. 2019;14:e0226496. https://doi.org/10.1371/journal.pone.0226496.
25. Heidelbaugh JJ, Kim AH, Chang R, Walker PC. Overutilization of proton-pump inhibitors: what the clinician needs to know. *Therap Adv Gastroenterol*. 2012;5:219. https://doi.org/10.1177/1756283X12437358.
26. Andersson T, Olsson R, Regårdh C-G, Skånberg I. Pharmacokinetics of [14C]omeprazole in patients with liver cirrhosis. *Clin Pharmacokinet*. 1993;24:71–78. https://doi.org/10.2165/00003088-199324010-00006.
27. Lawson EB, Wu JC, Baldwin RM, et al. Omeprazole limited sampling strategies to predict area under the concentration–time curve ratios: implications for cytochrome P450 2C19 and 3A phenotyping. *Eur J Clin Pharmacol*. 2012;68:407–413. https://doi.org/10.1007/s00228-011-1136-y.
28. Clooney AG, Bernstein CN, Leslie WD, et al. A comparison of the gut microbiome between long-term users and non-users of proton pump inhibitors. *Aliment Pharmacol Ther*. 2016;43:974–984. https://doi.org/10.1111/apt.13568.
29. Imhann F, Bonder MJ, Vila AV, et al. Proton pump inhibitors affect the gut microbiome. *Gut*. 2016;65:740–748. https://doi.org/10.1136/gutjnl-2015-310376.
30. Jackson MA, Goodrich JK, Maxan M-E, et al. Proton pump inhibitors alter the composition of the gut microbiota. *Gut*. 2016;65:749–756. https://doi.org/10.1136/gutjnl-2015-310861.
31. Maier L, Pruteanu M, Kuhn M, et al. Extensive impact of non-antibiotic drugs on human gut bacteria. *Nature*. 2018;555:623–628. https://doi.org/10.1038/nature25979.
32. Rosa PSL, Warner BB, Zhou Y, et al. Patterned progression of bacterial populations in the premature infant gut. *Proc Natl Acad Sci USA*. 2014;111:12522–12527. https://doi.org/10.1073/pnas.1409497111.
33. Korpela K, Blakstad EW, Moltu SJ, et al. Intestinal microbiota development and gestational age in preterm neonates. *Sci Rep*. 2018;8:1–9. https://doi.org/10.1038/s41598-018-20827-x.

34. Belenguer A, Duncan SH, Calder AG, et al. Two routes of metabolic cross-feeding between Bifidobacterium adolescentis and butyrate-producing anaerobes from the human gut. *Appl Environ Microbiol.* 2006;72:3593–3599. https://doi.org/10.1128/AEM.72.5.3593-3599.2006.
35. Duncan SH, Barcenilla A, Stewart CS, Pryde SE, Flint HJ. Acetate utilization and butyryl coenzyme a (CoA):acetate-CoA transferase in butyrate-producing bacteria from the human large intestine. *Appl Environ Microbiol.* 2002;68:5186–5190. https://doi.org/10.1128/AEM.68.10.5186-5190.2002.
36. Duncan SH, Louis P, Thomson JM, Flint HJ. The role of pH in determining the species composition of the human colonic microbiota. *Environ Microbiol.* 2009;11:2112–2122. https://doi.org/10.1111/j.1462-2920.2009.01931.x.
37. Ewaschuk JB, Diaz H, Meddings L, et al. Secreted bioactive factors from Bifidobacterium infantis enhance epithelial cell barrier function. *Am J Physiol Gastrointest Liver Physiol.* 2008;295:G1025–G1034. https://doi.org/10.1152/ajpgi.90227.2008.
38. Huda MN, Lewis Z, Kalanetra KM, et al. Stool microbiota and vaccine responses of infants. *Pediatrics.* 2014;134:e362–e372. https://doi.org/10.1542/peds.2013-3937.
39. Sonnenburg JL, Xu J, Leip DD, et al. Glycan foraging in vivo by an intestine-adapted bacterial symbiont. *Science.* 2005;307:1955–1959. https://doi.org/10.1126/science.1109051.
40. Round JL, Mazmanian SK. Inducible Foxp3+ regulatory T-cell development by a commensal bacterium of the intestinal microbiota. *Proc Natl Acad Sci USA.* 2010;107:12204–12209. https://doi.org/10.1073/pnas.0909122107.
41. Bäckhed F, Roswall J, Peng Y, et al. Dynamics and stabilization of the human gut microbiome during the first year of life. *Cell Host Microbe.* 2015;17:852. https://doi.org/10.1016/j.chom.2015.05.012.
42. Marques TM, Wall R, Ross RP, Fitzgerald GF, Ryan CA, Stanton C. Programming infant gut microbiota: influence of dietary and environmental factors. *Curr Opin Biotechnol.* 2010;21:149–156. https://doi.org/10.1016/j.copbio.2010.03.020.
43. Hourigan SK, Subramanian P, Hasan NA, et al. Comparison of infant gut and skin microbiota, resistome and virulome between neonatal intensive care unit (NICU) environments. *Front Microbiol.* 2018;9:1361. https://doi.org/10.3389/fmicb.2018.01361.
44. Ravi A, Estensmo ELF, Abée-Lund TML, et al. Association of the gut microbiota mobilome with hospital location and birth weight in preterm infants. *Pediatr Res.* 2017;82:829–838. https://doi.org/10.1038/pr.2017.146.
45. Martin CR, Osadchiy V, Kalani A, Mayer EA. The brain-gut-microbiome axis. *Cell Mol Gastroenterol Hepatol.* 2018;6:133–148. https://doi.org/10.1016/j.jcmgh.2018.04.003.
46. Borre YE, O'Keeffe GW, Clarke G, Stanton C, Dinan TG, Cryan JF. Microbiota and neurodevelopmental windows: implications for brain disorders. *Trends Mol Med.* 2014;20:509–518. https://doi.org/10.1016/j.molmed.2014.05.002.
47. Tracey KJ. Physiology and immunology of the cholinergic antiinflammatory pathway. *J Clin Invest.* 2007;117:289–296. https://doi.org/10.1172/JCI30555.
48. Chatterjee PK, Al-Abed Y, Sherry B, Metz CN. Cholinergic agonists regulate JAK2/STAT3 signaling to suppress endothelial cell activation. *Am J Physiol Cell Physiol.* 2009;297:C1294–C1306. https://doi.org/10.1152/ajpcell.00160.2009.
49. Yang I, Corwin EJ, Brennan PA, Jordan S, Murphy JR, Dunlop A. The infant microbiome: implications for infant health and neurocognitive development. *Nurs Res.* 2016;65:76–88. https://doi.org/10.1097/NNR.0000000000000133.
50. Sudo N, Chida Y, Aiba Y, et al. Postnatal microbial colonization programs the hypothalamic–pituitary–adrenal system for stress response in mice. *J Physiol.* 2004;558:263–275. https://doi.org/10.1113/jphysiol.2004.063388.

51. Konturek PC, Brzozowski T, Konturek SJ. Stress and the gut: pathophysiology, clinical consequences, diagnostic approach and treatment options. *J Physiol Pharmacol.* 2011;62:591–599.
52. Dada T, Rosenzweig JM, Shammary MA, et al. Mouse model of intrauterine inflammation: sex-specific differences in long-term neurologic and immune sequelae. *Brain Behav Immun.* 2014;38:142–150. https://doi.org/10.1016/j.bbi.2014.01.014.
53. Braniste V, Al-Asmakh M, Kowal C, et al. The gut microbiota influences blood-brain barrier permeability in mice. *Sci Transl Med.* 2014;6:263ra158. https://doi.org/10.1126/scitranslmed.3009759.
54. Desbonnet L, Clarke G, Shanahan F, Dinan TG, Cryan JF. Microbiota is essential for social development in the mouse. *Mol Psychiatry.* 2014;19:146–148. https://doi.org/10.1038/mp.2013.65.
55. Heijtz RD, Wang S, Anuar F, et al. Normal gut microbiota modulates brain development and behavior. *Proc Natl Acad Sci USA.* 2011;108:3047–3052. https://doi.org/10.1073/pnas.1010529108.
56. Neufeld KM, Kang N, Bienenstock J, Foster JA. Reduced anxiety-like behavior and central neurochemical change in germ-free mice. *Neurogastroenterol Motil.* 2011;23:255-e119. https://doi.org/10.1111/j.1365-2982.2010.01620.x.
57. Arrieta M-C, Walter J, Finlay BB. Human microbiota-associated mice: a model with challenges. *Cell Host Microbe.* 2016;19:575–578. https://doi.org/10.1016/j.chom.2016.04.014.
58. Ridaura VK, Faith JJ, Rey FE, et al. Gut microbiota from twins discordant for obesity modulate metabolism in mice. *Science.* 2013;341:1241214. https://doi.org/10.1126/science.1241214.
59. Yu Y, Lu L, Sun J, Petrof EO, Claud EC. Preterm infant gut microbiota affects intestinal epithelial development in a humanized microbiome gnotobiotic mouse model. *Am J Physiol Gastrointest Liver Physiol.* 2016;311:G521–G532. https://doi.org/10.1152/ajpgi.00022.2016.
60. Lu J, Lu L, Yu Y, Cluette-Brown J, Martin CR, Claud EC. Effects of intestinal microbiota on brain development in humanized gnotobiotic mice. *Sci Rep.* 2018;8:1–16. https://doi.org/10.1038/s41598-018-23692-w.

CHAPTER TWO

Microbiota in viral infection and disease in humans and farm animals

Lijuan Yuan*, Casey Hensley, Hassan M. Mahsoub, Ashwin K. Ramesh, Peng Zhou

Department of Biomedical Sciences and Pathobiology, Virginia-Maryland College of Veterinary Medicine, Virginia Polytechnic Institute and State University, Blacksburg, VA, United States
Integrated Life Science Building, Blacksburg, VA, United States
*Corresponding author: e-mail address: lyuan@vt.edu

Contents

Progress in Molecular Biology and Translational Science, Volume 171
ISSN 1877-1173
https://doi.org/10.1016/bs.pmbts.2020.04.005

Abstract

The influence of the microbiota on viral infection susceptibility and disease outcome is undisputable although varies among viruses. The purpose of understanding the interactions between microbiota, virus, and host is to identify practical, effective, and safe approaches that target microbiota for the prevention and treatment of viral diseases in humans and animals, as currently there are few effective and reliable antiviral therapies available. The initial step for achieving this goal is to gather clinical evidences, focusing on the viral pathogens—from human and animal studies—that have already been shown to interact with microbiota. The subsequent step is to identify mechanisms, through experimental evidences, to support the development of translational applications that target microbiota. In this chapter, we review evidences of virus infections altering microbiota and of microbiota enhancing or suppressing infectivity, altering host susceptibility to certain viral diseases, and influencing vaccine immunogenicity in humans and farm animals.

1. Introduction

The human and animal body surface and cavities are inhabited by a large number of commensal microorganisms, collectively referred to as the microbiota.[1] This network of microbial communities recently has been revealed to have significant impacts on host health and immunity against infectious diseases.[1] The microbiota has come to the forefront of infection and immunity research, as there are mounting evidence suggesting that microbiota plays an important role in the development of host immune system and immunity.[2] This discovery has provided many research opportunities regarding the role of microbiota in viral diseases, in which virologists can explore how the microbiota suppresses or enhances the infectivity of viruses, the host susceptibility, and how to harness this information for targeted antiviral therapeutics. This new field of research on microbiome-virus-host interaction is made possible by the massive advances in the nucleotide sequencing approaches, metagenomic analysis and data mining, and bioinformatics analytical software.[3,4]

Most available data on the microbiome describe the gut microbiota; however, microbiota is pertinent not only to enteric viral diseases, but also to viral diseases that affect every organ and system in the body. Although the majority of the microbiota resides in the gut, there are large populations that also reside in the mouth, urinary tract and on the skin.[5–7] Most principles learned from the study of gut microbiome apply to microbial habitats throughout the body. The study of microbiota-virus-host interactions in each specific location of the body allows the understanding of localized immunity and susceptibility to invading viruses and how the local microbiota factor into these processes.

In addition to bacteria, developments in sequencing techniques have led to the expansion of knowledge in the viral aspect of the microbiome, deemed the virome. The virome is comprised of all prokaryotic and eukaryotic viruses found in and on humans and animals. The role of the virome in health and disease is poorly understood at this time. Studies of the role of microbiome in viral diseases have focused almost exclusively on bacteria instead of viruses.

In this chapter, we review and discuss how the microbiota (bacteria) enhances or suppresses infectivity and alters host susceptibility to certain viral diseases and also how virus infection alters microbiota in humans and farm animals, which include poultry, swine, and ruminant species. Because of the exceedingly fast-growing literature in this field, we only chose a few major viral diseases as examples for each species (Table 1). We also provide brief background information on virome in general health in each species.

2. Microbiota in virus infection and diseases in humans

2.1 Influenza virus

Influenza is a common viral infection responsible for the seasonal pandemics that sweep the globe each year. Influenza viruses possess a segmented negative sense, single-stranded RNA genome and belong to the family *Orthomyxoviridae*. Influenza virus infection causes acute respiratory inflammation in humans with symptoms such as high fever, body aches, fatigue, and can result in death. Up to one-half million deaths occur annually, with close to five million reported cases.[45] The best defense thus far is the yearly strain-specific vaccine. The efficacy of this vaccine is low due to several factors, including continual antigenic drift (minor mutational changes that occur over time) and annual vaccine mismatches to circulating strains.[45] The low efficacy of the vaccine coupled with the ease of transmission and

Table 1 Microbiota associated with exacerbation or amelioration of viral diseases.

Viral pathogen	Host species	Bacterial Taxa associated with exacerbated disease	Bacterial Taxa associated with reduced disease severity
Influenza virus	Human	*Streptococcus, Neisseria, Alloprevotella, Prevotella* (Lee et al.[2]), *Dolosigranulum, Staphylococcus, Moraxella* (Ding et al.[8])	*Bacteroidetes* (Lee et al.[2]), *Corynebacterium* (Bomar et al.[9]; Ding et al.[8]), *Bifidobacteriaceae* (Trompette et al.[10])
Human immunodeficiency virus (HIV)	Human	*Gardnerella, Mobiluncus* (Fredricks et al.[11]), *Prevotella amnii, Sneathia amnii, Sneathia sanguinegens, Fusobacterium, Aerococcus, Gemella, Mobiluncus mulieris* (Anahtar et al.[12])	*Lactobacillus* (Gosmann et al.[13]), *Lactobacillus crispatus* (Fredricks et al.[11])
Human norovirus (HuNoV)	Human	*Escherichia coli* (Nelson et al.[14])	*Ruminococcaceae, Faecalibacterium* (Rodriguez-Diaz et al.[15])
	Gn pig	*Bacteroides, Lactobacillus* (Lei et al.[16])	–
Human rotavirus (HRV)	Human	Bacteroidetes, *Bacteroides, Prevotella* (Harris et al.[17])	*Streptococcus bovis* (Harris et al.[17]), Clostridium cluster XI, *Proteobacteria, Serratia, Escherichia coli* (Harris et al.[18])
	Gn pig	*Proteobacteria* (Twitchell et al.[19])	*Bacteroidetes* (Twitchell et al.[19]), *Lactobacillus rhamnosus GG, Bifidobacterium lactis Bb12* (Vlasova et al.[20])
Human hepatitis B (HBV)	Human	*Proteobacteria, Enterobacteriaceae, Veillonellaceae, Streptococcaceae* (Wei et al.[21])	*Bacteroidetes* (Wei et al.[21])
Human hepatitis C (HCV)	Human	*Bacteroidetes, Prevotella, Acinetobacter, Veillonella, Phascolarctobacterium* (Aly et al.[22])	*Proteobacteria, Bifidobacterium* (Aly et al.[22]), *Ruminococcus* (Bajaj et al.[23])

Avian influenza virus	Chicken	*Proteobacteria* (Yitbarek et al.[24]), *Enterobacteriaceae, Escherichia coli, Clostridium, Veillonella* (Li et al.[25])	*Vampirovibrio, Pseudoflavonifractor, Ruminococcus, Clostridium cluster XIVb, Isobaculum* (Yitbarek et al.[24]) *Novosphingobium, Sphingomonas, Bradyrhizobium, Bifidobacterium, Lactobacillus, Enterococcus, Streptococcus, Candidatus Arthromitus, Bacteroides, Parabacteroides, SMB53* (Li et al.[25])
	Waterfowl	*Veillonella dispar, Rothia mucilaginosa* (Ganz et al.[26])	*Firmicutes, Proteobacteria, Bacteroidetes, Fusobacteria, Actinobacteria, Tenericutes* (Hird et al.[26])
Newcastle disease virus (NDV)	Chicken	*Sinobacteraceae, Rhodoplanes, Xanthomonadaceae, Cytophagaceae* (Cui et al.[27])	*Proteobacteria, Bacteroidetes, Firmicutes, and Actinobacteria, Serratia, Escherichia, Chitinophagaceae, Saprospiraceae, Lactobacillus, Paenibacillus, Erwinia, Enterococcus faecium, Bifidobacterium bifidium* (Cui et al.[27]), *L. reuteri PIA16* (Gonmei et al.[28])
Infectious bursal disease virus (IBDV)	Chicken	*Escherichia coli, Shigella* (Li et al.[29]), *Campylobacter jejuni* (Li et al.[30]), *Salmonella enteritidis* (Phillips and Opitz[31], Arafat et al.[32])	–
Hemorrhagic enteritis virus (HEV)	Turkey	*Propionibacteriaceae, Clostridiaceae, Campylobacteriaceae* (D'Andreano et al.[33]), *Escherichia coli* (Fitzgerald et al.[34])	*Lactobacillaceae* (D'Andreano et al.[33])
African swine fever (ASFV)	Pig	–	*Plaudibacter, Anaeroplasma, Petrimonas, Moraxella* (Correa-Fiz et al.[35])

Continued

Table 1 Microbiota associated with exacerbation or amelioration of viral diseases.—cont'd

Viral pathogen	Host species	Bacterial Taxa associated with exacerbated disease	Bacterial Taxa associated with reduced disease severity
Porcine reproductive and respiratory syndrome virus (PRRSV)	Pig	*Methanobacteriaceae* spp. (Ober et al.[36])	*Ruminococcaceae,* Streptococcaceae (Ober et al.[36])
Porcine circovirus type 2 (PCV2)	Pig	*Prevotella* spp. (van Sambeek et al.[37]), *Methanobacteriaceae* spp. (Ober et al.[36])	Non-pathogenic *E. coli* (Niederwerder et al.[38]), *Ruminococcaceae, Streptococcaceae* (Ober et al.[36])
Porcine epidemic diarrhea virus (PEDV)	Pig	*Escherichia, Shigella, Enterococcus, Fusobacterium,* and *Veillonella* (Huang et al.[39])	*Rikenellaceae* (RC9 gut group), *Butyricimonas,* and *Alistipes* (Song et al.[40]), *Ruminococcaceae, Porphyromonadaceae* (Huang et al.[39])
Bovine rotavirus (BRV)	Cow	*Proteobacteria* (Singh et al.[41]), *Escherichia coli, Clostridium,* and *Streptococcus* (Margreiter et al.[42]; Jang et al.[43])	*Firmicutes, Bacteroidetes* (Jang et al.[43]), *Lactobacillus gasseri, Lactobacillus amylovorus* (Margreiter et al.[42]), *Gemmiger formicilis, Blautia glucerasei, Blautia coccoides, Blautia obeum* (Jang et al.[43])
Bovine leukemia virus (BLV)	Cow	*Sanguibacteroides* (Uchiyama et al.[44])	*Lachnospiraceae, Veillonellaceae* (Uchiyama et al.[44])
Lumpy skin disease virus (LSDV)	Cow	*Brucella* (de Macedo et al.[185]) and *Rickettsia genera* (Sazmand et al.[186]), *Bordetella* (Vordermeier et al.[187]) and *Neisseria* (Sneath and Barrett[188]), *Escherichia* (Wolf-Jackel et al.[189])	–

severity of disease among the very young, very old, and other immunocompromised individuals has made alternative therapies attractive.[46]

Lee et al. conducted a longitudinal study on the human nose/throat microbiome and the potential influence it may have on the susceptibility of the human host to influenza infection.[2] They demonstrated that the level of host susceptibility to influenza is associated with microbial community makeup within the nose/throat.[2] Nose/throat samples were assigned nasal/oropharyngeal (NOP) community state types (CST) and samples from study participants were assigned to one of five groups based on bacterial taxa present. All five of these groups differed significantly in their relative abundance of 15 different bacterial genera.[2] When looking at the implications of certain taxa within the NOP CSTs on susceptibility to influenza, results showed a negative association between *Bacteroidetes* and NOP CST 4, of which the highest relative abundance was both *Streptococcus* and *Neisseria*, on host susceptibility to influenza infection. A positive correlation was noted between *Alloprevotella* and *Prevotella* and infection susceptibility.[43]

Lee et al. also demonstrated notable differences in the microbiome of people of different ages in regards to concurrent influenza infection. Of the NOP CSTs defined in the study, NOP CST 4, which was associated with a decreased vulnerability to influenza infection, was not as abundant or stable in children and showed a consistent increase in prevalence with increased age of a given participant.[2] This suggests the greater occurrence of influenza among young children can be attributed to immature and underdeveloped microbial networks in the nose/throat.[2]

The microbiome of the nasopharynx may also play a role in severity of disease and the development of secondary bacterial infections related to influenza. Ding et al. showed that both influenza A and B infected individuals had a nasopharynx microbiome containing taxa such as *Dolosigranulum* and *Staphylococcus*.[8] These genera have been correlated with the development of pneumonia, a common sequela of influenza in elderly patients, and *Dolosigranulum* also has been identified as a causative agent in septicemia.[47–49] This study classified nasopharyngeal (NP) swabs from infected and healthy subjects into four different NP types. NP type A was most prevalent in influenza-infected patients, and consisted of the two bacteria mentioned above.[8] NP type B was significantly enriched in the healthy control group and was dominated by *Corynebacterium*, NP type C was dominated by *Moraxella*, and NP type D had a high prevalence of *Staphylococcus*.[8] *Corynebacterium* has been shown to hinder *Streptococcus pneumoniae* growth in vitro and has been demonstrated to be abundant in children lacking

S. pneumoniae nasal colonization.[9] *Corynebacterium* has also been shown to have antimicrobial effects against *S. aureus*.[50] The discoveries made in the study conducted by Ding et al. support the previously mentioned study's findings, with NP type B being dominated by *Corynebacterium* with relatively low levels of *Staphylococcus*.[8] These results reiterate the need for more focused studies on the therapeutic potentials of certain bacteria among microbiota and how they can alter vulnerable patient outcomes.

A recent study using a mouse model of influenza virus infection has shown that fermentable dietary fiber increased survival of influenza-infected mice by dampening airway inflammation through several immune modulating mechanisms.[10] Dietary fiber inulin altered the intestinal microbiota in mice and exhibited butyrogenic properties. In inulin-treated mice, the microbiome structure was significantly improved, and beneficial bacteria that produce short-chain fatty acids (SCFAs), including Bifidobacteriaceae, became dominant. Inulin or SCFAs intake prevented excessive neutrophil influx into the airways by blunting the levels of CXCL1 produced by lung monocytes and macrophages. In addition, inulin and SCFAs increased antiviral immunity through $CD8^+$ T cell activation.[10] Another study in a mouse model of severe respiratory syncytial virus (RSV) reported similar findings that fermentable dietary fiber can reduce virus replication, protect the lungs against inflammation and pathological changes induced by RSV infection through microbiota-derived acetate in mice.[51] Since fermentable dietary fiber play a role in regulating immunity by promoting the growth of SCFA-producing bacteria, taking fermentable dietary fiber and reducing the excessive immune response and lung pathology caused by the viruses are a causal relationship, not just a correlation. To date, there are no human clinical studies on using fermentable dietary fiber to prevent influenza or other respiratory viral infection or to reduce the disease severity. This area of translational research warrants further investigation.

2.2 Human immunodeficiency virus (HIV)

According to the World Health Organization, 1.7 million new HIV cases occurred globally in 2018, and nearly 38 million people currently are living with the infection.[52] This disease typically is spread through sexual contact, and the disruption of cervicovaginal microbiota has been correlated with increased susceptibility to HIV infection.[53] Investigation of the cervicovaginal microbiota's influence on HIV infection susceptibility could lead to the development of therapeutic agents for preventing HIV in especially at-risk communities, such as those in sub-Saharan Africa, where approximately 24 million people are infected, and young women are eight times more likely than men to acquire the virus.[13]

Though the majority of the human microbiota resides in the gut, the cervix and vagina maintain their own commensal microbial communities.[13] A cervicovaginal microbiota dominated by *Lactobacillus,* specifically *L. crispatus*, is associated with many health benefits, including reduced risk of contracting HIV.[52] When cervicovaginal dysbiosis occurs, *Lactobacillus* communities are disrupted and other resident microbes, such as *Gardnerella* or *Mobiluncus*, become the predominant species, resulting in bacterial vaginosis.[11] A study conducted by Fredricks et al. demonstrated increased bacterial diversity in women with bacterial vaginosis, and a relatively "normal" vaginal flora dominated by *L. crispatus* in women without the disease.[11] A HIV study by Gosmann et al. showed that none of the participants who contracted HIV had a cervicovaginal microbiota dominated by *Lactobacillus* prior to infection.[13] They also demonstrated that women who maintained the most diverse microbiome were significantly more likely to contract HIV than those with less diverse communities dominated by L. *crispatus.*[13] These results are consistent with the notion that a *Lactobacillus* dominated cervicovaginal microbiota for the prevention of HIV is critical. Several anaerobic bacteria also were implicated in the increased susceptibility to HIV infection in this study, including *Prevotella* and *Sneathia.*[13] These bacteria were shown to increase inflammation in the genital tract, further decreasing the participants' immune robustness against HIV infection.[13]

Bacterial vaginosis and high diversity/low *L. crispatus* presence in the vagina has been associated with increased susceptibility to HIV infection, partly due to resulting genital inflammation and the vulnerability of cell types that dominate the inflamed genital tract.[12,54] Genital inflammation has been shown to lead to disturbances in the vaginal epithelium, making women especially susceptible to HIV infection.[13,54] Anahtar et al. found that a diverse cervicovaginal microbiota is strongly linked to genital inflammation, including an increase in proinflammatory cytokines.[12] *Fusobacterium, Aerococcus, Sneathia, Gemella, Mobiluncus,* and *Prevotella* were all found to have a significant association with an increase in inflammatory cytokines.[12] These results were confirmed in vitro when human epithelial cells co-cultured with these bacteria, specifically *Prevotella amnii, Mobiluncus mulieris, Sneathia amnii,* and *Sneathia sanguinegens* were shown to have a marked increase in cytokine secretion than those co-cultured with *L. crispatus*, which showed no detectable levels of cytokine secretion.[12] Women with increased proinflammatory cytokine levels attributed to highly diverse cervicovaginal microbiota also had higher levels of CD4 + T cells, one of HIV's main target cells, establishing a strong link between increased bacterial diversity and an increased risk of contracting HIV.[12]

Along with cervicovaginal microbiota diversity and the implications it has on HIV status and pathogenesis, gut microbiota diversity has also been identified as playing a significant role in these areas in regards to HIV. Because HIV is known to replicate in gut lymphoid tissues, it is relevant to discuss the implications the disruption of the gut microbiota and subsequent inflammation may have on HIV pathogenesis and vice versa.[55] HIV has been previously associated with gut dysbiosis, and again represents a potential tool that can be used to mitigate the impact of HIV infection.[56]

In a meta-analysis performed by Tuddenham et al., it was demonstrated that diversity of the microbiome in HIV-1 positive individuals was significantly decreased prior to controlling using demographic categories.[56] When controlling for sexual preferences and gender, this held true for men who have sex with women (MSW) as well as men.[56] Men who have sex with men (MSM) showed no significant association with microbiota diversity and HIV status; however, when controlling for MSM, HIV positive men were more likely to have a decreased diversity than HIV positive women.[56] This study highlights the complicated relationship between many factors involved in HIV infection, including diversity of the microbiota. In future studies, it will be beneficial to investigate the role the other factors play in diversity and how therapeutics can be tailored based on factors like gender and sexual preference.

In a study by Qing et al., researchers demonstrated how diversity of specific short-chain fatty acid (SCFA) producing bacteria can influence HIV progression through intestinal injury.[57] Because SCFAs are known to contribute to immune homeostasis and the expression of antimicrobial peptides by epithelial cells in the gut, they have been implicated in the progression of HIV infection.[57] This study found that certain species in the Firmicutes and Bacteroidetes phyla, who produce various SCFAs such as butyric and valeric acid, were much less abundant in the HIV positive group.[57] Butyric acid has been shown to reduce colonic inflammation in patients with inflammatory bowel disease, and the reduction of bacterial producers of this metabolite may contribute to increased intestinal injury and permeability, ultimately leading to bacterial translocation from the gut resulting in systemic inflammation.[58–62]

Several factors contribute to the enhancement of HIV infection, including cervicovaginal and gut dysbiosis. The resulting inflammatory and immune-related sequelae can have significant impact on the progression and outcome of the disease. The prevention and management of HIV infection in developed countries has improved dramatically since the first cases were reported in the 1980s; however, there are still millions of people in underdeveloped areas who would benefit from microbiota research in regards to HIV prevention.[52] The findings in the above studies show promise for the development of accessible and personalized medicine alternative to help these areas recover from a several decades long HIV epidemic.

2.3 Human norovirus (HuNoV)

2.3.1 HuNoV infection disrupts human microbiome

The majority of the world's acute gastroenteritis cases are caused by HuNoV, an enteric viral pathogen spread primarily via the fecal-oral route.[63,64] HuNoVs are non-enveloped RNA viruses with a positive-sense single-stranded genome in the family *Caliciviridae*.[65]. Due to the ease of transmission, cases of HuNoV tend to be concentrated and typically are more problematic in children, especially in low-income regions.[63] In more developed countries, HuNoV is known to cause outbreaks of gastroenteritis in all age groups.[63] HuNoV is of great concern in public health practice, but due to the lack of reliable and well-established in vitro cell culture systems, the development of therapeutics has been slow.[66]

Certain microbial populations can positively contribute to gut health and microbial balance, and one of the most intriguing benefits of these bacteria is their ability to influence outcomes of infectious diseases.[67] When the gut microbiota is disrupted, the protection potential is diminished and the host becomes susceptible to numerous infections.[14,67] Nelson et al. conducted a study whose results suggest that HuNoV infections are capable of causing significant alterations in human gut microbiota.[14] In this study, 20% of HuNoV-infected subjects showed significant changes in their gut microbiota, including an increase in Proteobacteria populations, especially *Escherichia coli*, and a decreased bacterial diversity.[14] Proteobacteria are associated with gut microbiota disruption, further suggesting the study participants had a disrupted microbiota at the time of sampling.[14] These patients were identified as being at risk for developing postinfection disease, such as irritable bowel syndrome.[68]

2.3.2 Microbiome and secretor status in the immunity to HuNoV infection

Along with establishing the need for a stable gut microbiota as a natural defense against HuNoV infection, other studies have identified a potential link between human genetic signatures, such as histo-blood group antigen (HBGA), specifically secretor status, the gut microbiome, and their combined influence on immunity to HuNoV infection. Secretor status of an individual is determined by the functionality of the *FUT2* gene, which is responsible for producing an enzyme critical to generating glycans present in saliva and mucus.[69] These glycans can serve as an infection initiating binding site of certain HuNoV strains.[69] When both alleles of this gene are altered and nonfunctional, a person is deemed a non-secretor.[69] Non-secretors have been shown to be highly resistant to symptomatic HuNoV infections due to the mutation in the *FUT2* gene, which diminishes the norovirus's ability to bind to receptors in the intestine, thus preventing HuNoV from entering cells and causing infection.[69] Several studies also have shown secretor status to be linked to microbiota composition.[15,70]

In a study by Rodriguez-Diaz et al., human patients were shown to have a correlation between their gut microbiota diversity, secretor status and susceptibility to HuNoV infection.[15] Host anti-HuNoV salivary IgA titers were obtained from the participants and were used as an indicator of previous infection susceptibility.[15] Though no significant differences in phyla composition of the gut (obtained through stool samples) and secretor status were found, significant family-level differences were present between non-secretors and secretors; with significantly higher levels of *Prevotellaceae* and *Paraprevotellaceae* found in non-secretors.[15] The study then involved separating the obvious groupings into operational taxonomic groups (OTUs) and were then analyzed for diversity levels. Sixteen of the more than 5000 groups were found to have significant differences, and consisted of Firmicutes and Bacteroidetes phyla, found in non-secretors. There were also significant differences in richness of the sampled microbiota, though they were not significant in regards to secretors versus non-secretors.[15] Firmicutes, Bacteroidetes, Actinobacteria, and Verrucomicrobia were the most abundant phyla categorized from both secretors and non-secretors. This study also found anti-HuNoV IgA levels were significantly correlated with secretor versus non-secretor status, with secretors maintaining higher levels of the two groups.[15] The viral infectivity and host susceptibility to HuNoV was negatively correlated to the level of *Ruminococcaceae* and *Faecalibacterium* present in host samples. When these insights are combined with the negative correlation between anti-norovirus IgA titers and susceptibility to HuNoV infection, the notion that host microbiome, secretor status and susceptibility are all reliant on each other to some extent is apparent.[15] More thorough research is needed to evaluate other factors that simultaneously play a role in microbiome composition of all secretors and non-secretors and how this influences host immunity.[15] This is especially important for the determination of the best strategies to protect those in low-income areas in which HuNoV is of greatest public health concern, and the microbiome is likely an effective target for future personalized therapies.[63]

2.3.3 Study of microbiota and HuNoV interaction in the gnotobiotic (Gn) pig model

To study the role of microbiota in response to virus infection and vaccines, animal models that are without a preexisting microbiota and can be colonized with well-defined microbiota, (i.e., Gn animals) are an indispensable tool. Gn animals have played a critical role in the understanding of interactions among microbiota, viral pathogen or vaccine antigen, and the host.[71] The neonatal Gn pig model is well suited for the study of HuNoV and microbiome interaction because Gn pig model of HuNoV infection reflects HuNoV biology in terms of supporting the natural oral route of infection and results in diarrhea, transient viremia, and fecal virus shedding.[72,73] Recently, we established a human gut microbiota (HGM) transplanted Gn pig model of HuNoV infection and disease, using a well-characterized stool sample from a healthy infant as HGM transplant and a HuNoV GII.4/2006b strain for virus inoculation.[16] Compared to germ-free Gn pigs, HuNoV inoculation in HGM transplanted Gn pigs resulted in increased HuNoV infection, characterized by significantly higher shedding titers and significantly longer mean duration of virus shedding. HuNoV infection also dramatically altered intestinal microbiota at the phylum and genus levels in HGM transplanted Gn pigs. At the phylum level, Proteobacteria (95.6% versus 56.5%) and Firmicutes (3.6% versus 0.5%) significantly decreased, while Bacteroidetes (0.1% versus 42.9%) significantly increased; at the genus level, *Enterococcus*, *Bifidobacterium*, *Clostridium*, *Ruminococcus*, and *Anaerococcus* significantly decreased, while *Bacteroides* and *Lactobacillus* significantly increased in HGM transplanted pigs infected with HuNoV compared to HGM transplanted pigs without HuNoV infection. In summary, enhanced GII.4 HuNoV infection was observed in the presence of HGM, and significant intestinal microbiota alterations were observed under HuNoV infection in HGM transplanted Gn pigs.[16] The mechanisms underlying the enhancing effect of HGM on HuNoV infection are probed using transcriptome analysis of the ileum tissues of the Gn pigs. One of the most important impact of HGM is the down regulation of interferon (IFN)-λ gene IFN-1L expression. IFN-λ is known to play a critical role in the host defense against enteric viruses, including rotavirus, reovirus[74] and norovirus.[75]

2.4 Human rotavirus (HRV)

2.4.1 Microbiota and immunogenicity of HRV vaccines

Rotaviruses belong to the genus *Rotavirus* within the family *Reoviridae*, which are characterized by a segmented double-stranded RNA (dsRNA) genome. HRV is the leading cause of severe, dehydrating diarrhea among children under 5 years of age worldwide.[76] Annual mortality rates due to RV infections have declined from ~528,000 in 2000 to ~215,000 in 2013[76] owing to the implementation of HRV vaccines RotaTeq® and Rotarix® in the national immunization program in many countries. These two vaccine vaccines are highly efficacious (80–90%) in high income countries, but the efficacies are significantly reduced (40–60%) in low-income countries. Studies indicated that several factors, including microbiota dysbiosis,[77] concurrent uses of poliovirus oral vaccines,[78,79] enterovirus infections,[80] and malnutrition[81] contributed to the observed impaired efficacy of the oral RV vaccines,[79] likely by changing intestinal environments and homeostasis.[82,83] Harris et al.[17] analyzed the serum IgA antibody responses to Rotarix® in 78 Ghanaian infants and their fecal microbiome. Comparing between the 39 responder and non-responder pairs and to the Dutch infants, the overall microbiome composition was significantly different, with the responders more similar to healthy Dutch infants than non-responders. Responses to Rotarix® correlated with an increased abundance of *Streptococcus bovis* and the lack of response is correlated with the high abundance of Bacteroidetes phylum, especially several bacteria related to species from the *Bacteroides* and *Prevotella* genera.[17] Another study was reported for Pakistani infants.[18] IgA antibody responses to Rotarix® correlated with higher abundance of bacteria belonging to *Clostridium* cluster XI and Proteobacteria, including bacteria related to *Serratia* and *Escherichia coli*. Abundance of these Proteobacteria was also significantly higher in Dutch infants when compared to non-responders in Pakistan. These clinical studies showed that the intestinal microbiota composition correlated with seroconversion rate in infants after vaccination and identified microbiome signatures of vaccine responder versus non-responder infants. Identification of key bacterial phylotypes that correlate with vaccine-induced responses could be used as indicator to predict which infants are at risk for vaccine failure and to design an intervention through modulating gut microbiota to improve rotavirus vaccine efficacy.[18,77] A study of microbiome diversity in RataTeq vaccinated versus unvaccinated children in Spain did not find any differences.[84]

2.4.2 Study of microbiota and immunogenicity of HRV vaccine in Gn pig model

To clearly identify the role of microbiota in the immunogenicity of HRV vaccine, a Gn pig model of enteric dysbiosis was developed to evaluate the effects of different HGM on immune responses to HRV vaccination and protective efficacy.[19] Fecal samples collected from infants from Nicaragua were analyzed and classified as healthy (HHGM) or unhealthy (UHGM) based on enteropathy score and seroconversion status after vaccination with RotaTeq®. Newborn Gn pigs were transplanted with the HHGM or UHGM, followed by oral vaccination with the live attenuated Wa HRV and challenged with virulent HRV. Prior to HRV challenge, HHGM in Gn pigs was predominantly composed of bacteria belonging to the phyla Firmicutes, Bacteroidetes, and Actinobacteria, whereas UHGM was composed of Proteobacteria and Firmicutes. When challenged, the severity of illness among vaccinated HHGM pigs was much lower compared to vaccinated UHGM pigs. Analysis of fecal microbiome post-challenge showed a significant increase in the abundance of Bacteroidetes in HHGM transplanted Gn pigs and a significant decrease in the abundance of Firmicutes, Proteobacteria, and Verrucomicrobia abundance. Stronger vaccine-induced immune responses (HRV-specific IgG, IgA and VN antibodies; and CD4 + IFN-γ + and CD8 + IFN-γ + T cells) were facilitated by HHGM, hence conferring a stronger protection against HRV disease among vaccinated HHGM pigs than UHGM pigs.[19] Similar results were also observed in Gn pigs colonized with *Lactobacillus rhamnosus GG* and *Bifidobacterium lactis* Bb12, commensal bacteria commonly found as part of the microbiota of breast-fed infants.[20] Findings in Gn pigs are highly translatable to humans and as suggested by Dr. Harris "Correlate microbiome composition with vaccine effectiveness in appropriate experimental platforms will lead to the identification of safe, vaccine-supporting microbiota targets that are relevant to populations in need of improvement in vaccine-induced immunity."[18]

2.5 Hepatitis viruses

Hepatitis B virus (HBV) and hepatitis C virus (HCV) are important public health issues that can lead to chronic infections and together are the most common causes of liver cirrhosis and hepatocellular carcinoma (HCC). Due to the well-documented crosstalk between the gut and liver, the microbiota and its immunomodulatory effects on HBV and HCV infections have recently become a topic of interest in the research community.[85,85a]

As was demonstrated by the previously discussed studies, certain phyla and families of beneficial bacteria appear to be underrepresented in patients with certain viral diseases. In a study by Wei et al., it was shown that patients with chronic HBV infection that had progressed to cirrhosis were lacking in Bacteroidetes composition in the gut, as well as showing increased Proteobacteria concentration, a phylum known to contain several pathogenic bacteria.[21] There was also marked prevalence of *Enterobacteriaceae*, *Veillonellaceae*, and *Streptococcaceae*, which made up for less than 1% of the gut microbiota sampled from healthy patients.[21]

In two HCV studies by Aly et al.[22] and Bajaj[23] et al., low diversity of gut microbiome composition was associated with inflammation and disease. Interestingly, Aly et al.'s data showed chronic HCV patients had a significant increase in Bacteroidetes, as opposed to the previously discussed HBV patients, who had a marked decrease of this phylum in their composition.[21,22] In the HCV patients, *Prevotella* was significantly increased, perhaps explaining the increase in Bacteroidetes phylum composition in this group.[22] This study also demonstrated a higher abundance of Proteobacteria in the healthy control group, dissimilar to the HBV study, which showed a higher abundance in the HBV group.[21,22] In the HCV study, ill patients also displayed significant prevalence of *Acinetobacter*, *Veillonella*, and *Phascolarctobacterium*, as well, with no detection of *Bifidobacterium* genus, which was of relatively increased richness in two healthy patients.[22] *Ruminococcus* was also significantly enriched in healthy patients versus those with HCV, and this genus is considered to be largely beneficial in the gut.[22,23]

Though more research is needed in regards to the gut-liver axis and its role in immunity against HBV and HCV infections, the current literature has demonstrated a strong link between the gut microbiota and how the immune system functions in acute and chronic viral diseases. The disadvantage that remains is the microbiota's complexity, as is shown by the differences in compositions between patients with two different strains of the same virus. Fortunately, due to the development of vaccines, new hepatitis cases have dramatically decreased across the world, though chronic infections continue to pose a threat to those who do contract the virus. The gut microbiota may be a means to harness new therapies to relieve the increased mortality that comes with progression of chronic hepatitis infection.

2.6 Virome in humans

The human virome is constituted of viruses that infect our cells, virus-derived elements in our genome, and viruses that infect the broad array of other types of microorganisms that inhabit us.[86] The integration of endogenous retroviral DNA into the human genome accounts for 8% of the total genome, and is thought to promote development of certain diseases such as cancers and autoimmune diseases.[87] These sequences are also passed from mother to child, perpetuating the effects to offspring.[88] Another aspect of the virome is made up of prokaryotic viruses, or bacteriophages. Bacteriophages are the most numerous beings on Earth and have been found to promote health in humans, mostly through the development of phage therapy. Phage therapy has been utilized to combat antibiotic-resistant bacterial infections in humans.[89]

Similar to the bacterial component of the microbiome, the virome contains certain makeup that varies between individuals.[90] Some common inhabitants of the human gut include *Phycodnaviridae*, *Herpesviridae*, *Poxviridae*, *Mimiviridae*, *Iridoviridae*, *Adenoviridae*, *Anelloviridae*, *Astroviridae*, *Parvoviridae*, *Picornaviridae*, *Picobirnaviridae*, and *Reoviridae* which have been identified in infant stool samples.[91–93] Viromes tend to stabilize over time as well, with humans establishing the individualized virome by around the age of 2 years, similar to the bacteriome.[94]

3. Microbiota in viral infection and diseases in avian species

3.1 Introduction

Poultry contain the most economically important avian species because they provide a great source of animal protein. Poultry eggs and meat are crucial components to maintaining food security all over the globe. Fast-growing domestic poultry species are exposed to many environmental factors that can affect their health, performance, and productivity. To protect poultry from factors that may render them vulnerable to disease, scientists continuously investigate internal and external factors that may affect their health and productivity. The gastrointestinal tract has been a central part of these investigations as the gut health, function, and integrity are essential for overall wellbeing of birds. The gut microbiota, which contains tremendous numbers of bacterial species, has recently been extensively characterized in some avian species including chickens, turkeys, and ducks.[95–97] The influence of viral infections on the gut health and the homeostasis of commensal microbial populations of affected birds have been also investigated. Viral infections are also known risk factors for coinfection by other secondary, opportunistic pathogens as they cause damage to local and systemic immune defenses.[98] Here, we review the interplay between a number of viral pathogens and the gut microbiota of their respective natural avian hosts.

3.2 Avian influenza virus

Avian influenza viruses (AIV) belong to species Influenza A virus (IAV), genus *Influenza A*, family *Orthomyxoviridae*, order *Mononegavirales*, with a negative sense, single-stranded, segmented RNA genome (talk.ictvonline.org/ictv-reports). Based on their pathogenicity, AIV strains are classified into two major pathotypes: high pathogenicity avian influenza (HPAI) viruses, e.g., H5N1 and low pathogenicity avian influenza (LPAI) viruses, e.g., H9N2. However, AIV strains causing natural infections in different species can be categorized into four groups from highly virulent to avirulent. The morbidity, mortality, and lesions are highly dependent on the virulence of the strain, the bird's age, and the immune status. Highly virulent AIV infections cause morbidity and mortality nearing 100% and systemic, severe lesions. Mortality decreases substantially in mildly virulent AIV infections, which cause mild respiratory disease.[99]

3.2.1 Effect of IAV infection on microbiota composition

The effect of LPAI infection on the homeostasis of the gut microbiota of poultry has been investigated very recently. Experimental infection of young chickens with H9N2 virus caused alterations in the intestinal microbiota composition. In infected chickens, the bacterial numbers of the phylum Proteobacteria were increased and the bacteria belonging to the genera *Vampirovibrio*, *Pseudoflavonifractor*, *Ruminococcus*, *Clostridium* cluster XIVb, and *Isobaculum* were enriched differentially.[24] Proteobacteria contains important gram-negative pathogenic bacteria belonging to the genera *Escherichia*, *Shigella*, *Salmonella*, and *Helicobacter*.[100] In uninfected—apparently healthy—chickens, differential enrichment of the genera *Novosphingobium*, *Sphingomonas*, *Bradyrhizobium*, and *Bifidobacterium* was observed.[24] Comparing the same chicken group before and after infection, Yitbarek et al. observed notable changes in the relative abundance of bacteria at various taxonomic levels.[24] Pre-infection, chickens were characterized by the increased abundance of two families and three genera of bacteria, while postinfection differential enrichment of members from one class, two orders, and one genus of bacteria were recorded. In a study by Li et al.[25] at the phylum level, the relative abundance of Proteobacteria was increased—as observed in the study above—while that of Firmicutes was decreased in H9N2 virus-infected young chickens. The relative abundance of endogenous bacterial genera within the family *Enterobacteriaceae* was increased; these genera included *Escherichia*—especially *E. coli* which colonized the ileum—, *Clostridium*, and *Veillonella*. On the other hand, a severe reduction in the probiotic organisms including the lactic acid-producing bacteria (i.e., *Lactobacillus*, *Enterococcus*, and *Streptococcus)*, *Candidatus* Arthromitus, *Bacteroides*, *Parabacteroides*, and SMB53 was observed.[25] These findings clearly indicate that an influenza virus infection can disrupt the commensal microbial population in affected chickens.

Waterfowl species are natural reservoir of IAV and infections in these birds, unlike chickens, are asymptomatic. The composition and diversity of cloacal microbiome and their relationship with the active status of IAV infection and affected host have been studied in five waterfowl species: *A. acuta* (northern pintail), *A. americana* (American wigeon), *A. carolinensis* (green-winged teal), *A. clypeata* (northern shoveler), and *A. platyrhynchos* (mallard).[101] Six bacterial phyla, namely Firmicutes, Proteobacteria, Bacteroidetes, Fusobacteria, Actinobacteria, and Tenericutes (ordered based on proportions of abundance) were consistently represented in IAV-negative, i.e., healthy, ducks of the above-mentioned five species. These phyla composed an average of 97.1% of the ducks' microbiota and thus were further investigated in the study. The relative abundance of bacterial phyla was drastically different within the various host species, with no unique patterns across species. However, green-winged teals and mallards showed the same pattern of abundance for all phyla, while northern shovelers showed the opposite pattern. The relationship among the various groups of bacterial taxa was evaluated using a microbiome diversity measurement, bacterial operational taxonomic unit (OTU) co-occurrence patterns. The observed OTUs in IAV-positive and -negative birds showed no consistent patterns among host species. Moreover, the number of observed OTUs by IAV infection status was only significant in two out of five duck species.[101] Findings from this study indicates that host genetics (and possibly other environmental factors, e.g., feed and habitat) may play a crucial role in the response of waterfowl to influenza virus infections. Another study on juvenile wild mallards (*Anas platyrhynchos*) only analyzed the differences in the cloacal microbiome of IAV-infected birds as compared to healthy ones.[26] Overall, the cloacal microbiome in healthy mallards was more rich, diverse, robust, and uniform than that in IAV-infected birds. Among 41 identified OTUs, IAV-infected mallards were found to have $\leq$12 versus 24 in healthy birds. Thus, IAV-negative mallards were richer in OTUs and had greater OTU diversity and evenness than those of IAV-positive group. The same six bacterial phyla listed above were also represented in both IAV-positive and -negative birds at similar levels of relative abundance. The higher differences were observed within the phyla Firmicutes, Proteobacteria, and Bacteroidetes. Members of the genus *Streptococcus* and species *Veillonella dispar* and *Rothia mucilaginosa* were major contributors to these differences.[26]

3.2.2 Effect of IAV infection on intestinal integrity and immunity

In chickens with compromised microbiota, intestinal damage was caused by H9N2 virus infection, including reduced villus height, increased crypt depth, and reduced villus height:crypt depth ratio of the ileum, as well as lymphocytic infiltration of ileal mucosa.[25,102] The breach of the intestinal wall integrity was consistent with downregulation in mRNA expression of the epithelial cell tight junction proteins (ZO-1, claudin 3, and occludin) and mucus layer proteins (trefoil factor 2 and mucin); the latter two have essential roles in preventing the inflammation and infection of the intestinal epithelium caused by invading bacteria.[25,103,104] Furthermore, the intestinal damage in H9N2 virus-infected chickens was associated by upregulation of mRNA expression of the proinflammatory cytokines IFN-γ, IL-22, IFN-α, and IL-17A by the intestinal epithelial cells.[25]

The importance of balanced gut microbiota in modulating innate immune responses and protecting chickens against influenza virus infections was demonstrated by several studies. In intestinal microbiota-depleted chickens, H9N2 virus infection caused significant increase in oropharyngeal and cloacal virus shedding. Also, the mRNA expression of type I IFNs (i.e., INF-α and IFN-β) and IL-22 was suppressed in the respiratory and gastrointestinal tracts.[102]

3.3 Newcastle disease virus

Newcastle disease virus (NDV, aka avian paramyxovirus 1) belongs to species *Avian orthoavulavirus 1*, genus *Orthoavulavirus*, family *Paramyxoviridae*, with a negative sense, single-stranded, monopartite RNA genome.[105] NDV affects a wide variety of avian species, among them the most susceptible and economically important are chickens and turkeys,[106] which can virtually be infected at all ages. Depending on the strain and tropism of the virus, NDV can cause a variety of symptoms involving the gastrointestinal tract, the nervous system, the respiratory system, and the reproductive system of affected birds, leading to up to 100% morbidity and mortality.[107]

The adverse effect of NDV infection on the development of intestinal microbiota in chickens has been recently reported by Cui et al.[27] Although the study was performed in newly hatched chicks which were infected *in ovo* with NDV, it presented significant insight into the impact of such an infection on the intestinal microbial communities in chickens. In uninfected chicks, the duodenum had a richer (indicated by higher OTU counts and abundance indices) and more diverse microbiota than the cecum. In normal duodenum, the predominant bacterial phyla were Proteobacteria, Bacteroidetes, Firmicutes, and Actinobacteria, with six more less-dominant phyla. At the genus level, bacteria from *Serratia*, *Escherichia*, *Chitinophagaceae* unclassified, *Saprospiraceae* unclassified, and *Lactobacillus* were represented the most in the normal duodenum among other genera. In the cecum of uninfected chicks, the prevailing bacterial phyla were only Firmicutes and Proteobacteria, with five more marginally-represented phyla detected. At the genus level, *Enterococcus*, *Paenibacillus*, *Erwinia*, *Escherichia*, *Lactobacillus*, and *Serratia* constituted the great majority of the normal cecal microbiota.[27] Comparing NDV-infected with uninfected chicks, the richness and diversity of duodenal microbiota were lower in the former. The cecal microbiota was almost unaffected by the NDV infection. As for the microbial composition, both duodenal and cecal bacterial communities were adversely affected in a similar manner. An increase in *Sinobacteraceae* and *Rhodoplanes* family members was observed. Also, at the genus level, a massive loss was observed in members of the following families in both duodenum and cecum of NDV-infected chicks: *Xanthomonadaceae*, *SC-I-84*, *Cytophagaceae*, *Bacteroidales*, *Chitinophagaceae*, *Gemm-1*, *Saprospiraceae*, *Ignavibacteriaceae*, *Rhizobiales*, *Sphingobacteriales*, *Ellin6067*, *Acidimicrobiales*, and *Pseudomonas*. Moreover, a dramatic reduction was found in the abundance of *Xanthomonadaceae* and *Cytophagaceae* family members.[27] A significant reduction in the relative abundance of *Serratia* and *Clostridium* genera residing in the duodenum was recorded as opposed to greatly increased numbers in the ceca. This implied a possible translocation of these genera from duodenum to cecum of NDV-infected chicks. Moreover, authors proposed a translocation of several bacterial genera from the cecum to the duodenum of infected birds according to changes in their relative abundance.[27]

Talebi et al. studied the effect of a mixture of probiotics (*Lactobacillus acidophilus*, *Lactobacillus casei*, *Enterococcus faecium*, and *Bifidobacterium bifidium*), administered in water, on the antibody response and production performance in NDV-vaccinated broiler chickens.[108] The study showed enhancement in birds' body weight and feed conversion rate; however, the humeral immune response to NDV vaccination was similar to that of untreated, vaccinated groups. Similar results were later revealed by Bautista-Garfias et al.,[109] who studied the effect of oral administration of *L. casei* (ATCC7469 strain) on the humeral immune response to NDV vaccination and production performance in fighting roosters. Blood anti-NDV antibody levels in *L. casei*-treated group were comparable to those of the vaccine control birds. On the other hand, the natural mortality rate and body weight gain were significantly improved in the probiotic-treated group, compared with the untreated one.[109] The use of two (cecal and jejunal) bacterial isolates of *L. reuteri* PIA16 strain in feed, either alone or with a prebiotic (mannan oligosaccharide), demonstrated substantially enhanced humoral immune response to NDV vaccine in broiler chickens. The cell-mediated immune response; however, remained largely unaffected throughout the course of the experimental period.[28] Results from the probiotics studies indicate that these beneficial bacteria can boost the performance and overall health status in poultry. Although these studies showed generic positive impact on NDV-vaccinated chickens, the associated changes in the intestinal microbiota have not been examined. Moreover, whether the supplementation of these probiotics/prebiotics has an enhanced protective effect for NDV vaccines against the consequences of an NDV challenge is yet to be investigated.

3.4 Infectious bursal disease virus

Infectious bursal disease virus (IBDV) belongs to the genus *Avibirnavirus*, family *Birnaviridae*, with a double-stranded, segmented RNA genome.[110] IBDV affects chickens at young ages and causes an acute, immunosuppressive disease, infectious bursal disease. IBDV infection causes depression of the local and humoral immune responses (due to a direct effect on B lymphocytes); and induces damages in the primary lymphoid organs, mainly the bursa of Fabricius, in addition to the spleen, thymus, and cecal tonsils.[111,112]

The role of IBDV in modulating the commensal microbiota in chickens has recently been studied.[29] Young chickens were infected with IBDV and the changes in the cecal bacterial communities were analyzed in comparison with uninfected, healthy birds. Regardless of the infection status, nine bacterial phyla were represented in the chicken cecal microbiota, with Firmicutes,

Proteobacteria, Actinobacteria, and Bacteroidetes members forming more than 95% of total bacterial population in the cecum. At the family level, *Lachnospiraceae* and *Ruminococcaceae* composed the majority of cecal microbiota, with the relative abundance of *Lachnospiraceae* decreasing over time and that of *Ruminococcaceae* increasing over time for up to 7 weeks of age. At the genus level, an increase in the abundance of *Faecalibacterium* was observed till ∼4 weeks of age, and decreased thereafter. IBDV infection of young chickens caused immediate and obvious alteration in the composition of cecal microbiota. Changes were continuously seen over a three-week observation period postinfection,[29] though no certain pattern was observed. At the genus level, in comparison to uninfected group, IBDV-infected chickens had a lower abundance in the commensal *Clostridium XIVa* at day 3 postinfection, which became higher thereafter. On the other hand, after 7 days of IBDV infection chickens had a higher abundance of *Faecalibacterium* (Gram-positive bacteria) than uninfected ones, which became lower thereafter. A reduction in the abundance of *Escherichia/Shigella* (Proteobacteria) was observed in chickens until 3 weeks postinfection, which contributed to the overall decrease recorded for family *Enterobacteriaceae*.[29] These changes in the cecal microbiota were associated with IBDV-induced immunosuppression, the manifestations of which is described below. Another study showed that the infection of young chickens with IBDV or *Campylobacter jejuni*, either alone or combined, provoked significant modifications in the composition of gut microbiota, as compared with uninfected chickens.[30] Changes in the abundance of OTUs from the following bacterial genera were observed in the cecal contents of infected chickens: *Campylobacter*, *Clostridium XIVa*, *Eubacterium*, *Faecalibacterium*, *Lachnospiracea incertae sedis*, *Lactobacillus*, and *Roseburia*. The abundance of *Campylobacter* was increased in IBDV/*C. jejuni*-infected chickens and it was higher in birds inoculated with bacteria at 7 days versus 9 days postinfection. Obviously, the exposure time to a secondary pathogen following an immunosuppressive viral infection has a significant impact on its pathogenicity, colonization, and persistence, as is the case here with IBDV and *C. jejuni*. Similar observations have been reported for hemorrhagic enteritis of turkeys—a siadenovirus-based immunosuppressive disease—and secondary *E. coli* infections.[113]

The particular effect of immunosuppression-inducing IBDV infection on the colonization and shedding of *C. jejuni* and *Salmonella* in the gastrointestinal tract of chickens has been investigated. Unlike *C. jejuni*, *Salmonella* is pathogenic in poultry; however, both are important pathogens in human and can cause severe food-borne gastroenteritis.[114,115] Apart from *C. jejuni* and *Salmonella* infections, inoculation of young chickens with various strains of IBDV resulted in atrophy and pronounced lesions in the bursa of Fabricius, bursal B cell depletion, reduction or depletion in circulating B lymphocytes as well as those localized in the cecal lamina propria and cecal tonsils, and a reduced thickness of cecal mucosa.[27,29,32,116,117] Virus replication in the bursa, cecal tonsils, and cecum was detected for up to several weeks postinfection.[27,29,116] In IBDV-infected chickens with immunosuppression, unlike *C. jejuni* only-infected birds, the bacteria colonized the small intestines, and was detected at an earlier time point and at much higher titers in the cecum and cloaca (i.e., virus shedding).[27,116] These results were consistent with the reduced B lymphocytes and the lack of *C. jejuni*-specific IgG antibodies in the cecal lamina propria.[27] Likewise, IBDV infection of chickens exacerbated *Salmonella* multi-organ colonization, pathogenicity, fecal shedding, and persistence in chickens exposed to infections at early ages.[31,32,117] The highest *Salmonella*-mediated pathogenicity and mortality rates were recorded in chickens coinfected with IBDV and *S. enteritidis,*[31,32] but not *S. typhimurium.*[117] Both intestinal mucosal[41] and systemic (*Salmonella*-specific IgG) antibody responses were lower in chickens coinfected with IBDV and *Salmonella,* which coincided with the immunosuppressive effects described above. However, unlike *C. jejuni, Salmonella* clearance appeared not to be related to the antibody response. Numbers of localized T lymphocytes and cell-mediated immune responses did not seem to be affected by IBDV infections.[27,32]

3.5 Hemorrhagic enteritis virus/turkey adenovirus 3

Turkey adenovirus 3, commonly known as hemorrhagic enteritis virus (HEV), belongs to genus *Siadenovirus*, family *Adenoviridae*, with a double-stranded, linear DNA genome.[118] HEV is present naturally in two strains; virulent, which causes an immune-mediated, acute disease characterized by bloody diarrhea, enteric hemorrhage, and death; and avirulent, which is widely used as a vaccine strain. Both strains cause transient period of immunosuppression and splenomegaly and target the IgM^+ B cells in the spleen and peripheral blood. In addition to turkeys, HEV causes immune-mediated diseases in pheasants and chickens characterized by pulmonary edema and splenomegaly. Lesions in several organs have been also reported in affected birds.[113]

Changes in the abundance of microbial populations between HEV-infected and uninfected turkeys have recently been investigated. Analysis of OTUs detected in the gut of healthy turkeys revealed the presence of four abundant phyla: Actinobacteria, Bacteroidetes, Firmicutes, and Proteobacteria. In healthy turkeys, Firmicutes was the predominant phylum (80–100%) in the three small intestinal sections, while in the cecum a more heterogeneous population containing representatives from Firmicutes, Bacteroidetes, and Proteobacteria was observed.[33] The diversity of cecal microbial population was also observed throughout the developmental stages of turkeys up to 16 weeks of age. The bacterial populations were also dissected at the genus level in the different gastrointestinal tract sections and *Lactobacillus*, *Streptococcus*, and *Clostridium XI* were the prevailing genera forming nearly 80% of the total microbiota. An interesting finding was that *Campylobacter* spp. was most abundant in the cecum of turkey poults at 10 weeks of age.[96] Interestingly, similar to turkeys, both chickens and ducks have a greater microbial diversity in the cecum as compared to the rest of the small intestine.[96,119,120] The abundance of jejunal bacterial populations at the family level was altered by natural infections of HEV in turkeys. Compared to uninfected turkeys, infected birds with clinical signs and HEV detected in the intestines—indicating an early-phase infection—had a lower abundance in *Micrococcaceae* and a higher abundance of *Propionibacteriaceae,* while those turkeys with no clinical signs and virus detected in the spleen—indicating a late-phase infection—had a much lower abundance in both families. The abundance of family *Bacteroidaceae* was extremely increased in HEV-infected turkeys as compared to uninfected controls. For the phyla Firmicutes, a substantial reduction in *Lactobacillaceae* and an increase in *Clostridiaceae* was observed in intestinally-positive turkeys with clinical signs. An opposite trend was recorded for turkeys with positive spleens and no clinical signs. However, both groups of infected turkeys showed an increased abundance of three other families belonging to Firmicutes. Family *Campylobacteriaceae* increased in abundance by 18- to 26-folds in HEV-infected turkeys.[33] HEV infection of turkeys has been associated with secondary *E. coli* infections during the transient immunosuppression period.[34] HEV is transmitted through fecal-oral route and eventually reaches the blood stream. In the spleen, the virus replicates rapidly in B lymphocytes and to a much lower extent in macrophages. Hyperplasia of the splenic white pulp occurs as a result of infiltration with high numbers of macrophages and lymphocytes, including CD4+ T cells. The interplay among these various types of immune cells within the site of viral replication leads to the production of very high levels of pro-inflammatory cytokines, e.g., interleukin 6 (IL-6) and tumor necrosis factor (TNF) as well as the anti-inflammatory cytokines of type I and II IFNs. The latter activates macrophages, inducing the production of the antiviral and immunosuppressive nitric oxide (NO). Most likely, IFN-I and NO play a role in clearing the virus later from the spleen.[121–123] Moreover, it was suggested that the intestinal hemorrhage in the jejunum area occurs through inflammation-induced "diapedesis."[124] This might be associated with a compromised intestinal epithelial cell barrier of the gut as noticed with other enteric viral infections.

3.6 Virome in avian species

Viruses which can colonize a healthy poultry gut have been recently identified in chickens. Specific pathogen-free chickens were placed for a few days with age-matched peers in commercial and back yard farms with previous history of enteric and respiratory problems. Intestinal samples were then collected and the enteric virome and microbiome were analyzed.[125] The analysis revealed that the new viruses which colonized the gut of SPF birds belong to the families *Picornaviridae*, *Picobirnaviridae*, *Reoviridae*, and *Astroviridae*. This implies that members of these families are associated with enteric diseases in chickens. In these birds, alterations of the gut microbiome were also observed, specifically in the *Lachnospiracea* family and the *Clostridium* and *Lactobacillus* genera. Other virus families that were represented in the enteric virome of chickens included *Adenoviridae*, *Birnaviridae*, *Caliciviridae*, *Coronaviridae*, *Leviviridae*, *Siphoviridae*, and *Retroviridae*.[125] In another study, the fecal virome in healthy commercial chickens was analyzed by viral nucleic acid purification, illumina sequencing, and de novo assembly.[126] Analysis of the assembled viral genome sequences indicated the presence of viruses belonging to the families *Adenoviridae*, *Caliciviridae*, *Circoviridae*, *Parvoviridae*, *Picobirnaviridae*, *Picornaviridae*, and *Reoviridae*.[126] Additionally, novel, unclassified, viruses with circular single-stranded DNA genomes were identified. The fecal virome of wild waterfowls has been also recently analyzed in various species of migratory wild ducks.[127] The DNA and RNA viral genomes were isolated from purified viral particles, their nucleotide sequences obtained, and analyzed. The most abundant viral familied recognized in the duck's virome were *Herpesviridae*, *Alloherpesviridae*, *Adenoviridae*, *Retroviridae*, and *Myoviridae*.[127] In a more recent study on healthy waterfowl and shorebirds, the presence of 27 viral species was reported. These species represent the families *Picobirnaviridae*, *Reoviridae*, *Astroviridae*, *Caliciviridae*, *Picornaviridae*, *Hepadnaviridae*, and *Parvoviridae*. Variations in virus abundance and diversity were detected among the various species studied.[128] The results signify the potential of wild birds as reservoir of disease-causing poultry pathogens.

4. Microbiota in viral infection and diseases in swine

4.1 Introduction

Over 35% of agriculturally produced protein for consumption is produced in pigs globally.[129] China is the leading producer of pork generating over 54 million metric tons annually. The majority of the pork exported to the rest of the world comes from the European Union.[130] A constantly increasing human population demands for the availability of more animal-based protein for consumption around the world. This demand for more meat has led to the intensification of farming practices often causing overcrowding of animals allowing for rapid pathogen transmission.[129] Pork production and trade are primarily impacted by infectious diseases that affect swine. The domestication of pigs for the purpose of livestock and husbandry has increased their exposure to emerging and re-emerging viral diseases causing a significant impact on the worldwide pig populations.

The recent advancement of technology and overcoming the requirement of culture-based species identification has allowed for extensive research initiatives that have investigated the role of microbiome and its influence on host immunity.[131–134] This section will focus on different microbial (bacterial) communities that constitute the microbiome and their effects on viral infections in pigs that have a significant impact on domestic pig rearing.

4.2 African swine fever

African swine fever (ASF) is a severe hemorrhagic disease, with a near 100% mortality rate. The disease is characterized by high fever, loss of appetite, ataxia and depression.[135] The causative agent African swine fever virus (ASFV) is a double-stranded DNA virus belonging to the *Asfaviridae* family. ASF is a reportable disease to the World Health Organization (WHO), and is often accompanied with regional, national and international trade restrictions with the capacity to impact the global pig meat economy.[136] Pigs of all ages are susceptible to virulent strains of ASF with the exception of African wild pigs. No vaccines or therapeutics are available, with the control of disease spread often carried out by culling large numbers of domestic pigs.

Transmission between traditional ASFV hosts, warthogs (*Phacochoerus aethiopicus*) and bushpigs (*Potamchoerus porcus*), occurs through the bite of soft ticks belonging to the genus *Ornithodoros,* which are endemic to sub-Saharan Africa.[137] Ingestion of feed contaminated with biological fluid from wild boars and pigs, and fomites have been identified to be efficient modes of virus transmission in the absence of the arthropod vector.[135,138] The oral route of infection is considered to be a major portal of entry for ASFV among pigs outside the natural territory of its arthropod host. Little work has been carried out in order to characterize the bacterial composition among different pigs, based on their susceptibility status to ASFV.

Specific pathogen-free (SPF) pigs have been demonstrated to be highly susceptible to lethal infection by an attenuated strain of ASFV (E75CV1) that was previously tested to be safe for administering to domestic pigs as a vaccine.[139] Diversity analysis of bacterial communities belonging to SPF, and domestic pigs from the same breed, domestic pigs indigenous to Africa, and warthogs from Africa showed that SPF pigs had the lowest diversity among the different species of pigs analyzed. Firmicutes were found to be the most abundant phylum among all the animals. Bacteria belonging to the genera *Plaudibacter*, *Anaeroplasma*, *Petrimonas*, and *Moraxella* were found to be among the core communities in pig species that were naturally resistant to ASFV.[35] Naturally, pigs reared under strict SPF-conditions were observed to be the least diverse based on microbial composition which has been correlated with an impaired immune system development and the ASFV-resistant warthogs were identified to contain OTUs belonging to previously-uncharacterized bacterial genera.

4.3 Porcine circovirus type 2

Porcine circovirus type 2 (PCV2) is a single-stranded circular DNA virus and is the primary causative agent of porcine circovirus associated disease. PCV2 belongs to the genus *Circovirus* within the family *Circoviridae.*[140] PCV2 targets the lymphoid organs leading to lymphoid depletion and tissue damage.[141] Direct contact with an infected pig has been known to be the most efficient route of virus transmission along with other routes such as contaminated vectors and fomites.[142]

Barrows (pigs aged 27–40 days) infected with PCV2 were observed to show an increase in abundance of *Prevotella* spp. in their microbiota composition.[37] The presence of a more diverse microbiota and the presence of non-pathogenic *E. coli* in pigs have been associated with a reduced in clinical signs of PCV associated disease as well as improved growth in sub clinically infected pigs. Fecal microbiota transplant in 3-week old barrows reduced the number of pigs affected by PCV infection, including a reduction in viral load and increased viral-antigen specific antibodies.[38]

Coinfections with porcine reproductive and respiratory syndrome virus (PRRSV) and PCV2 occur regularly among overcrowded pig farms.[36] Bacterial colonization dynamics at the time of viral infection, that have been associated with a reduced PCV2/PRRSV burden include the presence of low abundance of *Methanobacteriaceae* spp., and an increased abundance of the species of *Ruminococcaceae*, and *Streptococcaceae*.[36] PRRSV is a single-stranded, positive-sense RNA virus belonging to the family *Arteriviridae* of the order Nidovirales.[143] Characteristics of PRRSV infection in pigs include reproductive failure in sows, and respiratory disease in growing and finishing pigs.[144] Each year, the pig industry in the US loses $664 million due to the respiratory disease and reduced weight gain among growing pigs infected with PRRSV and coinfected with other pathogens, including PCV2.[38]

4.4 Porcine epidemic diarrhea virus

Porcine epidemic diarrhea (PED) is a highly contagious and lethal enteric disease and is caused by PED virus (PEDV) which belongs to the *Alphacoronavirus* genus in the *Coronaviridae* family. Transmission occurs through the fecal-oral route with disease characterized by severe vomiting, diarrhea and dehydration.[145] Genomic sequencing of different PEDV isolates have shown the emergence of PEDV strains with insertion/deletions or multiple mutations in the spike (S) protein sequence allowing for the circulation PEDV among the global pig population.[146] The age of infected pigs, immune status and the virulence of the virus strain have all been attributed to the development of clinical signs of disease.[147] Pigs of all ages are susceptible to PEDV, with the highest mortality rate occurring among neonatal and suckling pigs.[148] A vaccine is available in the form of an adjuvanted attenuated virus that is administered to healthy pregnant sow. Antibodies generated by the sow during and after gestation have been successful to an extent but fail to induce sufficient protection in neonatal piglets farrowed from PEDV naïve sows.[149]

Bacterial composition of the neonatal gastrointestinal tracts[35] often begins with the colonization of aerobic or facultatively aerobic bacteria belonging to *Escherichia coli*, *Shigella flexneri*, and *Streptococcus* spp. The consumption of oxygen within the GIT of the newborn piglets allows for the subsequent colonization of anaerobes such as *Bacteroides*, *Clostridium*, *Oscillibacter*, *Bifidobacterium*, *Lactobacillus*, and *Ruminococcaceae* spp.[150,151] Analysis of the microbiota composition in the small and large intestines of PEDV infected piglets showed an increase in abundance of *Escherichia-Shigella*, *Enterococcus*, *Fusobacterium*, and *Veillonella*.[39] In parallel to this, a decrease in SCFA-producing bacteria such as *Rikenellaceae* (RC9 gut group), *Butyricimonas*, and *Alistipes* were also observed among infected piglets.[40] A reduction in SCFA production within the GIT has been linked with an increased state of dysbiosis in piglets implicating reasons for severity of PEDV infection.[152,153] Huang et al. showed that bacteria belonging to *Ruminococcaceae, Rikenellaceae, and Porphyromonadaceae,* were associated with healthy piglets as compared to PEDV infected piglets.[39]

4.5 Virome in swine

Investigative studies on viruses outside their pathogenic interaction with their swine hosts have been limited. With an increased access to advanced sequencing technology, the influences of virome on swine health can be investigated in greater detail. Liver biopsies taken from healthy pigs revealed a diverse range of small ssDNA viral genome sequences that were closely related to viruses belonging to *Anelloviridae*, *Circoviridae* and *Parvoviridae* families.[154] Another study also reported a rich viral abundance and diversity among blood-fed mosquitoes collected from pig farms. Apart from the presence of viruses belonging to families that were known to cause infection among pigs, humans, and mosquitoes, a considerable number of non-classified virus reads were also detected that were presumed to belong to unexplored novel virus species.[155]

4.6 Summary

Microbiota diversity is an integral part of pig health. As identified by Ober and colleagues, an increased microbial diversity together with shifts in the presence of several bacterial families prior to infection allowed pigs to recover faster after the infection fully resolved.[36] The use of antimicrobials in regular farming practices have been attributed to the reduction in microbiota diversity increasing their susceptibility to infections. Considering alternative therapy such as fecal microbiome transplantation for enhancing swine health would enhance the overall economic benefits associated with pig farming.

5. Microbiome and virus disease in ruminants

5.1 Introduction

Rumen microbes play a critical role in the development of digestive system in neonatal, nursing calves. They also play important roles in the nutritional acquisition (feed digestion), and physiological and immunological functions of the host. Generally, they support each other to ferment plant structural and nonstructural carbohydrates and proteins. Along with the large-scale application of high throughout sequences and metagenome techniques, literatures in characterization and variation of microbiota in ruminant animals increased explosively.[156–160] The rumen microbiota is very complex, and the diversity of ruminal microorganisms can be affected by diet composition, genetics and environmental factors. There are approximately 7000 bacterial species and 1500 archaeal species in the rumen.[161] Rumen protozoa is present when animals are fed high-grain diets, and rumen fungi represents approximately 10% of the total rumen microbiota at any given time.[162] Plenty of studies have found differences of micro-biological development in the digestive system of calves in relation to variations in management practices, such as housing, feeding, and antimicrobial administration during the neonatal period.[163,164] The following review focuses on the microbiota and how it changes and acts during virus infection in ruminants.

5.2 Virus infection changes microbiota of infected animals

5.2.1 Bovine rotavirus

Bovine rotavirus is the main pathogen associated with neonatal calf diarrhea. It infects young calves orally, replicates in the cytoplasm of intestinal epithelial cells, and destroys mature intestinal cells.[165] It has been reported that the incidence of rotavirus diarrhea was associated with a decrease in bacterial diversity in the gut, which is a well-known hallmark of dysbiosis in the early stages of life in the calves.[43] Alterations were found in the gut microbiota of neonatal diarrhea samples as a result of rotavirus infection. The relative abundances of Firmicutes and Bacteroidetes increased in healthy calves, while Proteobacteria was abundant in rotavirus diarrhea samples.[43] This was consistent with human rotavirus diarrhea cases that the phylum Proteobacteria is associated with gut dysbiosis and inflammation.[41] At the genus level, in rotavirus-infected calves, there was significant increase of the genera *Escherichia, Clostridium, and Streptococcus*. However,

Subdoligranulum, Blautia, Bacteroides, and Coprococcus were decreased. Notably, *Lactobacillus* was significantly decreased in the rotavirus diarrhea calves. At the species level, *Lactobacillus* species such as *L. gasseri* and *L. amylovorus* were found to be relatively abundant in healthy calves. *L. gasseri* is known to prevent the severity of acute self-limiting diarrhea in adults.[42] In rotavirus-infected calves, the relative abundance of *G. formicillis, B. glucerasea, B. coccoides*, and *B. obeum* were significantly higher in the healthy group. Two species of *C. perfringens* and *E. coli* associated with the diarrheic calves were identified.[43] This study demonstrated that rotavirus infection changed the structure of the gut microbiota, which would affect the follow up diagnosis and administration.

5.2.2 Bovine leukemia virus (BLV)

Bovine leukemia virus (BLV) is a retrovirus which causes enzootic bovine leukosis in cattle, and it is closely related to the human T-lymphotropic virus type 1.[166] BLV infection of domestic cattle is highly prevalent in several geographic regions.[167] Once BLV infects a cow, it cannot be eliminated.[168] Because the immune system of host cattle can be impaired during infection, BLV infection consistently results in the inability of cattle to maintain normal health.[169] Comparing to BLV infected dairy cows, *Lachnospiraceae* and *Veillonellaceae* families associated with ruminal fermentation were more abundant in the fecal microbiota of uninfected cows.[44] Higher titer BLV infection was associated with less diversity of microbiota in infected cows. Meanwhile, the virus propagation ability of BLV strains was negatively correlated with one taxon of *Sanguibacteroides*. Besides causing lymphoproliferation and leukemia, BLV infection may decrease energy production efficiency in the infected cows through modification of rumen and hindgut microbiota.[44] According to function speculation of the bacteria detected in the differential abundance analysis, energy production loss in the rumen and hindgut was assumed, which may explain the secondary negative effects such as increased susceptibility to other infections and decreased lifetime milk production and reproductive efficiency.[170–173] Further studies need to systematically investigated rumen microbiota with rumen nutrition status, such as the level of volatile fatty acids together with BLV strain genetics.

5.3 Opportunistic pathogen activities during virus infection

5.3.1 Escherichia coli *and rotavirus and coronavirus infection*

Escherichia coli strains comprise a group of bacteria with a huge genetic diversity that make them able to colonize different niches and to adapt as a component of intestinal commensal microbiota of animals and human beings.[174] *E. coli* are classified into eight different phylogenetic groups (A, B1, B2, C, D, E, F and Clado1).[175] Changes in compositions of commensal *E. coli* were observed during rotavirus and coronavirus intestinal infection in calves with diarrhea.[176] From this study with 30 calves each in the infected and control group, *E. coli* isolates identified in virus-infected calves were phylogenetically classified as B1 (70%), B2 (3.33%), C (3.33%), D (3.33%), E (13.33%) and unknown (6.7%), whereas *E. coli* isolates from the control group were classified only as B1 (83.3%), E (10%) and unknown (6.7%). B2, C and D groups were found only in samples from animals with diarrhea due to rotavirus and coronavirus infection.[176] According to Escobar-Paramo[177] and Clermont,[175] B2 and D are *E. coli* phylogenetic groups that comprise strains with pathogenic potential, being more commonly observed in extra-intestinal infections.

In *E. coli* isolated from diarrheic calves, F5 and F18 Fimbriae were found.[176] Fimbriae are virulence factors that allow *E. coli* colonization mainly in the small intestine, avoiding bacteria elimination along with the feces.[178] F5 is one of the most frequent factors in enterotoxigenic *E. coli* pathotype isolated from calves with diarrhea.[179] F18 is more commonly associated with post-weaning diarrhea and edema disease in swine.[178,180,181] In summary, rotavirus and coronavirus infection changes the homeostasis of intestinal microbiota which involves the participation of *E. coli* in the pathological progress. *E. coli* was the only species that positively correlated with the number of neutrophils, which may explain the inflammation during diarrhea.[43]

5.3.2 Lumpy skin disease virus and secondary bacterial infection

Lumpy skin disease (LSD) is a viral disease of cattle caused by lumpy skin disease virus (LSDV). LSDV is one of the most important animal poxviruses because of the serious economic consequences in cattle. The World Organization for Animal Health[182] categorizes LSD as a notifiable disease.[182] It is characterized by fever, reduced milk production and skin nodules. Mastitis, swelling of peripheral lymph nodes, loss of appetite, increased nasal discharge and watery eyes are also common. Temporary or permanent infertility occur among infected cows and bulls. The disease can cause high morbidity and low mortality.[183,184] Secondary bacterial infection in the affected skin lesions can increase the severity and prolong the course of the disease. The analysis of prevalent bacterial communities in affected lesion were carried out using 16s rRNA gene sequencing. Up to 98 species were found, most of them belonging to the phyla of *Proteobacteria*, followed by

Firmicutes, *Actinobacteria*, and *Bacteroidetes*. Many common mammalian pathogens are found in the *Proteobacteria* phylum. For example, the *Brucella*[185] and *Rickettsia*[186] genera belong to the *Alphaproteobacteria* class; *Bordetella*[187] and *Neisseria*[188] belong to the *Betaproteobacteria* class; and *Escherichia*,[189] *Shigella* and *Salmonella* belong to the *Gammaproteobacteria* class. All bacterial species found are known as opportunistic pathogens, but can withstand the inflammatory reaction.

5.4 Virome in ruminants

Ruminant virome is important in the nutrient and energy cycling,[190] development of immunity,[191] and a major source of genes through lysogenic conversion.[192] As metagenomic next-generation sequencing (mNGS) has been employed to identify uncommon and novel infectious etiologies, massive virus species have been identified in ruminants.[193]

Besides the common diarrhea-causing viral pathogens rotavirus and coronavirus, torovirus, norovirus, nebovirus, astrovirus, kobuvirus and enterovirus have been detected from calves' feces with diarrhea by mNGS.[194] Viral genomes of pestivirus A, Ungulate erythroparvovirus 1, bosavirus, and hypothetical circular Rep-encoding single-stranded DNA viruses were also identified from calf serum with mucosal disease.[195] Most studies have involved epidemiologic investigations seeking to show association with diarrhea for each virus alone or in combination with potential pathogens.

In bovine respiratory disease, 16 viruses were identified, of which bovine parvovirus 2 was the most prevalent (11.5%, 15/130) followed by ungulate tetraparvovirus 1 (UTPV1, 8.5%, 11/130) and bovine respiratory syncytial virus (BRSV, 8.5%, 11/130).[7] Unconventional viruses such as influenza D virus, bovine rhinitis A virus, bovine coronavirus and bovine rhinitis B virus had also been detected.[196,197]

Non-suppurative encephalitis is one of the most frequent pathological diagnosis in cattle with neurological disease. Six virus candidates: parainfluenza virus 5, bovine astrovirus CH13/NeuroS1 (BoAstV-CH13/NeuroS1), bovine polyomavirus 2 (BoPV-2 SF), bovine herpesvirus 2, bovine herpesvirus 6 (BoHV-6) and a novel bovine betaretrovirus termed BoRV-CH15 had been detected in neurologically-diseased cows. BoAstV-CH13, BoPV-2 SF and BoHV-6 were significantly associated with the disease. These data expanded our knowledge on encephalitis-associated pathogens in cattle and point to the value of NGS in resolving complex infection scenarios in a clinical disease setting.[198]

5.5 Summary

Most of the studies have focused on the variety of microbiome in ruminants. Limited work has been done to study the interaction between microbiota and ruminant viruses or the microbiota changes before and after different virus infection. Diarrhea due to rotavirus and coronavirus infection was associated with a decrease in bacterial diversity in the gut.[43,176] Diarrhea of any etiology may result in a variety of complications as acid-base and electrolyte imbalance and increased lumen-fluid volume with consequent dehydration.[199,200] These poor health condition encourages mass propagation of conditional pathogenic bacteria, which leads to the secondary bacterial infection. Meanwhile, the immunosuppression caused by BLV infection may decrease energy production efficiency in the cows through modification of rumen and hindgut microbiota. The mechanisms of interactions between microbiota and viruses in ruminant host in regard to pathological progress in viral diseases should be better explored in future research.

Like the bacterial microbiome studies, the metagenome sequencing opens a door for us to peek into the virome structure in ruminant diseases. The diarrhea, respiratory diseases, and encephalitis are the main health threats to ruminants. The detection of virome components will fill in the knowledge gap in identifying the disease-associated pathogens. However, determining the contribution of these viruses to health and diseases in ruminants and other animal species would be challenging and much uncertainty still remains concerning their roles as primary pathogens, coinfection agents, or commensals.

6. Concluding remarks

Microbiota is capable of modulating host immune systems and determining the susceptibility to various viral pathogens. The studies reviewed in this chapter have shown such significant implications of the microbiota on host immunity. The influence of the microbiota on viral infection susceptibility and viral disease outcome is undisputable although varies among viruses. These data are very important in terms of viral infection, as currently there are few effective and reliable antiviral therapies available. Further understanding of the interactions between viral pathogens, host species and their microbiota will eventually allow us to tailor the contents of the microorganisms to the extent that it is able to protect the host from certain life-threatening viral infections. More research is needed to establish concrete cause and effect relationships between the microbiota composition and viral infection, as many of these studies did not address other factors that also are known to influence microbiota composition, such as genetics, diet, and environment. It should also be noted that some of these studies focused on small cohorts of participants, and more work is needed to establish trends among larger populations. Despite the limitations of some of the presented works, the microbiota still prevails as a hopeful means of immunotherapy for prevention of viral infection in the future in both humans and farm animals.

References

1. Li N, Ma WT, Pang M, Fan QL, Hua JL. The commensal microbiota and viral infection: a comprehensive review. *Front Immunol.* 2019;10:1551.
2. Lee KH, Gordon A, Shedden K, et al. The respiratory microbiome and susceptibility to influenza virus infection. *PLoS One.* 2019;14(1), e0207898.
3. Nelson KE. Microbiomes. *Microb Ecol.* 2013;65(4):916–919.
4. Shang Y, Kumar S, Oakley B, Kim WK. Chicken gut microbiota: importance and detection technology. *Front Vet Sci.* 2018;5:254.
5. Gimblet C, Meisel JS, Loesche MA, et al. Cutaneous leishmaniasis induces a transmissible dysbiotic skin microbiota that promotes skin inflammation. *Cell Host Microbe.* 2017;22(1)13-24.e14.
6. Proctor DM, Relman DA. The landscape ecology and microbiota of the human nose, mouth, and throat. *Cell Host Microbe.* 2017;21(4):421–432.
7. Rothschild D, Weissbrod O, Barkan E, et al. Environment dominates over host genetics in shaping human gut microbiota. *Nature.* 2018;555(7695):210–215.
8. Ding T, Song T, Zhou B, et al. Microbial composition of the human nasopharynx varies according to influenza virus type and vaccination status. *mBio.* 2019;10(4), e01296-19.
9. Bomar L, Brugger SD, Yost BH, Davies SS, Lemon KP. Corynebacterium accolens releases antipneumococcal free fatty acids from human nostril and skin surface triacylglycerols. *mBio.* 2016;7(1), e01725-15.
10. Trompette A, Gollwitzer ES, Pattaroni C, et al. Dietary fiber confers protection against flu by shaping Ly6c(-) patrolling monocyte hematopoiesis and CD8(+) T cell metabolism. *Immunity.* 2018;48(5), 992-1005.e1008.
11. Fredricks DN, Fiedler TL, Marrazzo JM. Molecular identification of bacteria associated with bacterial vaginosis. *N Engl J Med.* 2005;353(18):1899–1911.
12. Anahtar MN, Byrne EH, Doherty KE, et al. Cervicovaginal bacteria are a major modulator of host inflammatory responses in the female genital tract. *Immunity.* 2015;42(5):965–976.
13. Gosmann C, Anahtar MN, Handley SA, et al. Lactobacillus-deficient cervicovaginal bacterial communities are associated with increased HIV acquisition in young South African women. *Immunity.* 2017;46(1):29–37.
14. Nelson AM, Walk ST, Taube S, et al. Disruption of the human gut microbiota following norovirus infection. *PLoS One.* 2012;7(10), e48224.
15. Rodriguez-Diaz J, Garcia-Mantrana I, Vila-Vicent S, et al. Relevance of secretor status genotype and microbiota composition in susceptibility to rotavirus and norovirus infections in humans. *Sci Rep.* 2017;7:45559.

16. Lei S, Twitchell EL, Ramesh AK, et al. Enhanced GII.4 human norovirus infection in gnotobiotic pigs transplanted with a human gut microbiota. *J Gen Virol.* 2019; 100(11):1530–1540.
17. Harris VC, Armah G, Fuentes S, et al. Significant correlation between the infant gut microbiome and rotavirus vaccine response in rural ghana. *J Infect Dis.* 2017; 215(1):34–41.
18. Harris V, Ali A, Fuentes S, et al. Rotavirus vaccine response correlates with the infant gut microbiota composition in Pakistan. *Gut Microbes.* 2018;9(2):93–101.
19. Twitchell EL, Tin C, Wen K, et al. Modeling human enteric dysbiosis and rotavirus immunity in gnotobiotic pigs. *Gut Pathog.* 2016;8:51.
20. Vlasova AN, Chattha KS, Kandasamy S, et al. Lactobacilli and bifidobacteria promote immune homeostasis by modulating innate immune responses to human rotavirus in neonatal gnotobiotic pigs. *PLoS One.* 2013;8(10):e76962.
21. Wei X, Yan X, Zou D, et al. Abnormal fecal microbiota community and functions in patients with hepatitis B liver cirrhosis as revealed by a metagenomic approach. *BMC Gastroenterol.* 2013;13:175.
22. Aly AM, Adel A, El-Gendy AO, Essam TM, Aziz RK. Gut microbiome alterations in patients with stage 4 hepatitis C. *Gut Pathog.* 2016;8(1):42.
23. Bajaj JS, Sterling RK, Betrapally NS, et al. HCV eradication does not impact gut dysbiosis or systemic inflammation in cirrhotic patients. *Aliment Pharmacol Ther.* 2016;44(6):638–643.
24. Yitbarek A, Weese JS, Alkie TN, Parkinson J, Sharif S. Influenza A virus subtype H9N2 infection disrupts the composition of intestinal microbiota of chickens. *FEMS Microbiol Ecol.* 2018;94(1):1–10.
25. Li H, Liu X, Chen F, et al. Avian influenza virus subtype H9N2 affects intestinal microbiota, barrier structure injury, and inflammatory intestinal disease in the chicken ileum. *Viruses.* 2018;10(5):270.
26. Ganz HH, Doroud L, Firl AJ, Hird SM, Eisen JA, Boyce WM. Community-level differences in the microbiome of healthy wild mallards and those infected by influenza A viruses. *mSystems.* 2017;2(1), e001188.
27. Cui N, Huang X, Kong Z, et al. Newcastle disease virus infection interferes with the formation of intestinal microflora in newly hatched specific-pathogen-free chicks. *Front Microbiol.* 2018;9:900.
28. Gonmei G, Sapcota D, Saikia GK, et al. Studies on immune response to Newcastle disease virus in broiler chickens fed with Lactobacillus reuteri PIA16 isolated from the gut of indigenous chicken of Assam, India. *Vet World.* 2019;12(8):1251–1255.
29. Li L, Kubasova T, Rychlik I, Hoerr FJ, Rautenschlein S. Infectious bursal disease virus infection leads to changes in the gut associated-lymphoid tissue and the microbiota composition. *PLoS One.* 2018;13(2), e0192066.
30. Li L, Pielsticker C, Han Z, et al. Infectious bursal disease virus inoculation infection modifies Campylobacter jejuni-host interaction in broilers. *Gut Pathog.* 2018;10:13.
31. Phillips RA, Opitz HM. Pathogenicity and persistence of Salmonella enteritidis and egg contamination in normal and infectious bursal disease virus-infected leghorn chicks. *Avian Dis.* 1995;39(4):778–787.
32. Arafat N, Eladl AH, Mahgoub H, El-Shafei RA. Effect of infectious bursal disease (IBD) vaccine on Salmonella Enteritidis infected chickens. *Vaccine.* 2017;35(29): 3682–3689.
33. D'Andreano S, Sanchez Bonastre A, Francino O, et al. Gastrointestinal microbial population of turkey (Meleagris gallopavo) affected by hemorrhagic enteritis virus. *Poult Sci.* 2017;96(10):3550–3558.

34. Fitzgerald SD, Rautenschlein S, Mahsoub HM, Pierson FW, Reed WM, Jack SW. Adenovirus infections. In: *Diseases of Poultry*. John Wiley & Sons, Inc.; 2020:321–363. 14th ed. vol 1.
35. Correa-Fiz F, Blanco-Fuertes M, Navas MJ, et al. Comparative analysis of the fecal microbiota from different species of domesticated and wild suids. *Sci Rep*. 2019; 9(1):13616.
36. Ober RA, Thissen JB, Jaing CJ, Cino-Ozuna AG, Rowland RRR, Niederwerder MC. Increased microbiome diversity at the time of infection is associated with improved growth rates of pigs after co-infection with porcine reproductive and respiratory syndrome virus (PRRSV) and porcine circovirus type 2 (PCV2). *Vet Microbiol*. 2017;208:203–211.
37. van Sambeek DM, Tran H, Fernando SC, Ciobanu DC, Miller PS, Burkey TE. Alteration of the pig intestinal microbiome when vaccinated against or inoculated with porcine circovirus 2 using a multivariate analysis model. *J Anim Sci*. 2016;94(Suppl. 3): 387–390.
38. Niederwerder MC, Constance LA, Rowland RRR, et al. Fecal microbiota transplantation is associated with reduced morbidity and mortality in Porcine Circovirus associated disease. *Front Microbiol*. 2018;9:1631.
39. Huang A, Cai R, Wang Q, Shi L, Li C, Yan H. Dynamic change of gut microbiota during porcine epidemic diarrhea virus infection in suckling piglets. *Front Microbiol*. 2019;10:322.
40. Song D, Peng Q, Chen Y, et al. Altered gut microbiota profiles in sows and neonatal piglets associated with porcine epidemic diarrhea virus infection. *Sci Rep*. 2017; 7(1):17439.
41. Singh P, Teal TK, Marsh TL, et al. Intestinal microbial communities associated with acute enteric infections and disease recovery. *Microbiome*. 2015;3:45.
42. Margreiter M, Ludl K, Phleps W, Kaehler ST. Therapeutic value of a Lactobacillus gasseri and Bifidobacterium longum fixed bacterium combination in acute diarrhea: a randomized, double-blind, controlled clinical trial. *Int J Clin Pharmacol Ther*. 2006;44(5): 207–215.
43. Jang JY, Kim S, Kwon MS, et al. Rotavirus-mediated alteration of gut microbiota and its correlation with physiological characteristics in neonatal calves. *J Microbiol*. 2019; 57(2):113–121.
44. Uchiyama J, Murakami H, Sato R, et al. Examination of the fecal microbiota in dairy cows infected with bovine leukemia virus. *Vet Microbiol*. 2020;240:108547.
45. Paules CI, Sullivan SG, Subbarao K, Fauci AS. Chasing seasonal influenza—the need for a universal influenza vaccine. *N Engl J Med*. 2018;378(1):7–9.
46. Iuliano AD, Roguski KM, Chang HH, et al. Estimates of global seasonal influenza-associated respiratory mortality: a modelling study. *Lancet*. 2018;391(10127):1285–1300.
47. Belibasakis GN, Hajishengallis G. Advances in oral mucosal immunity and the microbiome. *Adv Exp Med Biol*. 2019;1197:1–9.
48. Robinson KM, Kolls JK, Alcorn JF. The immunology of influenza virus-associated bacterial pneumonia. *Curr Opin Immunol*. 2015;34:59–67.
49. Lecuyer H, Audibert J, Bobigny A, et al. Dolosigranulum pigrum causing nosocomial pneumonia and septicemia. *J Clin Microbiol*. 2007;45(10):3474–3475.
50. Ramsey MM, Freire MO, Gabrilska RA, Rumbaugh KP, Lemon KP. Staphylococcus aureus shifts toward commensalism in response to corynebacterium species. *Front Microbiol*. 2016;7:1230.
51. Antunes KH, Fachi JL, de Paula R, et al. Microbiota-derived acetate protects against respiratory syncytial virus infection through a GPR43-type 1 interferon response. *Nat Commun*. 2019;10(1):3273.

52. Marrazzo JM. Biomedical prevention of HIV in women: challenges and approaches, with particular reference to the vaginal microbiome. *Trans Am Clin Climatol Assoc.* 2018;129:63–73.
53. Farcasanu M, Kwon DS. The influence of cervicovaginal microbiota on mucosal immunity and prophylaxis in the battle against HIV. *Curr HIV/AIDS Rep.* 2018; 15(1):30–38.
54. Passmore JA, Jaspan HB, Masson L. Genital inflammation, immune activation and risk of sexual HIV acquisition. *Curr Opin HIV AIDS.* 2016;11(2):156–162.
55. Dillon SM, Frank DN, Wilson CC. The gut microbiome and HIV-1 pathogenesis: a two-way street. *AIDS.* 2016;30(18):2737–2751.
56. Tuddenham SA, Koay WLA, Zhao N, et al. The impact of human immunodeficiency virus infection on gut microbiota alpha-diversity: an individual-level meta-analysis. *Clin Infect Dis.* 2020;70(4):615–627.
57. Qing Y, Xie H, Su C, et al. Gut microbiome, short-chain fatty acids, and mucosa injury in young adults with human immunodeficiency virus infection. *Dig Dis Sci.* 2019; 64(7):1830–1843.
58. Chevalier MF, Petitjean G, Dunyach-Remy C, et al. The Th17/Treg ratio, IL-1RA and sCD14 levels in primary HIV infection predict the T-cell activation set point in the absence of systemic microbial translocation. *PLoS Pathog.* 2013; 9(6), e1003453.
59. Correa-Oliveira R, Fachi JL, Vieira A, Sato FT, Vinolo MA. Regulation of immune cell function by short-chain fatty acids. *Clin Trans Immunol.* 2016;5(4), e73.
60. Gori A, Tincati C, Rizzardini G, et al. Early impairment of gut function and gut flora supporting a role for alteration of gastrointestinal mucosa in human immunodeficiency virus pathogenesis. *J Clin Microbiol.* 2008;46(2):757–758.
61. Tincati C, Douek DC, Marchetti G. Gut barrier structure, mucosal immunity and intestinal microbiota in the pathogenesis and treatment of HIV infection. *AIDS Res Ther.* 2016;13:19.
62. Luhrs H, Gerke T, Muller JG, et al. Butyrate inhibits NF-kappaB activation in lamina propria macrophages of patients with ulcerative colitis. *Scand J Gastroenterol.* 2002;37(4):458–466.
63. Ahmed SM, Hall AJ, Robinson AE, et al. Global prevalence of norovirus in cases of gastroenteritis: a systematic review and meta-analysis. *Lancet Infect Dis.* 2014;14(8):725–730.
64. Baldridge MT, Turula H, Wobus CE. Norovirus regulation by host and microbe. *Trends Mol Med.* 2016;22(12):1047–1059.
65. Zheng DP, Ando T, Fankhauser RL, Beard RS, Glass RI, Monroe SS. Norovirus classification and proposed strain nomenclature. *Virology.* 2006;346(2):312–323.
66. Lei S, Ramesh A, Twitchell E, et al. High protective efficacy of probiotics and rice bran against human norovirus infection and diarrhea in gnotobiotic pigs. *Front Microbiol.* 2016;7:1699.
67. Reeves AE, Theriot CM, Bergin IL, Huffnagle GB, Schloss PD, Young VB. The interplay between microbiome dynamics and pathogen dynamics in a murine model of Clostridium difficile infection. *Gut Microbes.* 2011;2(3):145–158.
68. Schwille-Kiuntke J, Frick JS, Zanger P, Enck P. Post-infectious irritable bowel syndrome—a review of the literature. *Z Gastroenterol.* 2011;49(8):997–1003.
69. Rydell GE, Kindberg E, Larson G, Svensson L. Susceptibility to winter vomiting disease: a sweet matter. *Rev Med Virol.* 2011;21(6):370–382.
70. Wacklin P, Tuimala J, Nikkila J, et al. Faecal microbiota composition in adults is associated with the FUT2 gene determining the secretor status. *PLoS One.* 2014;9(4), e94863.

71. Cram JA, Hager KW, Kublin JG. Utilizing gnotobiotic models to inform the role of the microbiome in vaccine response heterogeneity. *Curr Opin HIV AIDS*. 2018;13(1):1–8.
72. Bui T, Kocher J, Li Y, et al. Median infectious dose of human norovirus GII.4 in gnotobiotic pigs is decreased by simvastatin treatment and increased by age. *J Gen Virol*. 2013;94(Pt. 9):2005–2016.
73. Lei S, Samuel H, Twitchell E, et al. Enterobacter cloacae inhibits human norovirus infectivity in gnotobiotic pigs. *Sci Rep*. 2016;6:25017.
74. Pott J, Mahlakoiv T, Mordstein M, et al. IFN-lambda determines the intestinal epithelial antiviral host defense. *Proc Natl Acad Sci U S A*. 2011;108(19):7944–7949.
75. Rocha-Pereira J, Jacobs S, Noppen S, Verbeken E, Michiels T, Neyts J. Interferon lambda (IFN-lambda) efficiently blocks norovirus transmission in a mouse model. *Antiviral Res*. 2018;149:7–15.
76. Tate JE, Burton AH, Boschi-Pinto C, Parashar UD. Global, regional, and national estimates of rotavirus mortality in children <5 years of age, 2000-2013. *Clin Infect Dis*. 2016;62(Suppl. 2):S96–s105.
77. Harris VC, Haak BW, Handley SA, et al. Effect of antibiotic-mediated microbiome modulation on rotavirus vaccine immunogenicity: a human, randomized-control proof-of-concept trial. *Cell Host Microbe*. 2018;24(2), 197–207.e194.
78. Ramani S, Mamani N, Villena R, et al. Rotavirus serum IgA immune response in children receiving rotarix coadministered with bOPV or IPV. *Pediatr Infect Dis J*. 2016;35(10):1137–1139.
79. Parker EP, Ramani S, Lopman BA, et al. Causes of impaired oral vaccine efficacy in developing countries. *Future Microbiol*. 2018;13:97–118.
80. Taniuchi M, Platts-Mills JA, Begum S, et al. Impact of enterovirus and other enteric pathogens on oral polio and rotavirus vaccine performance in Bangladeshi infants. *Vaccine*. 2016;34(27):3068–3075.
81. Rytter MJ, Kolte L, Briend A, Friis H, Christensen VB. The immune system in children with malnutrition—a systematic review. *PLoS One*. 2014;9(8), e105017.
82. Desselberger U. Differences of rotavirus vaccine effectiveness by country: likely causes and contributing factors. *Pathogens*. 2017;6(4):65.
83. Desselberger U. The mammalian intestinal microbiome: composition, interaction with the immune system, significance for vaccine efficacy, and potential for disease therapy. *Pathogens*. 2018;7(3):57.
84. Garcia-Lopez R, Perez-Brocal V, Diez-Domingo J, Moya A. Gut microbiota in children vaccinated with rotavirus vaccine. *Pediatr Infect Dis J*. 2012;31(12):1300–1302.
85. Milosevic I, Vujovic A, Barac A, et al. Gut-liver axis, gut microbiota, and its modulation in the management of liver diseases: a review of the literature. *Int J Mol Sci*. 2019;20(2):395.
85a. WHO. World Health Organization Hepatitis Fact Sheets. https://www.who.int/hepatitis/en/. Published 2019. Updated July 2019. Accessed April 10, 2020.
86. Virgin HW. The virome in mammalian physiology and disease. *Cell*. 2014;157(1):142–150.
87. Griffiths DJ. Endogenous retroviruses in the human genome sequence. *Genome Biol*. 2001;2(6), REVIEWS1017.
88. Luzuriaga K. Introduction to retroviridae. In: Long SS, Pickering LK, Prober CG, eds. *Principles and Practice of Pediatric Infectious Diseases*. 4th ed. Philadelphia, PA, USA: Saunders; 2012:1196.
89. Dalmasso M, Hill C, Ross RP. Exploiting gut bacteriophages for human health. *Trends Microbiol*. 2014;22(7):399–405.
90. Reyes A, Haynes M, Hanson N, et al. Viruses in the faecal microbiota of monozygotic twins and their mothers. *Nature*. 2010;466(7304):334–338.

91. Kapusinszky B, Minor P, Delwart E. Nearly constant shedding of diverse enteric viruses by two healthy infants. *J Clin Microbiol.* 2012;50(11):3427–3434.
92. Lim ES, Zhou Y, Zhao G, et al. Early life dynamics of the human gut virome and bacterial microbiome in infants. *Nat Med.* 2015;21(10):1228–1234.
93. Pannaraj PS, Ly M, Cerini C, et al. Shared and distinct features of human milk and infant stool viromes. *Front Microbiol.* 2018;9:1162.
94. Carding SR, Davis N, Hoyles L. Review article: the human intestinal virome in health and disease. *Aliment Pharmacol Ther.* 2017;46(9):800–815.
95. Choi JH, Kim GB, Cha CJ. Spatial heterogeneity and stability of bacterial community in the gastrointestinal tracts of broiler chickens. *Poult Sci.* 2014;93(8):1942–1950.
96. Wilkinson TJ, Cowan AA, Vallin HE, et al. Characterization of the microbiome along the gastrointestinal tract of growing turkeys. *Front Microbiol.* 2017;8:1089.
97. Best AA, Porter AL, Fraley SM, Fraley GS. Characterization of gut microbiome dynamics in developing pekin ducks and impact of management system. *Front Microbiol.* 2016;7:2125.
98. Borda-Molina D, Seifert J, Camarinha-Silva A. Current perspectives of the chicken gastrointestinal tract and its microbiome. *Comput Struct Biotechnol J.* 2018;16:131–139.
99. Swayne DE, Suarez DL, Sims LD. Influenza. In: Swayne DE, Boulianne M, Logue CM, et al. *Diseases of Poultry.* John Wiley & Sons, Inc.; 2020:210–256. vol 1.
100. Rizzatti G, Lopetuso LR, Gibiino G, Binda C, Gasbarrini A. Proteobacteria: a common factor in human diseases. *Biomed Res Int.* 2017;2017:9351507.
101. Hird SM, Ganz H, Eisen JA, Boyce WM. The cloacal microbiome of five wild duck species varies by species and influenza A virus infection status. *mSphere.* 2018;3(5), e00382-18.
102. Yitbarek A, Taha-Abdelaziz K, Hodgins DC, et al. Gut microbiota-mediated protection against influenza virus subtype H9N2 in chickens is associated with modulation of the innate responses. *Sci Rep.* 2018;8(1):13189.
103. Hansson GC. Role of mucus layers in gut infection and inflammation. *Curr Opin Microbiol.* 2012;15(1):57–62.
104. Aihara E, Engevik KA, Montrose MH. Trefoil factor peptides and gastrointestinal function. *Annu Rev Physiol.* 2017;79:357–380.
105. Rima B, Balkema-Buschmann A, Dundon WG, et al. ICTV virus taxonomy profile: paramyxoviridae. *J Gen Virol.* 2019;100(12):1593–1594.
106. Wakamatsu N, King DJ, Kapczynski DR, Seal BS, Brown CC. Experimental pathogenesis for chickens, turkeys, and pigeons of exotic Newcastle disease virus from an outbreak in California during 2002–2003. *Vet Pathol.* 2006;43(6):925–933.
107. Suarez DL, Miller PJ, Koch G, Mundt E, Rautenschlein S. Newcastle disease, other avian paramyxoviruses, and avian metapneumovirus infections. *Dis Poult.* 2020;109–166.
108. Talebi A, Amirzadeh B, Mokhtari B, Gahri H. Effects of a multi-strain probiotic (PrimaLac) on performance and antibody responses to Newcastle disease virus and infectious bursal disease virus vaccination in broiler chickens. *Avian Pathol.* 2008; 37(5):509–512.
109. Bautista-Garfias CR, Rios-Flores E, Garcia-Rubio VG. Comparative effect of Lactobacillus casei and a commercial mangosteen dietary supplement on body weight gain and antibody response to Newcastle disease virus vaccine in fighting roosters. *J Med Food.* 2011;14(7–8):828–833.
110. Delmas B, Attoui H, Ghosh S, et al. ICTV virus taxonomy profile: Birnaviridae. *J Gen Virol.* 2019;100(1):5–6.
111. Hirai K, Shimakura S, Kawamoto E, et al. The immunodepressive effect of infectious bursal disease virus in chickens. *Avian Dis.* 1974;18(1):50–57.
112. Saif YM. Immunosuppression induced by infectious bursal disease virus. *Vet Immunol Immunopathol.* 1991;30(1):45–50.

113. Rautenschlein S, Mahsoub HM, Fitzgerald SD, Pierson FW. Hemorrhagic enteritis and related infections. In: Swayne DE, Boulianne M, Logue CM, et al. *Diseases of Poultry*. USA: John Wiley & Sons, Inc.; 2020:339–347. 14th ed. vol 1.
114. Pan D, Yu Z. Intestinal microbiome of poultry and its interaction with host and diet. *Gut Microbes*. 2014;5(1):108–119.
115. Barrow PA, Jones MA, Smith AL, Wigley P. The long view: Salmonella—the last forty years. *Avian Pathol*. 2012;41(5):413–420.
116. Subler KA, Mickael CS, Jackwood DJ. Infectious bursal disease virus-induced immunosuppression exacerbates Campylobacter jejuni colonization and shedding in chickens. *Avian Dis*. 2006;50(2):179–184.
117. Bautista DA, Elankumaran S, Heckert RA. Effect of a variant infectious bursal disease virus (E/Del) on Salmonella typhimurium infection in commercial broiler chickens. *Avian Dis*. 2004;48(2):361–369.
118. Davison A, Harrach B. Siadenovirus. In: Tidona C, Darai G, eds. *The Springer Index of Viruses*. New York: Springer; 2011:49–56.
119. Chen X, Zheng M, Lin F, et al. Impacts of novel duck reovirus infection on the composition of intestinal microbiota of Muscovy ducklings. *Microb Pathog*. 2019;137:103764.
120. Ma X, Wang Q, Li H, Xu C, Cui N, Zhao X. 16S rRNA genes Illumina sequencing revealed differential cecal microbiome in specific pathogen free chickens infected with different subgroup of avian leukosis viruses. *Vet Microbiol*. 2017;207:195–204.
121. Suresh M, Sharma JM. Hemorrhagic enteritis virus induced changes in the lymphocyte subpopulations in turkeys and the effect of experimental immunodeficiency on viral pathogenesis. *Vet Immunol Immunopathol*. 1995;45(1–2):139–150.
122. Suresh M, Sharma JM. Pathogenesis of type II avian adenovirus infection in turkeys: in vivo immune cell tropism and tissue distribution of the virus. *J Virol*. 1996;70(1):30–36.
123. Rautenschlein S, Suresh M, Sharma JM. Pathogenic avian adenovirus type II induces apoptosis in turkey spleen cells. *Arch Virol*. 2000;145(8):1671–1683.
124. Saunders GK, Pierson FW, Hurk JV. Haemorhagic enteritis virus infection in turkeys: a comparison of virulent and avirulent virus infections, and a proposed pathogenesis. *Avian Pathol*. 1993;22(1):47–58.
125. Day JM, Oakley BB, Seal BS, Zsak L. Comparative analysis of the intestinal bacterial and RNA viral communities from sentinel birds placed on selected broiler chicken farms. *PLoS One*. 2015;10(1), e0117210.
126. Lima DA, Cibulski SP, Finkler F, et al. Faecal virome of healthy chickens reveals a large diversity of the eukaryote viral community, including novel circular ssDNA viruses. *J Gen Virol*. 2017;98(4):690–703.
127. Ramírez-Martínez LA, Loza-Rubio E, Mosqueda J, González-Garay ML, García-Espinosa G. Fecal virome composition of migratory wild duck species. *PLoS One*. 2018;13(11), e0206970.
128. Wille M, Shi M, Klaassen M, Hurt AC, Holmes EC. Virome heterogeneity and connectivity in waterfowl and shorebird communities. *ISME J*. 2019;13(10):2603–2616.
129. VanderWaal K, Deen J. Global trends in infectious diseases of swine. *Proc Natl Acad Sci U S A*. 2018;115(45):11495.
130. Pork.org. *Top 10 Pork-Producing Countries*. 2019. https://www.pork.org/facts/stats/u-s-pork-exports/top-10-pork-producing-countries/. Accessed 1-2-20.
131. Sender R, Fuchs S, Milo R. Revised estimates for the number of human and bacteria cells in the body. *PLoS Biol*. 2016;14(8), e1002533.
132. Mackie RI, Sghir A, Gaskins HR. Developmental microbial ecology of the neonatal gastrointestinal tract. *Am J Clin Nutr*. 1999;69(5):1035s–1045s.
133. Canny GO, McCormick BA. Bacteria in the intestine, helpful residents or enemies from within? *Infect Immun*. 2008;76(8):3360–3373.

134. Qin J, Li R, Raes J, et al. A human gut microbial gene catalog established by metagenomic sequencing. *Nature*. 2010;464(7285):59–65.
135. Chenais E, Depner K, Guberti V, Dietze K, Viltrop A, Ståhl K. Epidemiological considerations on African swine fever in Europe 2014–2018. *Porcine Health Manag*. 2019;5(1):6.
136. Arias M, Jurado C, Gallardo C, Fernández-Pinero J, Sánchez-Vizcaíno JM. Gaps in African swine fever: analysis and priorities. *Transbound Emerg Dis*. 2018;65(S1):235–247.
137. Endris RG, Haslett TM, Hess WR. African swine fever virus infection in the soft tick, Ornithodoros (Alectorobius) puertoricensis (Acari: Argasidae). *J Med Entomol*. 1992;29(6):990–994.
138. Guinat C, Gogin A, Blome S, et al. Transmission routes of African swine fever virus to domestic pigs: current knowledge and future research directions. *Vet Rec*. 2016; 178(11):262–267.
139. Lacasta A, Ballester M, Monteagudo PL, et al. Expression library immunization can confer protection against lethal challenge with African swine fever virus. *J Virol*. 2014; 88(22):13322–13332.
140. Meng XJ. Circoviridae. In: Knipe DM, Howley P, eds. *Fields Virology*. 6th ed. Philadelphia, USA: Wolters Kluwer; 2013:1792–1801.
141. Meng XJ. Porcine circovirus type 2 (PCV2): pathogenesis and interaction with the immune system. *Annu Rev Anim Biosci*. 2013;1:43–64.
142. López-Lorenzo G, Díaz-Cao JM, Prieto A, et al. Environmental distribution of porcine circovirus type 2 (PCV2) in swine herds with natural infection. *Sci Rep*. 2019;9(1):14816.
143. Lunney JK, Fang Y, Ladinig A, et al. Porcine reproductive and respiratory syndrome virus (PRRSV): pathogenesis and interaction with the immune system. *Annu Rev Anim Biosci*. 2016;4:129–154.
144. Nan Y, Wu C, Gu G, Sun W, Zhang YJ, Zhou EM. Improved vaccine against PRRSV: current progress and future perspective. *Front Microbiol*. 2017;8:1635.
145. Song D, Park B. Porcine epidemic diarrhoea virus: a comprehensive review of molecular epidemiology, diagnosis, and vaccines. *Virus Genes*. 2012;44(2):167–175.
146. Diep NV, Norimine J, Sueyoshi M, Lan NT, Yamaguchi R. Novel porcine epidemic diarrhea virus (PEDV) variants with large deletions in the spike (S) gene coexist with PEDV strains possessing an intact S gene in domestic pigs in Japan: a new disease situation. *PLoS One*. 2017;12(1), e0170126.
147. Lv C, Xiao Y, Li X, Tian K. Porcine epidemic diarrhea virus: current insights. *Virus Adapt Treat*. 2016;8:1–12.
148. Mole B. Deadly pig virus slips through US borders. *Nature*. 2013;499(7459):388.
149. Lin C-M, Ghimire S, Hou Y, et al. Pathogenicity and immunogenicity of attenuated porcine epidemic diarrhea virus PC22A strain in conventional weaned pigs. *BMC Vet Res*. 2019;15(1):26.
150. Aluthge ND, Van Sambeek DM, Carney-Hinkle EE, Li YS, Fernando SC, Burkey TE. Board invited review: the pig microbiota and the potential for harnessing the power of the microbiome to improve growth and health. *J Anim Sci*. 2019;97(9):3741–3757.
151. Mach N, Berri M, Estelle J, et al. Early-life establishment of the swine gut microbiome and impact on host phenotypes. *Environ Microbiol Rep*. 2015;7(3):554–569.
152. Tan Z, Dong W, Ding Y, Ding X, Zhang Q, Jiang L. Porcine epidemic diarrhea altered colonic microbiota communities in suckling piglets. *Genes (Basel)*. 2019;11(1):44.
153. Tan Z, Dong W, Ding Y, Ding X, Zhang Q, Jiang L. Changes in cecal microbiota community of suckling piglets infected with porcine epidemic diarrhea virus. *PLoS One*. 2019;14(7), e0219868.
154. Da Silva MS, Budaszewski RF, Weber MN, et al. Liver virome of healthy pigs reveals diverse small ssDNA viral genomes. *Infect Genet Evol*. 2020;81:104203.

155. Hameed M, Liu K, Anwar MN, et al. A viral metagenomic analysis reveals rich viral abundance and diversity in mosquitoes from pig farms. *Transbound Emerg Dis*. 2020; 67(1):328–343.
156. Oikonomou G, Teixeira AG, Foditsch C, Bicalho ML, Machado VS, Bicalho RC. Fecal microbial diversity in pre-weaned dairy calves as described by pyrosequencing of metagenomic 16S rDNA. Associations of Faecalibacterium species with health and growth. *PLoS One*. 2013;8(4), e63157.
157. Mamun MAA, Sandeman M, Rayment P, et al. The composition and stability of the faecal microbiota of Merino sheep. *J Appl Microbiol*. 2020;128(1):280–291.
158. Debevere S, De Baere S, Haesaert G, Rychlik M, Fievez V, Croubels S. Development of an UPLC-MS/MS method for the analysis of mycotoxins in rumen fluid with and without maize silage emphasizes the importance of using matrix-matched calibration. *Toxins (Basel)*. 2019;11(9):519.
159. Meenatchi R, Thinesh T, Brindangnanam P, Hassan S, Kiran GS, Selvin J. Revealing the impact of global mass bleaching on coral microbiome through 16S rRNA gene-based metagenomic analysis. *Microbiol Res*. 2019;233:126408.
160. Stewart RD, Auffret MD, Warr A, Walker AW, Roehe R, Watson M. Compendium of 4,941 rumen metagenome-assembled genomes for rumen microbiome biology and enzyme discovery. *Nat Biotechnol*. 2019;37(8):953–961.
161. Chaucheyras-Durand F, Ossa F. The rumen microbiome: composition, abundance, diversity, and new investigative tools. *Prof Anim Sci*. 2014;30:1–12.
162. Walker R. *An Introduction to the Rumen Microbiome*. Noble News and Views; 2020. Accessed 03/23/2020.
163. Edrington TS, Dowd SE, Farrow RF, et al. Development of colonic microflora as assessed by pyrosequencing in dairy calves fed waste milk. *J Dairy Sci*. 2012;95(8):4519–4525.
164. Uyeno Y, Sekiguchi Y, Tajima K, Takenaka A, Kurihara M, Kamagata Y. An rRNA-based analysis for evaluating the effect of heat stress on the rumen microbial composition of Holstein heifers. *Anaerobe*. 2010;16(1):27–33.
165. Cho YI, Yoon KJ. An overview of calf diarrhea—infectious etiology, diagnosis, and intervention. *J Vet Sci*. 2014;15(1):1–17.
166. Lefkowitz EJ, Dempsey DM, Hendrickson RC, Orton RJ, Siddell SG, Smith DB. Virus taxonomy: the database of the International Committee on Taxonomy of Viruses (ICTV). *Nucleic Acids Res*. 2018;46(D1):D708–d717.
167. Aida Y, Murakami H, Takahashi M, Takeshima SN. Mechanisms of pathogenesis induced by bovine leukemia virus as a model for human T-cell leukemia virus. *Front Microbiol*. 2013;4:328.
168. Murakami H, Yamada T, Suzuki M, Nakahara Y, Suzuki K, Sentsui H. Bovine leukemia virus integration site selection in cattle that develop leukemia. *Virus Res*. 2011;156(1–2):107–112.
169. Frie MC, Coussens PM. Bovine leukemia virus: a major silent threat to proper immune responses in cattle. *Vet Immunol Immunopathol*. 2015;163(3–4):103–114.
170. Brenner J, Van-Haam M, Savir D, Trainin Z. The implication of BLV infection in the productivity, reproductive capacity and survival rate of a dairy cow. *Vet Immunol Immunopathol*. 1989;22(3):299–305.
171. Nekouei O, VanLeeuwen J, Stryhn H, Kelton D, Keefe G. Lifetime effects of infection with bovine leukemia virus on longevity and milk production of dairy cows. *Prev Vet Med*. 2016;133:1–9.
172. Norby B, Bartlett PC, Byrem TM, Erskine RJ. Effect of infection with bovine leukemia virus on milk production in Michigan dairy cows. *J Dairy Sci*. 2016;99(3): 2043–2052.

173. Polat M, Takeshima SN, Aida Y. Epidemiology and genetic diversity of bovine leukemia virus. *Virol J.* 2017;14(1):209.
174. Tenaillon O, Skurnik D, Picard B, Denamur E. The population genetics of commensal Escherichia coli. *Nat Rev Microbiol.* 2010;8(3):207–217.
175. Clermont O, Christenson JK, Denamur E, Gordon DM. The clermont Escherichia coli phylo-typing method revisited: improvement of specificity and detection of new phylo-groups. *Environ Microbiol Rep.* 2013;5(1):58–65.
176. Maciel JF, Matter LB, Tasca C, et al. Characterization of intestinal Escherichia coli isolated from calves with diarrhea due to rotavirus and coronavirus. *J Med Microbiol.* 2019;68(3):417–423.
177. Escobar-Paramo P, Le Menac'h A, Le Gall T, et al. Identification of forces shaping the commensal Escherichia coli genetic structure by comparing animal and human isolates. *Environ Microbiol.* 2006;8(11):1975–1984.
178. Dubreuil JD, Isaacson RE, Schifferli DM. Animal enterotoxigenic Escherichia coli. *EcoSal Plus.* 2016;7(1):10.
179. Kolenda R, Burdukiewicz M, Schierack P. A systematic review and meta-analysis of the epidemiology of pathogenic Escherichia coli of calves and the role of calves as reservoirs for human pathogenic E. coli. *Front Cell Infect Microbiol.* 2015;5:23.
180. Nagy B, Fekete PZ. Enterotoxigenic Escherichia coli in veterinary medicine. *Int J Med Microbiol.* 2005;295(6–7):443–454.
181. Tiels P, Verdonck F, Smet A, Goddeeris B, Cox E. The F18 fimbrial adhesin FedF is highly conserved among F18+Escherichia coli isolates. *Vet Microbiol.* 2005;110(3–4): 277–283.
182. OIE. Lumpy skin disease. In: *Manual of Diagnostic Tests and Vaccines for Terrestrial Animals.* World Organization for Animal Health; 2018:1158–1171.
183. Tageldin MH, Wallace DB, Gerdes GH, et al. Lumpy skin disease of cattle: an emerging problem in the Sultanate of Oman. *Tropl Anim Health Prod.* 2014;46(1):241–246.
184. Tuppurainen ES, Oura CA. Review: lumpy skin disease: an emerging threat to Europe, the Middle East and Asia. *Transbound Emerg Dis.* 2012;59(1):40–48.
185. de Macedo AA, Galvao NR, Sa JC, et al. Brucella-associated cervical bursitis in cattle. *Tropl Anim Health Prod.* 2019;51(3):697–702.
186. Sazmand A, Harl J, Eigner B, et al. Vector-borne bacteria in blood of camels in Iran: new data and literature review. *Comp Immunol Microbiol Infect Dis.* 2019;65:48–53.
187. Vordermeier HM, Simsova M, Wilkinson KA, et al. Recognition of mycobacterial antigens delivered by genetically detoxified Bordetella pertussis adenylate cyclase by T cells from cattle with bovine tuberculosis. *Infect Immun.* 2004;72(11): 6255–6261.
188. Sneath PH, Barrett SJ. A new species of Neisseria from the dental plaque of the domestic cow, Neisseria dentiae sp. nov. *Lett Appl Microbiol.* 1996;23(5):355–358.
189. Wolf-Jackel GA, Hansen MS, Larsen G, Holm E, Agerholm JS, Jensen TK. Diagnostic studies of abortion in Danish cattle 2015–2017. *Acta Vet Scand.* 2020;62(1):1.
190. Anderson CL, Sullivan MB, Fernando SC. Dietary energy drives the dynamic response of bovine rumen viral communities. *Microbiome.* 2017;5(1):155.
191. Seth RK, Maqsood R, Mondal A, et al. Gut DNA virome diversity and its association with host bacteria regulate inflammatory phenotype and neuronal immunotoxicity in experimental Gulf War illness. *Viruses.* 2019;11(10):968.
192. Ross EM, Petrovski S, Moate PJ, Hayes BJ. Metagenomics of rumen bacteriophage from thirteen lactating dairy cattle. *BMC Microbiol.* 2013;13(1):242.
193. Kwok KTT, Nieuwenhuijse DF, Phan MVT, Koopmans MPG. Virus metagenomics in farm animals: a systematic review. *Viruses.* 2020;12(1):107.
194. Gomez DE, Weese JS. Viral enteritis in calves. *Can Vet J.* 2017;58(12):1267–1274.

195. Weber MN, Cibulski SP, Silveira S, et al. Evaluation of the serum virome in calves persistently infected with Pestivirus A, presenting or not presenting mucosal disease. *Virus Genes*. 2018;54(6):768–778.
196. Zhang M, Hill JE, Godson DL, Ngeleka M, Fernando C, Huang Y. The pulmonary virome, bacteriological and histopathological findings in bovine respiratory disease from western Canada. *Transbound Emerg Dis*. 2020;67(2):924–934.
197. Zhang M, Hill JE, Fernando C, et al. Respiratory viruses identified in western Canadian beef cattle by metagenomic sequencing and their association with bovine respiratory disease. *Transbound Emerg Dis*. 2019;66(3):1379–1386.
198. Wuthrich D, Boujon CL, Truchet L, et al. Exploring the virome of cattle with non-suppurative encephalitis of unknown etiology by metagenomics. *Virology*. 2016; 493:22–30.
199. Gomez DE, Arroyo LG, Costa MC, Viel L, Weese JS. Characterization of the fecal bacterial microbiota of healthy and diarrheic dairy calves. *J Vet Intern Med*. 2017; 31(3):928–939.
200. Trefz FM, Lorenz I, Lorch A, Constable PD. Clinical signs, profound acidemia, hypoglycemia, and hypernatremia are predictive of mortality in 1,400 critically ill neonatal calves with diarrhea. *PLoS One*. 2017;12(8), e0182938.

CHAPTER THREE

Gut microbiota in reductive drug metabolism

Yukuang Guo, Hyunwoo Lee*, Hyunyoung Jeong*
Department of Pharmaceutical Sciences, Center for Biomolecular Sciences, College of Pharmacy, University of Illinois at Chicago, Chicago, IL, United States
*Corresponding authors: e-mail address: hlee31@uic.edu; yjeong@uic.edu

Contents

Abstract

Gut bacteria are predominant microorganisms in the gut microbiota and have been recognized to mediate a variety of biotransformations of xenobiotic compounds in the gut. This review is focused on one of the gut bacterial xenobiotic metabolisms, reduction. Xenobiotics undergo different types of reductive metabolisms depending on chemically distinct groups: azo (-N=N-), nitro (-NO_2), alkene (-C=C-), ketone (-C=O), *N*-oxide (-N—O), and sulfoxide (-S=O). In this review, we have provided select examples of drugs in six chemically distinct groups that are known or suspected to be subjected to the reduction by gut bacteria. For some drugs, responsible enzymes in specific gut bacteria have been identified and characterized, but for many drugs, only circumstantial evidence is available that indicates gut bacteria-mediated reductive metabolism. The physiological roles of even known gut bacterial enzymes have not been well defined.

1. Introduction

The human gut harbors around 100 trillion bacteria, whose genomes are predicted to contain about 100 times more protein-encoding genes than the human genome.[1,2] Such richness in the gene contents likely reflects

Progress in Molecular Biology and Translational Science, Volume 171
ISSN 1877-1173
https://doi.org/10.1016/bs.pmbts.2020.04.002

extremely diverse catalytic reactions that can be mediated by gut bacteria, and unsurprisingly, gut bacteria have repeatedly been identified to metabolize small xenobiotic molecules including therapeutic drugs.[3–6] Gut, either small or large intestine, provides an environment with low redox potentials and with no to little oxygen available.[7–10] Such anaerobic and reducing conditions in the gut likely position resident gut bacteria to readily mediate the reductive metabolism of incoming xenobiotics such as therapeutic drugs. This review is focused on drugs undergoing reductive metabolism by gut bacteria. Depending on chemically distinct groups, xenobiotics may undergo one or more types of six major reductive metabolisms: azo (-N=N-), nitro (-NO_2), alkene (-C=C-), ketone (-C=O), *N*-oxide (-N—O), and sulfoxide (-S=O).[3–6] We have provided select examples of drugs in six chemically distinct groups, for which gut bacteria are known or suspected to mediate the reduction (summarized in Table 1).

2. Reductive drug metabolism

2.1 Azoreduction

Azo compounds are a group of organic chemicals (R-N=N-R′), in which "-N=N-" is the azo bond, and R or R′ is either alkyl or aryl (aromatic) group.[34] Azo compounds with alkyl groups are colorless, whereas those (azo dyes) containing aromatic groups give characteristic colors. Because azo dyes are the most common coloring agents used in the textile, paper, cosmetics, food, and pharmaceutical industries, their (potentially toxic) metabolism in animals and humans, as well as their degradation in the environment, have been important health-related research subjects.[34,35] Earlier studies have well established that azo dyes are metabolized in the body via both oxidative and reductive reactions, with the former mediated by the host and the latter mainly by intestinal bacteria.[36]

2.1.1 Azo compounds as pro-drugs

2.1.1.1 Prontosil

The azo dye prontosil (Fig. 1A) was identified as an anti-infective agent active against *Streptococcal* infections in mice.[37] Prontosil exhibited its efficacy only in vivo mouse infection models, with no apparent antibacterial activity against *Streptococci* in vitro. Because *p*-aminobenznesulfonamide (sulfanilamide) (Fig. 1A), which can be generated upon the reductive cleavage of the azo bond of prontosil, was as effective as prontosil for controlling *Streptococcal* infections, prontosil was proposed to be reductively metabolized

Table 1 Gut bacteria identified to mediate reductive metabolisms.

Reductive metabolism	Drug	Gut bacteria identified	Cecal/ intestinal contents/ feces tested
Azoreduction	Prontosil	Many gut bacterial species	Rat cecal contents [11]
	Sulfasalazine	*Bacteroides fragilis*; *Clostridium perfringens*; *Corynebacterium acnes*; *Fusobacterium* sp.; *Streptococcus faecium*; *Streptococcus faecalis*; *Escherichia coli*; *Lactobacillus* sp. [12]	
Nitroreduction	Chloramphenicol	*Escherichia coli* [13,14]; *Haemophilus influenza* [15,16]; *Neisseria meningitides* [16]; *Bacteroides fragilis* [17]	
	Nitrobenzodiazepines	*Escherichia coli* [18]; *Clostridium leptum* [19]	
Alkene reduction	Deleobuvir	N/A	Rat feces; human feces [20]
	Digoxin	*Eggerthella lenta* [21]	Human feces [22]
Keto-reduction	Nabumetone	*Escherichia coli*; *Lactobacillus casei*; *Lactobacillus plantarum*; *Parabacteroides distasonis* [23]	
	Doxorubicin	*Raoultella planticola*; *Escherichia coli* [24]	
	Tacrolimus	*Ruminococcaceae* sp.; *Clostridium innocuum*; *Anaerostipes* sp.; *Dorea formicigenerans*; *Clostridium clostridioforme*; *Clostridium hathewayi*; *Blautia* sp.; *Clostridium aldenense*; *Clostridium symbiosum*; *Clostridium citroniae*; *Coprococcus* sp.; *Clostridium bolteae*; *Clostridium cadaveris*; *Ruminococcus gnavus*; *Erysipelotrichaceae* sp.; *Bacteroides cellulosilyticus*; *Bacteroides finegoldii*; *Bacteroides ovatus*; *Parabacteroides merdae*; *Parabacteroides johnsonii*; *Parabacteroides goldsteinii* [25]	Human feces [25]

Continued

Table 1 Gut bacteria identified to mediate reductive metabolisms.—cont'd

Reductive metabolism	Drug	Gut bacteria identified	Cecal/ intestinal contents/ feces tested
Sulfoxide reduction	Sulindac	*Escherichia coli* [26]	Human intestinal contents [27]; Human feces [28,29]
	Sulphinpyrazone	*Escherichia coli*; *Enterobacter* sp.; *Proteus* sp.; *Providencia* sp.; *Klebsiella* sp.; *Citrobacter* sp.; *Pseudomonas* sp.; *Enterococcus* sp.; *Bacteroides* sp.; *Fusobacterium* sp.; *Clostridium* sp.[29]	Rat feces [30]; Human feces [29]
N-Oxide reduction	Nicotine	N/A	Rat intestinal and cecal contents [31]
	Loperamide	N/A	Rat intestinal and cecal contents [32]
	Ranitidine	N/A	Human feces [33]

N/A: individual bacteria not tested.

in the body into the active component sulfanilamide.[38,39] As was the case for many other azo dyes,[40,41] prontosil was shown to be reduced by intestinal bacteria in rats and humans.[11,42] Moreover, antibiotic treatments of rats receiving oral prontosil reduced the excretion of sulfanilamide significantly in urine,[42] further suggesting that gut bacteria play a critical role in cleaving the azo bond and producing the active component of prontosil. With the finding of prontosil as a pro-drug that is converted to the active metabolite by gut bacteria, the azo linkage has been extensively employed in the development of in particular colon-specific pro-drugs that are activated by gut bacteria.[43]

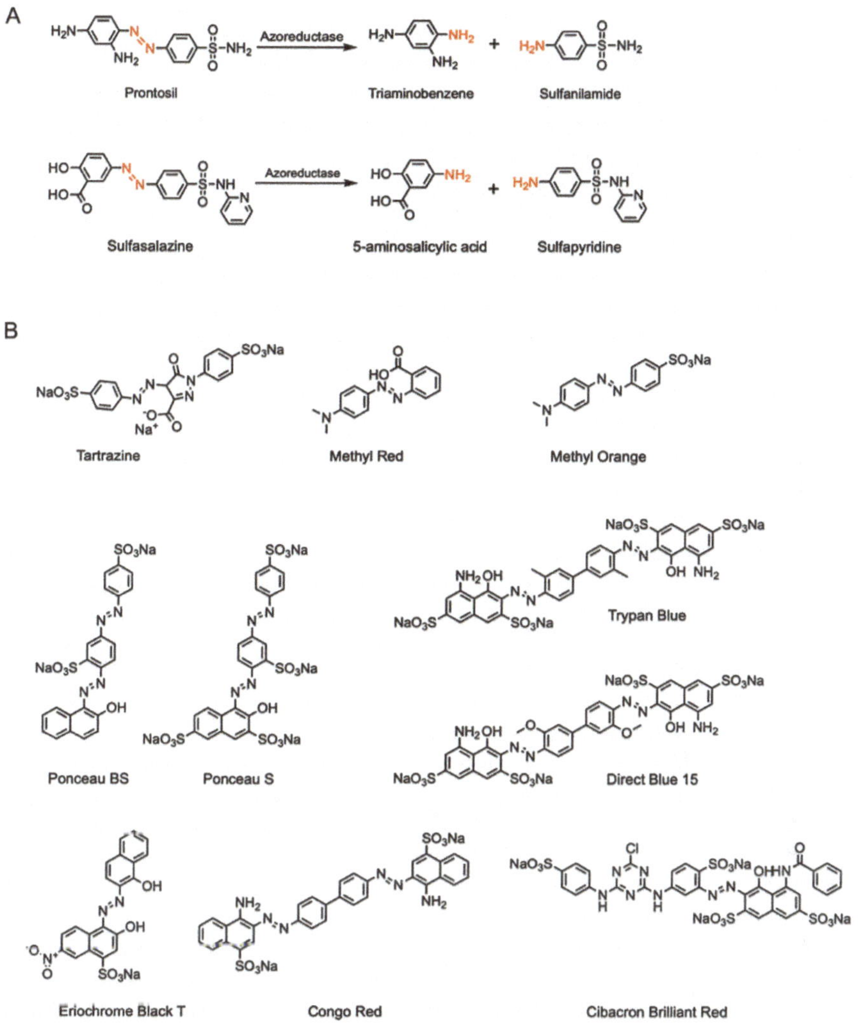

Fig. 1 Azoreduction by gut bacteria. Drugs (A) and dyes (B) that undergo azoreduction by gut bacteria are shown. See text for details.

2.1.1.2 Sulfasalazine

Sulfasalazine (salazopyrin; salicylazosulfapyridine) is a compound synthesized by combining between 5-aminosalicylic acid and sulfapyridine (Fig. 1A). Since sulfasalazine was identified to be active in ulcerative colitis and introduced into clinical practice in 1941,[44] it has been used as the main treatment of ulcerative colitis.[45]

Using conventional and germ-free rats and gut bacteria, sulfasalazine was shown to be reductively metabolized by gut bacteria, the result of which is the generation of 5-aminosalicylic acid and sulfapyridine.[12,46] A later study determined that 5-aminosalicylic acid is the therapeutically active component[47] while sulfapyridine was deemed to be responsible for the side effects of sulfasalazine.[48] Because the colon is the site where 5-aminosalicylic acid exerts its therapeutic effect and most of the 5-aminosalicylic acid liberated from sulfasalazine by gut bacteria in the colon is not absorbed for systemic circulation,[49] it is not surprising that antibiotic treatments affect the clinical outcomes of sulfasalazine treatments. For example, in healthy volunteers [taking sulfasalazine for 3 days, 6-day break, ampicillin (250 mg four times daily) for 5 days followed by 3 days of sulfasalazine (2 g twice daily)], the plasma concentration of sulfapyridine was significantly lower in post-treatment than in pre-treatment with ampicillin, indicating that ampicillin treatment impairs the azo reduction of sulfasalazine by gut bacteria and the amount of liberated 5-aminosalicylic acid is reduced.[50]

2.1.2 Azoreductases

Many earlier studies that investigated the reduction of azo dyes led to the identification of bacterial enzymes collectively called azoreductases.[51] The majority of these enzymes are identified in aerobic bacteria, a few azoreductases are characterized in facultative anaerobic bacteria, and only one enzyme is known in obligate anaerobic bacteria. Phylogenetic analysis based on primary amino acid sequences classifies most of the known and putative azoreductases into four distinct clades (I–IV), while the AzoC azoreductase in the strict anaerobic gut bacterium *Clostridium perfringens* is phylogenetically separated from these clades.[52]

Facultative and obligate anaerobic gut bacteria belonging to major phyla present in the human intestine appear to possess one or more (un)characterized azoreductases.[53–56] However, the physiological role of azoreductases in gut bacteria is unclear. Since the substrates of azoreductases, i.e., aromatic azo compounds, are extremely rare in nature,[57] it is difficult to imagine that these gut bacterial enzymes have evolved to catalyze azo reduction specifically. There is evidence that may suggest that azo reduction is an off-target activity of certain reductases; some azoreductases can catalyze the reduction of different types of substrates. For example, AzoR, the azoreductase of the facultative anaerobic gut bacterium *Escherichia coli*, catalyzes quinone (menadione) reduction with higher specific activity than the azo reduction of Methyl Red,[58] and it is also able to catalyze the reduction of nitro compounds.[59,60] Moreover, a Δ*azoR* mutant exhibits an aerobic growth defect in the presence of quinones.[61] More work will be required to define better the physiological roles of azoreductases in gut bacteria.

As mentioned above, AzoC in *C. perfringens* is the only azoreductase characterized in strictly anaerobic bacteria.[62,63] Though a physiological role of AzoC is not well defined, a study suggests that it might be involved in resistance to certain toxic azo dyes. It was shown that wild-type *C. perfringens* ATCC 3626 grows slower in the presence than the absence of an azo dye such as tartrazine, Methyl Red, Methyl Orange, Ponceau BS, Ponceau S, Trypan Blue, or Direct Blue 15 (Fig. 1B).[64] The growth inhibitory effects of these azo dyes appear to be specific since no such effects were observed with other azo dyes such as Eriochrome Black T, Congo Red, and Cibacron Brilliant Red (Fig. 1B). Interestingly, a Δ*azoC* mutant was partially or entirely resistant to the growth inhibitory effects of Ponceau BS, Ponceau S, Trypan Blue, and Direct Blue 15, suggesting that the AzoC-mediated reductive cleavage products of these azo dyes might be responsible for the impaired growth. Moreover, the Δ*azoC* mutant displayed a similar growth defect to the wild type in the presence of tartrazine, Methyl Red, and Methyl Orange, suggesting the existence of additional azoreductase(s).[64] Our unpublished search for AzoC homologs in gut microbial metagenomes using MetaQuery[65] indicate that gut bacteria in all major phyla possess AzoC homologs (>70% coverage, >30% identity, e-value >1e−5; data not shown). This gut microbiome search result, together with the reported phenotype of the Δ*azoC* mutant in *C. perfringens*, raises an intriguing question about the effects of azo dyes on the gut microbiota. Tartrazine is one of the FDA-approved azo dyes for use as food additives. The fact that both the wild-type *C. perfringens* and Δ*azoC* mutant strains show similar growth defects when grown in the presence of tartrazine might suggest the existence of as-yet-unknown azoreductase(s) that mainly mediates the azo reduction of tartrazine. To the best of our knowledge, studies investigating the effects of azo dyes on the composition and diversity of the gut microbiota have not been reported. However, since AzoC homologs are present in all major gut bacterial phyla, short- or long-term intake of certain azo dyes as food additives may have profound effects on the relative abundance and/or function of the gut microbiota, consequently affecting host physiology.

2.2 Nitroreduction

Similar to azo dyes, numerous aromatic compounds that contain nitro groups (-NO_2) have been synthesized for a variety of industrial purposes such as the synthesis of plastics, dyes, explosives, and drugs.

Nitroaromatics also include nitrated-polycyclic aromatic hydrocarbons formed by (photo)chemical transformation of polycyclic aromatic hydrocarbons in air and during combustion processes.[66] Since many nitroaromatic compounds have been found to have mutagenic and/or tumorigenic activity, their contamination in the environment and toxic effects on humans have been investigated extensively.[66–68] The mutagenicity and/or tumorigenicity of nitroaromatic compounds have been proposed to be due to nitroso and hydroxylamine reactive intermediates that are generated during the reduction of nitro groups to amines by enzymes collectively termed nitroreductases.[69] A large number of studies about nitroreduction have been directed toward the identification and characterization of (an)aerobic environmental bacteria, with their potential applications in biodegradation of nitroaromatic compounds.[66,70,71] However, nitroaromatic compounds also include therapeutic drugs used in the treatment of a variety of human diseases, and nitroreduction by intestinal bacteria has been implicated in modulating the efficacy and/or toxicity of some nitroaromatic drugs.

2.2.1 Chloramphenicol

Chloramphenicol is a natural nitroaromatic compound (Fig. 2) that was isolated from the culture of the soil Gram-positive bacterium *Streptomyces venezuelae* as an antibacterial agent active against Gram-positive and Gram-negative bacteria.[72] Although effective in treatment of certain bacterial infections, the clinical utility of chloramphenicol is limited due to side effects such as neurotoxicity and hematological toxicity.[73] The mechanism of hematological toxicity is unclear, but a potential involvement of an amine metabolite of chloramphenicol (Fig. 2) has been suggested based on the association of the toxicity with the oral (but not intravenous) route of chloramphenicol administration.[13] Commensal gut and pathogenic bacteria can reduce chloramphenicol to the amine metabolite in vitro;[15–17,74,75] and in animal models of anaerobic infections, the anaerobic gut bacterium *Bacteroides fragilis* was shown to metabolize chloramphenicol via nitroreduction.[17] Because the amine metabolite of chloramphenicol exhibits very weak, if any, antibacterial activity as compared to the parent compound,[16] nitroreduction has been suggested to be one of the bacterial mechanisms for chloramphenicol resistance.[14,16] This notion is supported by a recent finding that overexpression of a nitroreductase of *Haemophilus influenzae* in *E. coli* increased chloramphenicol resistance.[15] Of note, a mutant strain with deletion of a nitroreductase has not yet been tested for susceptibility to chloramphenicol in any gut bacterial species; and the

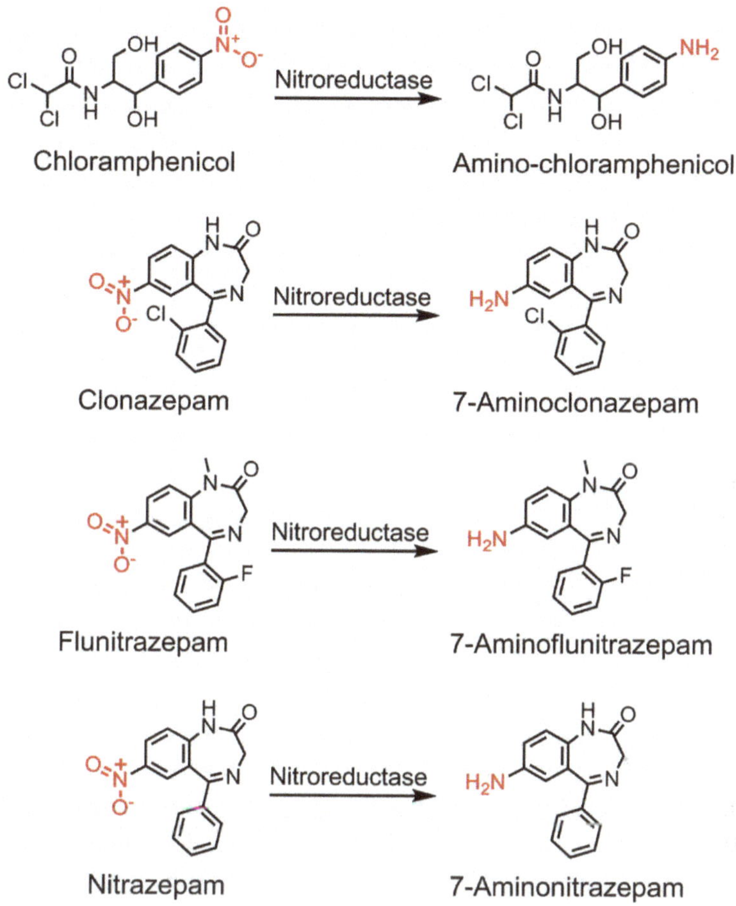

Fig. 2 Nitroreduction by gut bacteria. Gut bacterial nitroreduction of chloramphenicol and nitrobenzodiazepines is shown. See text for details.

contribution of nitroreduction to intrinsic and/or inducible chloramphenicol resistance in gut bacteria is unclear.

2.2.2 *Nitrobenzodiazepines*

Nitrobenzodiazepines (clonazepam, flunitrazepam, and nitrazepam) are sedative-hypnotic drugs that are widely used to treat sleeping disorders and relieve anxiety-related symptoms.[76,77] Although safe at therapeutic doses, an accidental or intentional overdose of nitrobenzodiazepines may cause adverse effects such as reproductive and hepatic toxicities.[78,79] These toxic effects have been suggested to be due at least in part to the amino metabolites of nitrobenzodiazepines (Fig. 2).

The reduction of a nitro group in nitrobenzodiazepines to an amino group has been well demonstrated to be mediated by intestinal bacteria in vitro and in vivo. For example, incubation of nitrobenzodiazepines with rat cecal contents or individual gut bacteria in vitro was shown to produce the amine metabolites.[19,80–85] In animal experiments, the levels of nitroreduction metabolites of nitrobenzodiazepines were significantly lower in germ-free and antibiotic-treated rats as compared to conventional rats.[78,83] The nitroreduction can also be catalyzed by host enzymes such as NADPH-cytochrome P450 reductase,[86] but its contribution to total amounts of amino metabolites produced in vivo appears to be minimal, and the nitroreduction by gut bacteria is suggested to be a primary determinant for nitrazepam-induced teratogenicity.[78]

Despite the observation that the gut bacterial conversion of nitrazepam to its amino metabolite is an essential step for the toxicity, only a few nitroreductases in gut bacteria have been identified and characterized to mediate the nitroreduction. To date, the NfsB protein of the facultative anaerobe *E. coli* and its homologous proteins (NfnB and NfsI) in closely-related enteric bacteria are the only nitroreductases experimentally shown to catalyze the nitroreduction of nitrobenzodiazepines to amino metabolites.[18] An earlier study, however, has shown that obligate anaerobic gut bacteria such as *B. fragilis* and *C. perfringens* can catalyze the nitroreduction of nitrobenzodiazepines at rates 50–70 times higher than that by *E. coli.*[85] In this respect, it is notable that four proteins have been purified from *B. fragilis* and determined to be nitroreductases,[87] although they have not been tested for the nitroreduction of nitrobenzodiazepines. In *C. perfringens*, no nitroreductase(s) has been identified, but an as-yet-uncharacterized azoreductase is shown to possess nitroreductase activity.[88] A study also indicates the presence of other anaerobic gut bacteria *Clostridial* and *Eubacterium* species with nitroreductase activity.[89]

It is clear from these studies that major nitroreductase(s) in gut bacteria that catalyzes the nitroreduction of nitrobenzodiazepines has not yet been identified. To better understand the role of gut bacteria in modulating the adverse effects of nitrobenzodiazepines, identification of responsible nitroreductase(s) and determination of their distribution in gut bacteria will be required.

2.3 Alkene reduction

Alkene reduction by the mammalian gut microbiotas was indicated in early studies on the metabolism of cinnamic acid derivatives.[90,91] For example,

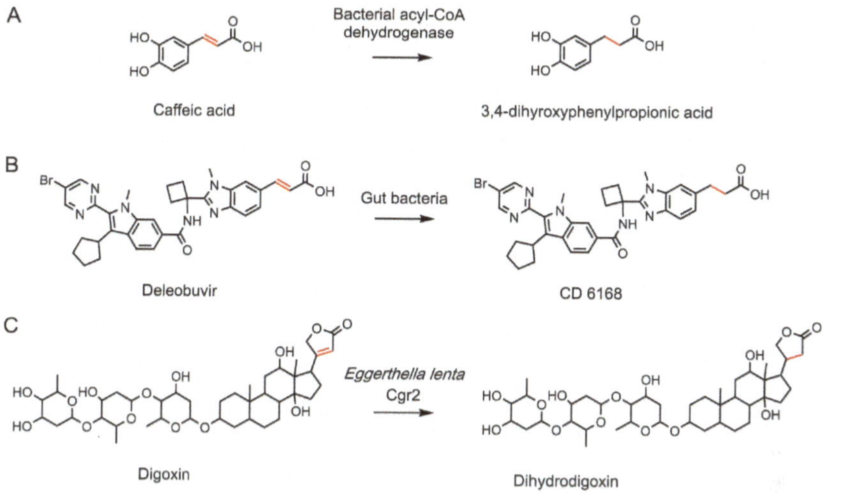

Fig. 3 Alkene reduction by gut bacteria. Gut alkene reduction of dietary compound caffeic acid (A) and drugs including deleobuvir (B) and digoxin (C) is shown. See text for details.

when caffeic acid (3,4-dihydroxycinnamic acid) was given orally to human, rat or rabbit, a reduced metabolite of caffeic acid, 3,4-dihydroxyphenylpropionic acid, was detected in urines (Fig. 3A). Later studies showed that antibiotic (neomycin) pre-treatment or germ-free condition significantly decreased or abrogated the urinary excretion of the reduced metabolite in rats, suggesting that the gut microbiota is involved in the alkene reduction of caffeic acid.[92] Several gut bacteria isolated from rat feces or cecal contents (*Pseudomonas* sp., *Lactobacillus* sp., and *Clostridium* sp.) have been known to mediate the caffeic acid reduction,[92–94] although responsible enzymes have not been identified. In a recent study, acyl-CoA dehydrogenase, one of the proteins encoded on a gene cluster for the reductive metabolism of aromatic amino acids, was shown to catalyze the alkene reduction of cinnamic acid to phenylpropionic acid in the gut bacterium *Clostridium sporogenes*,[95] suggesting a potential involvement of a gut bacterial metabolic pathway in the reductive metabolism of certain cinnamic acid derivatives. Below, we have shown two examples of therapeutic drugs that contain double bonds undergoing reduction by gut bacteria.

2.3.1 Deleobuvir

Deleobuvir (Fig. 3) is an NS5B RNA polymerase inhibitor used for the treatment of hepatitis C virus infection.[96] A study using radiolabeled deleobuvir in healthy volunteers revealed that the maximum plasma concentration of deleobuvir is reached at 3.5–5 h after oral administration.[97] The alkene reduction product (CD 6168, 10-fold lower antiviral activity than deleobuvir) and acyl glucuronide conjugate of deleobuvir were shown to be two major circulating metabolites, representing ~15% and ~20% of the total radioactivity in the blood, respectively. Excretion into feces was the major route of elimination such that >95% of the dose was found in feces after oral dose, and CD 6168 and its hydroxylated metabolites in feces accounted for a major portion of the oral dose (~60%). Of note, the amount of CD 6168 excreted into bile was minimal (less than 3% of the dose) in rats,[20] suggesting that CD 6168 detected in feces is produced mainly in the gastrointestinal tract. The in vitro incubation of deleobuvir with rat or human fecal samples (but not with human liver tissue homogenates) resulted in CD 6168 production, indicating that CD 6168 is a gut microbial metabolite of deleobuvir.[20] Consistently, in rats pretreated with antibiotics (streptomycin and neomycin), the plasma exposure of CD 6168 decreased 9-fold as compared to control rats after the oral administration of deleobuvir. In the antibiotic-treated rats, most of the deleobuvir dose was found in feces as the parent drug (as compared to only 26% of the dose in control rats), suggesting that gut bacteria determine the disposition of deleobuvir and its metabolites. To better define factors determining deleobuvir exposure after oral administration, the gut bacterial enzymes responsible for the alkene reduction of deleobuvir will need to be identified and their distribution in gut bacteria determined.

2.3.2 Digoxin

Digoxin (Fig. 3) is a cardiac glycoside that has been used to treat heart failure and atrial fibrillation for over 200 years.[98] Digoxin is poorly water-soluble and its oral bioavailability varies among different individuals, ranging from 50% to 90%.[99] Digoxin has a narrow therapeutic window (0.6–1.2 nM); supra-therapeutic concentrations may cause fatal cardiac arrhythmia.[100,101]

In the late 1960s, Luchi et al. reported a case of a patient who required an unusually high dose of digoxin (to control atrial fibrillation), which was accompanied by the presence of a large quantity of dihydrodigoxin, a ~20-fold less active reduction metabolite of digoxin, in the urine.[102] Later studies demonstrated significant inter-individual variations in the extent of urinary excretion of dihydrodigoxin.[103,104] For example, among 100 patients receiving digoxin, the fraction of dihydrodigoxin in the lipid-extractable cardiac glycosides in the urine ranged from 2% to 52%, and the fraction was greater than 30% in ~10% of the patients.[104]

The urinary excretion of dihydrodigoxin exhibited an inverse relationship with digoxin bioavailability; it is greatest after a poorly absorbed tablet was ingested, and least after intravenous administration, suggesting that the colonic gut microbiota is involved in dihydrodigoxin formation. Indeed, incubation of digoxin with stool samples from dihydrodigoxin "excretors" (i.e., subjects who excreted >15% dihydrodigoxin in the urine) showed the production of dihydrodigoxin, whereas the same metabolite was not observed upon incubation with non-excretor's stool samples.[22] Furthermore, the administration of an antibiotic (e.g., erythromycin, tetracycline, or clarithromycin) virtually eliminated the urinary excretion of dihydrodigoxin, with a concomitant increase (up to fourfold) in serum digoxin concentrations.[22,105]

Examination of human fecal bacteria grown in digoxin-containing media led to the identification of the gut Actinobacterium *Eggerthella lenta* that mediates digoxin reduction to dihydrodigoxin. *E. lenta* strains exhibited marked variations for digoxin reduction, as 10 out of 28 E. lenta strains tested were unable to produce dihydrodigoxin in vitro.[21] Interestingly, the fecal abundance of dihydrodigoxin-forming *E. lenta* was found to be similar between excretors and non-excretors, indicating that the mere presence of the bacterium in the gut does not account for dihydrodigoxin production in an individual. The finding partly explained this apparent discrepancy that the amino acid arginine stimulates *E. lenta* growth but represses digoxin reduction by *E. lenta.*[21] Using gene expression profiling, Haiser et al. identified a genetic locus, named cardiac glycoside reductase (*cgr*), whose expression is upregulated in the presence of digoxin in dihydrodigoxin-forming *E. lenta.*[106] The *cgr* locus was present only in *E. lenta* strains able to catalyze digoxin reduction, and its fecal abundance was greater in individuals highly producing dihydrodigoxin (as compared to low producers).[106] Furthermore, arginine repressed *cgr* expression in *E. lenta* in vitro; and in gnotobiotic mice monocolonized with *E. lenta* and fed protein-rich diet, digoxin concentrations in serum significantly increased, further demonstrating the role of *E. lenta* and diet (i.e., arginine) in determining digoxin exposure.

The *cgr* locus is a two-gene operon (*cgr1* and *cgr2*), and *cgr2* is shown to encode the enzyme catalyzing digoxin reduction. Heterologous expression of *cgr2* in a non-digoxin metabolizing Actinobacterium *Rhodococcus erythropolis* L88 was sufficient to provide metabolic ability for digoxin reduction.[107] The Cgr2 protein was characterized to be an oxygen-sensitive flavin- and [4Fe—4S] cluster-dependent reductase with narrow substrate specificity for cardenolides (including digoxin). The *cgr2* gene was detected in over 70% of fecal samples collected from >150 individuals, and its fecal abundances were well correlated with fecal *E. lenta* abundances (rho = 0.725). Among *E. lenta* strains carrying *cgr2*, a naturally occurring genetic polymorphism (Y333N) was identified to be associated with decreased digoxin reduction.[107] Together, these studies have illustrated multiple layers of factors modulating drug metabolism by the specific gut bacterium, namely gut bacterial abundance, bacterial gene regulation, strain-level genomic variation, genetic polymorphism, the nutritional milieu in the gut, and diet.[106,107]

2.4 Ketone reduction

The conversion of carbonyl groups to alcohol by host enzymes, such as carbonyl reductases (CBRs) and aldo-keto reductases (ABRs), has been relatively well characterized.[108,109] However, gut bacterial enzymes functionally equivalent to the host CBRs and ABRs are underexplored, and only a few examples of the keto-reduction of drugs by gut bacteria have been reported.

2.4.1 *Nabumetone*

Nabumetone (Fig. 4A) is a non-steroidal anti-inflammatory drug used for the management of arthritis and rheumatoid diseases to reduce pain and inflammation.[110] Orally administered nabumetone is completely metabolized, and a minimal amount of the parent drug was detected in the blood.[110] The major circulating nabumetone metabolite (accounting for ~35% of the dose) is the pharmacologically active 6-methoxy-2-naphtylacetic acid (6-MNA) whose formation is catalyzed by hepatic cytochrome P450 1A2.[111] Earlier studies have also found a minor pharmacologically inactive metabolite [4-(6-methoxy-2-naphthyl)-butan-2-ol], which is produced by the keto-reduction of nabumetone and is present in the urine samples at a low amount (~1% of total nabumetone metabolites excreted into the urine).[112] Microsomal and cytosolic fractions of human livers, as well as host CBRs and AKRs, were shown to produce the keto-reduction metabolite.[113] In a recent study, gut bacteria such as *E. coli* were also shown to mediate the keto-reduction of nabumetone under both aerobic and anaerobic conditions in vitro.[23] However, antibiotic (imipenem) pre-treatment or germ-free status had no effects on the plasma exposure of 6-MNA in rats after the oral administration of nabumetone.[23] These results suggest the minimal contribution, if any, of the gut microbiota to the metabolism of orally administered nabumetone.

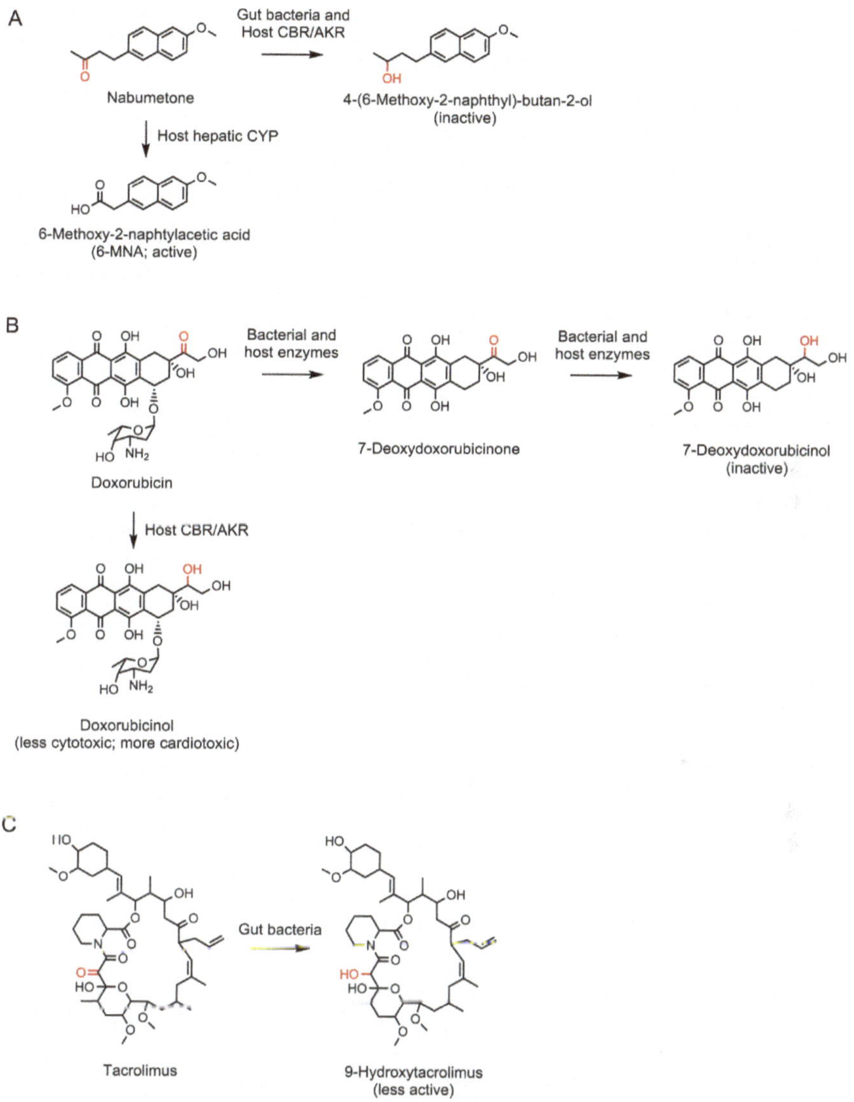

Fig. 4 Keto-reduction by gut bacteria. Keto-reduction of nabumetone (A) and doxorubicin (B) is mediated by gut bacteria as well as host enzymes. CBR, carbonyl reductases; ABR, aldo-keto reductases. Tacrolimus (C) undergoes keto-reduction by gut bacteria. See text for details.

2.4.2 Doxorubicin

Doxorubicin (Fig. 4B) is an anticancer agent that belongs to anthracycline antibiotics. Anthracyclines with anticancer activity inhibit cancer cell growth via multiple mechanisms, including DNA cross-linking and free radial formation.[114] When doxorubicin is administered intravenously, about half of the dose is eliminated into feces after biliary excretion.[115] The major circulating metabolite is doxorubicinol, the C7-reduction product of doxorubicin, produced by host CBRs and AKRs.[116,117] Doxorubicinol is less cytotoxic than the parent drug, but it inhibits the calcium pump at cardiac muscle sarcoplasmic reticulum, potentially responsible for the dose-limiting cardiotoxicity of doxorubicin.[117,118] Considering significant biliary excretion of doxorubicin, a gut microbial metabolism of doxorubicin was hypothesized, and a screen for gut bacteria that may metabolize doxorubicin was conducted.[24] This screen led to the identification of the gut bacterium *Raoultella planticola* that deglycosylates doxorubicin and produces 7-deoxydoxorubicinone, which is subsequently converted to 7-deoxydoxorubicinol via keto-reduction.[24] Further, by screening a collection of defined mutants of *E. coli* BW25113 (another bacterium that metabolizes doxorubicin, producing 7-deoxydoxorubicinone and 7-deoxydoxorubicinol), authors identified a group of mutants unable to metabolize doxorubicin. Genes inactivated in the identified *E. coli* mutants encode enzymes required for the biosynthesis of the molybdenum cofactor (called Moco), suggesting that Moco-requiring enzymes are responsible for doxorubicin metabolism.[24] On the other hand, the isolated perfused rat liver was also able to metabolize doxorubicin to 7-deoxydoxorubicinone and 7-deoxydoxorubicinol, although responsible host enzymes were unidentified.[119] The extent to which the gut microbiota contributes to the overall keto reductive metabolism of doxorubicin in vivo remains to be determined.

2.4.3 Tacrolimus

Tacrolimus (Fig. 4C) is an immunosuppressant used to prevent organ rejection after transplantation.[120] Tacrolimus has a narrow therapeutic window, and supra- and under-therapeutic concentrations of tacrolimus are associated with drug toxicity (nephrotoxicity and neurotoxicity) and graft rejection, respectively.[121] Maintaining tacrolimus concentrations within the therapeutic range is difficult, because of its low and variable oral bioavailability (ranging from 5% to 93%; average 25%).[121] The activity of CYP3A4 and the drug efflux pump P-glycoprotein in the liver and intestine had been proposed to explain the tacrolimus bioavailability. However, drug-drug interaction studies of tacrolimus and ketoconazole (an inhibitor of both hepatic and intestinal CYP3A4 and P-glycoprotein) showed that even after CYP3A4 and P-glycoprotein inhibition, a significant (>50%) fraction of the dose failed to reach systemic circulation,[122,123] suggesting the presence of additional route(s) of tacrolimus elimination.

In a pilot study of 19 adult kidney transplant recipients, stool samples and tacrolimus dosing information were longitudinally collected for one-month after transplantation, and the relationship between the gut microbiota and tacrolimus dosing was examined.[124] Analysis of the gut microbial composition revealed that the fecal abundance of *Faecalibacterium prausnitzii* in the first week of transplantation is positively correlated with the tacrolimus dose required for achieving therapeutic concentrations. In vitro incubation of tacrolimus with *F. prausnitzii* revealed the conversion of tacrolimus to a C-9 keto-reduction metabolite that exhibits ~15-fold lower immunosuppressive activity than the parent compound.[25] When additional human gut bacteria were tested for tacrolimus metabolism, most gut bacteria belonging to *Firmicutes*, a major phylum in the gut microbiota, were able to catalyze the same keto-reduction, suggesting that tacrolimus metabolism may be prevalent in the gut microbiota. Indeed, mouse cecal contents and human feces readily converted tacrolimus to the reductive metabolite.[25] The keto-reduction of tacrolimus was found to be oxygen-sensitive and not mediated by mouse hepatic microsomes, perhaps suggesting that tacrolimus keto-reduction is unique to as-yet-unidentified gut bacterial enzymes.

Orally administered antibiotics, including erythromycin, clarithromycin, chloramphenicol, and clindamycin, were shown to increase tacrolimus exposure.[121,125] Since these antibiotics inhibit the function of CYP3A4 and/or P-glycoprotein, it is unclear whether the antibacterial activity (i.e., depletion of gut bacteria) is solely responsible for the effect on tacrolimus exposure. On the other hand, results from a few drug-drug interaction studies, in which antibiotics with minimal effects on CYP3A4 and P-glycoprotein expression/activity were used, may support the involvement of the gut microbiota in determining tacrolimus disposition. For example, levofloxacin increased tacrolimus exposure by ~25% in five kidney transplant recipients,[126] and ertapenem decreased the tacrolimus dosage required to achieve therapeutic concentrations by ~2-fold.[127] Also, a recent retrospective chart review of subjects who had received kidney transplants ($n=260$) revealed that antibiotic treatment (for 7 days) increased the ratio of tacrolimus trough concentration to dosage by ~20%.[128] Moreover, significantly higher intra-individual variations in tacrolimus trough concentrations were observed in patients who received antibiotics as compared to non-antibiotic users. The most prominent effects on tacrolimus exposure were observed with antibiotics belonging to the β-lactam (penicillin or cephalosporin) class that is mainly eliminated via renal excretion and thus has minimal effects on host CYP3A4 and P-glycoprotein expression/activity. While these results suggest that the gut microbiota is involved in tacrolimus metabolism and may be one of the determinants of tacrolimus exposure, its clinical impact remains to be clarified via the measurement of systemic exposure to the keto-reduction metabolite as well as the estimation of oral dose that is subject to gut bacterial metabolism in vivo.

2.5 Sulfoxide reduction

Gut bacterial reduction of the sulfoxide group was most extensively studied for sulphinpyrazone and sulindac (Fig. 5A and B). Gut bacterial metabolism of other compounds containing sulfoxide group such as the proton pump inhibitor omeprazole (Fig. 5C) in the colon has also been reported; aerobic incubation of omeprazole with rat cecal or colonic contents led to ~50%

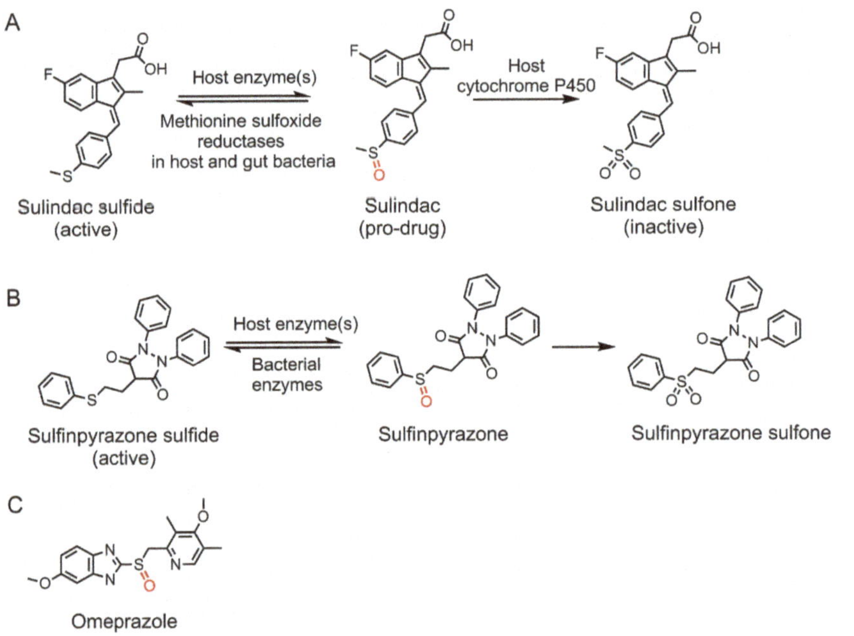

Fig. 5 Sulfoxide reduction by gut bacteria. Sulindac (A) and sulfinpyrazone (B) are known to undergo sulfoxide reduction by gut bacteria as well as host enzymes. Omeprazole (C) is a potential substrate of bacterial sulfoxide reductase. See text for details.

decrease in amounts within 30 min.[129] Based on the vast body of supporting evidence, our discussion of gut bacteria-mediated sulfoxide reduction will be focused on sulindac and sulphinpyrazone.

2.5.1 Sulindac

Sulindac (Fig. 5A) is a non-steroidal anti-inflammatory drug used to treat inflammatory diseases such as rheumatoid arthritis and osteoarthritis.[27] Mass balance studies using radiolabeled sulindac in humans have demonstrated that sulindac is extensively and rapidly absorbed after oral administration; ~90% of the dose is absorbed, and the maximum plasma concentration reached within 1 h.[28] Sulindac metabolites, sulindac sulfone (called "SSone" hereafter) and sulindac sulfide (called "SSide" hereafter) (Fig. 5A) provide similar plasma exposures as that of the parent drug sulindac. Upon intra-articular injection into the inflamed synovial fluid space, SSide was shown to be >60-fold more active as an anti-inflammatory agent than sulindac whereas SSone is inactive.[130] After ingestion, sulindac (but not SSide) undergoes significant enterohepatic recycling as unmodified or conjugated form (acyl glucuronide), exhibiting prominent double peaks on plasma concentration vs. time profiles.[28,131] Interestingly, cumulative recovery analysis found significant amounts of SSide but not the parent drug in human fecal samples, raising the possibility that sulindac may be converted to SSide in the gut. Consistent with this idea, when sulindac was incubated in vitro with rat cecal contents or human fecal samples, it was rapidly and completely converted to SSide.[28]

In addition to putative gut microbial enzymes, host enzymes also appear to catalyze the reduction of sulindac to SSide. After intravenous administration of sulindac or SSide in rats, sulindac was readily converted to SSide or vice versa in the body.[130] Animal tissue samples, including hepatic microsomes, mitochondria, and cytosol fractions, were shown to convert sulindac to SSide under both aerobic and anaerobic conditions.[26,132,133] Consistent with these in vitro results, patients with ileostomies (where the colon is bypassed in drug elimination process into feces) and healthy control groups exhibited similar plasma exposures to SSide up to 12 h after oral administration of sulindac.[134] On the other hand, SSide exposure after 12 h (which constitutes 55% of the total AUC in healthy subjects) decreased drastically in the ileostomy patients, suggesting that about half of the circulating SSide originates from the colonic metabolism of sulindac in healthy individuals. Supporting the notion, incubation of sulindac with ileostomy effluent showed negligible SSide production, whereas extensive SSide formation was observed when sulindac was incubated with the fecal samples of healthy subjects.[134]

Reactive nitrogen or oxygen species can cause chemical modifications of biological molecules. For example, the oxidation of the amino acid methionine produces methionine sulfoxides that exhibit altered biological activities.[135,136] Most, if not all, eukaryotic and prokaryotic cells have methionine sulfoxide reductases that can reduce proteinous methionine sulfoxide residues back to methionine in a thioredoxin-dependent manner. Methionine sulfoxide reductases, MsrA in *E. coli* and MsrB1 in the rat liver, were shown to catalyze sulindac reduction to SSide under aerobic conditions.[26,137] On the other hand, microsomal enzymes such as flavin-containing monooxygenases in the human liver and kidney catalyzed the oxidation of SSide back to sulindac,[138] and hepatic cytochrome P450 enzymes further oxidized sulindac to SSone.[137] These findings suggest that multiple host enzymes and (putative) gut bacterial enzymes govern the extent of host exposure to bioactive SSide.

2.5.2 Sulphinpyrazone

Sulphinpyrazone (Fig. 5B) is a uricosuric and anti-platelet agent.[139] The major circulating metabolites detected in systemic circulation after oral administration of radiolabeled sulphinpyrazone are its sulfide and sulfone (Fig. 5B), representing 13% and 6% of the radioactivity, respectively.[140] Similar to sulindac, the sulfide metabolite of sulphinpyrazone exhibits more potent anti-platelet activity than the parent drug. However, unlike sulindac that is also reduced by host enzymes to SSide, sulphinpyrazone does not appear to be the substrate of host enzymes.[140,141] Anaerobic incubation of sulphinpyrazone with rat or rabbit liver and kidney homogenates showed minimal production of the sulfide metabolite.[30,142] Furthermore, after oral administration of sulphinpyrazone, the sulfide metabolite was not detected in the blood and cecal contents of germ-free rats or rats treated with a cocktail of antibiotics (bacitracin, neomycin, and tetracycline).[30] In addition to the difference between sulindac and sulphinpyrazone as being the substrates of host enzymes, gut bacteria and their enzymes for catalyzing sulphinpyrazone and sulindac reduction appear to differ. When sulphinpyrazone or sulindac was incubated anaerobically with one of over 200 bacterial strains isolated from human feces, sulphinpyrazone sulfide production was observed mainly with aerobic bacteria, whereas sulindac sulfide production was observed with both anaerobic and aerobic bacteria.[29] Sulphinpyrazone is discontinued from the US market due to its renal side effects and unfavorable pharmacological interactions with other drugs.[143] However, a large body of information available on sulphinpyrazone reduction to the sulfide metabolite provides a rare opportunity to investigate and compare different substrates that undergo sulfoxide reduction by gut bacteria.

2.6 *N*-oxide reduction

Alkaloid *N*-oxides constitute a significant part of natural *N*-oxidation products in plants.[144] In particular, pyrrolizidine alkaloids exist mostly as *N*-oxides. Metabolism of *N*-oxides by gut contents was first reported for the pyrrolizidine alkaloid heliotrine *N*-oxide.[145] The plant *Heliotropium europaeum* producing heliotrine *N*-oxide was known to cause chronic liver diseases in sheep after multiple seasons of grazing pastures. Considering the relatively low levels of acute liver toxicity observed in sheep that grazed on the plants for only one season, potential inactivation of toxic heliotrine *N*-oxide before intestinal absorption was postulated. Incubation of the alkaloid heliotrine *N*-oxide with sheep rumen contents produced an *N*-oxide reduction metabolite heliotrine (Fig. 6), suggesting that the gut microbiota mediates the *N*-oxide reduction. In vivo, heliotrine appeared to be further metabolized to nontoxic 7-hydroxy-1-methylene pyrrolizidine in sheep rumen. Below, we have presented the xenobiotic nicotine and two therapeutic drugs undergoing *N*-oxide reduction.

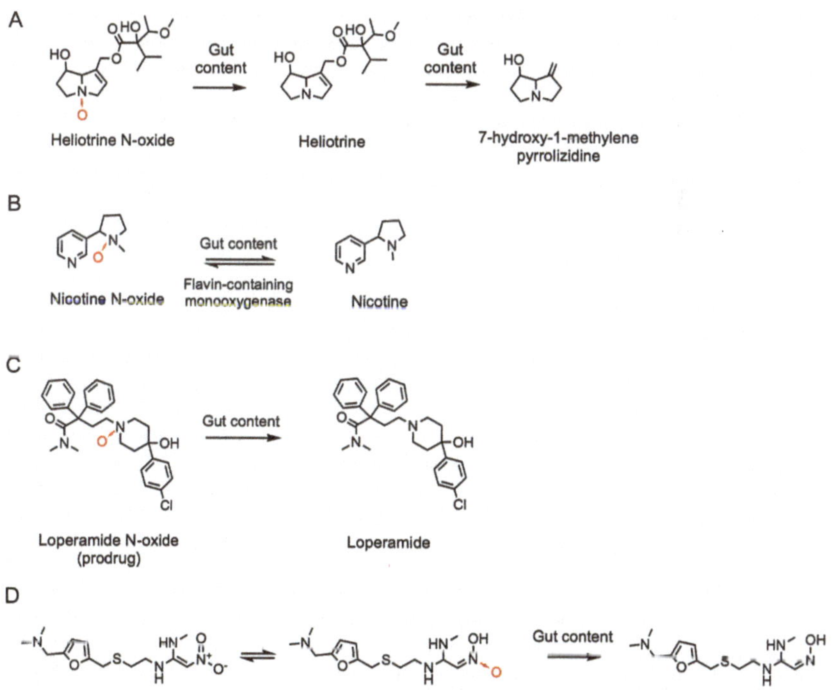

Fig. 6 *N*-oxide reduction by gut contents. A plant alkaloid heliotrine *N*-oxide (A) and drugs (B-D) that potentially undergo *N*-oxide reduction by gut contents are shown. See text for details.

2.6.1 Nicotine

Nicotine is the primary psychoactive component in tobacco, and the recent introduction of electronic nicotine delivery systems has dramatically increased nicotine consumption.[146] Nicotine is converted to multiple metabolites, including nicotine *N*-oxide by hepatic flavin monooxygenase.[147] When nicotine *N*-oxide was orally administered to humans or rats, significant amounts of nicotine were excreted in the urine.[31,148] This was not observed after intravenous administration,[148] suggesting that the conversion of nicotine *N*-oxide to nicotine is mediated by intestinal tissues and/or gut contents. Indeed, rat intestine homogenates, as well as gut contents, were shown to catalyze nicotine *N*-oxide reduction under anaerobic conditions.[149] However, the oral administration of nicotine *N*-oxide to germ-free or antibiotics-treated rats led to urinary nicotine excretion to the extents similar to those in conventional or non-antibiotic-treated rats.[149] These results indicate that nicotine *N*-oxide reduction is mediated mainly by host enzymes and/or nonenzymatic processes, with a minimal contribution by the gut microbiota.

Multiple microsomal and cytosolic *N*-oxide reductases in host tissues, including NADP- or NADPH-supported enzymes, have been reported.[144,149,150] Notably, these experiments were performed under anaerobic conditions,[149] and *N*-oxide reductase activities in the tissue fractions were highly oxygen-sensitive, i.e., 50% inhibition concentration of oxygen ranging from 2 to 4 μM.[150,151] Physiological oxygen concentrations in the liver range from 46 μM (perivenous blood) and 88 μM (periportal blood),[152] and it is plausible that *N*-oxide reduction by host enzymes does not occur in physiological settings.

Nonenzymatic reduction of *N*-oxide compounds has been reported. *N*-oxides of aliphatic tertiary amines, including nicotine *N*-oxide, are chemically labile as compared to the *N*-oxides of heteroaromatic amines such as nicotinamide *N*-oxide.[153] Reducing agents such as EDTA complex of iron readily catalyzed *N*-oxide reduction of indicine in vitro under aerobic conditions as well as in vivo in rats.[154] *N*-oxide in *N,N*-dimethylaniline was reduced to the amine in the aerobic solution of hemoglobin.[155] Despite the evidence indicating the nonenzymatic reduction of *N*-oxides in the presence of reducing agents, nicotine was not detected in the urine of human subjects administered nicotine *N*-oxide intravenously.[148] Whether nicotine *N*-oxide is susceptible to reduction by hemoglobin in the blood remains to be determined. However, the currently available data support the notion that the nonenzymatic process (likely mediated by reducing agents in diet) is the major contributor to nicotine *N*-oxide reduction after oral administration.

2.6.2 Loperamide

Loperamide is an antidiarrheal agent that activates the opioid receptor in the large intestine.[156] Given that *N*-oxides are more hydrophilic than the parent amine compound (and thus less likely to be absorbed in the small intestine), loperamide *N*-oxide was developed as a colon-targeting pro-drug of loperamide with limited systemic availability and thus reduced central nervous system side effects of loperamide. After the oral administration of loperamide *N*-oxide to dogs, the pro-drug was reduced to loperamide, exhibiting significant loperamide plasma exposure (~50% of exposure observed after administering an equal oral dose of loperamide itself).[32] Furthermore, the mass balance of the metabolites, as well as the excretion pattern, were similar to those obtained after the oral dose of loperamide, indicating almost complete conversion of loperamide *N*-oxide to loperamide in the body. Anaerobic incubation of the pro-drug with small intestinal or cecal contents of rats and dogs revealed significant *N*-oxide reductase activities in gut contents, with the highest activity in the cecum. The *N*-oxide reductase activity was oxygen-sensitive (i.e., ~90% activity loss upon air exposure of gut contents). Of note, the amount of loperamide *N*-oxide reduced by small intestinal contents was similar between germ-free and conventional rats,[32] likely due to nonenzymatic *N*-oxide reduction. On the other hand, dramatic differences in loperamide *N*-oxide reductase activities were observed between cecal contents of germ-free and conventional rats(i.e., minimal enzyme activity in germ-free rat cecum), suggesting that the role of the gut microbiota is critical for the bioactivation of pro-drug and for the antidiarrheal effects of loperamide.

2.6.3 Ranitidine

Ranitidine is a histamine H_2-receptor antagonist that is used to decrease gastric secretion and for treating gastrointestinal ulcers.[157] In earlier pharmacokinetic studies of orally administered ranitidine, apparent "second peak" on concentration vs. time profile was observed for the majority of subjects (i.e., the first and second peak concentrations at 0.5–1.5 and 3–4 h after an oral dose).[158] A similar phenomenon was also found for cimetidine, another H_2-receptor antagonist developed and marketed before ranitidine.[159] Considering minimal enterohepatic recycling of these drugs,[160] it was postulated that discontinuous intestinal absorption of these drugs along the gastrointestinal tract might explain the second peak. To determine the origin of the second peak, ranitidine was injected directly into the stomach, jejunum, or cecum via a nasoenteric tube in healthy volunteers, and blood samples were collected for drug analysis.[161] There was no difference in pharmacokinetic parameters or the presence of second peaks between gastric and jejunal dosing. On the other hand, ranitidine exposure after cecal administration was less than 15% of that after gastric or jejunal administration, suggesting that absorption windows for ranitidine may be located between jejunum and cecum. While the reason for the second peak phenomenon remained unclear, the low bioavailability after cecal administration raised the possibility of colonic ranitidine metabolism. Indeed, incubation of ranitidine with human feces revealed extensive disappearance of the drug over time, which was saturated at high drug concentrations.[33] The major metabolite was identified as a product of oxygen loss at the diaminonitroalkene moiety and proposed to be an *N*-oxide reduction product of ranitidine tautomer. A similar metabolism by gut contents was observed for nizatidine, an H_2-receptor antagonist that shares the same structural moiety of diaminonitroalkene as ranitidine, but not for cimetidine and famotidine that do not have the diaminonitroalkene moiety.[162] Of note, the *N*-oxide reduction metabolites of ranitidine and nizatidine have not been identified in the systemic circulation, suggesting that the contribution of the metabolic pathway to overall ranitidine disposition may be minor. The clinical significance of ranitidine metabolism by the gut microbiota remains to be determined.

3. Concluding remarks

Orally administered drugs can be metabolized by the gut microbiota (e.g., via reduction) and/or by host drug-metabolizing enzymes (via oxidation), consequently being bioactivated, eliminated or inactivated. On the other hand, medication use is shown to be associated with the largest inter-individual variability in the gut microbiota,[163] and many non-antibiotic drugs are determined to have anti-commensal activity,[164] suggesting that certain drugs drive compositional changes in the gut microbiota. Considering the emerging roles of the gut microbiota in maintaining host health or causing diseases, certain prescription drugs (and other xenobiotics) may have an unintended activity of impacting host physiology via their effects on gut microbiota composition/structure (Fig. 7).

Xenobiotic metabolism by gut microbiota has been reported over decades and supported by a large body of evidence encompassing drugs in diverse chemical classes and therapeutic areas. Despite the accumulating evidence, biochemical and genetic bases of the chemical reactions remain largely unknown. This is due in part to perceived difficulties in culturing

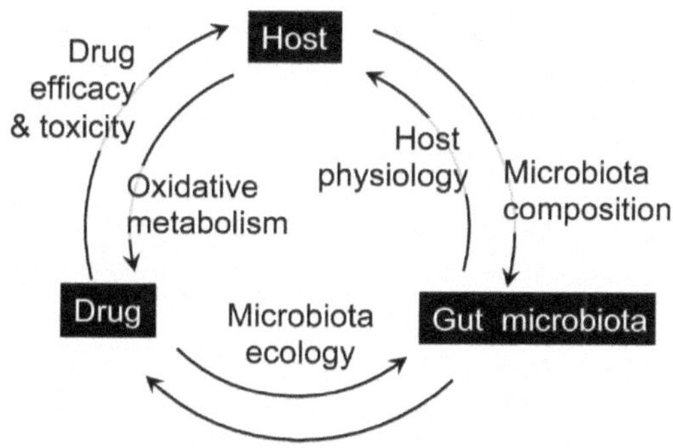

Fig. 7 Reciprocal interactions among host, gut microbiota, and drug. Orally administered drugs (and other xenobiotics) can be metabolized by the gut microbiota (via reduction) and/or by host drug-metabolizing enzymes (via oxidation), with different consequences: bioactivation and/or elimination. On the other hand, xenobiotics may alter gut microbial composition and community structure, consequently affecting host physiology.

some gut bacteria and extreme complexity in the gut microbiota community. With advances in metagenomics, metabolomics, and bioinformatics tools, as well as the collection of thousands of gut bacteria and better culturing techniques, enormous amounts of resources are now available that allow the identification of specific gut bacteria and their enzymes mediating the biotransformation of drugs. For most of the reductive metabolisms described in this review, responsible gut bacteria and enzymes mediating those reactions are largely unknown. Identification and characterization of these gut bacteria and enzymes will not only lay a foundation to integrate the gut microbiota into optimizing drug therapy outcomes in the host, but also provide an opportunity for defining their physiological roles in bacteria and within the gut community.

References

1. Sender R, Fuchs S, Milo R. Revised estimates for the number of human and bacteria cells in the body. *PLoS Biol.* 2016;14(8):e1002533.
2. Gill SR, Pop M, DeBoy RT, et al. Metagenomic analysis of the human distal gut microbiome. *Science.* 2006;312(5778):1355–1359.
3. Clarke G, Sandhu KV, Griffin BT, Dinan TG, Cryan JF, Hyland NP. Gut reactions: breaking down xenobiotic-microbiome interactions. *Pharmacol Rev.* 2019;71(2):198–224.
4. Spanogiannopoulos P, Bess EN, Carmody RN, Turnbaugh PJ. The microbial pharmacists within us: a metagenomic view of xenobiotic metabolism. *Nat Rev Microbiol.* 2016;14(5):273–287.

5. Sousa T, Paterson R, Moore V, Carlsson A, Abrahamsson B, Basit AW. The gastrointestinal microbiota as a site for the biotransformation of drugs. *Int J Pharm.* 2008;363(1):1–25.
6. Scheline RR. Metabolism of foreign compounds by gastrointestinal microorganisms. *Pharmacol Rev.* 1973;25(4):451–532.
7. Friedman ES, Bittinger K, Esipova TV, et al. Microbes vs. chemistry in the origin of the anaerobic gut lumen. *Proc Natl Acad Sci U S A.* 2018;115(16):4170–4175.
8. Espey MG. Role of oxygen gradients in shaping redox relationships between the human intestine and its microbiota. *Free Radic Biol Med.* 2013;55:130–140.
9. Circu ML, Aw TY. Redox biology of the intestine. *Free Radic Res.* 2011;45(11–12):1245–1266.
10. Celesk RA, Asano T, Wagner M. The size pH, and redox potential of the cecum in mice associated with various microbial floras. *Proc Soc Exp Biol Med.* 1976;151(2): 260–263.
11. Gingell R, Bridges JW, Williams RT. Gut flora and the metabolism of prontosils in the rat. *Biochem J.* 1969;114(1):5–6.
12. Peppercorn MA, Goldman P. The role of intestinal bacteria in the metabolism of salicylazosulfapyridine. *J Pharmacol Exp Ther.* 1972;181(3):555–562.
13. Holt R. The bacterial degradation of chloramphenicol. *Lancet.* 1967;1(7502): 1259–1260.
14. Merkel JR, Steers E. Relationship between "chloramphenicol reductase activity" and chloramphenicol resistance in *Escherichia coli. J Bacteriol.* 1953;66(4):389.
15. Crofts TS, Sontha P, King AO, et al. Discovery and characterization of a nitroreductase capable of conferring bacterial resistance to chloramphenicol. *Cell Chem Biol.* 2019;26(4):559–570.e556.
16. Smith AL, Erwin AL, Kline T, et al. Chloramphenicol is a substrate for a novel nitroreductase pathway in *Haemophilus influenzae. Antimicrob Agents Chemother.* 2007;51(8):2820–2829.
17. Onderdonk AB, Kasper DL, Mansheim BJ, Louie TJ, Gorbach SL, Bartlett JG. Experimental animal models for anaerobic infections. *Rev Infect Dis.* 1979;1(2): 291–301.
18. LinWu S-W, Syu C-J, Chen Y-L, Wang AH-J, Peng F-C. Characterization of *Escherichia coli* nitroreductase NfsB in the metabolism of nitrobenzodiazepines. *Biochem Pharmacol.* 2009;78(1):96–103.
19. Rafii F, Sutherland JB, Hansen EB, Jr., Cerniglia CE. Reduction of nitrazepam by *Clostridium leptum*, a nitroreductase-producing bacterium isolated from the human intestinal tract. *Clin Infect Dis.* 1997;25(Suppl 2):S121–S122.
20. McCabe M, Sane RS, Keith-Luzzi M, et al. Defining the role of gut bacteria in the metabolism of deleobuvir: In vitro and in vivo studies. *Drug Metab Dispos.* 2015;43(10):1612–1618.
21. Saha BV, Neu H, Lindenbaum J. Digoxin-inactivating bacteria: identification in human gut flora. *Science.* 1983;220(4594):325–327.
22. Lindenbaum J, Rund DG, Butler VPJ, Tse-Eng D, Saha JR. Inactivation of digoxin by the gut flora: reversal by antibiotic therapy. *N Engl J Med.* 1981;305(14):789–794.
23. Jourova L, Anzenbacher P, Matuskova Z, et al. Gut microbiota metabolizes nabumetone in vitro: Consequences for its bioavailability in vivo in the rodents with altered gut microbiome. *Xenobiotica.* 2019;49(11):1296–1302.
24. Yan A, Culp E, Perry J, et al. Transformation of the anticancer drug doxorubicin in the human gut microbiome. *ACS Infect Dis.* 2018;4(1):68–76.
25. Guo Y, Crnkovic CM, Won KJ, et al. Commensal gut bacteria convert the immunosuppressant tacrolimus to less potent metabolites. *Drug Metab Dispos.* 2019;47(3): 194–202.

26. Etienne F, Resnick L, Sagher D, Brot N, Weissbach H. Reduction of Sulindac to its active metabolite, sulindac sulfide: assay and role of the methionine sulfoxide reductase system. *Biochem Biophys Res Commun.* 2003;312(4):1005–1010.
27. Duggan DE. Sulindac: therapeutic implications of the prodrug/pharmacophore equilibrium. *Drug Metab Rev.* 1981;12(2):325–337.
28. Duggan DE, Hare LE, Ditzler CA, Lei BW, Kwan KC. The disposition of sulindac. *Clin Pharmacol Ther.* 1977;21(3):326–335.
29. Strong HA, Renwick AG, George CF, Liu YF, Hill MJ. The reduction of sulphinpyrazone and sulindac by intestinal bacteria. *Xenobiotica.* 1987;17(6):685–696.
30. Renwick AG, Evans SP, Sweatman TW, Cumberland J, George CF. The role of the gut flora in the reduction of sulphinpyrazone in the rat. *Biochem Pharmacol.* 1982;31(16):2649–2656.
31. Dajani RM, Gorrod JW, Beckett AH. Reduction in vivo of (-)-nicotine-1'-*N*-oxide by germ-free and conventional rats. *Biochem Pharmacol.* 1975;24(5):648–650.
32. Lavrijsen K, van Dyck D, van Houdt J, et al. Reduction of the prodrug loperamide oxide to its active drug loperamide in the gut of rats, dogs, and humans. *Drug Metab Dispos.* 1995;23(3):354–362.
33. Basit AW, Lacey LF. Colonic metabolism of ranitidine: implications for its delivery and absorption. *Int J Pharm.* 2001;227(1-2):157–165.
34. Chung KT. Azo dyes and human health: a review. *J Environ Sci Health C Environ Carcinog Ecotoxicol Rev.* 2016;34(4):233–261.
35. Feng J, Cerniglia CE, Chen H. Toxicological significance of azo dye metabolism by human intestinal microbiota. *Front Biosci (Elite Ed).* 2012;4:568–586.
36. Levine WG. Metabolism of azo dyes: implication for detoxication and activation. *Drug Metab Rev.* 1991;23(3-4):253–309.
37. Domagk G. Twenty-five years of sulfonamide therapy. *Ann N Y Acad Sci.* 1957;69(3):380–384.
38. Fuller AT. Is *p*-aminobenzenesulphonamide the active agent in prontosil therapy? *Lancet.* 1937;229(5917):194–198.
39. Colebrook L, Buttle GAH, O'Meara RAQ. The mode of action of *p*-aminobenzenesulphonamide and prontosil in hemolytic Streptococcal infections. *Lancet.* 1936;228(5910):1323–1326.
40. Ryan AJ, Roxon JJ, Sivayavirojana A. Bacterial azo reduction: a metabolic reaction in mammals. *Nature.* 1968;219(5156):854–855.
41. Radomski JL, Mellinger TJ. The absorption, fate and excretion in rats of the water-soluble azo dyes, FD&C Red No. 2, FD&C Red No. 4, and FD&C Yellow No. 6. *J Pharmacol Exp Ther.* 1962;136:259–266.
42. Gingell R, Bridges JW. Intestinal azo reduction and glucuronide conjugation of prontosil. *Biochem J.* 1971;125(2):24p.
43. Ali Y, Hamid SA, Rashid U. Biomedical applications of aromatic azo compounds. *Mini Rev Med Chem.* 2018;18(18):1548–1558.
44. Svartz N. Salazopyrin, a new sulfanilamide preparation. *Acta Med Scand.* 1942;110(6):577–598.
45. Hauso O, Martinsen TC, Waldum H. 5-Aminosalicylic acid, a specific drug for ulcerative colitis. *Scand J Gastroenterol.* 2015;50(8):933–941.
46. Schroder H, Gustafsson BE. Azo reduction of salicyl-azo-sulphapyridine in germ-free and conventional rats. *Xenobiotica.* 1973;3(4):225–231.
47. Azad Khan AK, Piris J, Truelove SC. An experiment to determine the active therapeutic moiety of sulphasalazine. *Lancet.* 1977;2(8044):892–895.
48. Das KM, Eastwood MA, McManus JP, Sircus W. The metabolism of salicylazosulphapyridine in ulcerative colitis. I. The relationship between metabolites and the response to treatment in inpatients. *Gut.* 1973;14(8):631–641.

49. Azadkhan AK, Truelove SC, Aronson JK. The disposition and metabolism of sulphasalazine (salicylazosulphapyridine) in man. *Br J Clin Pharmacol.* 1982;13(4):523–528.
50. Houston JB, Day J, Walker J. Azo reduction of sulphasalazine in healthy volunteers. *Br J Clin Pharmacol.* 1982;14(3):395–398.
51. Mahmood S, Khalid A, Arshad M, Mahmood T, Crowley DE. Detoxification of azo dyes by bacterial oxidoreductase enzymes. *Crit Rev Biotechnol.* 2016;36(4):639–651.
52. Suzuki H. Remarkable diversification of bacterial azoreductases: primary sequences, structures, substrates, physiological roles, and biotechnological applications. *Appl Microbiol Biotechnol.* 2019;103(10):3965–3978.
53. Xu H, Heinze TM, Paine DD, Cerniglia CE, Chen H. Sudan azo dyes and Para Red degradation by prevalent bacteria of the human gastrointestinal tract. *Anaerobe.* 2010;16(2):114–119.
54. Rafii F, Franklin W, Cerniglia CE. Azoreductase activity of anaerobic bacteria isolated from human intestinal microflora. *Appl Environ Microbiol.* 1990;56(7):2146–2151.
55. Brown JP. Reduction of polymeric azo and nitro dyes by intestinal bacteria. *Appl Environ Microbiol.* 1981;41(5):1283–1286.
56. Chung KT, Fulk GE, Egan M. Reduction of azo dyes by intestinal anaerobes. *Appl Environ Microbiol.* 1978;35(3):558–562.
57. Blair LM, Sperry J. Natural products containing a nitrogen-nitrogen bond. *J Nat Prod.* 2013;76(4):794–812.
58. Nakanishi M, Yatome C, Ishida N, Kitade Y. Putative ACP phosphodiesterase gene (*acpD*) encodes an azoreductase. *J Biol Chem.* 2001;276(49):46394–46399.
59. Mercier C, Chalansonnet V, Orenga S, Gilbert C. Characteristics of major *Escherichia coli* reductases involved in aerobic nitro and azo reduction. *J Appl Microbiol.* 2013;115(4):1012–1022.
60. Prosser GA, Copp JN, Syddall SP, et al. Discovery and evaluation of *Escherichia coli* nitroreductases that activate the anti-cancer prodrug CB1954. *Biochem Pharmacol.* 2010;79(5):678–687.
61. Liu G, Zhou J, Fu QS, Wang J. The *Escherichia coli* azoreductase AzoR is involved in resistance to thiol-specific stress caused by electrophilic quinones. *J Bacteriol.* 2009;191(20):6394–6400.
62. Morrison JM, John GH. Non-classical azoreductase secretion in *Clostridium perfringens* in response to sulfonated azo dye exposure. *Anaerobe.* 2015;34:34–43.
63. Morrison JM, Wright CM, John GH. Identification, Isolation and characterization of a novel azoreductase from *Clostridium perfringens*. *Anaerobe.* 2012;18(2):229–234.
64. Morrison JM, John GH. Growth and physiology of *Clostridium perfringens* wild-type and Δ*azoC* knockout: an azo dye exposure study. *Microbiology.* 2016;162(2):330–338.
65. Nayfach S, Fischbach MA, Pollard KS. MetaQuery: a web server for rapid annotation and quantitative analysis of specific genes in the human gut microbiome. *Bioinformatics.* 2015;31(20):3368–3370.
66. Ju KS, Parales RE. Nitroaromatic compounds, from synthesis to biodegradation. *Microbiol Mol Biol Rev.* 2010;74(2):250–272.
67. Rosenkranz HS, Mermelstein R. Mutagenicity and genotoxicity of nitroarenes. All nitro-containing chemicals were not created equal. *Mutat Res.* 1983;114(3):217–267.
68. Peres CM, Agathos SN. Biodegradation of nitroaromatic pollutants: from pathways to remediation. *Biotechnol Annu Rev.* 2000;6:197–220.
69. Roldan MD, Perez-Reinado E, Castillo F, Moreno-Vivian C. Reduction of polynitroaromatic compounds: the bacterial nitroreductases. *FEMS Microbiol Rev.* 2008;32(3):474–500.
70. Symons ZC, Bruce NC. Bacterial pathways for degradation of nitroaromatics. *Nat Prod Rep.* 2006;23(6):845–850.

71. Rieger PG, Meier HM, Gerle M, Vogt U, Groth T, Knackmuss HJ. Xenobiotics in the environment: present and future strategies to obviate the problem of biological persistence. *J Biotechnol.* 2002;94(1):101–123.
72. Ehrlich J, Bartz QR, Smith RM, Joslyn DA, Burkholder PR. Chloromycetin, a new antibiotic from a soil *Actinomycete. Science.* 1947;106(2757):417.
73. Dinos GP, Athanassopoulos CM, Missiri DA, et al. Chloramphenicol derivatives as antibacterial and anticancer agents: Historic problems and current solutions. *Antibiotics (Basel).* 2016;5(2):20.
74. Smith GN, Worrel CS. Reduction of chloromycetin and related compounds by *Escherichia coli. J Bacteriol.* 1953;65(3):313–317.
75. Smith GN, Worrel CS. The decomposition of chloromycetin (chloramphenicol) by microorganisms. *Arch Biochem.* 1950;28(2):232–241.
76. Pinder RM, Brogden RN, Speight TM, Avery GS. Clonazepam: a review of its pharmacological properties and therapeutic efficacy in epilepsy. *Drugs.* 1976;12(5):321–361.
77. Mattila MA, Larni HM. Flunitrazepam: a review of its pharmacological properties and therapeutic use. *Drugs.* 1980;20(5):353–374.
78. Takeno S, Sakai T. Involvement of the intestinal microflora in nitrazepam-induced teratogenicity in rats and its relationship to nitroreduction. *Teratology.* 1991;44(2): 209–214.
79. Gidai J, Acs N, Banhidy F, Czeizel AE. Congenital abnormalities in children of 43 pregnant women who attempted suicide with large doses of nitrazepam. *Pharmacoepidemiol Drug Saf.* 2010;19(2):175–182.
80. Hewick DS, Shaw V. The importance of the intestinal microflora in nitrazepam metabolism in the rat [proceedings]. *Br J Pharmacol.* 1978;62(3):427 pp.
81. Golovenko NY, Bogatskii AV, Orlyuk EI, Kurushin AA, Karaseva TL. Metabolism of nitrazepam in the intestine of albino rats. *Bull Exp Biol Med.* 1977;84(1):981–984.
82. Levin AA, Dent JG. Comparison of the metabolism of nitrobenzene by hepatic microsomes and cecal microflora from Fischer-344 rats in vitro and the relative importance of each in vivo. *Drug Metab Dispos.* 1982;10(5):450–454.
83. Elmer GW, Remmel RP. Role of the intestinal microflora in clonazepam metabolism in the rat. *Xenobiotica.* 1984;14(11):829–840.
84. Colburn WA, Bekersky I, Min BH, Hodshon BJ, Garland WA. Contribution of gut contents, intestinal wall and liver to the first-pass metabolism of clonazepam in the rat. *Res Commun Chem Pathol Pharmacol.* 1980;27(1):73–90.
85. Robertson MD, Drummer OH. Postmortem drug metabolism by bacteria. *J Forensic Sci.* 1995;40(3):382–386.
86. Peng FC, Chang HH, Tang SH, Chen PC, Lu SC. NADPH-cytochrome P-450 reductase is involved in flunitrazepam reductive metabolism in Hep G2 and Hep 3B cells. *J Toxicol Environ Health A.* 2004;67(2):109–124.
87. Kinouchi T, Ohnishi Y. Purification and characterization of 1-nitropyrene nitroreductases from *Bacteroides fragilis. Appl Environ Microbiol.* 1983;46(3):596–604.
88. Rafii F, Cerniglia CE. Comparison of the azoreductase and nitroreductase from *Clostridium perfringens. Appl Environ Microbiol.* 1993;59(6):1731–1734.
89. Rafil F, Franklin W, Heflich RH, Cerniglia CE. Reduction of nitroaromatic compounds by anaerobic bacteria isolated from the human gastrointestinal tract. *Appl Environ Microbiol.* 1991;57(4):962–968.
90. Deeds F, Booth AN, Jones FT. Methylation and dehydroxylation of phenolic compounds by rats and rabbits. *J Biol Chem.* 1957;225(2):615–621.
91. Booth AN, Emerson OH, Jones FT, Deeds F. Urinary metabolites of caffeic and chlorogenic acids. *J Biol Chem.* 1957;229(1):51–59.
92. Perez-Silva G, Rodriguez D, Perez-Silva J. Dehydroxylation of caffeic acid by a bacterium isolated from rat faeces. *Nature.* 1966;212(5059):303–304.

93. Soleim HA, Scheline RR. Metabolism of xenobiotics by strains of intestinal bacteria. *Acta Pharmacol Toxicol (Copenh)*. 1972;31(5):471–480.
94. Whiting GC, Carr JG. Metabolism of cinnamic acid and hydroxy-cinnamic acids by *Lactobacillus pastorianus* var. quinicus. *Nature*. 1959;184(Suppl 18):1427–1428.
95. Dodd D, Spitzer MH, Van Treuren W, et al. A gut bacterial pathway metabolizes aromatic amino acids into nine circulating metabolites. *Nature*. 2017;551(7682):648–652.
96. Cheng R, Tu T, Shackel N, McCaughan GW. Advances in and the future of treatments for hepatitis C. *Expert Rev Gastroenterol Hepatol*. 2014;8(6):633–647.
97. Chen LZ, Sabo JP, Philip E, et al. Mass balance, metabolite profile, and in vitro-in vivo comparison of clearance pathways of deleobuvir, a hepatitis C virus polymerase inhibitor. *Antimicrob Agents Chemother*. 2015;59(1):25–37.
98. Ehle M, Patel C, Giugliano RP. Digoxin: clinical highlights: a review of digoxin and its use in contemporary medicine. *Crit Pathw Cardiol*. 2011;10(2):93–98.
99. Winter ME. *Basic clinical pharmacokinetics*. Lippincott Williams & Wilkins; 2004.
100. Smith TW, Haber E. Digoxin intoxication: the relationship of clinical presentation to serum digoxin concentration. *J Clin Invest*. 1970;49(12):2377–2386.
101. Vamos M, Erath JW, Hohnloser SH. Digoxin-associated mortality: a systematic review and meta-analysis of the literature. *Eur Heart J*. 2015;36(28):1831–1838.
102. Luchi RJ, Gruber JW. Unusually large digitalis requirements. A study of altered digoxin metabolism. *Am J Med*. 1968;45(2):322–328.
103. Clark DR, Kalman SM. Dihydrodigoxin: a common metabolite of digoxin in man. *Drug Metab Dispos*. 1974;2(2):148–150.
104. Peters U, Falk LC, Kalman SM. Digoxin metabolism in patients. *Arch Intern Med*. 1978;138(7):1074–1076.
105. Hirata S, Izumi S, Furukubo T, et al. Interactions between clarithromycin and digoxin in patients with end-stage renal disease. *Int J Clin Pharmacol Ther*. 2005;43(1):30–36.
106. Haiser HJ, Gootenberg DB, Chatman K, Sirasani G, Balskus EP, Turnbaugh PJ. Predicting and manipulating cardiac drug inactivation by the human gut bacterium *Eggerthella lenta*. *Science*. 2013;341(6143):295–298.
107. Koppel N, Bisanz JE, Pandelia ME, Turnbaugh PJ, Balskus EP. Discovery and characterization of a prevalent human gut bacterial enzyme sufficient for the inactivation of a family of plant toxins. *Elife*. 2018;7:e33953.
108. Barski OA, Tipparaju SM, Bhatnagar A. The aldo-keto reductase superfamily and its role in drug metabolism and detoxification. *Drug Metab Rev*. 2008;40(4):553–624.
109. Hoffmann F, Maser E. Carbonyl reductases and pluripotent hydroxysteroid dehydrogenases of the short-chain dehydrogenase/reductase superfamily. *Drug Metab Rev*. 2007;39(1):87–144.
110. Hedner T, Samulesson O, Wahrborg P, Wadenvik H, Ung KA, Ekbom A. Nabumetone: therapeutic use and safety profile in the management of osteoarthritis and rheumatoid arthritis. *Drugs*. 2004;64(20):2315–2343. [discussion 2344–2315].
111. Turpeinen M, Hofmann U, Klein K, Murdter T, Schwab M, Zanger UM. A predominate role of CYP1A2 for the metabolism of nabumetone to the active metabolite, 6-methoxy-2-naphthylacetic acid, in human liver microsomes. *Drug Metab Dispos*. 2009;37(5):1017–1024.
112. Davies NM. Clinical pharmacokinetics of nabumetone. The dawn of selective cyclo-oxygenase-2 inhibition? *Clin Pharmacokinet*. 1997;33(6):404–416.
113. Skarydova L, Nobilis M, Wsol V. Role of carbonyl reducing enzymes in the phase I biotransformation of the non-steroidal anti-inflammatory drug nabumetone in vitro. *Xenobiotica*. 2013;43(4):346–354.
114. Gewirtz DA. A critical evaluation of the mechanisms of action proposed for the antitumor effects of the anthracycline antibiotics adriamycin and daunorubicin. *Biochem Pharmacol*. 1999;57(7):727–741.

115. Speth PA, van Hoesel QG, Haanen C. Clinical pharmacokinetics of doxorubicin. *Clin Pharmacokinet.* 1988;15(1):15–31.
116. Jacquet JM, Bressolle F, Galtier M, et al. Doxorubicin and doxorubicinol: intra- and inter-individual variations of pharmacokinetic parameters. *Cancer Chemother Pharmacol.* 1990;27(3):219–225.
117. Piska K, Koczurkiewicz P, Bucki A, Wojcik-Pszczola K, Kolaczkowski M, Pekala E. Metabolic carbonyl reduction of anthracyclines - role in cardiotoxicity and cancer resistance. Reducing enzymes as putative targets for novel cardioprotective and chemosensitizing agents. *Invest New Drugs.* 2017;35(3):375–385.
118. Olson RD, Mushlin PS, Brenner DE, et al. Doxorubicin cardiotoxicity may be caused by its metabolite, doxorubicinol. *Proc Natl Acad Sci U S A.* 1988;85(10):3585–3589.
119. Ballet F, Vrignaud P, Robert J, Rey C, Poupon R. Hepatic extraction, metabolism and biliary excretion of doxorubicin in the isolated perfused rat liver. *Cancer Chemother Pharmacol.* 1987;19(3):240–245.
120. Gardiner KM, Tett SE, Staatz CE. Multinational evaluation of mycophenolic acid, tacrolimus, cyclosporin, sirolimus, and everolimus utilization. *Ann Transplant.* 2016;21:1–11.
121. Staatz CE, Tett SE. Clinical pharmacokinetics and pharmacodynamics of tacrolimus in solid organ transplantation. *Clin Pharmacokinet.* 2004;43(10):623–653.
122. Floren LC, Bekersky I, Benet LZ, et al. Tacrolimus oral bioavailability doubles with coadministration of ketoconazole. *Clin Pharmacol Ther.* 1997;62(1):41–49.
123. Tuteja S, Alloway RR, Johnson JA, Gaber AO. The effect of gut metabolism on tacrolimus bioavailability in renal transplant recipients. *Transplantation.* 2001;71(9): 1303–1307.
124. Lee JR, Muthukumar T, Dadhania D, et al. Gut microbiota and tacrolimus dosing in kidney transplantation. *PLoS One.* 2015;10(3) e0122399.
125. Paterson DL, Singh N. Interactions between tacrolimus and antimicrobial agents. *Clin Infect Dis.* 1997;25(6):1430–1440.
126. Federico S, Carrano R, Capone D, Gentile A, Palmiero G, Basile V. Pharmacokinetic interaction between levofloxacin and ciclosporin or tacrolimus in kidney transplant recipients: ciclosporin, tacrolimus and levofloxacin in renal transplantation. *Clin Pharmacokinet.* 2006;45(2):169–175.
127. Bora F, Aliosmanoglu I, Kocak H, et al. Drug interaction between tacrolimus and ertapenem in renal transplantation recipients. *Transplant Proc.* 2012;44(10): 3029–3032.
128. Zheng Y, Masand A, Wagner M, et al. Identification of antibiotic administration as a potentially novel factor associated with tacrolimus trough variability in kidney transplant recipients: a preliminary study. *Transplant Direct.* 2019;5(9) e485.
129. Watanabe K, Yamashita S, Furuno K, Kawasaki H, Gomita Y. Metabolism of omeprazole by gut flora in rats. *J Pharm Sci.* 1995;84(4):516–517.
130. Duggan DE, Hooke KF, Risley EA, Shen TY, Arman CG. Identification of the biologically active form of sulindac. *J Pharmacol Exp Ther.* 1977;201(1):8–13.
131. Dobrinska MR, Furst DE, Spiegel T, et al. Biliary secretion of sulindac and metabolites in man. *Biopharm Drug Dispos.* 1983;4(4):347–358.
132. Kitamura S, Tatsumi K, Yoshimura H. Metabolism in vitro of sulindac. Sulfoxide-reducing enzyme systems in guinea pig liver. *J Pharmacobiodyn.* 1980;3(6):290–298.
133. Tatsumi K, Kitamura S, Yamada H. Sulfoxide reductase activity of liver aldehyde oxidase. *Biochim Biophys Acta.* 1983;747(1 2):86–92.
134. Strong HA, Warner NJ, Renwick AG, George CF. Sulindac metabolism: the importance of an intact colon. *Clin Pharmacol Ther.* 1985;38(4):387–393.
135. Moskovitz J, Flescher E, Berlett BS, Azare J, Poston JM, Stadtman ER. Overexpression of peptide-methionine sulfoxide reductase in Saccharomyces cerevisiae and human

T cells provides them with high resistance to oxidative stress. *Proc Natl Acad Sci U S A*. 1998;95(24):14071–14075.
136. Achilli C, Ciana A, Minetti G. The discovery of methionine sulfoxide reductase enzymes: an historical account and future perspectives. *Biofactors*. 2015;41(3):135–152.
137. Brunell D, Sagher D, Kesaraju S, Brot N, Weissbach H. Studies on the metabolism and biological activity of the epimers of sulindac. *Drug Metab Dispos*. 2011;39(6): 1014–1021.
138. Hamman MA, Haehner-Daniels BD, Wrighton SA, Rettie AE, Hall SD. Stereoselective sulfoxidation of sulindac sulfide by flavin-containing monooxygenases. Comparison of human liver and kidney microsomes and mammalian enzymes. *Biochem Pharmacol*. 2000;60(1):7–17.
139. Margulies EH, White AM, Sherry S. Sulfinpyrazone: a review of its pharmacological properties and therapeutic use. *Drugs*. 1980;20(3):179–197.
140. Dieterle W, Faigle JW, Moppert J. New metabolites of sulfinpyrazone in man. *Arzneimittelforschung*. 1980;30(6):989–993.
141. Wallis RB. Mechanisms of action of sulphinpyrazone. *Thromb Res Suppl*. 1983;4: 31–38.
142. Strong HA, Renwick AG, George CF. The site of reduction of sulphinpyrazone in the rabbit. *Xenobiotica*. 1984;14(10):815–826.
143. Strilchuk L, Fogacci F, Cicero AF. Safety and tolerability of available urate-lowering drugs: a critical review. *Expert Opin Drug Saf*. 2019;18(4):261–271.
144. Hlavica P. Biological oxidation of nitrogen in organic compounds and disposition of N-oxidized products. *CRC Crit Rev Biochem*. 1982;12(1):39–101.
145. Dick AT, Dann AT, Bull LB, Culvenok CCJ. Vitamin B_{12} and the detoxification of hepatotoxic pyrrolizidine alkaloids in rumen liquor. *Nature*. 1963;197:207–208.
146. Moerke MJ, McMahon LR, Wilkerson JL. More than smoke and patches: The quest for pharmacotherapies to treat tobacco use disorder. *Pharmacol Rev*. 2020;72(2): 527–557.
147. Cashman JR, Park SB, Yang ZC, Wrighton SA, Jacob 3rd P, Benowitz NL. Metabolism of nicotine by human liver microsomes: stereoselective formation of *trans*-nicotine *N'*-oxide. *Chem Res Toxicol*. 1992;5(5):639–646.
148. Beckett AH, Gorrod JW, Jenner P. Absorption of (-)-nicotine-1'-*N*-oxide in man and its reduction in the gastrointestinal tract. *J Pharm Pharmacol*. 1970;22(9):722–723.
149. Dajani RM, Gorrod JW, Beckett AH. In vitro hepatic and extra-hepatic reduction of (-)-nicotine-1'-*N*-oxide in rats. *Biochem Pharmacol*. 1975;24(1):109–117.
150. Sugiura M, Iwasaki K, Kato R. Reduction of tertiary amine N-oxides by liver microsomal cytochrome P-450. *Mol Pharmacol*. 1976;12(2):322–334.
151. Sugiura M, Iwasaki K, Kato R. Reduced nicotinamide adenine dinucleotide-dependent reduction of tertiary amine *N*-oxide by liver microsomal cytochrome P-450. *Biochem Pharmacol*. 1977;26(6):489–495.
152. Jungermann K, Kietzmann T. Oxygen: modulator of metabolic zonation and disease of the liver. *Hepatology*. 2000;31(2):255–260.
153. Bickel MH. The pharmacology and biochemistry of *N*-oxides. *Pharmacol Rev*. 1969;21(4):325–355.
154. Powis G, Svingen BA, Degraw C. Iron-EDTA stimulated reduction of indicine *N*-oxide by the hepatic microsomal fraction, isolated hepatocytes, and the intact rat. *Biochem Pharmacol*. 1982;31(3):293–299.
155. Kiese M, Renner G, Schlaeger R. Mechanism of the autocatalytic formation of ferrihemoglobin by *N,N*-dimethylaniline-N-oxide. *Naunyn Schmiedebergs Arch Pharmakol*. 1971;268(3):247–263.
156. Stanciu CN, Gnanasegaram SA. Loperamide, the "Poor Man's Methadone": brief review. *J Psychoactive Drugs*. 2017;49(1):18–21.

157. Morgan KA, Ranitidine AR. In: *StatPearls*. Treasure Island (FL): StatPearls Publishing LLC.; 2019
158. Woodings EP, Dixon GT, Harrison C, Carey P, Richards DA. Ranitidine—a new H_2-receptor antagonist. *Gut*. 1980;21(3):187–191.
159. Bodemar G, Norlander B, Fransson L, Walan A. The absorption of cimetidine before and during maintenance treatment with cimetidine and the influence of a meal on the absorption of cimetidine—studies in patients with peptic ulcer disease. *Br J Clin Pharmacol*. 1979;7(1):23–31.
160. Klotz U, Walker S. Biliary excretion of H_2-receptor antagonists. *Eur J Clin Pharmacol*. 1990;39(1):91–92.
161. Williams MF, Dukes GE, Heizer W, et al. Influence of gastrointestinal site of drug delivery on the absorption characteristics of ranitidine. *Pharm Res*. 1992;9(9): 1190–1194.
162. Basit AW, Newton JM, Lacey LF. Susceptibility of the H_2-receptor antagonists cimetidine, famotidine and nizatidine, to metabolism by the gastrointestinal microflora. *Int J Pharm*. 2002;237(1-2):23–33.
163. Falony G, Joossens M, Vieira-Silva S, et al. Population-level analysis of gut microbiome variation. *Science*. 2016;352(6285):560–564.
164. Maier L, Pruteanu M, Kuhn M, et al. Extensive impact of non-antibiotic drugss on human gut bacteria. *Nature*. 2018;555(7698):623–628.

CHAPTER FOUR

Relationship between T cells and microbiota in health and disease

Takahiro Kageyama, Takashi Matsuo, Reina Kurakake, Teruyuki Sano*
Department of Microbiology and Immunology, University of Illinois at Chicago, College of Medicine, Chicago, IL, United States
*Corresponding author: e-mail address: tsano1@uic.edu

Contents

Progress in Molecular Biology and Translational Science, Volume 171
ISSN 1877-1173
https://doi.org/10.1016/bs.pmbts.2020.03.007

Abstract

In the past decades, the fields of microbiology and immunology have largely advanced by using germ-free animals and next-generation sequencing. Many studies revealed the relationship among gut microbiota, activation of immune system, and various diseases. Especially, some gut commensals can generate their antigen-specific T cells. It is becoming clear that commensal bacteria have important roles in various autoimmune and inflammatory diseases, such as autism, rheumatoid arthritis (RA), and inflammatory bowel diseases (IBD). Recently, it was reported that commensals contribute to the cancer immune therapy. However, how commensal-specific T cells contribute to the disease development and cancer treatment are not fully understood yet. In this chapter, we will summarize the decade history of the studies associated with commensal-induced T cells and commensal-causing diseases.

1. Introduction

In these days, not only scientists/physicians but also non-specialists who are interested in our health have recognized that microbiota is very important in our bodies. Since the sequencing technology has been improved, we have known that the number of commensal bacteria is greater than the number of cells in our bodies. Moreover, improvement in the germ-free (GF) technologies has moved microbiome studies much forward. Studies using GF animals and gnotobiotic approaches have revealed two important roles of microbiome in our health; protective and/or harmful. The immune system of GF mice is immature. In particular, GF mice lack IL-17-producing $CD4^+$ helper T cells (Th17) and have a few inducible regulatory T cells (iTreg) and a small number of IFN-γ producing $CD4^+$ helper T cells (Th1) compared with specific pathogen free (SPF) mice. In addition, numbers of intestinal intraepithelial lymphocytes (IELs) in GF mice are also fewer than in SPF mice, but some IEL subsets are largely maintained in GF mice.[1] Thus, GF mice have weak immune responses to pathogen infection. However, once protective commensals are used to colonize GF mice, Th17, iTreg, and Th1 cells are significantly induced and maintain tissue homeostasis. On the other hand, mouse models of rheumatoid arthritis (RA) and inflammatory bowel diseases (IBD) develop severe tissue inflammation under conventional or SPF conditions, whereas these mice show no or minor tissue inflammation under GF conditions. Therefore, gut commensal bacteria are recognized as important players in autoimmune and inflammatory diseases. It is becoming clear that commensal bacteria have key roles for various autoimmune and inflammatory diseases, such as RA, systemic lupus erythematosus (SLE), autism, multiple sclerosis (MS), Type 1 diabetes (T1D), and IBD. These diseases are caused by dysregulated T cells. Thus, there would be strong relationships among microbiome, T cells, and diseases. The commensal bacteria responsible for promoting these diseases can be new drug targets giving a clue to develop pharmaceutical medicines for various autoimmune diseases. Moreover, some bacteria provide beneficial aspects in our bodies by promoting protective immune responses to be natural drugs to enhance our immune systems. To understand the microbiome-associated protective immunity and autoimmune diseases, in this chapter, we will summarize the roles of commensal bacteria in our health and diseases mediated by T cells and discuss the possibilities of how commensal-induced T cells contribute to the tissue homeostasis and commensals-associated diseases. First of all, we will introduce T cell subsets and the history of GF animals as a general introduction.

2. T cell subsets

T cells mainly regulate adaptive immune responses, and can be divided into $CD4^+$ T cells and $CD8^+$ T cells. $CD4^+$ T cells play crucial roles in vertebrate adaptive immune responses against invading pathogenic microbes. Expansion of $CD4^+$ T cells requires T cell receptor (TCR) stimulation by antigen-presenting cells (APCs) expressing MHC class II bound with cognate antigen in the form of small peptides. During antigen presentation, APCs also produce specific cytokines, which induce differentiation programs for the diverse functions of the activated T helper cells.[2] The effector responses of T helper cells are determined by the expression of key transcription factors and the production of different types of cytokines.[3] Type 1 T helper (Th1) cells express the key transcription factor Tbet (*Tbx21*),[4] and produce interferon-γ (IFN-γ), which controls infection with viral and other intracellular pathogens.[5] Type 2 T helper (Th2) cells express GATA3 and make interleukin 4 (IL-4), IL-5, and IL-13, which are required for the clearance of parasites such as helminths.[6] Dysregulation of Th1 and Th2 contribute on autoimmune diseases and asthma, respectively.[7] Th17 cells are defined by the expression of the transcription factor RORγt^{+}[8] and are characterized by the production of the effector cytokines IL-17A,

IL-17F, and IL-22. Th17 cells are critical for mucosal epithelial homeostasis and host defense against bacteria and fungi at various mucosal sites.[9] Dysregulated Th17 cell responses, however, promote several chronic inflammatory and autoimmune diseases such as inflammatory bowel disease (IBD), rheumatoid arthritis (RA), and multiple sclerosis (MS).[10] Regulatory T cells express key transcription factor, Foxp3, produce IL-10, and have an immunosuppressive function.[11] The presence of regulatory T cells is critical to suppress the various autoimmune diseases.[12] On the other hand, $CD8^+$ T cells can attack directly to virus-infected cells and tumor cells.[13] The expansion of $CD8^+$ T cells requires T cell receptor (TCR) stimulation by APCs expressing MHC class I bound with the antigen.

3. GF animals

3.1 History of GF animals

Our bodies possess trillions of microorganisms, consisting of bacteria, fungi, and viruses, which outnumber the total host cells in our bodies. GF animals lack all microorganisms, and are housed in tightly controlled and monitored isolators to prevent the contamination of all microorganisms. Using these animals, we can simply examine what is happening if there is no microorganism in their bodies. Many review articles and books indicated that the concept of a GF animal was proposed more than a century ago by Pasteur (1885). The first GF animals were guinea pigs generated by Nuttall and Thierfelder.[13a,13b] Almost 50 years later, the first GF rat colonies were established in the late 1940s.[13c,13d] The first GF mice were successfully developed by Pleasants.[13e,14,15] In the review article written by Al-Asmakh et al.[16] and Gensollen et al.,[17] the history of GF animals and the differences between GF and SPF mice were well summarized. Kostic et al. summarized the differences in anatomical and physiological features among various organisms not only mammals but also squid and fly.[18] Briefly, their body size and weight were less than their SPF counterparts. Moreover, the size of cecum is enlarged with full of cecum contents, due to the accumulation of mucus and undigested foods and fibers. Moreover, the small intestine of GF mice is observed due to a smaller surface area, abnormal villi, and reduced renewal of epithelial cells. The lungs are typically populated with commensal bacteria. We can find phenotypes in lungs due to the lack of commensals in GF mice; thinner alveolar and capsular walls, and fewer reticuloendothelial elements are observed in GF mice. The liver size is also reduced and the levels of ferritin and cholesterol increased in GF animals. In the circulation of GF mice, the total volume of blood is reduced and the blood flow to major tissues such as skin, liver, and lungs are also reduced.[16]

3.2 Current stage of using germ-free model in research

In these days, re-derivation of experimental animals carrying gene mutations and transgenic gene in conventional or SPF condition into GF condition has been general approaches. Using the approaches, whether microbiota is necessary for the phenotype in specific animals can be confirmed. To understand the physiological roles of one bacteria species, gnotobiotic approach is often used. Investigators will introduce a single organism into GF animals. These gnotobiotic animals must have this single bacteria species in their bodies. Moreover, various research groups have generated a humanized gnotobiotic mouse model, in which GF mice are colonized with human microbiota from healthy individuals and diseased patients.[19,20]

3.3 Advantages and limits of the germ-free models

Importantly, it often remains unclear whether microbiota distribution in patients with diseases is a cause or consequence of the disease state. GF animal model provides a lot of useful information and clues to uncover the role of microbiome in the immune system and the pathogenesis of diseases. However, the information from GF animal model cannot be directly applied to humans, because several bacterial species are found in limited animals. It has been widely recognized that some human commensal bacteria cannot colonize in the other host species. There would be strong host-bacteria interaction within host species level. Furthermore, the different physiology and anatomy in animals influence their microbial groups. Even though there are some limitations in GF animal model, GF animal model is definitely an important tool to understand the host-microbe interactions.

This chapter focuses on the other surprising observation; GF animals are very weak against pathogen infections and have immature immune systems. The differences in the immune cell population between GF and SPF mice have been characterized and analyzed deeper and deeper. The biggest changes have been found in T cell population.

4. Microbiota and Th17

4.1 Gut Th17 cells are controlled by commensal bacteria

Since the GF animals were established in 1885, it has been known that GF animals possess an underdeveloped immune system. Introduction of commensal bacteria leads to an increase in $CD4^+$ T cell numbers, induction of secretory IgA, and development of organized gut-associated lymphoid tissues (GALT).[21] Ivanov et al. first described that Th17 cells in the intestinal lamina propria (LP) are not detected in GF mice.[22] The group first realized that the appearance of Th17 cells in the LP correlated with the colonization of the intestines with normal commensal bacteria upon weaning, as indicated by the change in the size of the cecum in the examined mice. They further examined Th17 cell representation in the small intestines of GF mice. GF 129S6 and C57BL/6 had almost no detectable Th17 cells, in the intestinal LP. In addition, GF Swiss-Webster (SW) mice from Taconic Farms have no Th17 cells in the small intestine LP. Importantly, Th17 cells in GF mice are restored within 2 weeks following the oral gavage of fecal homogenates from regular SPF mice. Thus, fecal microbiota from SPF mice are necessary and sufficient to induce intestinal Th17 cells in the mice. These results indicated that there was something in the fecal contents to induce gut Th17 cells, and very likely it must be commensal bacteria in the gut based on their experiments using different antibiotics treatment. In the process of exploring the bacteria responsible for the induction of gut Th17 cells, the group found two perfect mouse models. First, C57BL/6 mice obtained from the Jackson Laboratory consistently had a quite limited number of Th17 cells (only 1–2% small intestine (SI) LP Th17 cells among $TCR\beta^+$ $CD4^+$ T cells), whereas C57BL/6 mice from Taconic farm had more Th17 cells (typically 10–15% of total $CD4^+$ T cells). These suggested that commensal flora of C57BL/6 mice from Taconic farm can possess Th17-inducible bacteria in their gut. Second, the elimination of Gram-positive or negative bacteria by the treatment with several antibiotics allowed them to reduce the number of candidate bacteria for Th17 induction in LP. Actually, ampicillin- or vancomycin-treated C57BL/6 mice lost Th17 cells but not Neomycin/Metronidazole-treated C57BL/6 mice. Taken together, Gram-positive commensals in the Taconic mice were the strong candidate bacteria for the induction of Th17 cells.[22]

4.2 Identification of Th17-inducible commensal bacterium, SFB in the mouse gut

By the 16s ribosomal RNA (rRNA) PhyloChip using luminal bacteria of C57BL/6 mice from Taconic farm and Jackson Lab, they finally identified segmented filamentous bacteria (SFB) as a Th17-inducible bacterium[23] (Fig. 1B). Mono-colonization of the SFB was sufficient to induce Th17 cells in SILP. SFB was originally identified in 1974.[24] SFB, which was

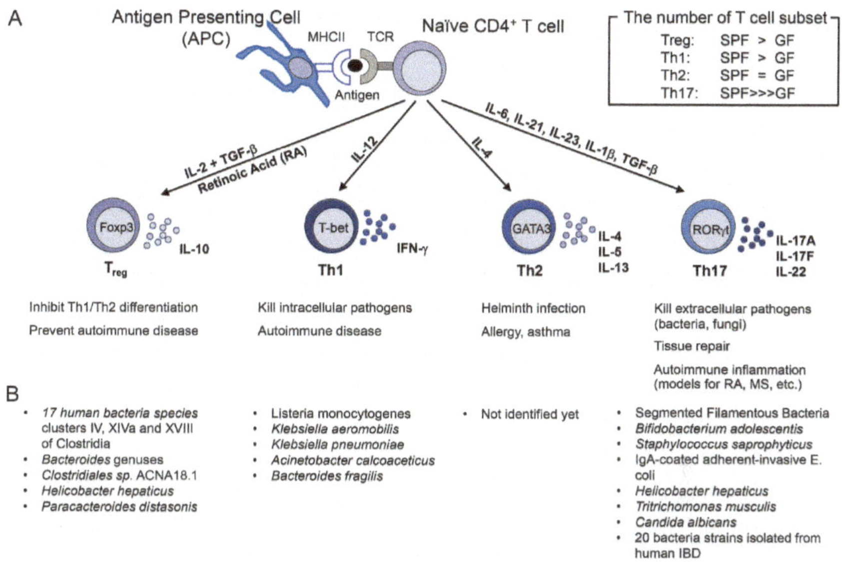

Fig. 1 Development of effector CD4$^+$ T cells. (A) Naïve T cells are educated by antigen presenting cells (APCs) through MHC classII (MHCII) with specific antigens. During the antigen presentation, stimulating Naïve T cells will be received different cytokines from APCs and the environment to be determined their fate of differentiation. (B) The list of CD4$^+$ T cell-inducible bacteria isolated from mice and humans.

biologically classified as *Candidatus* Savagella, is Gram-positive, anaerobic, and spore-forming commensal bacterium located in the terminal ileum of small intestine.[24,25] Each SFB is a quite small but makes filament-like structure, is comprised of multiple segments, and adheres tightly to epithelial cells in villi.[26] The presence of SFB has been reported in many vertebrate species, including rat,[24] fish, and chicken.[27,28] However, whether the SFB colonize in human is in the big argument. SFB is known to actively interact with the immune system.[29] Colonization of GF mice with pure SFB promotes the stimulation of IgA secretion and increases the number of intraepithelial lymphocytes (IELs) in the gut villi.[30,31] Another group reported that colonization of SFB in GF mice induces a wide spectrum of intestinal CD4$^+$ T cell responses.[32] In SFB monocolonized mice, they detected strong IL-10 production as well as IL-17 production. Moreover, IFN-γ production in SFB monocolonized mice also increased more in SILP and PPs compared with GF mice. They used C3H/HeN mice, thus understanding the difference between C57BL/6 and C3H/HeN among the mouse background can be important to fully understand the mechanisms for the bacteria-host interaction by the colonization of SFB.

4.3 SFB provides their antigens for the expansion of Th17 cells in the gut

Although SFB makes a big change in $CD4^+$ T cell population, how SFB induces Th17 cells and how these cells contribute to self-reactive pathological responses remain key unanswered questions. Yang et al. undertook an examination of the repertoire and specificity of naturally-arising Th17 cells.[33] In the paper, they showed that intestinal Th17 cells are specific for SFB- and other microbiota-derived antigens, but the most TCRs on Th17 cells recognized antigens from proteins harboring SFB. They further sought to identify epitopes recognized by TCRs of Th17 cells using a whole genome shotgun cloning and in vitro screening using *E. coli* library expressing about 1500 SFB protein. Finally, they identified various SFB-derived antigens for Th17 cells and established several SFB-specific TCR transgenic mice and tetramers which can recognize SFB-specific Th17 cells. In their paper, they also proposed an important tip for Th17 cell differentiation; antigen-itself cannot determine the fate of T cell differentiation. Oral gavage of *Listeria monocytogenes* can induce Th1 response in the intestine[34] (Fig. 1B). Moreover, Listeria overexpressed with SFB antigens also induced Th1 response but not Th17 cell response. These data strongly suggested that SFB is endowed with the ability to direct a dominant signal specialized for induction of Th17 cells, but not *Listeria monocytogenes*. These knowledge and tools accelerated the studies to understand how SFB can induce Th17 cells to modulate the entire immune response in mouse intestine.[35–38]

4.4 How does SFB induce Th17 cells in the gut?

It has been suggested a two-step process for the functional differentiation of intestinal Th17 cells following SFB colonization during homeostasis.[39] Following the SFB gavage into SFB negative mice, SFB antigen-specific $CD4^+$ T cells were primed and polarized to express RORγt in the intestine draining mesenteric lymph nodes (MLN), but these Th17 cells do not produce effector cytokines such as IL-17A in the MLN.[40] Following the migration of recently induced $ROR\gamma t^+$ Th17 cells to the intestinal LP, Th17 cells modestly upregulated effector cytokines. However, robust IL-17A producing Th17 cells were enriched in the ileum, where SFB makes direct contact with the epithelium. Serum amyloid A1 and 2, which are produced from ileum epithelial cells in response to SFB colonization, contribute the induction of IL-17A in Th17 cells.[40] Atarashi et al. nicely showed that the epithelial adhesion of SFB is important for Th17 cell induction in the intestine.[41] In the study, the group also isolated 20 candidate bacteria strains from human stools, for the Th17 cell induction (Fig. 1B). Although they did not show that all of 20 strains can attach to the epithelial cells, it was shown that numerous bacteria adhering to the colonic epithelial cells in mice when 20 strains were orally introduced into GF mice.

4.5 Other Th17-inducing bacteria

Although a single bacterial strain, SFB was identified as strong Th17 cell inducer and provide their antigens for the SFB-inducing Th17 cells in mouse, single Th17-inducing microbes derived from the human gut had not yet been identified. Many research groups appreciated these pioneer studies and started to hunt the unique immunomodulatory and immunogenic commensal bacteria in humans. Tan et al. collected a large set of microbes (38 strains) available from American Type Culture Collection, Biological and Emerging Infections Resources, Deutsche Sammlung von Mikroorganismen und Zellkulturen GmbH, or laboratory collections.[42] To test their immunomodulatory functions, Tan et al. orally introduced each bacterial strain into GF mice and checked the immune cell population such as Th17 cells. They found that two out of 37 human bacterial strains, *Bifidobacterium adolescentis* and *Staphylococcus saprophyticus* elicited SILP Th17 cells as large as SFB (Fig. 1B). Interestingly, *Bifidobacterium adolescentis* induced another T cell subset, inducible regulatory T cells defined as Foxp3 and RORγt double positive and Helios negative CD4+ T cells, but not IFN-γ-producing Th1 cells following the oral inoculation as similar as SFB. Consistent with the findings by Atarashi et al.,[41] Th17-incucible *Bifidobacterium adolescentis* colonizes closely associating with the ileum epithelial cells. It was also reported that IgA-coated adherent-invasive *E. coli* (AIEC) isolated from Crohn's disease spondyloarthritis patients (CD-SpA isolate 2A) promote Th17 cell immunity in the mouse intestine[43] (Fig. 1B). Moreover, the colonization with CD-SpA isolate 2A also induces systemic Th17 immunity and worsened the development of inflammation in large intestine and joints in mouse models.[43] Chudnovskiy et al. found that a murine commensal protest, protozoan named *Tritrichomonas musculis* activates Th1 and Th17 immunity using the circuit among epithelial cells, dendritic cells and T helper cells in the colon[44] (Fig. 1B). In addition, Atarashi et al. reported that *Candida albicans* increases the frequency of Th17 cells in the colon of monocolonized mice[41] (Fig. 1B). The next question is whether these Th17-inducible bacteria can contribute to the inflammatory diseases or not. In the next topic, we will focus on the inflammatory bowel diseases (IBD) and microbiota.

5. Microbiota and T cells in IBD

IBD is caused by the un-controllable activation of intestinal mucosal immune responses and/or decreasing the protective immune responses.[45] Dysregulated immune cells are hyperactivated in the intestine and these induce severe inflammation in the intestine.[46] Although human IBD is complex diseases, there are many IBD mouse models available for studying of IBD.[47] It has been indicated that gut microbiome is highly associated with IBD. In fact, several IBD mouse models develop intestinal inflammation under the SPF condition, whereas those under the GF condition do not develop any inflammation in the intestine. For example, one of the most well-studied models is IL-10 signaling deficient mice. IL-10 deficient mice in the SPF condition spontaneously show multifocal and transmural inflammation with epithelial hyperplasia following their weaning.[48]

5.1 Mouse model of IBD: IL-10 deficient mouse

On the other hand, IL-10 deficient mice under the GF condition do not develop any gut inflammation.[49] During IL-10 deficient colitis, strong TNFα and IL-17 production are observed in the inflamed gut. Another well-established IBD mouse model is transfer colitis.[50] Transferring naïve T cells ($CD44^{low}$, $CD62L^{high}$, $CD25^{-}$ $CD4^{+}$ T cells or $CD45RB^{high}$ $CD4^{+}$ T cells) into immune deficient host such as Rag deficient or SCID mice induces T cell mediated inflammation in colon. In this model, transferred T cells are dysregulated due to the lack of regulatory T cells (Treg) and produce Th1 and Th17-related cytokines such as IFN-γ and IL-17.[51]

5.2 Mouse model of IBD: Transfer colitis

As well as IL-10 deficient colitis model, the transfer colitis is also microbiota-dependent.[52] Jiang et al. reported that transfer colitis did not develop when the host mice were kept under the GF condition. In the paper, they reconstituted GF SCID host with conventional flora or several single bacterial strains and transferred naïve T cells into the gnotobiotic SCID mice. The colonization of conventional flora or *Helicobacter muridarum* restored the colitis phenotypes; however, *Ochrobactrum anthropi* and nonpathogenic mutant of *Listeria monocytogenes* showed very mild colitis and SFB did not induce any colitis. But, these tested bacteria activated immune response in the gut and gut-draining mesenteric lymph node (MLN).[52] These observations strongly suggested two important things; (1) it is not correct that all immunomodulatory bacterial strains can induce the inflammation under the disease condition, and (2) there are certain unknown rules to induce/initiate the tissue inflammation.

5.3 Mouse model of IBD: SAMP1/YitFc mouse

In this chapter, we cannot introduce many IBD mice models, thus we add one more unique IBD mouse model. It is the SAMP1/YitFc mouse strain represents a model of Crohn's disease (CD)-like ileitis.[53] SAMP1/YitFc mice are derived from a substrain of AKR/J mice. Originally, a colony of AKR/J mice was developed in Japan from several pairs of founder mice purchased from The Jackson Laboratory. After a number of generations of brother-sister breeding, the litters (SAMP1 mice) expressed an accelerated senescent phenotype, spontaneously.[54] A SAMP1/Yit substrain was subsequently inbred the SAMP1 line by selective breeding of the littermates that exhibited skin lesions and ileitis.[53] Following multi-times inbreeding, this substrain (SAMP1/YitFc mice) had lost the senescent phenotype, but consistently developed spontaneous inflammation in the terminal ileum. Some ileitis-susceptibility gene loci have been suggested for their ileitis phenotype.[55] In SAMP1/YitFc mice, high amount of IFN-γ and TNFα were detected in the MLN between 4 and 7 weeks of age, but strong visible inflammation was not shown in that time point. During the initial stage, Th1 polarization is further amplified in the SILP, but Th2 cells and Th2-related cytokines such as IL-4, IL-5, and IL-13 were not induced. The treatment of blocking antibody against IFN-γ significantly suppresses the ileitis at this stage.[56] However, interestingly, at the chronic phase (between 9 and 16 weeks of age) in which inflammation is expanded and become stronger and stronger, Th2-related cytokines were highly activated and Th1 response and TNFα production were comparable to those at the early phase. As similar as IL-10 deficient colitis, the ileitis pathology in SAMP1/YitFc was significantly decreased when the mice were re-derived into GF condition,[57] but not completely diminished as well as IL-10 deficient and transfer colitis.

5.4 Can Th17-inducible bacterial strains induce inflammation in the gut?

Since the sequence technologies have been improved, many research groups have been reported that the composition of the microbiome changes dramatically in diseased patients and these changes are one of the important factors in the pathogenesis of multiple diseases.[58] The condition is called dysbiosis or dysbacteriosis.[59] Dysbiosis is defined as a state of imbalance between the types of organism present in a person's natural microflora in the gut, and contribute to our health. Dysbiosis is associated with several diseases including obesity, diabetes, autoimmune diseases, and inflammatory bowel disease.[58] Especially, the abundance of bacterial species in the feces from IBD patients decrease. It is very likely that there are key bacteria which can directly affect diseases. For example, some protective bacteria can be removed under such disease condition and hosts lose protective immune response, or some pathogenic bacteria can be expanded in such condition, resulting in tissue inflammation. In the review article written by Tamboli et al., the relationship between IBD and dysbiosis was discussed deeply.[59] As in fecal culture studies, mucosal analyses in IBD have found increased concentrations of anaerobes, E coli, and decreased concentrations of *Bifidobacteria* species, including in the terminal ileum following ileocolonic resections for CD. In autoimmune diseases, the host immune system attacks and harms our own tissues, which is usually mediated by autoreactive and dysregulated T cells. Although process by which the development of many autoimmune diseases occurs remains unknown, these abnormal T cells are critical players in autoimmune diseases. However, how bacteria-induced T cells and the other immune cells activated by bacteria can contribute to disease development is still unknown for the most part. In the WT mice, any Th17-inducible bacteria isolated from mice and humans do not induce obvious gut inflammation. However, under the IBD-susceptible condition, such as IBD-associated gene deficient mice, some of the bacteria isolated from IBD patients induce gut inflammation. This kind of bacteria has been classified as "pathobiont." Although the entire mechanisms remain largely unknown, it appears that genetic defects and/or environmental factors may predispose hosts to immune-mediated diseases triggered by potentially pathogenic symbionts of the microbiota.[60] One example is that the colonization with Th17-incudible AIEC isolated from Crohn's disease and spondyloarthritis patients induces systemic Th17 immunity and worsened the development of inflammation in large intestine and joints in mouse models of IBD and RA, respectively.[43] However, it is still unclear whether bacteria-specific T cells directly induce tissue inflammation or the environmental changes such as metabolites mediated by the pathobiont provide the inflammatory condition.

6. Microbiota and Th1 cells

T helper type 1 (Th1) cells express key transcription factor, Tbet (*Tbx21*) and produce IFN-γ, which have an important role for control of infection with viruses and other intracellular pathogens.[4] Th1-inducing cytokines are induced and regulated by bacterial DNA, lipopolysaccharide, and dead bacteria such as heat-inactivated *Brucella abortus*.[61] Innate immune response could modulate the function of effector T cells. *Salmonella* are large group of enteric bacteria, not commensal gut bacteria, cause significant human and veterinary disease, and are associated with significant mortality in the world.[62–64] The immune response against *Salmonella* has been studied in both humans and mice; IFN-γ-inducing cytokines such as IL-12, IL-18 and IL-15 play essential roles in the host defense against *Salmonella* infection. In *Salmonella* infected mice, the expansion of splenic $CD4^+$ and $CD8^+$ T cells are induced and these cells work as a major fraction of the T cell pool until infected *Salmonella* are completely removed.[65] The majority of $CD4^+$ T cells responding to *Salmonella* infection expressed the Th1-master transcription factor Tbet and produced IFN-γ.[66] As described above in Section 4.3, *Listeria monocytogenes* is also enteric pathogenic bacteria and induce Th1 immunity in the gut (Fig. 1B). These enteric bacteria usually induce their antigen-specific Th1 responses and are utilized to clearance of these bacterial infections. In 2017, Atarashi et al. found that salivary microbiota strains of *Klebsiella* spp isolated from IBD patients are the strong inducers of IFN-γ producing Th1 cells in the intestine[67] (Fig. 1B). In their paper, *Klebsiella aeromobilis* and *Klebsiella pneumoniae* induced massive IFN-γ production from $CD4^+$ T cells when these bacterial isolates were orally inoculated into GF mice. *Klebsiella* spp generally colonize in the mouth; however, they tend to colonize in the intestine when the intestinal microbiota is significantly impaired by antibiotics treatment or inflammation in the gut. *Klebsiella pneumoniae* is in the family of *Enterobacteriaceae*, which has been implicated in IBD pathogenesis.[68,69] However, the colonization of the isolate, Kp-2H7 did not induce any inflammation in the intestine of WT host. On the other hand, under the disease-mimicked condition, such as in IL-10 deficient mice, Kp-2H7 inoculated IL-10 KO mice developed gut inflammation and had more increased number of $IL\text{-}17^+$ $IFN\text{-}\gamma^+$ $CD4^+$ T cells compared with Kp-2H7 colonized WT mice. Injection of *Klebsiella* intratracheally into the lung can induce Th17 cell response rather than Th1 response in the mouse lung and result in severe lung inflammation.[70,71] Therefore, the effect of Kp-2H7 on Th1 induction might be a specific response only in the intestine. Furthermore, persistence of *Klebsiella* spp. is also observed in several other diseases.[72] The comparison among the pathogenic strains (disease-causing isolates), Th1-inducible *Klebsiella* species, and non-Th1-inducible ones might provide deep insights of what the pathogenic and immunogenic signatures for the differentiation of T helper cells are. Recently, Zimmermann et al. suggested that the difference of intestinal microbiota can determine the types of colitis; Th1-associated or Th17-associated colitis.[73] When transferring $CD4^{+}$ $CD45RB^{high}$ $CD25^{-}$ naïve T cells into Rag1 deficient hosts purchased from the Jackson Laboratory (USA) and bred at a Charles River facility (Germany), WT but not T-bet-deficient donor cells induced colitis. When Rag1 deficient hosts were transplanted with fecal microbiota from the DRFZ Berlin animal facility prior to T cell transfer, T-bet-deficient donor cells induced colitis as similar as WT donors. They further analyzed the composition of microbiota between these two microbiomes. They did not conclude which bacteria are responsible for this phenotype in the paper. Further studies are needed to identify the bacteria and evaluate how the microbiome can contribute to the phenotype.

7. Microbiota and regulatory T cells

$CD4^{+}$ regulatory T cells (Treg) express the master transcription factor forkhead box P3 (Foxp3) and are abundant at higher frequencies in the intestinal lamina propria (LP), particularly in the colon than in other organs.[74] From early 2000, it has been believed that the number and function of mucosal Treg are affected by the presence of intestinal bacteria such as probiotic mixture. Notably, GF mice have fewer numbers and frequencies of Treg as compared to SPF-housed mice.[75] It is reported that probiotic administration increase number of Treg cells and the production of IL-10.[76–78] Indeed, daily treatment of mice with mixture of probiotic strains including *bifidobacteria*, *lactobacilli*, and *S. salivarius* modifies the inflammatory status of mice by the induction of Treg.[76] In other organs, such as lung, *Lactobacillus reuteri* protect against allergic airway response by the induction of regulatory T cells in the mice.[78] Atarashi et al. showed that GF mice inoculated with chloroform-treated feces have an increased number of Treg cells comparable to that in SPF mice. They further found that *Clostridium* clusters

IV and XIVa (also known as *Clostridium leptum* and *coccoides* groups, respectively) are the most abundant bacteria in the cecum and proximal colon of the mice inoculated with chloroform-treated feces (Fig. 1B). Importantly, they showed that the colonization of *Clostridium* is sufficient to increase the number of Treg cells in the colon LP by generating gnotobiotic mice with fecal material obtained from mice colonized with a cocktail of 46 strains of *Clostridium*. They further tried to select these Treg-inducible bacteria using human microbiota and finally succeeded isolating and selecting 17 strains of human bacteria on the basis of their potency in enhancing Treg cell abundance and inducing important regulatory cytokines, such as IL-10 upon inoculation into GF mice[79] (Fig.1B). More importantly, the treatment with the mixture of 17 human bacteria species suppresses experimental colitis. These pioneer studies have opened new doors;

(1) Can each of 17 strains of human bacteria induce Treg expansion in the mouse colon?
(2) Can these bacteria provide the antigen-specificities to induce Treg cells in response to these bacteria?
(3) If there are bacteria-specific Treg cells induced, these bacteria-specific Treg cells can be maintained and work in the other tissues as regulatory T cells?

As an independent research, Lathrop et al. identified multiple TCRs in colonic Treg cells, which can induce Foxp3$^+$ Treg cells in the colon LP.[80] Because Treg cells are abundant in the colon of SPF mice and less in the colon of GF mice, they speculated that these colonic Treg TCRs can be stimulated by colon colonized bacteria. They further tried to identify specific bacterial antigens recognized by these Treg TCRs by screening small pools (mixture of 2–3 bacterial isolates) of heat-killed bacteria isolated in pure culture from the colonic contents of the mice. Testing individual bacterial isolates from these pools revealed that one of the TCR clones, CT6 responded to isolated *Clostridiales* sp. ACNA18.1 (Fig. 1B). Using the strategy, they also found that another TCR clone, CT7 recognize *Parabacteroides distasonis* (Fig. 1B). CT7 broadly recognizes commensal species within the *Parabacteroides/Bacteroides* genuses. These results strongly suggested that these TCRs do not recognize self-antigens in the host, and the activation of these TCRs are differentially induced by these closely related bacteria. Their efforts also suggested that these colonic Treg TCR activation is not due to non-specific stimulation by generic immune-stimulatory bacterial components or superantigens. In the transfer colitis mouse model setting, transferred naïve T cells can be pathogenic T cells in the immune deficient animals such as Rag deficient mice and SCID mice, due to the lack of regulatory T cells in the host mice. In the presence of Treg cells by the co-transfer of naïve T cells with isolated Treg cells, the colitis is attenuated. Although the stimulation of Treg TCR clones, CT2 and CT6 by bacterial antigen is predicted to induce Treg induction, the transfer of naïve T cells from CT2 and CT6 mice into Rag1 deficient mice induced more severe colitis. They concluded that if T cells expressing these TCRs fail to undergo Treg development and instead become effector cells, they have the potential to induce colitis. They also concluded that these data support a model in which an individual's T cell population is not only instructed by classic self/non-self discrimination mechanisms during thymic development but is also educated in the periphery to accommodate the variety of non-self antigens derived from the commensal microbiota at mucosal sites.

In 2018, Other Treg-inducible TCR transgenic was established.[81] Colonization of *Helicobacter hepaticus* induced their antigen-specific Treg cells which express both Foxp3 and RORγt (Fig. 1B). The population has been considered as inducible Treg cells and has stronger immunosuppressive function, and is abundant in the colon. Interestingly, these *Helicobacter hepaticus* induced T cells became pathogenic T cells in IBD mouse models such as IL-10 deficient mice. Under the inflamed condition, presumably, *Helicobacter hepaticus*-specific iTreg cells cannot maintain their suppressive functions and trans-differentiate into pathogenic Th17 cells, which express RORγt, not Foxp3. They nicely showed that T cell specific and Treg cell specific c-Maf deficient mice lost the iTreg cells, had increased number of pathogenic Th17 cells, and develop severe spontaneous colonic inflammation. It indicated that c-Maf can contribute on the maintenance of iTreg cells and prevent from gut inflammation. Adoptive transfer of *Helicobacter hepaticus*-specific naïve T cells from HH7-2tg or HH5-1tg mice into *Helicobacter hepaticus*-colonized Rag1 deficient mice led to colitis due to the lack of Treg cells. Taking the advantage of this system, they compared the suppressive function of iTreg cells differentiated in vitro from naive HH7-2tg, HH5-1tg and polyclonal T cells. Interestingly, Helicobacter-specific iTreg suppressed colonic inflammation of transfer colitis more than polyclonal Tregs. They concluded that epitope-specific iTreg were better at suppressing colitis, providing a potential explanation for why pathobiont-specific iTreg cells are required in addition to nTreg cells to maintain gut homeostasis. However, it is still unclear how such epitope-specificity can provide the stronger immunosuppressive function on iTreg cells.

Some commensal bacteria in the gut induce the differentiation and activation of various T cell subsets. Dysregulation of these commensal-induced T cells sometimes causes gut inflammation. In the past decades, several strong evidences have been reported; there are strong links between commensals and autoimmune and inflammatory diseases occurred in other tissues far from gut. In the chapter 8 and 9, we will discuss the relationships among microbiota, T cells, and other autoimmune diseases such as rheumatoid arthritis, inflammation in central nervous systems.

8. Relationships among microbiota, T cells, and rheumatoid arthritis (RA)

RA is a heterogeneous disease in which various factors are involved in the development of RA. The main factors are classified into genetic and environmental factors. Environmental factors are smoking, sex hormones, dust exposure (silica, etc.), and periodontal disease, which are well known risk factors for RA.[82,83] Regarding the immunological pathogenesis in RA, Th1 cells were thought to play a major role in various animal models of RA. However, it was found that deficiency of IFN-γ and IL-12 paradoxically exacerbated the disease in animal model of RA.[84] However, since Th17 was discovered in 2005,[85,86] it has been realized that Th17 has critical roles in RA. IL-17-producing Th17 cells have already been discovered in the synovial membrane of RA,[87,88] and the involvement of T cells in the pathogenesis of RA has been focused, again. Th17 induces joint inflammation, and osteoclast differentiation, and bone destruction by acting on neutrophils, macrophage, fibroblasts, mesenchymal cells and osteoblasts.[89–92] There are many mouse models in which Th17 is involved in pathological conditions of RA. In recent years, the relationship between Th17 and microbiota has been reported more and more. In this section, thus we will introduce several well-established mouse models of RA and discuss the relationship among genetic abnormality, microbiome, and T cells.

8.1 Mouse models of RA

In human RA, treatment with antibiotics such as tetracycline is sometimes used as a therapeutic agent for RA.[93] In several mouse models of RA, the disease onset and pathogenesis of arthritis are suppressed under GF conditions, suggesting the involvement of commensal bacteria. In some mouse models, specific bacteria are shown to be important for the development of arthritis using gnotobiotic method. IL-1 receptor antagonist (IL-1Ra) deficient mice are known to be spontaneous RA mouse model.

In this model, excessive IL-1 signaling induces spontaneous arthritis and activates Th17, which are indispensable for the development of arthritis.[94] Interestingly, IL-1Rα deficient mice do not develop arthritis under GF conditions, whereas the mice develop arthritis if single commensal bacterium, *Lactobacillus bifidus* is colonized. In this model, TLR2,4 is involved in the development of arthritis and arthritis is caused by changes in the balance of Th17 and Treg.[95,96]

Collagen induced arthritis (CIA) is induced by the injection of type2 collagen in complete Freund's adjuvant, into the genetically susceptible mouse. In this model, CD4+ T cells promote the production of pathological autoantibody and secrete IL-17 which contribute on arthritis.[97] It was reported that CIA-susceptible mice have more *Lactobacillus* genera than CIA resistant-mice.[98]

K/BxN mice which develop spontaneous arthritis are generated by NOD mice mated with KRN T cell receptor transgenic mice. The TCR specifically recognizes host protein, Glucose-6-phosphate isomerase (GPI). As a result, the mice produce antibodies against GPI and show arthritogenicity.[99] In DBA/1 mice, immunization of GPI induces arthritis and Th17 play a crucial role in induction phase. Blocking IL-17 ameliorate GPI-induced arthritis.[100] In K/BxN mice, Th17 cells in the SILP and the spleen are reduced in GF condition, and arthritis is suppressed. When SFB is monocolonized in this GF mouse, dendritic cells present SFB antigen, Th17 cells increase, and arthritis develops.[36,101] The latest report also shows that SFB induces peyer's patch Tfh in this model and contributes to autoantibody production and arthritis.[102] This SFB was found to be involved in the activation of autoreactive T cells by changing the TCR signal threshold in the T cell transfer model.[103] Moreover, SFB triggers autoimmune pathology in the lung during pre-arthritic phase.[104] In this mouse model, T cells responsible for arthritis, express dual TCRs; one TCR is specific against SFB and second TCR is specific to antigen from host (GPI). This dual-specificity can contribute on the induction of lung autoimmunity and result in arthirits.[104,105] Recently, three independent research groups confirmed the presence of SFB in human intestine ([106–108]). Therefore, further studies are needed to be elucidated the roles of SFB in the development of human RA.

SKG mice that have a single base substitution in ZAP-70, a major molecule related to TCR signal (W163C).[109] In SKG mice, the negative selection in the thymus is abnormal due to the ZAP70 mutation in T cells. As a result, the autoreactive T cells escape into periphery and cause autoimmune arthritis. Transfer SKG CD4 T cells into Rag KO mice cause

Th17-dependent arthritis.[110] It has been reported that when this mouse is bred in a conventional environment, arthritis spontaneously develops but does not occur when bred in an SPF and GF environment.[109,111] In addition, when this mouse is injected with β-glucan (curdlan, zymosan, etc.), which is a component of fungal cell wall, the injected mouse develops arthritis through the activation of innate immunity in the SPF environment. It has been reported that β-glucan injection does not cause arthritis in GF condition, which indicated commensal bacteria are involved in the development of arthritis in SKG mice.[111,112] ZAP70 mutation by SKG causes lupus-like nephritis in the C57BL/6 background (B6SKG).[113] In this mouse model, as well, it is important to clarify whether symbiotic bacteria are involved in the development of lupus.[114]

8.2 RA in human

In human RA research, heterogeneous background such as the race, gender, and age, affects the microbiome outcomes Other factors also affect the microbiome results; whether it is early or established RA, whether it is being treated, whether control group is a patient or a healthy person. Thus, Bacteria species related to RA pathogenesis tend to differ in various reports. Even under such conditions, there are several consistent reports suggesting that *Prevotella* species are associated to RA patients. In 2008, Vaahtovuo et al. compared 51 early RA patients who had used no anti-rheumatic drugs and fibromyalgia by flow cytometry using 16S rRNA-specific probe hybridization DNA staining and they compared the gut microbiota of 50 patients.[114] The amount of three groups; (1) *Bifidobacteria*, *Bacteroides-Porphyromonas-Prevotella* group, (2) *Eubacterium rectale-Clostridium coccoides* group, and (3) *Bacteroides fragilis* subgroup were significantly less in the RA patients than those of fibromyalgia patients. In 2013, a comparative study of intestinal microbiota was conducted using next-generation sequencer methods targeting 16S rRNA and following biospecimens; 44 untreated RA patients with less than 6 months from the onset, 26 RA treated patients, 16 patients with psoriatic arthritis, and 28 healthy control in the United States were analyzed.[115] In untreated RA patients, the genus *Prevotella* (especially *Prevotella copri*) increased compared to other disease groups including healthy subjects. In addition, HLA-DRB1 shared epitope (SE) is known as one of the genetic factors for the onset of RA, it was also shown that *Prevotella copri* is more common in RA patients who do not have this SE. Maeda et al., used 16SrRNA deep sequencing, compared 17 early RA patients with 14 healthy individuals, and confirmed that there were many *Prevotella copri* in RA patients.[116] The stool of RA patients with a lot of *Prevotella copri* and the stool of healthy subjects (no *Prevotella copri)* were orally introduced into GF SKG mice and arthritis was induced by the injection of zymosan, which is a component of βD glucan. As a result, Th17 were significantly increased more in the intestine and arthritis became more severe in the zymosan-treated SKG mice fed with RA patients' stool. Moreover, SKG mice monocolonized with *Prevotella copri* developed arthritis upon the administration of zymosan. More importantly, when *Prevotella copri*-specific $CD4^+$ T cells were transferred to immune deficient SCID mice, the transferred mice develop arthritis. Although these *Prevotella copri*-specific $CD4^+$ T cells were generated in vitro system by co-culturing antigen-presenting cells, naïve T cells, and these results strongly suggested that commensal bacteria-specific T cells play pathogenic roles in arthritis. Research on microbiota has focused on analysis methods; targeting on 16S rRNA of some genes of bacteria previously. But in these days, a shotgun whole genome sequence is recommended, and part of the pathogenic mechanism of the microbiota is being elucidated. In 2015, Zhang investigated the intestinal flora and oral flora of RA patients using the metagenomic shotgun sequencing method.[117] For gut microbiota, 77 untreated RA patients and 80 healthy subjects were evaluated. Intestinal bacteria increased in patients with RA include *Clostridium paragiforme*, *Gordonibacter pamelaeae*, *Eggerthella lenta*, and *Lachnospiraceae bacterium*. As a whole, intestinal bacteria showed an increase in the proportion of Gram-positive bacteria and a decrease in the proportion of Gram-negative bacteria. It has not yet been clarified whether there is a relationship between these increased intestinal bacteria and pathological T cells or regulatory T cells in RA. In 2019, Kishikawa et al. conducted a comprehensive metagenome-wide association analysis using the shotgun sequence for the gut microbiota of the Japanese population (82 RA patients, 42 healthy subjects).[118] In comparison of the bacterial species composition between RA patients and healthy subjects, it showed the metagenome derived from RA includes more species belonging to the genus *Prevotella* (*P. denticola, P. marshii, P. disiens, P. corporis, P. amnii*) than the healthy group. Kishikawa et al. performed a pathway analysis and various pathways centered on those in metabolic response areas such as fatty acid biosynthesis and glycosaminoglycan degradation showed significant differences between patients and healthy subjects. Regarding these differences, it is possible that the gut microbiota including *Prevotella* may affect the pathology of RA. They also compared several pathways between gut metagenome and host genome data and showed significant enrichment pathways in (e.g., Th17 differentiation, IL-17 signaling). It is thought that *Prevotella* species also promote inflammation in diseases such as periodontal disease, bacterial vaginitis, HIV, and metabolic syndrome through the T cell activation.[119] However, *Prevotella histicola* was found to suppress inflammation by increasing regulatory T cells in the CIA model of Humanized mice.[120] *Prevotella histicola* also reduced arthritis score in DBA/1 mice with CIA as well, and the effect was shown not to depend on the mouse genetic background. In the future, it is expected that various types of *Prevotella* should be evaluated whether it is beneficial for human immunity, and the effects on T cells will be elucidated.

Even in research on diet therapy and probiotics targeting the microbiome, the effect for RA has not been concluded, and further research can be expected.[121] Finally, the number of reports about microbiome and RA are increasing, but there is a report mice with deficient reactive oxygen develop arthritis even under GF condition.[122] At present, microbiome should be considered as an important factor in the development of RA, a heterogeneous disease.

9. Inflammation in central nerve system (CNS) and microbiota

The bidirectional interaction between the gut microbiome and brain function has been focused on recently.[123] This crosstalk involves immune, neural, neuroactive, and endocrine pathways. The gut microbiome can modulate immune responses, stimulate the vagus nerve, induce cells to produce neurotransmitters, and secrete neuroactive metabolites.[124,125] Via the microbiota-gut-brain axis, the gut microbiome also plays an important role in neurological disorders, including multiple sclerosis (MS) and autism spectrum disorder (ASD).[126,127] In this section, the role of the microbiome, primarily related to an immune pathway, is reviewed in MS and ASD.

9.1 Multiple sclerosis (MS)

MS is the most common autoimmune disease of the central nervous system, characterized pathologically by multifocal areas of demyelination with loss of oligodendrocytes and astroglial scarring. Patients with MS have variable clinical and pathologic features reflecting different pathways to tissue injury. It has been widely considered that autoreactive lymphocytes induce an inflammatory-mediated CNS disorder in the development of MS. Inflammatory lymphocytes and macrophages are typically seen on histopathologic examination of MS lesions.[128] Also, myelin reactive T cells are found in patients with MS.[129] In an animal model of MS, experimental autoimmune encephalomyelitis (EAE), the autoreactive $CD4^+$ T cells in CNS may be skewed toward a Th1 or Th17 lineage.[130] Therefore, it is not surprising that the gut microbiome, which can modulate immune systems, is related to the pathogenesis of MS.

The clinical symptoms and pathology of EAE in GF mice were decreased compared to SPF mice.[131] The colonization of SFB in ileum promoted the differentiation of Th17 in both the lamina propria and CNS, and affected the onset and severity of EAE.[131] Furthermore, some human MS studies showed alterations of the gut microbiome in patients with MS compared to control.[132,133] Cekanaviciute et al. reported that *Akkermansia muciniphila* and *Acinetobacter calcoaceticus* increased and *Paracacteroides distasonis* decreased in gut microbiome from untreated patients with MS compared to control.[133] Both *Akkermansia muciniphila* and *Acinetobacter calcoaceticus* induced the differentiation of IFN-γ^+ Th1 lymphocytes in human peripheral blood mononuclear cells (PBMC). On the other hand, *Paracacteroides distasonis* stimulated IL-10^+ regulatory T cells in human PBMC. Furthermore, colonization of mice with microbiota from MS donor exacerbates disease severity of EAE and reduced proportions of IL-10^+ regulatory T cells compared with mice with microbiota from healthy control.[133] Berer et al. compared the gut microbiota in 34 pairs of monozygotic twins that were discordant for MS and reported a significant increase in some taxa, such as *Akkermansia muciniphila,* in the twin with MS.[132] This study also showed that transplantation of MS twin-derived microbiota to a transgenic mouse model of spontaneous EAE induced a higher EAE incidence than healthy-twin. Also, splenocytes from mice recolonized with MS twin samples produced less IL-10 compared to healthy-twin samples. Moreover, the treatment with human commensal *Prevotella histicola* also suppressed the development of EAE by upregulating regulatory T cells.[134] As mentioned above, the gut microbiome may be involved in the development of MS by affecting the differentiation of $CD4^+$ T cells.

9.2 Autism spectrum disorder (ASD)

ASD is a heterogeneous neurodevelopmental disorder characterized by persistent deficits in social communication and social interaction, and restricted, repetitive patterns of behavior. It has been considered that ASD is caused by both genetic factors and environmental factors.[135] Environmental factors include maternal immune activation (MIA) and microbiome.[135,136] Choi et al. reported that Th17 cells induced cortical abnormalities and ASD-like behaviors in the offspring of a mouse model of MIA.[137] Neutralization of IL-17A in pregnant mice prevented the development of MIA-induced behavioral abnormalities in the offspring. It uncovered the relationship between immunological and neurological pathways in the development of MIA. Furthermore, the maternal gut microbiome was reported to be involved in the development of MIA.[138] Kim et al. compared MIA model in C57BL/6 mice purchased from Taconic Biosciences (Tac), which have abundant gut Th17 cells owing to the presence of SFB, with C57BL/6 mice purchased from Jackson Laboratory (Jax), which lack SFB and gut Th17 cells and SFB. Consistent with previous reports, only the offspring from Tac showed MIA-associated behavioral phenotypes. Besides, the offspring from Jax mothers of MIA that had been co-housed with Jax or gavaged with SFB displayed MIA-associated phenotypes. This study also demonstrated that oral administration of a mix of 20 different commensal bacteria isolated from human fecal samples,[41] which was previously shown to induce Th17 cells in the large intestines of mice, into MIA model of Jax mice induced high levels of IL-17A in the maternal plasma and phenotypes in the offspring.[138] It has been suggested that human commensal bacteria, inducing the differentiation of Th17, play an essential role in the development of MIA-associated abnormality in mouse offspring. On the other hand, it has been also reported that the metabolite contributes to the pathology of ASD.[139,140] Hsiao et al. revealed that oral treatment with *Bacteroides fragilis* improved autism-related behaviors by normalizing metabolite alterations in MIA offspring.[139] Moreover, Sharon et al. reported transplantation of gut microbiota from individuals with ASD into GF mice was sufficient to induce ASD-like behaviors.[140] Importantly, this study suggested the amino acid metabolism may have a key function in the mechanism under which the microbiota regulates behavior.

10. Cancer immune therapy and microbiota

It has been considered that the development of cancer is a result of the defects of immunosurveillance.[141] Immune checkpoint molecules such as CTLA4 and PD1 are expressed on activated T cells and serve as the negative regulators to prevent from further and unnecessary T cell activation.[142,143] Therefore, Immune checkpoint inhibitors (ICIs) induce the activation of the immune system and then promote immune responses to the tumor.[144]

During this decade, the ICIs significantly have been expanded as the treatment of cancer patients. Recently, many reports suggested that gut microbiome plays an important role in determining the efficacy and toxicity of the ICIs therapies. The mechanisms of gut microbiome regulating the efficacy of ICIs treatments mainly can be divided into three parts describing below.

10.1 Microbiota and Th1 immune response on cancer immune therapy

The first point is to enhance T helper 1 immune responses.[145,146] In a mice model of sarcoma, the effect of CTLA-4 blockade was remarkably reduced in GF mice and SPF mice treated with broad-spectrum antibiotics.[146] Furthermore, this defect was overcome by oral administration with *Bacteroides fragilis* or by adoptive transfer of *Bacteroides fragilis*-specific T cells.[146] Oral feeding of *Bacteroides fragilis* promotes anti-CTLA-4 efficacy via upregulating Th1 immune responses in the tumor-draining lymph nodes and maturation of intratumoral dendritic cells. Besides, the correlation between clinical response to immune check inhibitors and the relative abundance of *Akkermansia muciniphila* was reported.[145] Oral administration of *Akkermansia muciniphila* to GF C57BL/6 mice restored the compromised efficacy of PD-1 blockade observed after the fecal microbiome transplantation from cancer patients who didn't respond to immune check inhibitors.[145] Moreover, the mechanism of immune modulation by *Akkermansia muciniphila* was suggested to induce dendritic cell secretion of IL-12, a Th1 cytokine.[145]

10.2 Microbiota and IFN-γ-producing CD8$^+$ T cells on cancer immune therapy

The second point is to induce IFN-γ-producing CD8$^+$ T cells.[147,148] Sivan et al. described that commensal *Bifidobacterium* was shown to promote the effect of anti-tumor by anti-PD-L1 in murine melanoma. *Bifidobacterium*-derived signals modulate the activation of DCs in the steady-state and enhance effector function of CD8$^+$ T cells.[147] Moreover, a consortium of 11 bacterial strains from human feces could induce IFN-γ-producing CD8$^+$ T cells.[148] Although the 11 strains localized to the caecum and colon, an increase in IFN-γ-producing CD8$^+$ T cells frequency was observed in several other organs. The 11-strain mixture enhanced the efficacy of immune checkpoint inhibitors in syngeneic tumor mice models by an increase in the frequency of IFN-γ$^+$ CD8 + tumor-infiltrating lymphocytes.

10.3 Microbiota and regulatory T cells on cancer immune therapy

The third point is to lead to decreased peripherally derived regulatory T cells.[149] *Enterococcus faecium, Collinsella aerofaciens, Bifdobacterium adolesentis,* and *Parabacteroides merdae* were increased in good responders to PD-1 blockade and reported to result in a decreased frequency of peripherally derived colonic regulatory T cells.[149] It is easy to expect that decreased Treg cells in host will promote stronger immune response and result in the promotion of anti-tumor immunity. However, the loss of Treg cells raises the possibility of increasing autoreactive T cells. Although the treatment with ICIs has shown great clinical impact as a new class of cancer therapeutics, 68.7% of patients showed immunotoxic effects in various organs following ICI treatment in a key trial.[150] Notably, 36.4% of these patients discontinued the therapy due to treatment-related immune adverse events such as autoimmune diseases and tissue inflammation. Because the systemic exposure to the ICI highly activates the immune system, resulting in anti-tumor immunity and the reduction of cancer, this stimulation might lead to undesirable non-specific T cell activation and causes inflammatory diseases, such as colitis, hepatitis, and pneumonitis.[150,151] Once cancer patients show immunotoxicity in response to treatment, the ICI administration must be terminated or postponed.[150] As we described in previous chapter, it has been reported that gut microbiota is associated with various autoimmune and inflammatory diseases such as RA, autism, MS, and IBD. Some of the commensal bacteria in our bodies can induce their antigen-specific T cell response. Thus, there would be a strong correlation among the microbiota, T cells, and ICI-induced toxicities.

11. How commensal-specific T cells generated in the gut contribute on the inflammation in tissues far away from intestine?

Many studies demonstrated the effect of gut microbiome on various autoimmune and inflammatory diseases. However, most of the mechanisms remain unclear, especially how microbiome induced immune cell can contribute the inflammation were not yet fully understood. Some of the microbiota can provide their antigens to induce $CD4^+$ and $CD8^+$ T cells, whereas many bacteria can change the metabolite profiling of hosts, foods, and microorganisms including bacteria. There are various interesting observations.

O'Donnell et al. reported that Salmonella-induced Th1 cells were further stimulated in response to the stimulation of Toll like receptors (TLR) and inflammasome pathways in the mice.[152] In the paper, O'Donnell summarized important history. *Brucella*-infected mice display transient cross-reactive protection against other intramacrophage pathogens.[153] It has reported that after clonal expansion of pathogen-specific T cells, its' effector T cells can be activated by a variety of noncognate stimuli.[154–159] It was also demonstrated that SFB-specific Th17 cells are further stimulated by the local epithelial production of serum amyloid A (SAA) 1/2. To induce the expansion and differentiation of SFB-specific Th17 cells, initial antigen presentation occurred in MLN. In the MLN, SFB-specific naïve T cells were expanded and express RORγt, which is master transcription factor for Th17 cells. However, the production of effector cytokines, such as IL-17A were significantly low. Following the migration from the MLN, $ROR\gamma t^{+}$ $IL\text{-}17A^{low}$ Th17 cells were distributed in intestinal LP and other tissues such as spleen. However, in the ileum where SFB mainly colonize, SFB initiated signaling pathway requiring IL-23 and type 3 innate lymphoid cells (ILC3) derived IL-22 and epithelial cells produced SAA1/2 in response to IL-22. As a result, only $ROR\gamma t^{+}$ $IL\text{-}17A^{low}$ Th17 cells were further stimulated by SAA1/2 and became fully matured Th17 cells, which produce robust effector cytokines such as IL-17A.[40] These findings suggested that local factors in the tissue microenvironment could contribute to the balance of homeostatic Th17 cells important for host protection and pathogenic Th17 cells implicated in autoimmunity. To evaluate the hypothesis that gut microbiota-specific T cells generated in the intestine and MLN directly affect to the inflammation in various tissues such as brain, lung, and joints, we must show the presence of microbiota-specific T cells in the inflamed tissues. If the hypothesis is true, microbiome-specific T cells generated in the gut will migrate to various tissues then promote the tissue inflammation in response to microenvironment.

Another possibility is through TCR cross-recognition. TCR recognized by commensal antigens may be also stimulated by the antigens from host proteins. The review article written by Rojas et al. summarized this phenomenon. Molecular mimicry occurs when similarities between foreign and self-peptides favor an activation of autoreactive T or B cells.[160]

It will be almost impossible to test all bacterial strains/species to examine their pathogenic roles in tissue inflammation. However, if we can understand the critical factors in disease-causing bacteria, which contribute to the pathogenesis, these features allow us to categorize the bacterial strains.

In the next 5 years, understanding the molecular mechanisms by which commensal bacteria contribute to the tissue inflammation will be one of the hot topics.

References

1. Umesaki Y, Setoyama H, Matsumoto S, Okada Y. Expansion of alpha beta T-cell receptor-bearing intestinal intraepithelial lymphocytes after microbial colonization in germ-free mice and its independence from thymus. *Immunology*. 1993;79:32–37.
2. Guermonprez P, Valladeau J, Zitvogel L, Thery C, Amigorena S. Antigen presentation and T cell stimulation by dendritic cells. *Annu Rev Immunol*. 2002;20:621–667.
3. Zhu J, Yamane H, Paul WE. Differentiation of effector CD4 T cell populations (*). *Annu Rev Immunol*. 2010;28:445–489.
4. Szabo SJ, Sullivan BM, Stemmann C, Satoskar AR, Sleckman BP, Glimcher LH. Distinct effects of T-bet in TH1 lineage commitment and IFN-gamma production in CD4 and CD8 T cells. *Science*. 2002;295:338–342.
5. Lazarevic V, Glimcher LH, Lord GM. T-bet: a bridge between innate and adaptive immunity. *Nat Rev Immunol*. 2013;13:777–789.
6. Agnello D, Lankford CS, Bream J, et al. Cytokines and transcription factors that regulate T helper cell differentiation: new players and new insights. *J Clin Immunol*. 2003;23:147–161.
7. Dardalhon V, Korn T, Kuchroo VK, Anderson AC. Role of Th1 and Th17 cells in organ-specific autoimmunity. *J Autoimmun*. 2008;31:252–256.
8. Ivanov II, McKenzie BS, Zhou L, et al. The orphan nuclear receptor RORgammat directs the differentiation program of proinflammatory IL-17+ T helper cells. *Cell*. 2006;126:1121–1133.
9. Korn T, Bettelli E, Oukka M, Kuchroo VK. IL-17 and Th17 cells. *Annu Rev Immunol*. 2009;27:485–517.
10. Maddur MS, Miossec P, Kaveri SV, Bayry J. Th17 cells: biology, pathogenesis of autoimmune and inflammatory diseases, and therapeutic strategies. *Am J Pathol*. 2012;181:8–18.
11. Josefowicz SZ, Lu LF, Rudensky AY. Regulatory T cells: mechanisms of differentiation and function. *Annu Rev Immunol*. 2012;30:531–564.
12. Sakaguchi S, Wing K, Onishi Y, Prieto-Martin P, Yamaguchi T. Regulatory T cells: how do they suppress immune responses? *Int Immunol*. 2009;21:1105–1111.
13. Ahmadzadeh M, Johnson LA, Heemskerk B, et al. Tumor antigen-specific CD8 T cells infiltrating the tumor express high levels of PD-1 and are functionally impaired. *Blood*. 2009;114:1537–1544.

13a. Nuttall GHF, Thierfelder H. ThierischesLeben ohne Bacterien im VerdauungskanalII. *Ztschr. Physiol. Chem*. 1895a;22:62–73.

13b. Nuttall GHF, Thierfelder H. ThierischesLeben ohne Bacterien im VerdauungskanalIII. *Ztschr. Physiol. Chem*. 1895b;23:231–235.

13c. Gustafsson BE. Germfree rearing of rats. *A Preliminary report, Acta Anat (Basel)*. 1947;2:376–391.

13d. Reyniers JA, Trexler PC, Erivin RF. *Rearing germfree albimo rats, Lobund Report # 1*. Notre Dame, IN: Notre Dame University Press; 1946.

13e. Pleasants JR. Rearing germfree cesarean-born rats, mice, and rabbits through weaning. *Ann N Y Acad Sci*. 1959;8(78):116–126.

14. Wostmann BS. The germfree animal in nutritional studies. *Annu Rev Nutr*. 1981;1:257–279.

15. Yi P, Li L. The germfree murine animal: an important animal model for research on the relationship between gut microbiota and the host. *Vet Microbiol.* 2012;157:1–7.
16. Al-Asmakh M, Zadjali F. Use of germ-free animal models in microbiota-related research. *J Microbiol Biotechnol.* 2015;25:1583–1588.
17. Gensollen T, Iyer SS, Kasper DL, Blumberg RS. How colonization by microbiota in early life shapes the immune system. *Science.* 2016;352:539–544.
18. Kostic AD, Howitt MR, Garrett WS. Exploring host-microbiota interactions in animal models and humans. *Genes Dev.* 2013;27:701–718.
19. Britton GJ, Contijoch EJ, Mogno I, et al. Microbiotas from humans with inflammatory bowel disease alter the balance of gut Th17 and RORgammat(+) regulatory T cells and exacerbate colitis in mice. *Immunity.* 2019;50(212–224):e214.
20. Nagao-Kitamoto H, Shreiner AB, Gillilland 3rd MG, et al. Functional characterization of inflammatory bowel disease-associated gut dysbiosis in gnotobiotic mice. *Cell Mol Gastroenterol Hepatol.* 2016;2:468–481.
21. Macpherson AJ, Harris NL. Interactions between commensal intestinal bacteria and the immune system. *Nat Rev Immunol.* 2004;4:478–485.
22. Ivanov II, Frutos Rde L, Manel N, et al. Specific microbiota direct the differentiation of IL-17-producing T-helper cells in the mucosa of the small intestine. *Cell Host Microbe.* 2008;4:337–349.
23. Ivanov II, Atarashi K, Manel N, et al. Induction of intestinal Th17 cells by segmented filamentous bacteria. *Cell.* 2009;139:485–498.
24. Davis CP, Savage DC. Habitat, succession, attachment, and morphology of segmented, filamentous microbes indigenous to the murine gastrointestinal tract. *Infect Immun.* 1974;10:948–956.
25. Thompson CL, Vier R, Mikaelyan A, Wienemann T, Brune A. '*Candidatus* Arthromitus' revised: segmented filamentous bacteria in arthropod guts are members of *Lachnospiraceae. Environ Microbiol.* 2012;14:1454–1465.
26. Koopman JP, Stadhouders AM, Kennis HM, De Boer H. The attachment of filamentous segmented micro-organisms to the distal ileum wall of the mouse: a scanning and transmission electron microscopy study. *Lab Anim.* 1987;21:48–52.
27. Klaasen HL, Koopman JP, Van den Brink ME, Bakker MH, Poelma FG, Beynen AC. Intestinal, segmented, filamentous bacteria in a wide range of vertebrate species. *Lab Anim.* 1993;27:141–150.
28. Ley RE, Hamady M, Lozupone C, et al. Evolution of mammals and their gut microbes. *Science.* 2008;320:1647–1651.
29. Klaasen HL, Van der Heijden PJ, Stok W, et al. Apathogenic, intestinal, segmented, filamentous bacteria stimulate the mucosal immune system of mice. *Infect Immun.* 1993;61:303–306.
30. Talham GL, Jiang HQ, Bos NA, Cebra JJ. Segmented filamentous bacteria are potent stimuli of a physiologically normal state of the murine gut mucosal immune system. *Infect Immun.* 1999;67:1992–2000.
31. Umesaki Y, Okada Y, Matsumoto S, Imaoka A, Setoyama H. Segmented filamentous bacteria are indigenous intestinal bacteria that activate intraepithelial lymphocytes and induce MHC class II molecules and fucosyl asialo GM1 glycolipids on the small intestinal epithelial cells in the ex-germ-free mouse. *Microbiol Immunol.* 1995;39:555–562.
32. Gaboriau-Routhiau V, Rakotobe S, Lecuyer E, et al. The key role of segmented filamentous bacteria in the coordinated maturation of gut helper T cell responses. *Immunity.* 2009;31:677–689.
33. Yang Y, Torchinsky MB, Gobert M, et al. Focused specificity of intestinal TH17 cells towards commensal bacterial antigens. *Nature.* 2014;510:152–156.

34. Hsieh CS, Macatonia SE, Tripp CS, Wolf SF, O'Garra A, Murphy KM. Development of TH1 CD4+ T cells through IL-12 produced by Listeria-induced macrophages. *Science*. 1993;260:547–549.
35. Esterhazy D, Canesso MCC, Mesin L, et al. Compartmentalized gut lymph node drainage dictates adaptive immune responses. *Nature*. 2019;569:126–130.
36. Goto Y, Panea C, Nakato G, et al. Segmented filamentous bacteria antigens presented by intestinal dendritic cells drive mucosal Th17 cell differentiation. *Immunity*. 2014;40:594–607.
37. Lecuyer E, Rakotobe S, Lengline-Garnier H, et al. Segmented filamentous bacterium uses secondary and tertiary lymphoid tissues to induce gut IgA and specific T helper 17 cell responses. *Immunity*. 2014;40:608–620.
38. Mao K, Baptista AP, Tamoutounour S, et al. Innate and adaptive lymphocytes sequentially shape the gut microbiota and lipid metabolism. *Nature*. 2018;554:255–259.
39. Sano T, Yang Y, Diehl G, Chen A, Kaplan DH, Littman DR. Multi-step Th17 differentiation in response to segmented filamentous bacteria in the mouse intestine. *J Immunol*. 2016;196.
40. Sano T, Huang W, Hall JA, et al. An IL-23R/IL-22 circuit regulates epithelial serum amyloid A to promote local effector Th17 responses. *Cell*. 2015;163:381–393.
41. Atarashi K, Tanoue T, Ando M, et al. Th17 cell induction by adhesion of microbes to intestinal epithelial cells. *Cell*. 2015;163:367–380.
42. Tan TG, Sefik E, Geva-Zatorsky N, et al. Identifying species of symbiont bacteria from the human gut that, alone, can induce intestinal Th17 cells in mice. *Proc Natl Acad Sci U S A*. 2016;113:E8141–E8150.
43. Viladomiu M, Kivolowitz C, Abdulhamid A, et al. IgA-coated *E. coli* enriched in Crohn's disease spondyloarthritis promote TH17-dependent inflammation. *Sci Transl Med*. 2017;9:eaaf9655.
44. Chudnovskiy A, Mortha A, Kana V, et al. Host-protozoan interactions protect from mucosal infections through activation of the inflammasome. *Cell*. 2016; 167(444–456):e414.
45. Duchmann R, Kaiser I, Hermann E, Mayet W, Ewe K, Meyer zum Buschenfelde, K.H. Tolerance exists towards resident intestinal flora but is broken in active inflammatory bowel disease (IBD). *Clin Exp Immunol*. 1995;102:448–455.
46. Xu XR, Liu CQ, Feng BS, Liu ZJ. Dysregulation of mucosal immune response in pathogenesis of inflammatory bowel disease. *World J Gastroenterol*. 2014;20: 3255–3264.
47. Kiesler P, Fuss IJ, Strober W. Experimental models of inflammatory bowel diseases. *Cell Mol Gastroenterol Hepatol*. 2015;1:154–170
48. Kuhn R, Lohler J, Rennick D, Rajewsky K, Muller W. Interleukin-10-deficient mice develop chronic enterocolitis. *Cell*. 1993;75:263–274.
49. Sellon RK, Tonkonogy S, Schultz M, et al. Resident enteric bacteria are necessary for development of spontaneous colitis and immune system activation in interleukin-10-deficient mice. *Infect Immun*. 1998;66:5224–5231.
50. Powrie F, Leach MW, Mauze S, Caddle LB, Coffman RL. Phenotypically distinct subsets of CD4+ T cells induce or protect from chronic intestinal inflammation in C. B-17 scid mice. *Int Immunol*. 1993;5:1461–1471.
51. Ostanin DV, Bao J, Koboziev I, et al. T cell transfer model of chronic colitis: concepts, considerations, and tricks of the trade. *Am J Physiol Gastrointest Liver Physiol*. 2009;296: G135–G146.
52. Jiang HQ, Kushnir N, Thurnheer MC, Bos NA, Cebra JJ. Monoassociation of SCID mice with Helicobacter muridarum, but not four other enterics, provokes IBD upon receipt of T cells. *Gastroenterology*. 2002;122:1346–1354.

53. Matsumoto S, Okabe Y, Setoyama H, et al. Inflammatory bowel disease-like enteritis and caecitis in a senescence accelerated mouse P1/Yit strain. *Gut*. 1998;43:71–78.
54. Takeda T, Hosokawa M, Takeshita S, et al. A new murine model of accelerated senescence. *Mech Ageing Dev*. 1981;17:183–194.
55. Kozaiwa K, Sugawara K, Smith Jr MF, et al. Identification of a quantitative trait locus for ileitis in a spontaneous mouse model of Crohn's disease: SAMP1/YitFc. *Gastroenterology*. 2003;125:477–490.
56. Bamias G, Martin C, Mishina M, et al. Proinflammatory effects of TH2 cytokines in a murine model of chronic small intestinal inflammation. *Gastroenterology*. 2005;128: 654–666.
57. Pizarro TT, Pastorelli L, Bamias G, et al. SAMP1/YitFc mouse strain: a spontaneous model of Crohn's disease-like ileitis. *Inflamm Bowel Dis*. 2011;17:2566–2584.
58. Carding S, Verbeke K, Vipond DT, Corfe BM, Owen LJ. Dysbiosis of the gut microbiota in disease. *Microb Ecol Health Dis*. 2015;26:26191.
59. Tamboli CP, Neut C, Desreumaux P, Colombel JF. Dysbiosis in inflammatory bowel disease. *Gut*. 2004;53:1–4.
60. Chow J, Tang H, Mazmanian SK. Pathobionts of the gastrointestinal microbiota and inflammatory disease. *Curr Opin Immunol*. 2011;23:473–480.
61. Huang L, Krieg AM, Eller N, Scott DE. Induction and regulation of Th1-inducing cytokines by bacterial DNA, lipopolysaccharide, and heat-inactivated bacteria. *Infect Immun*. 1999;67:6257–6263.
62. Grassl GA, Finlay BB. Pathogenesis of enteric Salmonella infections. *Curr Opin Gastroenterol*. 2008;24:22–26.
63. Parry CM, Hien TT, Dougan G, White NJ, Farrar JJ. Typhoid fever. *N Engl J Med*. 2002;347:1770–1782.
64. Rabsch W, Tschape H, Baumler AJ. Non-typhoidal salmonellosis: emerging problems. *Microbes Infect*. 2001;3:237–247.
65. Srinivasan A, Foley J, McSorley SJ. Massive number of antigen-specific CD4 T cells during vaccination with live attenuated Salmonella causes interclonal competition. *J Immunol*. 2004;172:6884–6893.
66. Griffin AJ, McSorley SJ. Development of protective immunity to Salmonella, a mucosal pathogen with a systemic agenda. *Mucosal Immunol*. 2011;4:371–382.
67. Atarashi K, Suda W, Luo C, et al. Ectopic colonization of oral bacteria in the intestine drives TH1 cell induction and inflammation. *Science*. 2017;358:359–365.
68. Gevers D, Kugathasan S, Denson LA, et al. The treatment-naive microbiome in new-onset Crohn's disease. *Cell Host Microbe*. 2014;15:382–392.
69. Morgan XC, Tickle TL, Sokol H, et al. Dysfunction of the intestinal microbiome in inflammatory bowel disease and treatment. *Genome Biol*. 2012;13:R79.
70. Chen K, McAleer JP, Lin Y, et al. Th17 cells mediate clade-specific, serotype-independent mucosal immunity. *Immunity*. 2011;35:997–1009.
71. Xiong H, Keith JW, Samilo DW, Carter RA, Leiner IM, Pamer EG. Innate lymphocyte/Ly6C(hi) monocyte crosstalk promotes *Klebsiella Pneumoniae* clearance. *Cell*. 2016;165:679–689.
72. Ebringer A, Rashid T, Tiwana H, Wilson C. A possible link between Crohn's disease and ankylosing spondylitis via Klebsiella infections. *Clin Rheumatol*. 2007;26:289–297.
73. Zimmermann J, Durek P, Kuhl AA, et al. The intestinal microbiota determines the colitis-inducing potential of T-bet-deficient Th cells in mice. *Eur J Immunol*. 2018;48:161–167.
74. Hall JA, Bouladoux N, Sun CM, et al. Commensal DNA limits regulatory T cell conversion and is a natural adjuvant of intestinal immune responses. *Immunity*. 2008;29:637–649.

75. Atarashi K, Tanoue T, Shima T, et al. Induction of colonic regulatory T cells by indigenous Clostridium species. *Science*. 2011;331:337–341.
76. Di Giacinto C, Marinaro M, Sanchez M, Strober W, Boirivant M. Probiotics ameliorate recurrent Th1-mediated murine colitis by inducing IL-10 and IL-10-dependent TGF-beta-bearing regulatory cells. *J Immunol*. 2005;174:3237–3246.
77. Jeon SG, Kayama H, Ueda Y, et al. Probiotic Bifidobacterium breve induces IL-10-producing Tr1 cells in the colon. *PLoS Pathog*. 2012;8:e1002714.
78. Karimi K, Inman MD, Bienenstock J, Forsythe P. Lactobacillus reuteri-induced regulatory T cells protect against an allergic airway response in mice. *Am J Respir Crit Care Med*. 2009;179:186–193.
79. Atarashi K, Tanoue T, Oshima K, et al. Treg induction by a rationally selected mixture of Clostridia strains from the human microbiota. *Nature*. 2013;500:232–236.
80. Lathrop SK, Bloom SM, Rao SM, et al. Peripheral education of the immune system by colonic commensal microbiota. *Nature*. 2011;478:250–254.
81. Xu M, Pokrovskii M, Ding Y, et al. c-MAF-dependent regulatory T cells mediate immunological tolerance to a gut pathobiont. *Nature*. 2018;554:373–377.
82. Deane KD, Demoruelle MK, Kelmenson LB, Kuhn KA, Norris JM, Holers VM. Genetic and environmental risk factors for rheumatoid arthritis. *Best Pract Res Clin Rheumatol*. 2017;31:3–18.
83. McInnes IB, Schett G. The pathogenesis of rheumatoid arthritis. *N Engl J Med*. 2011;365:2205–2219.
84. Vermeire K, Heremans H, Vandeputte M, Huang S, Billiau A, Matthys P. Accelerated collagen-induced arthritis in IFN-gamma receptor-deficient mice. *J Immunol*. 1997; 158:5507–5513.
85. Harrington LE, Hatton RD, Mangan PR, et al. Interleukin 17-producing CD4+ effector T cells develop via a lineage distinct from the T helper type 1 and 2 lineages. *Nat Immunol*. 2005;6:1123–1132.
86. Park H, Li Z, Yang XO, et al. A distinct lineage of CD4 T cells regulates tissue inflammation by producing interleukin 17. *Nat Immunol*. 2005;6:1133–1141.
87. Chabaud M, Durand JM, Buchs N, et al. Human interleukin-17: a T cell-derived proinflammatory cytokine produced by the rheumatoid synovium. *Arthritis Rheum*. 1999;42:963–970.
88. Kotake S, Udagawa N, Takahashi N, et al. IL-17 in synovial fluids from patients with rheumatoid arthritis is a potent stimulator of osteoclastogenesis. *J Clin Invest*. 1999;103: 1345–1352.
89. Fossiez F, Djossou O, Chomarat P, et al. T cell interleukin-17 induces stromal cells to produce proinflammatory and hematopoietic cytokines. *J Exp Med*. 1996;183: 2593–2603.
90. Kolls JK, Linden A. Interleukin-17 family members and inflammation. *Immunity*. 2004;21:467–476.
91. Sato K, Suematsu A, Okamoto K, et al. Th17 functions as an osteoclastogenic helper T cell subset that links T cell activation and bone destruction. *J Exp Med*. 2006;203:2673–2682.
92. Yao Z, Fanslow WC, Seldin MF, et al. Herpesvirus Saimiri encodes a new cytokine, IL-17, which binds to a novel cytokine receptor. *Immunity*. 1995;3:811–821.
93. Stone M, Fortin PR, Pacheco-Tena C, Inman RD. Should tetracycline treatment be used more extensively for rheumatoid arthritis? Metaanalysis demonstrates clinical benefit with reduction in disease activity. *J Rheumatol*. 2003;30:2112–2122.
94. Koenders MI, Devesa I, Marijnissen RJ, et al. Interleukin-1 drives pathogenic Th17 cells during spontaneous arthritis in interleukin-1 receptor antagonist-deficient mice. *Arthritis Rheum*. 2008;58:3461–3470.

95. Abdollahi-Roodsaz S, Joosten LA, Koenders MI, et al. Stimulation of TLR2 and TLR4 differentially skews the balance of T cells in a mouse model of arthritis. *J Clin Invest*. 2008;118:205–216.
96. Scher JU, Abramson SB. The microbiome and rheumatoid arthritis. *Nat Rev Rheumatol*. 2011;7:569–578.
97. Murphy CA, Langrish CL, Chen Y, et al. Divergent pro- and antiinflammatory roles for IL-23 and IL-12 in joint autoimmune inflammation. *J Exp Med*. 2003;198: 1951–1957.
98. Liu X, Zeng B, Zhang J, et al. Role of the gut microbiome in modulating arthritis progression in mice. *Sci Rep*. 2016;6:30594.
99. Matsumoto I, Staub A, Benoist C, Mathis D. Arthritis provoked by linked T and B cell recognition of a glycolytic enzyme. *Science*. 1999;286:1732–1735.
100. Iwanami K, Matsumoto I, Tanaka-Watanabe Y, et al. Crucial role of the interleukin-6/interleukin-17 cytokine axis in the induction of arthritis by glucose-6-phosphate isomerase. *Arthritis Rheum*. 2008;58:754–763.
101. Wu HJ, Ivanov II, Darce J, et al. Gut-residing segmented filamentous bacteria drive autoimmune arthritis via T helper 17 cells. *Immunity*. 2010;32:815–827.
102. Teng F, Klinger CN, Felix KM, et al. Gut microbiota drive autoimmune arthritis by promoting differentiation and migration of Peyer's patch T follicular helper cells. *Immunity*. 2016;44:875–888.
103. Chappert P, Bouladoux N, Naik S, Schwartz RH. Specific gut commensal flora locally alters T cell tuning to endogenous ligands. *Immunity*. 2013;38:1198–1210.
104. Bradley CP, Teng F, Felix KM, et al. Segmented filamentous bacteria provoke lung autoimmunity by inducing gut-lung axis Th17 cells expressing dual TCRs. *Cell Host Microbe*. 2017;22(697–704):e694.
105. Deng J, Wei Y, Fonseca VR, Graca L, Yu D. T follicular helper cells and T follicular regulatory cells in rheumatic diseases. *Nat Rev Rheumatol*. 2019;15:475–490.
106. Chen B, Chen H, Shu X, et al. Presence of segmented filamentous bacteria in human children and its potential role in the modulation of human gut immunity. *Front Microbiol*. 2018;9:1403.
107. Jonsson H, Hugerth WL, Sundh J, Andersson FA. Genome sequence of segmented filamentous bacteria present in the human intestine. *BioRxiv*. 2019. https://doi.org/10.1101/813196.
108. Yin Y, Wang Y, Zhu L, et al. Comparative analysis of the distribution of segmented filamentous bacteria in humans, mice and chickens. *ISME J*. 2013;7:615–621.
109. Sakaguchi N, Takahashi T, Hata H, et al. Altered thymic T-cell selection due to a mutation of the ZAP-70 gene causes autoimmune arthritis in mice. *Nature*. 2003; 426:454–460.
110. Hirota K, Hashimoto M, Yoshitomi H, et al. T cell self-reactivity forms a cytokine milieu for spontaneous development of IL-17+ Th cells that cause autoimmune arthritis. *J Exp Med*. 2007;204:41–47.
111. Rehaume LM, Mondot S, Aguirre de Carcer D, et al. ZAP-70 genotype disrupts the relationship between microbiota and host, leading to spondyloarthritis and ileitis in SKG mice. *Arthritis Rheumatol*. 2014;66:2780–2792.
112. Yoshitomi H, Sakaguchi N, Kobayashi K, et al. A role for fungal β-glucans and their receptor Dectin-1 in the induction of autoimmune arthritis in genetically susceptible mice. *J Exp Med*. 2005;201:949–960.
113. Matsuo T, Hashimoto M, Sakaguchi S, et al. Strain-specific manifestation of lupus-like systemic autoimmunity caused by Zap70 mutation. *J Immunol*. 2019;202: 3161–3172.
114. Vaahtovuo J, Munukka E, Korkeamaki M, Luukkainen R, Toivanen P. Fecal microbiota in early rheumatoid arthritis. *J Rheumatol*. 2008;35:1500–1505.

115. Scher JU, Sczesnak A, Longman RS, et al. Expansion of intestinal Prevotella copri correlates with enhanced susceptibility to arthritis. *Elife*. 2013;2:e01202.
116. Maeda Y, Kurakawa T, Umemoto E, et al. Dysbiosis contributes to arthritis development via activation of autoreactive T cells in the intestine. *Arthritis Rheumatol*. 2016;68:2646–2661.
117. Zhang X, Zhang D, Jia H, et al. The oral and gut microbiomes are perturbed in rheumatoid arthritis and partly normalized after treatment. *Nat Med*. 2015;21:895–905.
118. Kishikawa T, Maeda Y, Nii T, et al. Metagenome-wide association study of gut microbiome revealed novel aetiology of rheumatoid arthritis in the Japanese population. *Ann Rheum Dis*. 2020;79:103–111.
119. Larsen JM. The immune response to Prevotella bacteria in chronic inflammatory disease. *Immunology*. 2017;151:363–374.
120. Marietta EV, Murray JA, Luckey DH, et al. Suppression of inflammatory arthritis by human gut-derived Prevotella histicola in humanized mice. *Arthritis Rheumatol*. 2016; 68:2878–2888.
121. Badsha H. Role of diet in influencing rheumatoid arthritis disease activity. *Open Rheumatol J*. 2018;12:19–28.
122. Wing K, Klocke K, Samuelsson A, Holmdahl R. Germ-free mice deficient of reactive oxygen species have increased arthritis susceptibility. *Eur J Immunol*. 2015;45: 1348–1353.
123. Sharon G, Sampson TR, Geschwind DH, Mazmanian SK. The central nervous system and the gut microbiome. *Cell*. 2016;167:915–932.
124. Cox LM, Weiner HL. Microbiota signaling pathways that influence neurologic disease. *Neurotherapeutics*. 2018;15:135–145.
125. Sgritta M, Dooling SW, Buffington SA, et al. Mechanisms underlying microbial-mediated changes in social behavior in mouse models of autism spectrum disorder. *Neuron*. 2019;101(246–259):e246.
126. Cryan JF, O'Riordan KJ, Sandhu K, Peterson V, Dinan TG. The gut microbiome in neurological disorders. *Lancet Neurol*. 2020;19:179–194.
127. Jangi S, Gandhi R, Cox LM, et al. Alterations of the human gut microbiome in multiple sclerosis. *Nat Commun*. 2016;7:12015.
128. Frohman EM, Racke MK, Raine CS. Multiple sclerosis—the plaque and its pathogenesis. *N Engl J Med*. 2006;354:942–955.
129. Van Noort JM, van Sechel AC, Bajramovic JJ, et al. The small heat-shock protein alpha B-crystallin as candidate autoantigen in multiple sclerosis. *Nature*. 1995;375:798–801.
130. Carbajal KS, Mironova Y, Ulrich-Lewis JT, et al. Th cell diversity in experimental autoimmune encephalomyelitis and multiple sclerosis. *J Immunol*. 2015;195: 2552–2559.
131. Lee YK, Menezes JS, Umesaki Y, Mazmanian SK. Proinflammatory T-cell responses to gut microbiota promote experimental autoimmune encephalomyelitis. *Proc Natl Acad Sci U S A*. 2011;108(suppl 1):4615–4622.
132. Berer K, Gerdes LA, Cekanaviciute E, et al. Gut microbiota from multiple sclerosis patients enables spontaneous autoimmune encephalomyelitis in mice. *Proc Natl Acad Sci U S A*. 2017;114:10719–10724.
133. Cekanaviciute E, Yoo BB, Runia TF, et al. Gut bacteria from multiple sclerosis patients modulate human T cells and exacerbate symptoms in mouse models. *Proc Natl Acad Sci U S A*. 2017;114:10713–10718.
134. Mangalam A, Shahi SK, Luckey D, et al. Human gut-derived commensal bacteria suppress CNS inflammatory and demyelinating disease. *Cell Rep*. 2017;20:1269–1277.
135. Bai D, Yip BHK, Windham GC, et al. Association of genetic and environmental factors with autism in a 5-country cohort. *JAMA Psychiatry*. 2019;76(10):1035–1043. https://doi.org/10.1001/jamapsychiatry.2019.141.

136. Kang DW, Adams JB, Gregory AC, et al. Microbiota transfer therapy alters gut ecosystem and improves gastrointestinal and autism symptoms: an open-label study. *Microbiome*. 2017;5:10.
137. Choi GB, Yim YS, Wong H, et al. The maternal interleukin-17a pathway in mice promotes autism-like phenotypes in offspring. *Science*. 2016;351:933–939.
138. Kim S, Kim H, Yim YS, et al. Maternal gut bacteria promote neurodevelopmental abnormalities in mouse offspring. *Nature*. 2017;549:528–532.
139. Hsiao EY, McBride SW, Hsien S, et al. Microbiota modulate behavioral and physiological abnormalities associated with neurodevelopmental disorders. *Cell*. 2013;155: 1451–1463.
140. Sharon G, Cruz NJ, Kang DW, et al. Human gut microbiota from autism spectrum disorder promote behavioral symptoms in mice. *Cell*. 2019;177. 1600–1618.e1617.
141. Dunn GP, Old LJ, Schreiber RD. The immunobiology of cancer immunosurveillance and immunoediting. *Immunity*. 2004;21:137–148.
142. Agata Y, Kawasaki A, Nishimura H, et al. Expression of the PD-1 antigen on the surface of stimulated mouse T and B lymphocytes. *Int Immunol*. 1996;8:765–772.
143. Walunas TL, Lenschow DJ, Bakker CY, et al. CTLA-4 can function as a negative regulator of T cell activation. *Immunity*. 1994;1:405–413.
144. Leach DR, Krummel MF, Allison JP. Enhancement of antitumor immunity by CTLA-4 blockade. *Science*. 1996;271:1734–1736.
145. Routy B, Le Chatelier E, Derosa L, et al. Gut microbiome influences efficacy of PD-1-based immunotherapy against epithelial tumors. *Science*. 2018;359:91–97.
146. Vetizou M, Pitt JM, Daillere R, et al. Anticancer immunotherapy by CTLA-4 blockade relies on the gut microbiota. *Science*. 2015;350:1079–1084.
147. Sivan A, Corrales L, Hubert N, et al. Commensal Bifidobacterium promotes antitumor immunity and facilitates anti-PD-L1 efficacy. *Science*. 2015;350:1084–1089.
148. Tanoue T, Morita S, Plichta DR, et al. A defined commensal consortium elicits CD8 T cells and anti-cancer immunity. *Nature*. 2019;565:600–605.
149. Matson V, Fessler J, Bao R, et al. The commensal microbiome is associated with anti-PD-1 efficacy in metastatic melanoma patients. *Science*. 2018;359:104–108.
150. Larkin J, Chiarion-Sileni V, Gonzalez R, et al. Combined nivolumab and ipilimumab or monotherapy in untreated melanoma. *N Engl J Med*. 2015;373:23–34.
151. Robert C, Long GV, Brady B, et al. Nivolumab in previously untreated melanoma without BRAF mutation. *N Engl J Med*. 2015;372:320–330.
152. O'Donnell H, Pham OH, Li LX, et al. Toll-like receptor and inflammasome signals converge to amplify the innate bactericidal capacity of T helper 1 cells. *Immunity*. 2014;40:213–224.
153. Mackaness GB. The immunological basis of acquired cellular resistance. *J Exp Med*. 1964;120:105–120.
154. Beadling C, Slifka MK. Differential regulation of virus-specific T-cell effector functions following activation by peptide or innate cytokines. *Blood*. 2005;105:1179–1186.
155. Berg RE, Cordes CJ, Forman J. Contribution of CD8+ T cells to innate immunity: IFN-gamma secretion induced by IL-12 and IL-18. *Eur J Immunol*. 2002;32: 2807–2816.
156. Freeman BE, Hammarlund E, Raue HP, Slifka MK. Regulation of innate CD8+ T-cell activation mediated by cytokines. *Proc Natl Acad Sci U S A*. 2012;109: 9971–9976.
157. Guo L, Wei G, Zhu J, et al. IL-1 family members and STAT activators induce cytokine production by Th2, Th17, and Th1 cells. *Proc Natl Acad Sci U S A*. 2009;106: 13463–13468.

158. Kupz A, Guarda G, Gebhardt T, et al. NLRC4 inflammasomes in dendritic cells regulate noncognate effector function by memory CD8(+) T cells. *Nat Immunol.* 2012;13:162–169.
159. Souja SM, Ruiz AL, Marie JC, Lauvau G. Inflammatory monocytes activate memory $CD8^+$ T and innate NK lymphocytes independent of cognate antigen during microbial pathogen invasion. *Immunity.* 2012;37:549–562.
160. Rojas M, Restrepo-Jimenez P, Monsalve DM, et al. Molecular mimicry and autoimmunity. *J Autoimmun.* 2018;95:100–123.

CHAPTER FIVE

Oral and intestinal bacterial exotoxins: Potential linked to carcinogenesis

Matthew Silbergleit[a], Adrian A. Vasquez[b], Carol J. Miller[b], Jun Sun[c], Ikuko Kato[d,e,*]

[a]Wayne State University School of Medicine, Detroit, MI, United States
[b]Department of Civil and Environmental Engineering, Wayne State University, Healthy Urban Waters, Detroit, MI, United States
[c]Department of Medicine, College of Medicine, University of Illinois at Chicago, Chicago, IL, United States
[d]Department of Oncology, Wayne State University School of Medicine, Detroit, MI, United States
[e]Department of Pathology, Wayne State University School of Medicine, Detroit, MI, United States
*Corresponding author: e-mail address: katoi@karmanos.org

Contents

Progress in Molecular Biology and Translational Science, Volume 171
ISSN 1877-1173
https://doi.org/10.1016/bs.pmbts.2020.02.004

Abstract

Growing evidence suggests that imbalances in resident microbes (dysbiosis) can promote chronic inflammation, immune-subversion, and production of carcinogenic metabolites, thus leading to neoplasia. Yet, evidence to support a direct link of individual bacteria species to human sporadic cancer is still limited. This chapter focuses on several emerging bacterial toxins that have recently been characterized for their potential oncogenic properties toward human orodigestive cancer and the presence of which in human tissue samples has been documented. These include cytolethal distending toxins produced by various members of gamma and epsilon Proteobacteria, Dentilisin from mammalian oral *Treponema*, *Pasteurella multocida* toxin, two Fusobacterial toxins, FadA and Fap2, *Bacteroides fragilis* toxin, colibactin, cytotoxic necrotizing factors and α-hemolysin from *Escherichia coli*, and *Salmonella enterica* AvrA. It was clear that these bacterial toxins have biological activities to induce several hallmarks of cancer. Some toxins directly interact with DNA or chromosomes leading to their breakdowns, causing mutations and genome instability, and others modulate cell proliferation, replication and death and facilitate immune evasion and tumor invasion, prying specific oncogene and tumor suppressor pathways, such as p53 and β-catenin/Wnt. In addition, most bacterial toxins control tumor-promoting inflammation in complex and diverse mechanisms. Despite growing laboratory evidence to support oncogenic potential of selected bacterial toxins, we need more direct evidence from human studies and mechanistic data from physiologically relevant experimental animal models, which can reflect chronic infection in vivo, as well as take bacterial-bacterial interactions among microbiome into consideration.

1. Introduction

1.1 Overview of association between infectious agents, microbiome and cancer

Infections have been recognized as a major preventable cause of human cancer.[1] Plummer et al.[2] estimate that 15.4% of the worldwide cancer cases are attributable to infection, amounting to approximately 2.2 million cases per year. A variety of infectious agents have been linked to pathogenesis of human cancers, including parasites, such as *Schistosoma haematobium* (urinary bladder cancer) and *Opisthorchis viverrini* (cholangiocarcinoma), bacteria, such as, *Helicobacter (H) pylori* (stomach cancer and gastric lymphoma) and viruses, such as hepatitis virus types B and C (liver cancer), human papillomavirus (HPV) (anogenital and oropharyngeal cancer), Epstein-Barr (EB) virus (Burkitt lymphoma and nasopharyngeal cancer), human herpes virus type 8 (Kaposi's sarcoma and non-Hodgkin's lymphoma) and human T cell leukemia/lymphoma virus type 1.[2,3] Etiological roles of these oncoviruses have been appreciated more unequivocally, as viral genomes can be integrated to the host genome and transcribed, producing own oncoproteins and manipulating host signaling pathways critical in oncogenesis. On the other hand, precise mechanisms of bacteria-induced carcinogenesis have been less clear, except a well characterized oncoprotein from *H. pylori*, CagA,[4–9] despite the fact that bacteria are the most ubiquitous microorganisms that collectively make up to 100 trillion cells in our bodies, 10-fold the number of human cells.[10] Bacteria are primarily extracellular organisms and thus their oncogenic effects may be rather indirect. Growing evidence also suggests that imbalances in resident microbes (dysbiosis) can promote chronic inflammation, immune-subversion and production of carcinogenic metabolites, thus leading to neoplasia.[1,11–13] Yet, evidence to support a direct link of individual bacteria species to human sporadic cancer is still limited, except the aforementioned *H. pylori* producing a cytotoxin, CagA. There are other bacterial toxins that have recently been characterized for their potential oncogenic properties and the presence in human tissue samples. This book chapter focuses on those emerging bacterial toxins in order to expand the repertoire of bacteria to be interrogated further to establish a causal association.

1.2 Overview of bacterial toxins

Toxins are a wide range of substances biologically produced within living cells or organisms that are harmful to other organisms while benefiting own survival. They are inanimate and not capable of reproducing themselves. Toxins can be small molecules, peptides, or proteins that are capable of causing disease on contact with or absorption by body tissues interacting with biological macromolecules such as enzymes or cellular receptors. Bacterial toxins are broadly grouped into two main types, lipopolysaccharides (LPS), which are associated with the cell wall of Gram-negative bacteria, and proteins, which are released from bacterial cells and may act at tissue sites removed from the site of bacterial growth. The cell-associated toxins are referred to endotoxins and the extracellular diffusible toxins are referred to as exotoxins, or protein toxins.[14] This book chapter focuses on exotoxins produced by distinctive groups of bacteria, as LPS is not bacteria specific, produced by a virtually all Gram-negative bacteria and has been known for a long time causing sepsis (endotoxemia) and a myriad of innate inflammatory responses in infected hosts.[15–17] While exotoxins can be named after organisms generating the toxin or after susceptible tissue/organ types, protein toxins can be further divided according to modes of actions; (1) targeting host membrane receptors; (2) targeting host membrane lipid bilayer, and (3) targeting host cytosolic proteins, either through uptake by host endocytosis of paired bacterial proteins called AB toxins, or by direct delivery through molecular syringe of a bacterial injection apparatus.[18] Protein toxins are also characterized based in their secretion systems.[19] To exert virulence, bacterial toxins produced in cytosol must be transported across the bacterial membranes through co-translational and post-translational mechanisms to reach their targets. These transports are mediated through various secretion systems, which now include nine different mechanisms, types I–IX.[20,21] Gram-positive bacteria utilize only types I and VII,[19] Gram-negative bacteria exploit all but type VII,[20] while one bacterial species can possess multiple systems. Finally, toxins are also classified based on virulence mechanisms, i.e., targeting cytoskeleton components, targeting ubiquitin/proteasomal system, targeting cell translational machinery, targeting neurotransmission system, targeting cAMP production and MAP kinase signaling, and targeting DNA and inducing endoplasmic reticulum stress responses,[18] some of which may be highly relevant to carcinogenesis. Subsequent sections provide more details about individual bacterial toxins, with which potential links to human cancers have been increasingly recognized.

2. Cytolethal distending toxins (CDTs)

2.1 Biochemical properties of CDTs

CDTs belong to the AB toxin family produced by several groups of Gram-negative bacteria and is composed of an active subunit (CdtB) and two binding subunits (CdtA and CdtC) (AB_2 toxin).[22] The crystal structure solved by Nesic et al. indicates that this holotoxin is a tripartite complex,[23] and thus it is distinguished from binary CDT toxin secreted by *Clostridium difficile* in its structure and functions. CDT consists of an enzyme of the DNase-I family (CdtB) bound to two ricin-like lectin domains (CdtA and CdtC), which are required for host cell attachment and translocation of CdtB into host cells.[22] It forms a ternary complex with three interdependent molecular interfaces, characterized by globular, as well as non-globular extensions from CdtA and CdtC that facilitate interactions among the three subunits. CdtA and CdtB have also been reported to contain disordered amino acid sequences pointing to dynamic properties of this molecule.[23,24] An overall surface area of interaction is estimated to be over 10,700 $Å^2$ for the entire holotoxin, with relative molecular mass ~70,000 (M_r 70 K).[23]

The CdtA and CdtC subunits are lectin-type molecules, sharing structural homology with the B-chain repeats of the plant toxin ricin and form a deeply grooved, highly aromatic surface that is critical for toxicity.[23] The CdtB subunit is functionally and structurally homologous to mammalian DNase I, is translocated into the nucleus, and acts as a genotoxin leading to DNA damage.[25] After cell surface binding via CdtA and CdtC, CdtB has to cross the plasma membrane through endocytosis to reach the nucleus, traveling through the trans-Golgi, where sulfation occurs, and then it is retrograde translocated via the endoplasmic reticulum (ER), where it is glycosylated.[26] Further confocal microscopic analysis indicates that most likely the active subunit of CDT is directly translocated from the ER to the nucleus.[27] CdtB is a highly potent DNase, inducing DNA at an extremely low concentration (50 pg/mL), leading to both single-strand and double-strand breaks.[25] CdtB also possesses lipid phosphatase activity, converting PI3,4,5,P3 to PI,4,5P2.[22,28,29] This leads to the blockade of PI-3K signaling, which is required for an extremely diverse array of cellular functions, most notably cellular proliferation and survival and frequently dysregulated in many cancers.[22,30] These biochemical characteristics of this toxin are indicative of its potential involvement in human cancers.

2.2 Microbiological characteristics of CDT-producing bacteria

CDTs are produced by a small but diverse group of bacterial pathogens that primarily belong to gamma and epsilon Proteobacteria.[28] To date, CDT have been documented in *Aggregatibacter actinomycetemcomitans* (*Aa*), *Campylobacter coli/fetus/hyointestinalis/jejuni* (*Cj*)/*lari/upsaliensis*, *Escherichia albertii/coli* (*Ea/Ec*), *Haemophilus ducreyi* (*Hd*), *Helicobacter canis/cinaedi/fennelliae/hepaticus/pullorum/winghamensis*, *Salmonella enterica* (*Se*) and *Shigella dysenteriae/boydii*,[5,25,28,31] which were isolated from human samples. However, it is important to note that this toxin is not necessarily produced by all subspecies/strains that belong to these specific bacterial groups. The higher carriage rates (>80%) have been reported for *Aa*, *Hd*, *Se* and enterohepatic *Campylobacter* and *Helicobacter* species, while it is generally low in *Ec* and *Shigella*.[28,31–33]

In nearly all bacteria, CDTs are encoded by the *cdtABC* gene cluster, a constitutively expressed operon on the chromosome, which yields molecular masses of 23–30, 28–29 and 19–21 kDa of CdtA, CdtB and CdtC subunits, respectively, according to bacterial species.[28] Among these three genes, *cdtB* is the most conserved with homology across species of 50% or higher, while that of *cdtA* and *cdtC* is generally less than 30%.[22,31,34] Nonsynonymous mutation in any conserved residue that is critical for the catalytic activity or the Mg^{2+} binding abolishes the ability of CdtB to cleave DNA.[22,23,35] Single nucleotide polymorphisms at amino acid 281 (H/R) in *Aa cdtB* and at 95 (P/S) in *Cj cdtB* have been shown to change its biological activity by several log orders.[36,37] Strains carrying the high activity allele were reported to be dominant in clinical isolates.[36,37] A notable exception for the structure of this gene cluster is *Se*, which is an intracellular pathogen and expresses CDT only after bacterial internalization into host cells.[22,38] *Se* does not carry any homologous genes to *cdtA* and *cdtC*, but possesses two other genes called, *pltA* and *pltb* (persussis-like toxin A and B), located upstream of *cdtB* in the same pathogenicity island, the products of which form with CdtB a tripartite.[39,40] These two subunits have been projected to facilitate the transport of CdtB from its production site within infected host cells to extracellular medium, from where CdtB can intoxicate other non-infected cells.[22,39]

Each of the translated CdtA, CdtB and CdtC subunits is independently translocated across the cytoplasmic membrane, through a standard mechanism for the movement of secreted proteins across the cytoplasmic membrane of bacteria, where precursor proteins are exported by a common sec-dependent pathway that involves both signal peptidase I and signal peptidase II.[41,42] The CdtB and CdtC proteins processed by signal peptidase I and the CdtA processed by signal peptidase II accumulate in the periplasmic space and all three proteins self-assemble at the outer membrane. Catalyzed by an unidentified processing enzyme to remove the hydrophobic glycerolipid of CdtA, the holotoxin is translocated across the outer membrane.[41,42] The hydrophilic holotoxin accumulates in the aqueous environment outside the bacterium until it recognizes a specific receptor on the target cell surface.

An alternative mode of bacterial toxin secretion is the release of holotoxins into enclosed outer membrane vesicles (OMVs). Bacterial OMVs are nano-sized compartments consisting of a lipid bilayer that encapsulates periplasm-derived, luminal content. OMVs produced by of Gram-negative bacteria are now recognized as a generalized secretion pathway that provides a means to transfer cargo to the host cells.[43] In the context of intestinal colonization, packaging and release of CDT into outer membrane vesicles may provide the toxin with extra protection against enzymatic digestion, thus increasing the likelihood of the uptake of intact toxin by host cells. Although a mechanism of delivering the toxin from the vesicles to the target cell has not been well understood, several potential routes, including clathrin or caveolin mediated or non-mediated endocytosis and membrane fusion, have been demonstrated.[43] OMVs-like vesicles have been also found to be synthesized by internalized *Se*, transported and released to the extracellular medium through host microtubule and actin tracks.[44]

2.3 In vitro cellular responses to CDTs

Biological consequences from CDT intoxication in mammalian cells have been evaluated in vitro using various normal and tumor cell lines that are exposed to bacterial lysates or purified toxins or are transfected with *cdtB* or CdtB, of different bacterial origins. DNA damage caused by CDT has been shown to induce the activation of DNA damage response (DDR), leading to an ATM (Ataxia telangiectasia mutated)-dependent cell cycle arrest at G2/M and/or G1/S transition, and initiation of multiple DNA repair pathways. However, the DDR system sometimes fails to properly repair DNA damage, leading to cell death by apoptosis or to a long-term cell cycle arrest (senescence).[25] This cellular senescence may alternatively lead to a senescence-associated secretory phenotype (SASP), which is characterized by secretion of a large number of growth factors and pro-inflammatory cytokines and can promote survival and proliferation of transformed cells.[45]

Cell fate following CDT-induced DNA damage depends on types of cells.[46,47] Hematopoietic lineages, particularly lymphocyte, are known to be greatly more sensitive to CDT than most other cells, inducing rapid move toward apoptosis after a brief cell cycle arrest.[29,46] Accordingly, CDT is considered an immunotoxin. It is capable of inhibiting phagocytosis in murine macrophages[48] as well of inhibiting functions of both T- and B-cells.[29,46] Furthermore, CDT has other immunomodulatory properties as it has been documented to induce the expression of pro-inflammatory cytokines/chemokines (such as IL-1β, IL-6, IL-8) in human macrophages and peripheral mononuclear cells.[25,49,50]

Epithelial and mesenchymal lineages mainly undergo cell cycle arrest accompanied by cytoplasmic elongation and distension due to accumulation of F-actin assemblies.[51] DNA double-strand breaks induced by *Hd*CDT led to the formation of actin stress fibers and the rearrangement of the actin cytoskeleton in epithelial cells and fibroblasts, via activation of the small GTPase RhoA. Activation of RhoA was part of the ATM-induced cellular responses to genotoxic stresses to support cell survival.[52] Moreover, in colorectal cancer cell lines, exposure to *Hd* CDT stimulated a marker of autophagy, a pro-survival pathway, in an ATM and p53-dependent manner.[53] CDT also induced remodeling of adherens junctions in gingival tissue and normal epithelial cells and compromised mucosal barrier function, which was accompanied by a pronounced increase in the expression and cytosolic distribution of E-cadherin and β-catenin.[54]

Investigators have demonstrated in both fibroblast and normal and cancer cell lines the enhanced senescence-associated β-galactosidase activity, expansion of promyelocytic leukemia nuclear compartments and induced expression of several cytokines (especially interleukins IL-6, IL-8 and IL-24), which were overall features shared by cells undergoing replicative or premature cellular senescence.[55] Furthermore, a long-term (30 weeks) of exposure of fibroblast and intestinal epithelial cells to *Hd* CDT led to increased frequency of mutations (mainly transversions), accumulation of chromosomal aberrations and enhanced anchorage-independent growth.[56] These observations suggest that CDT-induced DNA damage promotes genomic instability, favoring tumor initiation and progression.[57] Graillot et al. postulate that CDT does not initiate colorectal cancer by itself, but may have promoting effects in premalignant epithelial cells. They found that chronic exposure (3 weeks) to *Ec* CDT resulted in enhanced anchorage-independent growth and genetic instability measured by the micronucleus formation in colonic epithelial cells defective in APC and p53 in comparison to normal cells, which were considered hallmarks of malignant transformation.[58] Frisan et al. also postulate that CDT may act synergistically with an oncovirus, specifically EBV. They found that exposure of gastric cancer cell lines to *Aa* CDT reactivated latent EBV infection and promoted latent infection in non-infected cells via genomic instability.[59]

2.4 CDT toxicity in animal models

Potential carcinogenic effects of CDTs in vivo have been evaluated in a few animal models. The vast majority of the studies were based on inoculation of murine enterohepatic *Helicobacter* species, either *H. hepaticus* or *H. bilis*, while a minority used enteric *Cj*. In these studies, natural or engineered CDT-deficient strains were often used for comparison.[60,61] In both liver and colon carcinogenesis, a progressive pathological sequence from chronic active inflammation (hepatitis/colitis), dysplasia, benign tumor to cancer has been acknowledged.[60,61] *Hh* inoculation to A/J, C57BL/6 and their F1 mice induced hepatitis, some of which further progressed to hepatocellular carcinomas or dysplastic liver lesions, in a gender-specific manner, resulting in 69% incidence in F1 male mice. *Hh* was isolated from hepatic tissues of all F1 mice with liver tumors, which were developed in the presence of hepatic steatosis and active hepatitis.[62] Likewise, perinatal intraperitoneal inoculation of *Hh* to A/J female mice resulted in increased incidence of hepatitis and hepatocellular tumors in their male (not female) offspring.[63] However, these experiments did not rule out the effects of other virulence factors co-expressed in CDT-positive *Hh*. Thus, others have tried to address the independent effects of CDT. While infection of A/JCr mice with wild-type (WT) *Hh* or an isogenic mutant lacking CDT activity induced comparable chronic hepatitis, the CDT mutant-infected mice did not develop any hepatic dysplastic nodules in contrast to the WT-infected male (not female) mice that developed dysplasia. Male mice infected with WT also exhibited significantly enhanced hepatic transcription of pro-inflammatory TNF-α, IFN-γ and Cox-2, growth mediators IL-6 and TGF-α, anti-apoptotic Bcl-2 and Bcl-XL, and increased hepatocyte proliferation compared with the CDT mutant-infected male mice, confirming that CDT plays a key role in promoting the dysplastic changes in the *Hh*-infected mouse livers.[64] Fox et al. further investigated the potential interactions of *Hh* with chemical or virial liver carcinogens. In both models, *Hh* intestinal colonization was sufficient to promote the development of hepatocellular carcinoma (HCC), via transactivation of the nuclear factor kappa

N (NF-κB) network in the liver. Thus bacterial translocation to the liver was not a requisite and, in the presence of a chemical carcinogen, aflatoxin, both male female mice equally developed HCC.[65] Tumor promoting effects of *Hh* on hepatitis C virus transgenic mice shown in this study also suggest potential bacterial involvement in viral activation as in the case of *Aa* CDT and EBV.[59]

Orogastric administration of culture supernatants of CDT-producing isolates to suckling mice induced overt inflammation in stomach, small intestine and large intestine, which was not seen in the experiments with CDT-negative *Cj* strains,[66] demonstrating that CDT plays a crucial role in gastrointestinal pathology in this model. In NF-κB-deficient mice, inoculation of the *cdtB* mutant strain did not induce gastroenteritis and gastric hyperplasia/dysplasia comparable to that developed in the WT-infected mice.[67] Furthermore, inoculation of CDT-negative Hh mutant showed a significantly diminished capacity to induce colitis lesions in an interleukin-10 deficient mouse model for inflammatory bowel disease.[68] Shen et al. tested mutant and WT CDT from zoonotic *H. cinaedi* using the same mouse model. Both mutant and WT colonized to a similar extent, but the mice infected with the mutant showed attenuated colitis and hyperplasia but no dysplasia or intra-mucosal carcinoma, compared to those infected with WT, which exhibited elevated mRNA expression of tumor necrosis factor alpha, inducible nitric oxide synthase, and gamma interferon in the cecum.[69] Another newly found CDT-positive *H. japonicum* induced the histopathology characterized by moderate to severe inflammation, mild edema, epithelial defects, mild to severe hyperplasia, dysplasia and carcinoma in the lower bowel of the same mouse model.[70] In an immunocompromised mouse model, inoculation of CDT-deficient *Hh* mutant led to less severe cecal pathology without inducing pre-neoplastic lesions, while WT-infected animals developed pre-neoplastic dysplasia and cancer, with concomitant upregulation of cecal Il-6 and Tnfα.[71] In a mouse xenograft model with *Hh cdtB*-transfected human hepatic and intestinal cancer cell lines, both types of the CdtB-derived tumors showed increased apoptosis, senescence, p21 and Ki-67 nuclear antigen expression, an overexpression of cytokeratins at the invasive front of the tumor and an increase in ploidy, while no difference in proliferating cells undergoing mitosis. All these features were considered hallmarks of endoreplication and aggressiveness in cancer.[72] More recently, oral infection of the $Apc^{Min/+}$ mice with CDT-producing *Cj* together with dextran sulfate sodium administration has been reported to induce significantly larger number and size of tumors when compared with a *ctdB* mutant *Cj*. WT *Cj* infection also induced expression of hundreds of colonic genes, with 22 genes dependent on the presence of *cdtB*, as well as altered microbial gene expression, which could also modulate colorectal carcinogenesis.[73]

2.5 Clinical and epidemiological studies concerning CDT

The direct evidence to link CDTs or CDT-positive bacteria to human cancer are still limited. To our knowledge, the presence of CDT itself in human tumor tissue or in surrounding mucosa has not been demonstrated. Alternatively, CDT-positive (by PCR) *Ec* has been reported to be more commonly present in colorectal cancer tissue than in colorectal mucosa from diverticulosis patients (16% vs 0%).[33] However, these CDT-positive strains also harbored other known virulence genes and thus it was difficult to disentangle the effect of CDT itself. Another routinely used method to assess the past exposure is serological assays to detect specific antibodies. Although anti-CDT antibody assays have been available, demonstrating increased titers in diarrhea type of irritable bowel syndrome (IRS) compared with control subjects,[74,75] their association with cancer or premalignant pathologies has not been studied. More consistent evidence from human populations has been complied with regard to hepatobiliary tract cancers. It should be noted that these studies evaluated the exposure to bacterial species with known high carriage rates of CDT, without confirmation of CDT status. The best studied is gallbladder cancer and its predisposing condition, cholelithiasis, in relation to infection with *Se* serovar Typhi or *Hh*. *Se* serovar Typhi status has been most often assessed by antibodies against a capsular polysaccharide (Vi). The summary odds ratio (OR) from a meta-analysis based on such serological studies was 4.6, consistent with the summary ORs based on stool or bile culture (4.7 and 5.5, respectively). A recent study in India also reported an equivalent difference in serology using the same assay between cancer and benign conditions, but more positive samples (74%) were detected using PCR for *S. typhi fliC* genes, whereas PCR-positivity in healthy gallbladder samples was very rare.[76] Serological assays against a *Hh* antigen have revealed a significantly higher antibody prevalence in bile duct/gallbladder cancer patients (39%) and even higher prevalence in patients with viral hepatitis (60%) or viral liver cirrhosis (68%), compared with cholelithiasis (13%) or healthy donors (27%), respectively.[77,78] Given low antibody prevalence in non-viral hepatitis and cirrhosis reported at the same time,[78] these findings are also be indicative of bacterial-viral interaction. Alternatively, using PCR, in situ hybridization for16S rRNA and Western blot for *Hh* antigens Hamada et al. reported high *Hh* carriage rates (40%) in cholelithiasis patients with or without gastric cancer, compared to the patients with other benign conditions (13%).[79]

3. Dentilisin (chymotrypsin-like protease (CTLP))

3.1 Biochemical properties of CTLP

Dentilisin belongs to only two known lipoproteins in the subtilase family of subtilisin-like serine proteases. Its activity is suggested to be based on an active seryl residue, on an active imidazole group, and on an active carboxyl group but not on metal cations.[80] The enzyme hydrolyzes specifically N-succinyl-L-Ala-L-Ala-L-Pro-L-Phe-p-nitroaniline (SAAPFNA, a typical chymotrypsin substrate) and various host proteins and peptides, including basement membrane components (type IV collagen, laminin, and fibronectin), serum proteins (transferrin, fibrinogen, IgG, IgA, and α1-antitrypsin), and bioactive peptides (cytokines, and chemokines).[81,82] These biochemical properties are considered to be instrumental in bacterial invasion, immune evasion and cytotoxicity.[81,82] The matured dentilisin is present as a complex, composed of an active enzyme, PrtP (~65 kDa), and two auxiliary proteins, PrcA (~70 kDa) and PrcB (~22 kDa).[83] The two auxiliary proteins are postulated to be required for stabilization of the active enzyme unit.[84,85] PrtP protease activity facilitates maturation of PrtP and PrcA, by releasing a 16 kDa N-terminal of the premature PrtP and cleaving PrcA to A1 (~30 kDa) and A2 (~40 kDa).[85,86]

3.2 Microbiological characteristics of dentilisin synthesis

Dentilisin is produced primarily by mammalian oral *Treponema* species,[83,87,88] represented by *T. denticola* (*Td*), a key pathogen of periodontitis, which are Gram-negative obligate anaerobic bacteria, spirochaetes uniquely lacking LPS. It is important to note that substantial inter-strain variability exist in this enzymatic activity, which is not entirely explained by the presence or sequence variabilities of encoding genes.[83,87,88] Typically, the dentilisin complex is encoded by a three-gene operon, consisting of *prcB*, *prcA* and *prtP*, with a putative promoter upstream of *prcB*.[85,86] These gene sequences are recently annotated,[83] demonstrating that *prcB* is the most conserved (89% out of 202 amino acids identical among the strains). *prcA* encoding 639 amino acids was 77% identical among the strains, and much of the variability was in three regions (151–180, 328–362 and 512–545). In *prtP* that transcribes 766 amino acids, the predicted catalytic triad (Asp^{203}, His^{258}, Ser^{447}) was conserved in all strains, and most *prtP* inter-strain variations were concentrated in the C-terminal 270 residues (up to 20%).[83] However, variability in proteolytic activities was not correlated with overall homology of *prtP*, suggesting that other unidentified sensing and signaling mechanisms modulate expression of these operons.

In *Td*, the dentilisin complex is located on bacterial outer membrane and extracellularly.[89] However, the secretion system by which the three subunits are transported through inner and outer membranes has not been well understood, Yet, studies have demonstrated that this complex is co-localized with another major *Td* virulence factor, Msp (major outer sheath protein),[83,85] encoded by *msp* and playing a major role in host adhesion.[90] MSP locates within the outer membrane, but is predominantly periplasmic, with only limited surface exposure.[91] Importantly, there is close interplay between PrtP and outer membrane segment of Msp,[92] and, as a result, *msp* mutants lack CTLP activity while *prtP* mutants show aberrant Msp expression.[93] In addition to cell surface expression, a recent mass-spectrometry-based study revealed that the three dentilisin subunits were enriched in outer membrane vesicles,[89] which enable remote delivery of the toxin.

3.3 Dentilisin in vitro activities

Cytopathic effects of isolated dentilisin, including membrane blebbing, vacuolization, inhibition of motility, loss of epithelial cell contacts and release of a cytosolic enzyme have been demonstrated in multi-layer epithelial cell culture system.[94,95] In this culture model, *Td* did not invade the cells of epithelial multilayers, but evidently shed dentilisin, which rapidly penetrated the cell layers, was transported into large intracellular vacuoles and led to increased epithelial permeability.[94] Using reconstructed basement membrane (Matrigel),[96] Grenier et al. also reported the migration of *Td* through the basement membrane, which was facilitated by protease activity of dentilisin to degrade the basement membrane components. Other investigators employed dentilisin mutant strains and confirmed that dentilisin was responsible for the increase in epithelial permeability and also found an evidence that dentilisin facilitated degradation of the tight junction protein, ZO-1.[97,98]

In addition, higher O_2^- production from polymorphonuclear leukocytes (PMNs) was observed in a dentilisin-positive wild-type strain in comparison to a dentilisin-negative mutant, suggesting another possible mechanism mediating the cytotoxicity of this enzyme.[99] The O_2^- production was medicated through C3 complement activation, which was associated with matrix metalloproteinase (MMP) 9 release from PMNs. On the other hand, gingival epithelial cells co-cultured with wild-type or dentilisin mutant *Td* showed dentilisin-dependent preferential degradation of tumor necrosis factor α, which resulted in substantial reduction in both IL-8 protein and mRNA levels.[100] These results were indicative of host immune evasion.

An additional potentially oncogenic property of dentilisin is its ability to induce MMP-2, a key enzyme both in tissue homeostasis and tissue destruction, which has been linked to tumor invasion, progression as well as tumorigenesis.[101–103] In periodontal ligament cell (PDL) culture, purified dentilisin protease induced MMP-2 activation, mediated through fragmentation of fibronectin in extracellular matrix.[104] Subsequently, it was found that *Td* plays a key role in the transcriptional regulation of MMP-2 and its activating complex MT1-MMP/TIMP-2 through downregulation of several chromatin modification enzymes, specifically histone phosphorylases (aurora kinases) and histone deacetylase. Inhibition of these enzymes that mediated epigenetic modifications prevented *Td*-mediated increases in MMP-2, MT1-MMP, and TIMP-2 in PDL cells.[105] While specific mechanisms by which dentilisin modulates the expression of these chromatin modification enzymes remain to be studied, this has substantial implication in transcriptional control of a wide range of tumor suppressor and oncogenes. Finally it is intriguing that a low dose exposure to *Td* (5×10^7 cells/mL) increased DNA synthesis (^{3}H-thymidine uptake) via activation of MAP kinase signal pathways in PDL cells, while infection at higher concentrations reduced DNA synthesis, exerting cytotoxic effects. Western blot analysis showed that *Td* strongly but transiently activated ERK1 and ERK2, signals mediating cell proliferation, and JNK and p38, kinases mediating apoptosis.[106] These results underscore the ability of *Td* to modulate cell proliferation and cell survival.

3.4 Dentilisin activities in animal models and clinical studies

To our knowledge, there have been no animal carcinogenesis models using dentilisin itself or *Td* inoculation, as the primary endpoint in animal models have been periodontal disease. However the effects of wild-type and dentilisin mutant strains have been tested in a mouse subcutaneous abscess model, which showed a significantly reduction in size of the lesion (abscess) with the mutant.[107] Despite the lack of in vivo data in animals, several recent clinical studies suggest potential carcinogenic effects of dentilisin in humans. IgG antibodies against CTLP have been shown to be detectable in human sera,[108] but no seroepidemiological studies have examined the associations between orodigestive cancers and *Td* or dentilisin. In saliva of periodontitis patients and healthy controls, the concentration of *Td* was correlated with activated forms of MMP-8, which was cleaved from its inactive precursor form and which exerts both tumorigenic and anti-tumorigenic effects.[109] In 2004, based on 16S rRNA gene sequencing, Narikiyo et al. reported that *Td* was significantly enriched in esophageal cancer tissue compared with adjacent normal mucosa and it was absent in saliva from healthy volunteers.[110] More recently, Nieminen et al. demonstrated using immunohitochemical staining the presence of CTLP in several types of orodigestive cancer tissue. CTLP was positive in 19 out of 29 tongue, 20 out of 25 tonsillar, 3 out of 3 esophageal, 21 out of 32 gastric, 6 out of 6 pancreatic and 25 out of 54 colon cancer cases.[111] The investigators also reported that CTLP degraded several MMP inhibitors, including TIMP-1, TIMP-2, hypothesizing that CTLP may contribute to carcinogenesis through immunomodulation.[111] Subsequently, the same group of investigators analyzed more tissue samples of mobile tongue squamous cell carcinoma. CTLP was present in 95% out of 141 tumors, of which many (40.4%) showed high immunopositivity. CTLP positivity was significantly associated with invasion depth, tumor size and the expression of TLR-7, TLR-9 and c-Myc. Furthermore high CTLP immunopositivity in younger patients ($\leq$60 years old) predicted early relapse.[112] Kylmä et al. also evaluated the presence of CTLP by immunohistochemistry in a series of 201 unselected consecutive oropharyngeal squamous cell carcinoma patients and reported 81% overall positivity, which was significantly more pronounced in HPV-negative than positive tumors. Among those HPV-negative, higher TLR 5 and lower TLR 7 expression associated with high CTLP expression, which resulted in poor disease-specific survival. No similar association emerged among HPV-positive subgroup.[113] In a prospective cohort study that characterized microbial compositions of oral rinse samples of cancer-free individuals, the risk of subsequent colorectal cancer statistically significantly increased in those who were *Td*-positive, yielding the OR of 1.8, which was highest among ORs for all other periodontal pathogens.[114] In order to substantiate oncogenic potentials of *Td*-CTLP, intrigued by the recent clinical observations, more mechanistic studies are warranted to delineate specific molecular pathways.

4. *Pasteurella multocida* toxin (PMT)

4.1 Biochemical properties of PMT

PMT belongs to a large group of deamidating toxins/effectors, deamidases, which remove the amide functional group from a key glutamine residue of the protein substrate and provoke significant pathophysiological changes of target cells. Other bacterial toxins structurally and functionally related to PMT include the cytotoxic necrotizing factors from *E. coli* and *Yersinia pseudotuberculosis* and the dermonecrotic toxin from *Bordetella* spp.[115] PMT selectively targets the α-subunit of host heterotrimeric G proteins and stimulate its substrates, $G\alpha_q$, $G\alpha_{13}$ and the $G\alpha_i$-family proteins, leading to activation of various host signal transduction pathways.[116,117] As a result, it acts as a highly potent mitogen at picomolar concentrations that stimulates quiescent cells to grow and divide.[118,119]

PMT is an AB toxin, consisting of a receptor binding and translocation domain (B) and a biologically active (A) domain. This AB toxin binds to host cell receptors through their binding B domains and facilitate the cellular uptake and delivery (translocation) of their toxic activity A domains into the host cell cytosol, where the A domains then interact with and modify their cellular G protein targets to cause cellular toxicity.[120] PMT consists of 1285 amino acid residues resulting in a mass of 146 kDa. The receptor binding domain is located in the N-terminal part of the protein, including the amino acid residues 1–580. Within this domain a putative translocation domain is located between residues 402–457 in the C-terminus.[116]

The C-terminal part of PMT further contains three domains designated C1, C2 and C3. Of these, the C3 is the major biologically active domain which harbors the catalytic activity of the toxin to modify intracellular targets. The C1-domain (feet), encompassing amino acid residues 575–719, consists of seven helices, four of which serve as a plasma membrane targeting signal in toxin B. Because the primary target proteins of PMT are plasma membrane-bound heterotrimeric G proteins, dysfunctional C1 impairs localization to the substrate and diminishes PMT toxicity. The largest domain in the C-terminal part of PMT is the so-called body or C2 domain (amino acid residues 720–1104), consisting of 18 helices and nine beta-strands and exhibit a structure typical of nucleotide-binding proteins.

This C3 domain (amino acid residues 1105–1285) is separated into two subdomains, providing the catalytic cleft for the enzymatic function of the toxin. It has been further clarified that Cys-1165, His-1205, His-1223, Asp-1220 and Gln-1225 are essential for PMT activity.[116]

4.2 Microbiological characteristics of PMT

PMT is produced by toxinogenic strains of *Pasteurella multocida*, mostly serogroups D and some A, which are classified based on capsular polysaccharides. These strains contain a unique 18-kbp region carrying 14 genes (PMCN06_2106 to PMCN06_2119), including the *toxA* gene for PMT flanked with several phage-related genes. *Pm* is a small, pleomorphic, Gram-negative, nonflagellated facultatively anaerobic coccobacillus and a multihost animal, zoonotic and opportunistic pathogen that is capable of causing respiratory and multisystemic diseases, bacteremia, and bite wound infections, while it is part of the normal flora in many animals.[121] The sequence analysis of *toxA*, indicating lower GC content than the rest of the genome, suggests its acquisition through horizontal transmission. Further sequence analysis has revealed that PMT is encoded within a lysogenic bacteriophage and that phage regulatory elements are present upstream and downstream of *toxA*, which control *toxA* expression.[122]

Specific secretion systems that translocate PMT from cytoplasm through bacterial external membrane surface have not been identified in *Pm*. Instead, it has been speculated that stress response induced by environmental factors encountered during host infections leads to induction of the phage lytic cycle, with resultant bacterial cell lysis and toxin release.[122] Knowledge of the uptake of PMT into eukaryotic cells has been still limited. Three major steps are generally acknowledged: (i) Binding of the toxin to a host cell membrane receptor; (ii) Internalization by endocytosis; and (iii) Release of the biologically active toxin into the host cytosol. To date specific host PMT receptors have not been well characterized. However, surface plasmon resonance analysis indicates that PMT initially binds with low affinity to a wide range of abundant membrane lipid components, which is followed by a more specific binding to sphingomyelin and possibly additional putative proteinaceous receptors, which would induce endocytosis.[116,120] Further trafficking within host cells is facilitated by interaction with transferrin receptors and a small regulatory G protein Art6.[116,120] It has been also clarified that PMT N-terminal mediates cytosolic delivery of its native C-terminal cargo as a single polypeptide (C1-C2-C3), without cleavage between subdomains.[123] However, cellular uptake process may be cell-type specific. While no endocytic uptake was detectable in enterocytes, mast cells in the lamina propria distinctly accumulated PMT in their secretory granules.[124]

4.3 Cellular activities of PMT

The incubation of mammalian cells with purified recombinant PMT induces strong mitogenic and anti-apoptotic effects in the cell-type specific manner. Myogenic effects have been reported in fibroblasts, preadipocytes, osteoblasts and embryonic kidney cells, while cytopathic responses have been noted in lung cells and cardiomyocytes.[115] Mitogenic responses are mediated through increased intracellular Ca^{2+} and inositol phosphate levels as a result of activation of phospholipase Cβ (PLCβ) and Rho-dependent cytoskeletal signaling.[115] PMT induces Ras-dependent activation of extracellular signal-regulated serine/threonine protein kinase[125] mitogen-activated protein kinase (MAPK) via G_q-dependent, protein kinase C (PKC)-independent transactivation of the epidermal growth factor receptor in a human embryonic kidney cell line.[118,126] The MAPK pathway, also known as the RAS-RAF-MEK-ERK cascade, is often activated in many human cancers.[127] Additionally, PMT has been shown in fibroblast culture to stimulate the mammalian target of rapamycin complex 1, which is also known to be often activated in human cancer,[128] through the $G\alpha_{q/11}$/PLCβ/PKC pathway.[129] PMT induces anchorage-independent growth in rat fibroblast culture with greater potency than that achieved by epidermal growth factor or platelet-derived growth factor.[119] PMT also activates anti-apoptotic pathways, e.g., protein kinase D signaling in both cardiac fibroblasts and cardiomyocytes, which leads to the phosphorylation of the transcription factor, cAMP response element binding protein (CREB), and upregulation of CREB target genes, including the anti-apoptotic Bcl-2 protein.[130,131] These cells exposed to PMT also exhibit increased cell migration capacity,[130,131] which may be instrumental for tumor progression. Additional evidence pointing to the oncogenic potential of PMT is the finding that PMT stimulates Janus kinases (JAK)[132]/signal transducer and activator of transcription (STAT) signaling in embryonic kidney cells and fibroblasts,[133,134] through G_q-dependent phosphorylation and activation of the JAK1 and JAK2 and increased expression of proto-oncogene serine/threonine kinase Pim-1. This is followed by activation of STAT1, STAT3 and STAT5 transcriptional factors, the upregulation of cyclooxygenase 2, a pro-inflammatory protein that is upregulated in many cancers and downregulation of the transcription factor suppressor of cytokine signaling-3 (SOCS-3).[115] Similarly to NF-κB, it has been acknowledged that STAT3 can act as a non-classical oncogene and plays a role in inflammation-associated cancer.[135] Through the STAT activation, PMT exposure in vitro induces proliferation of native T cells and the differentiation of Th17 cells/IL-17 production,[136,137] which is linked to excessive inflammation such as inflammatory bowel disease. PMT also initiates cytoskeletal rearrangements, including focal adhesion assembly and actin stress fiber development in the RhoA-dependent manner,[115] which may underlie increased motility and tumor-aggressive behavior.[138] In summary, these in vitro data support PMT's ability to manipulate host signaling pathways just like oncoviruses.

4.4 Animal, epidemiological and clinical data of PMT and *Pm*

There have been no animal carcinogenesis models using PMT to our knowledge. However, a limited number of experimental animal studies for PMT-induced atrophic rhinitis have reported the development of epithelial squamous metaplasia in some of the PMT-injected animals.[139,140] This seems compatible to *H. pylori*-induced gastric carcinogenesis where atrophic gastritis and intestinal metaplasia precede gastric cancer.[141,142]

Likewise, a few studies have reported carriage of *Pm* in human populations using serology or oropharyngeal swab culture. Serological studies were designed to target capsular antigen types A and D or non-specific non-PMT somatic antigens. These studies have revealed that *Pm* carriage is quite common in people routinely exposed to animals, such as meat industry workers and pet owners and the highest rate was found in pig breeders (~77%, serology and culture combined).[143–145] Nevertheless, information concerning prevalence of toxigenic *Pm* has been limited, but it is estimated to be very low in human populations (<10%).[146] Because toxigenic activities were tested biochemically in earlier studies, it was not clear whether strains negative to PMT did not carry the *toxA* gene or whether its expression was transcriptionally repressed. There have also been case reports of *Pm* infection diagnosed in various types of cancer patients.[147,148] Yet, as *Pm* was primarily isolated from non-cancerous infectious lesions or blood, it is difficult to infer any causal associations from these reports.

5. *Fusobacterium nucleatum* toxins

5.1 *Fusobacterium nucleatum* background

The oral microbiota contains one of the highest diversities of bacteria in the human body. Many of these bacteria are believed to contribute to the development of cancer. *Fusobacterium nucleatum* (*F. nucleatum*) is one such bacteria and is considered an oncobacterium.[149] *F. nucleatum* is a Gram-negative anaerobic bacteria that is found primarily in the oral mucosa under normal conditions. However, *F. nucleatum* has been implicated in several disease states including colorectal cancer.[150] *F. nucleatum* is more abundantly found in diseased lesions.[150] Multiple studies have shown that *F. nucleatum* is found to be present in potentially high numbers in cancerous tissues.[149] There has been a renewed interest in studying the pathogenicity of *F. nucleatum* given its prominent placement in diseased states. *F. nucleatum* featured prominently in a review of the bacteria associated with the oral microbiome and their potential contribution to systemic disease and cancer.[16] Two *F. nucleatum* exotoxins with potential links to carcinogenesis are FadA and Fap2. In this review we will focus on the FadA and Fap2 adhesins which may play significant roles in tumor formation and cancer.

5.2 *Fusobacterium* adhesin A (FadA)

5.2.1 FadA molecular and chemical characteristics

Fusobacterium adhesin A (FadA) was discovered and found to be conserved among *Fusobacterium* genera that inhabit the oral mucosa and is important for cell binding.[151] FadA is a 129 amino acid protein with an 18 amino acid signaling peptide.[151] It has a secreted form that was shown to cause an upregulation of the β-catenin pathway and Wnt gene expression, both important developmental pathways that when dysregulated are leading causes of carcinogenesis.[152] The crystal structure of FadA revealed a unique "leucine chain" structure which, when mutated, abrogated the binding of host cells.[153] Although they did not identify the receptor binding site for FadA they suggest that FadA functions as a filament and therapeutic targets in this region might be suitable for future clinical importance.[153]

5.2.2 Microbial characteristics of FadA

FadA is coded by the fadA gene and shown to be expressed on the cell surface.[151,153] FadA contains a signaling peptide that would be used for its secretory pathway out of the cell and studies have shown that it also has a secreted form.[151]

5.2.3 FadA in vitro studies

Colorectal cancer cell lines HCT116, DLD1, SW480 and HT29 exposed to wild-type *F. nucleatum* demonstrated proliferation but those exposed to deleted FadA mutants did not have the same result.[152] FadA was shown to promote the activation of the β-catenin pathway and Wnt signaling seen by Western blotting suggesting a pathway that potentially contributes to carcinogenesis.[152] When *F. nucleatum* is introduced to E-cadherin expressing CRC cells there is an increase of cell proliferation but not in CRC cells that do not express E-cadherin indicating that FadA promotes tumor growth.[152] Thus, FadA acts like a ligand of E-cadherin and when it binds E-cadherin it reduces its tumor suppressor activity.[152] Work by this same group showed that Annexin A1, a modulator of the Wnt/β-catenin molecular pathway, is also upregulated by FadA when acting on the E-cadherin molecular pathway.[154] This upregulation enters into a positive feedback loop thereby promoting cancer progression by this microbe.[154]

FadA was also shown to stimulate inflammatory responses in these cells to the same extent as *F. nucleatum* pointing to its significance as an oncogenic microorganism.[152] FadA might represent a pathway whereby *F. nucleatum* might attach and invade. When FadA was deleted it was shown that *F. nucleatum* was unable to attach and invade but FadA was shown to trigger a signaling cascade inside HCT116 cell lines promoting lymphoid enhancing factor expression.[152] Collectively, these results suggest a strong contribution to colorectal cancer progression by FadA.

5.2.4 FadA activities in animal models

Both FadA and *F. nucleatum* were studied in vivo by subcutaneously injecting them into nude mice.[152] When HCT116 cells with FadA were injected subcutaneously in nude mice this promoted growth of tumors.[152] Tumor growth increased in nude mice 20% when compared to controls after treatment with FadA protein.[152] Rubinstein et al. also reported positive correlation between fadA gene copy numbers and *ANXA1* mRNA levels in a F. *nucleatum*-induced *Apc*$^{min/+}$ mouse model.[154] Further studies in mice also conclusively demonstrated that *F. nucleatum*, resulted in increased tumor growth and enriched populations in cancer tissues.[155]

5.2.5 FadA clinical and epidemiological results

Expression of the FadA gene was statistically significantly higher in colorectal cancer tissues than in normal mucosa or adenoma tissues.[152] Presence of FadA in colon tissue from patients with colorectal cancer showed that it was 10–100 times the normal level.[152] However, we are just starting to understand the mechanistic roles that *F. nucleatum* may be contributing to the progression of cancer or if it is merely a "hitch hiker" found at tumor tissues. Rubinstein et al. also demonstrated co-localization of FadA and Annexin A1 proteins as well as positive correlation between fadA gene copy numbers and *ANXA1* mRNA levels in human colon cancer tissue.[154] This group suggests that Annexin A1 presents as a novel marker of cancer prognosis based on the results of disease-free survival of 466 CRC cancer patients.[154] On the other hand, a recent study measured the antibody responses to 11 *F. nucleatum* antigens, including FadA, in pre-diagnostic serum samples of colorectal cancer patients and did not find that FadA was statistically significant as an indication of colorectal cancer risk.[156]

5.3 *Fusobacterium* apoptosis protein (Fap2)

5.3.1 Fap2 molecular and chemical characteristics

Fap2 was shown to have a predicted molecular mass of 389.8 kDa and a size of 11.3 kb.[157] This corresponds to a 3692 amino acid protein with no cysteine residues.[157] Using site-directed mutagenesis it was shown that Fap2 contains three domains: the C-terminal autotransporter domain, the central domain containing the repeats seen in other filamentous hemagglutinins and adhesins, and the N-terminal domain.[158] The 3596th to 3606th amino acids have been identified to be an immunogenic epitope of the Fap2 protein.[159] Using an ELISA-based analysis; the immunogenic region: "TELAYKHYFGT," in the Fap2 protein was identified as the specific region.[159]

5.3.2 Microbial characteristics of Fap2

Fap2 was previously identified as a potential apoptosis inducing protein that was treated with proteases and heat which suggested that Fap2 (unnamed at that time) was probably a membrane-bound protein.[160] Further studies by the same group elucidated that the gene product of fap2 codes for the large outer membrane protein with similar parallel β-helix structures and highly conserved C-terminal barrel autotransporter domains seen in other homologues.[161] The fap2 gene codes for the membrane-bound Fap2 protein and was identified as an important binding and galactose sensitive adhesin that was not able to hemagglutinate.[157]

5.3.3 Fap2 in vitro studies

Fusobacterium apoptosis protein (Fap2) was initially isolated as a potentially important membrane-bound protein that could play a role in apoptosis.[160]

F. nucleatum was shown to activate apoptosis in human peripheral blood mononuclear cells (PBMCs) and polymorphonuclear cells (PMNs).[160] More recently it was shown that Fap2 has the capacity to bind to galactose residues and fap2 mutants failed to coaggregate with human embryonic kidney[92] 293 T cells presumably due to lack of galactose binding.[157] Fap2 is able to bind to TIGIT, an inhibitory receptor present on all human natural killer cells that is then able to abrogate the immune cell action against *F. nucleatum*.[162] Human cell lines HCT116, RKO and HT29 were used to demonstrate the binding of *F. nucleatum* to colorectal cancer cells via Fap2, while the binding was absent with Fap2 deletion mutants.[163] When colorectal cancer cells (CRC) were treated with O-glycanase, the binding of *F. nucleatum* and CRC cells was reduced.[163] This could be as a result of loss of sugar residues Gal-Gal-NAc which are needed to bind the Fap2 lectin expressed by *F. nucleatum*. It was also shown that *F. nucleatum* strains that lacked Fap2 or had Fap2 mutant proteins showed reduced binding to Gal-Gal-NAc expressing CRC cells.[163] The binding between Fap2 and Gal-Gal-NAc provides an important mechanism by which *F. nucleatum* is able to alter the tumor microenvironment to promote cancer. Given the capacity for binding to a widely present sugar residue, Fap2 presents an important target of studies aimed at uncovering its contribution to cancer.[159,163]

5.3.4 Fap2 activities in animal models

Use of Fap2 mutants that lack Fap2 show decreased colorectal cancer cells in mice.[163] The mice model showed that established colorectal cancer tumors when inoculated with *F. nucleatum* had more colonization than with the Fap2 deletion mutant and *F. nucleatum* localized to mouse tumor tissues in a Fap2-dependent manner.[163] Consistent with the observations in human tissue described below, Gal-GalNAc (measured using FITC-labeled PNA) was overexpressed in the mouse CRC sections compared with sections prepared from adjacent normal colon tissues.[163]

5.3.5 Fap2 clinical and epidemiological studies

Fap2 and Gal-GalNAc was shown to potentially give rise to the presence of *F. nucleatum* in cancer and its precursor lesions and metastasis in human.[163] When Gal-GalNAc levels on healthy human colorectal tissues, human colonic adenomas, and human colorectal adenocarcinomas was assessed, its levels were significantly higher in adenocarcinomas compared with adenomas.[163] Within the adenoma group, there were statistically significant trends by histology, i.e., the highest levels of Gal-GalNAc expression on villous adenomas, followed by tubulous villous adenomas.[163]

When 37 human patient colorectal cancer cases were tested immuno- and seroreactivity to Fap2 protein, investigators found a 100% seroreactivity with the colorectal cancer samples and a 32% seroreactivity with healthy controls.[159] The Fap2 peptide mimotope proved to be immunogenic in both animal and human models of colorectal cancer giving some indication of its potential use as a biomarker for cancer progression. However, in a study using 11 *F. nucleatum* antigens (including Fap2 and FadA) in pre-diagnostic serum samples from 485 colorectal cancer patients the results indicated that these proteins did not implicate a statistically significant risk to contract colorectal cancer.[156] The approach used differs from other studies in that they had access to pre-diagnostic serum samples, which were not taken before cancer diagnosis and matched to controls who did not develop colorectal cancer.[156]

5.4 FadA and Fap2 potential mechanistic roles in carcinogenesis

FadA and Fap2 are two adhesins produced by *F. nucleatum* that are of significant interest to cancer researchers. *F. nucleatum* has the ability to adhere to multiple cell types creating a platform for invasion and persistence of other cell types. *F. nucleatum* is considered an important "bridge" organism paving the way for early colonizers of the oral microbiome such as *Streptococcus*.[132,149] In this regard *F. nucleatum* might be looked at as an ecosystem engineer, a term first coined by Jones.[164] An ecosystem engineer has the ability to transform the environment to suit its need for survival. Studies looking at cancer formation has already implicated cancer cells as being ecosystem engineers.[165] This understanding could help form future research decisions and possible medical interventions. The human oral microbiome contains a very diverse collection of bacteria in addition to other constituents such as viruses and fungi.[16] The mouth has been described as an island with many biogeographical features which result in multiple unique ways for bacteria to colonize the different micro niches.[132] These adaptations may include the ability for *F. nucleatum* to signal to other microbial and host cells (as in the case of FadA which acts as a ligand to E-cadherin-expressing colorectal cancer cell types) and adhere to other cells (as Fap2 can bind to sugar residues found on colorectal cancer cells). Under dysbiosis, bacteria, including their toxins, can make their way to other parts of the body and

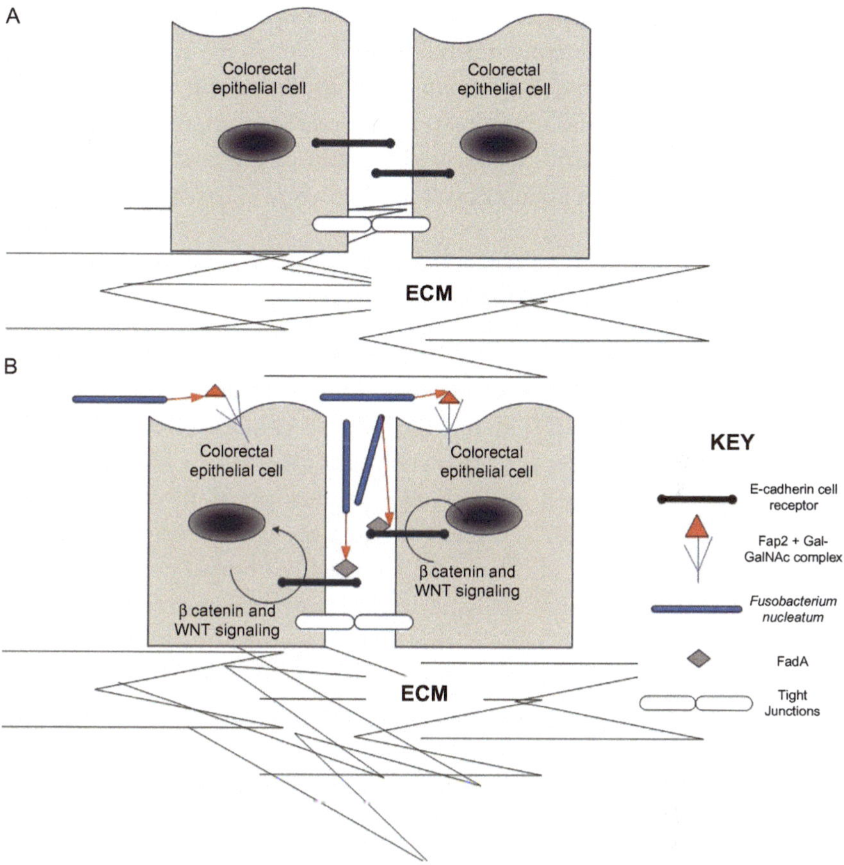

Fig. 1 *F. nucleatum's* putative role as an ecosystem engineer. (A) Normal state without the presence of *F. nucleatum*. ECM: extracellular matrix. (B) The presence of rod shaped *F. nucleatum* releases FadA which triggers a signaling cascade in colorectal epithelial cells leading to proliferation.[152,154] Fap2 binds to sugar residues on colorectal epithelial cells which leads to co-aggregation and increased recruitment of *F. nucleatum* thus modifying the tumor microenvironment and behaving as an ecosystem engineer.[163]

result in many diseased states.[16] These same adaptations might serve to disrupt homeostasis leading to severe conditions such as cancer. The role of *F. nucleatum* in adhering to other cells and creating a stable habitat for other colonizers has already been shown but we are suggesting that it plays an additional role as an ecosystem engineer in altering the tumor environment to facilitate proliferation and colonization. In Fig. 1B *F. nucleatum* functions as a putative ecosystem engineer by signaling to other cell types such as colorectal cancer cells via FadA which in turns leads to proliferation and tumor growth. This would have significant impacts to the tumor microenvironment such as invasion and increase in size. *F. nucleatum* also has the ability to adhere to other cell types via Fap2 binding to sugar residues on the cell surface and this would result in an increased recruitment of *F. nucleatum*. This contributes to the modification of the tumor microenvironment thus implicating *F. nucleatum* as an ecosystem engineer.

6. *Bacteroides fragilis* toxin (Bft)

6.1 Chemical and molecular characteristics of Bft

Bacteroides fragilis toxin (Bft) is a zinc-dependent metalloprotease toxin with three isotypes (BFT-1, BFT-2, and BFT-3).[166] While this enterotoxin has been known for several decades, it was not until the early 1990s that Bft was purified and characterized extensively. At this point it was established through the use of SDS-PAGE that Bft had a molecular weight of approximately 19,000,[167] a value similar to previous estimates that predicted a value of 19,500.[168] Bft was further characterized to have an isoelectric point in the range of 4.4–4.6 and to be stable within a pH range of 5–10.[167] Interestingly, Bft appears to be stable at temperatures ranging from −20 °C to 20 °C, but shows marked inactivation at higher temperatures. Specifically when incubated for 1 h at 55 °C over 90% of the toxin activity was lost, and when incubated for 1 h at 65 °C over 99% of the toxin activity was lost.[167]

At this time the enzyme susceptibility of Bft was also tested, revealing that the toxin is resistant to trypsin, chymotrypsin, dispase, bacterial amylase and bacterial lipase, yet sensitive to proteinase K and *Streptomyces* protease.[167] Antiserum specific to the toxigenic strain of *B. fragilis* has been shown to neutralize the activity of the toxin, while exposure to antiserum specific to nontoxigenic strains does not produce the same effect.[167] In addition to these physical characteristics, the molecular structure of Bft was also established at this time, with the first 20 amino acids from the N-terminal known to be Ala-Val-Pro-Ser-Glu-Pro-Lys-Thr-Val-Tyr-Val-Ile-Xxx-Leu-Arg-Glu-Asn-Gly-Ser-Thr.[167]

6.2 Microbial characteristics of Bft

Bacteroides Fragilis is an obligate anaerobe, Gram-negative bacteria that is a part of the normal biome of the human colon. Functionally, there are two distinct groups of *Bacteroides fragilis*. Strains that secrete a zinc-dependent metalloprotease toxin (Bft) which is encoded from the *B. fragilis* pathogenicity island are referred to as enterotoxigenic *B. fragilis* (ETBF). Strains that lack this ability are called nontoxigenic *B. fragilis* (NTBF).[169] While *Bacteroides fragilis* normally comprises only 1–2% of colon flora,[170] a value 10–100 times smaller than some other species of *Bacteroides* found in human intestines, it is the species most frequently isolated from clinical specimens, and as such is often viewed as the most virulent species of *Bacteroides*.[171] *B. fragilis* is so clinically prevalent due to its propensity to cause abscesses, soft-tissue infections, and bacteremias.[172]

Moncrief et al. generated a *Bft* recombinant PCR product that, when cloned into pUC19 and sequenced was determined to be 538 nucleotides in length,[173] providing insight into *B. fragilis*' enterotoxin encoding gene. This PCR product, which they designated as rBF107, was further analyzed and found to contain sequence identity to a number of metalloproteases, including *Pseudomonas aeruginosa* alkaline protease, and matrix metalloproteases.[173] Additionally, the sequence HELGHILGAEH, found in rBF107, exhibits consistency with the zinc-binding motif HEXX HXXGXXH, which is characteristic of the metzincin family.[173] This discovery gave strong support to the theory that Bft is a zinc-dependent metalloprotease.

Kling et al. did additional analysis of the encoding gene for the enterotoxin belonging to *B. fragilis*. This gene, termed *BftP* was found to be 1191 nucleotides long and code for a 44.4 kDa protein that is comprised of 397 amino acids.[174] BftP has a signal peptide 18 amino acids in length that is immediately succeeded by a protoxin that is 379 amino acids in length, while the protoxin found in stool samples begins at the 212th amino acid.[174] Three potential AUG start codons have been identified in the promoter region of the *B. fragilis* enterotoxin, with the middle start codon being followed by the amino acid sequence typical of a signal peptide. The two additional start codons are located 24 nucleotides upstream and 25 nucleotides downstream relative to the middle start codon. BftP has a signal I peptidase cleavage site located between two alanine residues at amino acid 17 and lacks an obvious ribosome-binding site.[174]

B. fragilis does not appear to utilize types III, IV, autotransporter or two-partner secretion systems,[175] yet does employ the type VI secretion system.[176] More specifically, *B. fragilis* expresses the GA3 subtype of T6SS (type VI secretion system), which is not highly conserved among other bacterial species.[177] The type VI secretion system exports the Bft exotoxin directly into eukaryotic host cells or other co-growing bacterial cells. This is accomplished through the T6 apparatus, a multiprotein, cell envelope spanning complex comprised of a core of Tss proteins.[177] This complex is characterized by a needle-like structure that becomes inundated with toxins in the cytoplasm.[178] The needle is then driven from the cell via contraction of the surrounding sheath in order to implant the toxins into extracellular matter.[177] Additionally, *B. fragilis* also appears to use type I secretory system, as it was determined to have three Tolc-like proteins that are encoded adjacent to genes that have previously been shown to be associated with type I secretion systems.[175]

6.3 In vitro toxicity of Bft

In vitro experiments have helped shed light on the potential mechanisms of action used by Bft. On a biochemical level Bft has been shown to utilize its zinc-dependent protease activity to cleave the extracellular domain of E-cadherin, which is followed by further intracellular degradation.[179] This is relevant because E-cadherin is not only the primary protein of zonula adherens, which function to maintain tight epithelial junctions, but also has tumor suppressor functions. Bft has been shown to provoke cell rounding, increase in volume, and effacement of microvilli and apical junctional complexes in intestinal and renal cells in culture.[180] This supports the idea that cadherins are functionally involved. This is thought to lead to actin rearrangement, and consequently pro-inflammatory cytokines, which diminish epithelial barrier function.[180] In addition, this leads to an increase in cellular proliferation mediated by elevated expression of the c-Myc oncogene.[181] Interestingly, these consequences appear to be reversible, as cells appear normal 2–3 days after toxin treatment.[180]

Additional experiments have shown that exposure to 15 ng of Bft per mL resulted in formation of blebs on the cell surface and complete disappearance of stress fibers along with F-actin condensation at the cell periphery.[182] These results were more pronounced at increased doses.[182] Cell organelles (including mitochondria and nucleus) appear to be unaffected by exposure to Bft,[182] as the toxin's effects seem to be localized to the actin cytoskeleton and associated structures. Additionally, Bft has been shown to stimulate intestinal epithelial shedding, which is dependent, in part, on E-cadherin degradation and β-catenin–T cell-factor nuclear signaling[183] and to stimulate proliferation and migration of human colon cancer cells in vitro.[181]

Incubation of human colorectal cancer cells with BFT led to upregulation of spermine oxidase, a polyamine catabolic, resulting in increased reactive oxygen species (ROS) and DNA damage measured by

γ-H2A induction.[184] Exposure to Bft has also been shown to induce changes in gene expression and chromatin availability in colonic epithelial cells.[185] Eight genes have been identified in colonic epithelial cells that are significantly differentially expressed 24 h post-Bft exposure, but not 48 h post-exposure. It has also been demonstrated that while Bft does induce changes in chromatin accessibility in colonic epithelial cells, these changes are transient, and greatly decline by 48 h after Bft treatment.[185] Specifically, there is a noted increase in chromatin accessibility at transcription factor binding sites, particularly those belonging to the AP-1/ATF family, many of which function downstream of mitogen-activated protein kinase (MAPK) pathways. Critically, while an increase in chromatin accessibility has been shown to have an association with differential DNA methylation and DNA mutation, these differences are not statistically significant.[185]

6.4 Toxicity of Bft in animal models

Mouse models have been used to demonstrate a link between Bft and colorectal cancer.[186] Mice colonized by the enterotoxigenic *B. fragilis* experienced a rapid onset colitis followed by tumors of the distal colon.[186] This was in contrast to mice exposed to nontoxigenic *B. fragilis*, which did not induce tumor formation at a higher rate than the control. Mice inoculated with the toxigenic strain developed observable tumors as early as 4 weeks after exposure, and microadenomas as early as 1 week after exposure.[186] Bft also induced a prominent lamina propria intrusion of IL-17-producing $CD4^{+}$ T cells (Th17) and $\gamma\delta$-T cells with a dominant signal transducer activator of transcription-3 (STAT-3) pathway.[186] The importance of IL-17 in the tumor-inducing response to Bft was demonstrated by the decrease in colonic tumors when the mice were treated with IL-17 blocking antibodies.[186] IL-17 has been shown to induce tumor growth both in vitro and in vivo via an IL-6-stat3 signaling pathway.[187] *B. fragilis* has also been found in increased quantity in the intestinal lumen of the 1,2-dimethylhydrazine (DMH)-induced rat colorectal cancer model.[188]

6.5 Clinical and epidemiological evidence to support in vivo toxicity in humans

In addition to animal models, there is now substantial indication that Bft is involved in the development of colorectal cancer in humans. In addition to being the most commonly isolated Bacteroides species found in clinical specimens,[171] enterotoxigenic *B. fragilis* has been shown to have an increased prevalence in samples from cases involving colorectal cancer when compared to a control group.[166] There has also been shown to be a difference in Bft detection depending on the stage in the course of the disease, with late-stage colorectal cancer patients having an increased rate of Bft detection in comparison to colorectal cancer patients at earlier stages of the disease.[166] This observation becomes more pronounced as the disease progresses, to the point where Bft was identified in 100% of late-stage colorectal cancer patients in one study.[166] It should be noted that this study still found Bft in approximately 54% of the controls when one-sided colon samples were obtained, yet this number was notably lower that the approximately 87% of colorectal cancer cases that tested positive for Bft.[166]

This link is reinforced by additional studies, which have demonstrated that stools from colorectal cancer patients were shown to test positive for Bft at significantly higher rates than controls.[189] Interestingly, *B. fragilis* has been reported to be found in more abundance in mucosal as opposed to luminal samples taken from colorectal cancer patients.[190] Here it is once again important to differentiate between toxigenic *B. fragilis* and nontoxigenic *B. fragilis*. One study determined that *B. fragilis* as a whole was found in 77% of colorectal cancer patients yet 68% of healthy controls, a difference that was not significant.[189] However, the Bft gene was detected in 38% of the *B. fragilis* isolated from the cancer patients, but only 12% from the control, a statistically significant difference.[189] Additionally, individuals with familial adenomatous polyposis, a disease that leads to digestive tract polyp formation at a young age due to mutation of tumor suppression gene APC, has been shown to be associated with enterotoxic *B. fragilis*.[191] In one study 60% of patients with familial adenomatous polyposis had gut mucosa with enterotoxic *B. fragilis* present, compared to only 30% of controls.[191]

7. *Escherichia (E) coli* toxins

E. coli secretes a number of protein exotoxins, some of them are lineage-specific, and among those, three toxins: colibactin, cytotoxic necrotizing factors (CNF) and α-hemolysin, which are pertinent to human cancer, are reviewed here in Sections 7.1–7.3.

7.1 Colibactin

7.1.1 Chemical and molecular characteristics

Colibactin is a genotoxin causing DNA double strand breaks in a contact-dependent manner. Colibactin is synthesized by a hybrid non-ribosomal peptide synthetase-polyketide synthase (NRPS-PKS) assembly line,[192] a sequence of tailoring and editing enzymes, and a critical phosphopantetheinyl transferase that mediates NRPS and PKS activation.[193] Despite multiple attempts, colibactin had not been able to be isolated successfully due to its' extreme instability.[194] Finally, in September 2019, through the interdisciplinary approach encompassing chemical synthesis, metabolomics, and probe-mediated natural product capture, a group of scientists from Yale University claimed a complete elucidation of colibactin structure.[195] It has a nearly symmetrical structure that contains two electrophilic spirocyclopropyldihydro-2-pyrrolone and the hydrolytically labile C36–C37 α-dicarbonyl.[195] These two electrophilic cyclopropane residues undergoes ring-opening through nucleotide addition,[195] consistent with earlier observation that colibactin-producing bacteria cross-link DNA.[196–198] Using genome-editing techniques, the group also showed the production of colibactin's precursor, precolibactin 1489, that requires every biosynthetic gene in the colibactin gene cluster, demonstrating it as being the first reported product derived from the long-elusive and completed biosynthetic pathway.[195] However, precolibactin 1489 was considered a stable product arising from oxidation and macrocyclization of a putative linear direct precursor of colibactin, precolibactin 1491.[195]

7.1.2 Microbial characteristics of colibactin

Colibactin is encoded by the *pks* genomic island, and is often found on *E. coli* strains belonging to the B2 phylogenetic group.[194] The bacteria *Klebsiella pneumoniae* is also known to carry the colibactin-producing *pks* island, with one study finding 16.7% of *K. pneumoniae* isolates were *pks*+.[190] It is estimated that roughly 20% of healthy individuals are colonized by *pks*+ *E. coli.*[199] In terms of *E. coli* prevalence, the *pks* genomic island is found in 30–40% of all B2 strains.[200] Colibactin-producing *E. coli* has also been shown to be relatively resistant to human macrophages and to not influence COX-2 expression.[201]

The *pks* genomic island is 54-kb in length and directs the synthesis of colibactin via specific cellular machinery.[202] It consists of 19 protein producing genes named *clbA* to *clbS*, which are arranged in sequence as *clbA*, *clbR*, *clbC* to *clbQ* and *clbS*.[192]

The first step in the assembly line of colibactin synthesis consists of the ClbA protein, a phosphopantetheinyl (PPant) transferase, activating the NRPS and PKS enzymes via the addition of a PPant onto the NRPS and PKS carrier protein domains.[192] ClbN and ClbB serve as the NRPS enzymes, with ClbN producing *N*-myristoyl-D-Asn, which is then accepted by ClbB. ClbB then proceeds to add either Ala or Val and incorporate a malonyl-CoA. ClbH and ClbI have been shown to be essential for the formation of cyclopropane, which is responsible for colibactin-induced DNA alkylation.[192] ClbsD-G are required for the synthesis of the unique PKS extender aminomalonyl unit (AM), which is necessary for genotoxicity. ClbG transfers AM to ClbK, which then incorporates AM into colibactin. While less is knows about the off-loading process, it is thought that ClbQ is involved in this process, and therefore controlling the flux of colibactin production.[192] Finally, ClbM (a MATE transporter) is responsible for precolibactin release into the periplasmic space, where ClbP generates the mature colibactin via removal of the *N*-myristoyl-D-Asn side chain.[192] ClbS is also produced by *pks*+ *E. coli* where it is thought to sequester colibactin and therefore protect the *E. coli* from damage from its own product.[192] Additionally, clbM has been shown to play a role in the transport of colibactin,[203] but to date the precise mechanism of how colibactin is secreted and transported to host cell cytosol is poorly understood.

7.1.3 In vitro toxicity data of colibactin

Colibactin induces cellular damage in eukaryotic cells via introduction of a double-stranded break into the DNA of the host cell.[193] This action requires the presence of live bacteria, not being observed with bacteria culture supernatants or cell lysates. This damage serves to activate a signaling pathway that eventually leads to cell cycle arrest, enlargement of cell bodies, and ultimately death of the cell.[193] Cells exposed to *pks*+ *E. coli* strains have been shown to demonstrate many of the defining characteristics of cell death, including chronic DNA double strand breaks, enhanced senescence-associated β-galactosidase (SA-β-Gal) activity, expansion of promyelocytic leukemia nuclear foci and senescence-associated heterochromatin foci.[204] Additionally, exposed cells increase their reactive oxygen species production and pro-inflammatory cytokines, chemokines and proteases secretion.[204] Importantly, the presence of colibactin-producing *E. coli* strains in fibroblasts in vitro has been shown to promote the growth of epithelial human lung adenocarcinoma and colon carcinoma tumor cells.[204] *Pks*+ *E. coli* infection has also been associated with the induction of IL-6 production, which is known to induce epithelial cell invasion, and MMP production, which has been shown to have direct tumor-promoting effects.[204] Taken together this suggests a mechanism for *pks+ E. coli* reprogramming fibroblast expression profiles in a way that promotes growth of cancer cells.[204]

Brief exposure to low doses of *pks+ E. coli* strains has been shown to transiently damage DNA in mammalian epithelial cells, followed by cellular division with incomplete DNA repair.[205] Increased rates of chromosomal aberrations, instability, and gene mutations were also observed in cells post-exposure.[205] Due to the difficulty in isolating colibactin, alternative methods have been turned to in order to study this toxin's exact mechanism of action. Production and observation of synthetic adenine-colibactin adducts mimicking the results of *pks+ E. coli* incubation in human cells demonstrated that colibactin breaks DNA strands by utilizing alkylation via a cyclopropane "warhead."[197] The degradation products observed in this experiment consisted of two diastereomeric adducts that both contain a 5-hyfroxypyrrolidin-2-one ring system with an attached N3-substituted adenine ring. Due to the small size of these adducts, it is likely that they are actually derived from a larger, yet unstable colibactin-DNA interstrand cross-link.[197]

7.1.4 Colibactin toxicity in animal models

In vivo experimentation has shown that *pks+ E. coli* strains induced the formation of phosphorylated H2AX foci in mouse enterocytes and indicates DNA damage to colonic epithelial cells.[205] *Pks+ E. coli* strains have also been found to induce a higher degree of lymphopenia in septicemic mice as compared to strain that did not produce colibactin.[206] Perhaps most notably, *Pks+ E. coli* has been observed to promote tumorigenesis in the $Apc^{Min/+}$; $Il10^{-/-}$ mouse model in a colibactin-dependent manner, suggesting colibactin is a driver of carcinogenesis.[207] A rat model has also demonstrated the link between *pks+ E. coli* strains and DNA damage, with observable signs of genotoxin stress including anaphase bridges, higher occurrence of crypt fission and accelerated renewal of the mature epithelium detected in adult rats.[208] This same model revealed that colibactin-producing *E. coli* strains increased intestinal epithelial cell apoptosis and proliferation, as well as amplified goblet, enteroendocrine and paneth cells.[208] This serves to support the connection between *Pks+ E. coli* and intestinal tumorigenesis.[208]

Notably, even transient exposure to *Pks+ E. coli* has been shown to increase tumor growth.[209] This was shown when intestinal epithelial cells

HCT116 that had been mixed with *Pks*+ *E. coli* were injected into nude mice. 3 h post injection the mice were given a broad-spectrum antibiotic (imipenem) in order to kill the bacteria. This short exposure resulted in significantly increased tumor growth, while injection of epithelial cells that had been mixed with *Pks*− *E. coli* had no effect on tumor growth, similarly to uninfected cells.[209] This tumor-promoting effects was mediated by induction of a senescence-associated secretory phenotype via over-expression of microRNA 20a-5p, which led to accumulation of SUMO conjugated p53.[209] The previously mentioned adenosine-colibactin adducts that was shown in vitro mechanistic study of colibactin action have also been observed in the colonic epithelial cells of mice that have been monocolonized with *pks*+ *E. coli*.[197]

7.1.5 Clinical and epidemiological data to support in vivo toxicity in humans

Among B2 *E. coli* isolated from colorectal cancer and diverticulosis patients, colibactin-producing variants were found to be prevalent at a high rate in the colorectal cancer patients (55.3%), especially when compared with the 19.3% seen among the diverticulosis patients.[33] This observation corroborates additional data showing colibactin-producing *E. coli* strains are associated with colorectal tumors, with 55–66.7% of cancer patient biopsies testing positive while only 19–21% of intestinal control tissue did.[199] Recently, PCR analysis has been used to demonstrate that the *clbA* gene is found at an increased prevalence in the stools of patients with colorectal cancer.[210] In this study 56.4% of stool from colorectal cancer patients was found to have *clbA*, as compared to only 18.5% seen in the controls.[210]

The presence of the *pks* island has also been shown to enhance the ability of *E. coli* to persist in the human gut microbiota.[211] One study on this subject showed that long-term colonizers of the gut were significantly more likely to have the *pks* island than either intermediate-term colonizers or transient strains.[211] This may help explain why B2 *E. coli* strains are represented to a large degree in the gut microbiome.[211] Additionally, individuals with familial adenomatous polyposis, a disease that leads to digestive tract polyp formation at a young age due to mutation of tumor suppression gene APC, has been shown to be associated with *pks*+ *E. coli*.[191] In one study 68% of patients with familial adenomatous polyposis had gut mucosa with *pks*+ *E. coli* present, compared to only 22% of controls.[191]

7.2 CNFs (cytotoxic necrotizing factors)

7.2.1 Chemical and molecular characteristics of CNFs

CNFs belong to a family of deamidating toxins that deamidate glutamine 63/61 in the switch II region of Rho GTPases, leading to constitutive activation of Rho GTPases.[212] There are four identified variants of cytotoxic necrotizing factor (CNF), CNF1, CNF2 and CNF3, which are secreted by *E. coli* and CNFY, which is secreted by *Yersinia*. CNF1 is the best characterized of all CNFs, while CNF2 shares 85% sequence identity with CNF1, CNF3 shares 70% sequence identity with CNF1 and CNFY shares 61% sequence identity with CNF1.[212] While there is a great deal of similarity among the CNFs, they differ in terms of their substrate specificity. CNFY exclusively acts on RhoA, B and C and has no activity for Rac or Cdc42.[213] CNF2 preferentially modifies RhoA and Rac,[214] whereas CNF1 and CNF3 deamidate RhoA, Rac and Cdc42.[212]

All CNFs are identical in length at 1013/1014 amino acids long and posses an N-terminal receptor binding domain.[212] Amino acids 710–1014 in CNF1 have been shown to be responsible for the catalytic activity of this protein, while amino acids 53–190 are responsible for cellular uptake.[212]

CNF1 is a protein with a molecular weight of 115 kDa.[215] Experimentation has determined that CNF is stable within a pH range of 6–10.5.[216] The toxic activity of CNF was destroyed when exposed both to a temperature of 75 °C for 15 min and 60 °C for 1 h.[216] As CNF is primarily a protein, it has a low carbohydrate content of no more than 25 μg.[216]

7.2.2 Microbial characteristics of CNFs

Compared to the *pks* island, *CNF1* is relatively less common, with one study determining that, of 78 *E. coli* strains isolated from adults with urinary tract infections, 13% were positive for the *CNF* gene.[217] The nucleotide sequence of the *CNF* gene has been identified, and notably the start codon is located at nucleotide 856 while the TGA termination codon is located at nucleotide 3900.[218] Interestingly, the C+G content of the *CNF1* is lower than that of the *E. coli* genome as a whole at 36.5%.[218]

For CNF1 specifically, ferredoxin is required for secretion at the step of crossing the cytoplasmic membrane.[219] CNF1 is transported in its active form by outer membrane vesicles.[220] This is supported by the fact that the histone-like nucleoid structuring protein H-NS has been shown to down regulate CNF1 production by affecting outer membrane vesicle release.[220] *CNFY* utilizes the type III secretory pathway, which supports *CNF1* using the same pathway.[221]

7.2.3 In vitro toxicity data of CNFs

CNF induces actin assembly and multinucleation in mouse embryonic fibroblasts that were incubated with the purified toxin.[222] CNF has been shown to inhibit the intrinsic GTPase activity of RhoA and completely blocks GTPase activity stimulated by the Rho-GTPase-activating protein (rhoGAP).[222] CNF has been observed to reduce function at the basolateral membrane of intercellular epithelial junctions.[125] This is accomplished through displacement of the proteins occludin and zonula occludens-1 (ZO-1), and reorganization of junction adhesion molecule-1 (JAM-1).[125] CNF1 has also been shown to increase F-actin content in cytoskeletons of polymorphonuclear leukocytes.[223] CNF1 also increases superoxide generation and adherence on epithelial monolayers, but significantly decreases their phagocytic function.[223] This implies a possible role of CNF1 in intestinal infections by way of by stimulating polymorphonuclear leukocyte cytotoxicity.[223]

It has also been shown that treatment with CNF1 initially inhibits mitosis and cytokinesis and elicits endoreplication and polyploidization in human colon cancer cells in culture.[224] These outcomes were found to be followed by depolyploidization-associated survival of cells. An important finding from this study was the discovery that progeny derived from cells treated with CNF1 exhibited marked genomic instability, which was clearly seen by an increased rate of aneuploidy.[224] These progeny cells were noted to be more resistant to CNF1, but not to commonly used chemotherapy agents 5-fluorouracil and oxaliplatin.[224] There is also evidence that CNF1 induces COX2 and NF-κB expression in mouse fibroblasts and a human squamous cancer cell line, respectively.[225,226] CNF1 has also been shown to act in endothelial cells to both inhibit apoptosis and promote pro-inflammatory cytokine release, with the later effect observed in both epithelial and endothelial cells.[227]

7.2.4 CNF toxicity in animal models

The majority of CNF-related research using animal models to date has focused primarily on inflammation rather than cancer, yet is still relevant for the purposes of this chapter. *CNF* positive *E. coli* has been shown to cause more inflammation-mediated morphological and histological tissue damage in a rat prostate model than an isogenic mutant.[228] Interestingly, this occurred despite the fact that total bacterial counts for each strain were nearly identical. A mouse bladder model demonstrated that mice infected with *CNF1*-positive strains consistently showed deeper and more extensive inflammation than those infected with the isogenic mutants.[229]

CNF1 has also been shown to have an association with spread of prostate cancer in a mouse model.[230] In this experiment induced overexpression of CNF1 was observed to promote both cellular invasion and migration.[230] Injection of CNF1 expressing cells into the tails of mice was observed to lead to a higher level of lung metastasis when compared to mice receiving a control vector.[230]

7.2.5 Clinical and epidemiological data to support carcinogenic potential in humans

Similar to colibactin, *cnf*+ B2 *E. coli* strains have been shown to have an association with colorectal cancer.[33] Of total *E. coli* strains isolated from colorectal cancer patients, *cnf*+ B2 *E. coli* strains comprised 18%.[231] Once again, this is more prevalent that what is seen in patients with diverticulitis.[231] One study found that 39.5% of biopsies from patients with colorectal cancer contained CNF1-producing *E. coli*, whereas 12.9% of biopsies from diverticulitis patients were positive for CNF1-producing *E. coli*.[33] Notably, All *cnf1*-harboring strains belonged to the B2 phylogroup.[33]

It is important to note that an individual *E. coli* strain is not limited to producing only one exotoxin. In fact, the previous study found that 33% of B2 strains possessed both the *pks* island and the *cnf1* gene.[33] Additional studies have shown that both the *cnf*1 gene and the *hlyC* (discussed below) are detectable in significantly lager numbers from *E. coli* blood isolates from cancer patients than in fecal isolates from healthy volunteers.[33] Cnf1-producing *E. coli* has also been associated with pathological conditions in other body systems. Strains of *E. coli* that possess the cnf1 gene have been shown to be associated with urosepsis,[232] possible indicating that the cnf1 gene is indicative of higher bacterial virulence. This higher virulence may lead to more severe inflammation and increases risk of carcinogenesis.

7.3 α-Hemolysin

7.3.1 Chemical and molecular characteristics of α-hemolysin

α-Hemolysin is a cytotoxic protein that is a member of the repeats in toxin (RTX) group. This group is characterized by glycine and aspartate-rich repeats located at the C-terminus of these toxin proteins.[233]

This nonapeptide sequence is GGXGXDXUX, where U represents a large, hydrophobic residue.[233] α-Hemolysin requires calcium to function optimally, and it is this repeating domain that is responsible for calcium binding.[233] Specifically, Asp-863 has been shown to be of special importance in calcium binding, and therefore HlyA conformational change and ability to breakdown the bilayer permeability layer.[233] HlyA possesses 15 of these nonapeptide repeats, implying the presence of a large number of calcium binding sites.[233]

HlyA consists of both protein (95% of weight) and carbohydrates, but lipids have not been detected.[234] The majority of hlyA is not heat-stable as it is inactivated at 56 °C,[234] but when exposed to temperatures of 37 °C inactivation did not occur for upwards of 18 h.[235] HlyA appears to be generally more stable at a lower pH, as it is stable at pH 3.0 for 6 h, but inactivates at pH 7.0–10.0 within 6 h.[236] This is not an unexpected result due to the acidity of the gut, a frequent location of *E. coli* colonization. Functionally, α-hemolysin of *E. coli* is a pore-forming toxin that acts by generating small cation-permeable channels in the membrane of a host cell.[237]

7.3.2 Microbial characteristics of α-hemolysin

The hemolysin gene (*hly*) was identified on 23% of 78 *E. coli* strains taken from adults with UTI's.[217] This study also concluded that *hly* was more prevalent than *CNF* in this population, with *hly* being present on 18 *E. coli* strains while *CNF* was present on 10.[217] Four defined genes encode for α-hemolysin. *HlyA* codes for a 107,000-kDa protein that is thought to be the inactive precursor of hemolysin.[238] *HlyC*, codes for a 20,000 polypeptide that is crucial for post-translational modification of the protein encoded by *hlyA*, rendering this protein hemolytically active. *HlyB* and *hlyD* are necessary for hemolysin secretion.[239] The gene for all four *hly* citrons has been localized to an 8211-base-pair region.[240] Of note, there is an ATG codon preceded by a potential ribosome-binding site (GCGG) at base-pair 5317.[240]

The secretory pathway of hly is well known, as it is both the prototypical and best characterized type I secretion system.[241] HlyA reaches the extracellular space in a soluble form without the formation of periplasmic intermediates.[242] Interestingly, a fraction of hlyA appears to be very tightly associated with outer membrane vesicles.[242] This corresponds with the observation that hlyA localized in outer membrane vesicles is frequently found among hemolytic *E. coli* strains.[242]

7.3.3 In vitro toxicity data of α-hemolysin

E. coli hemolysin has a pronounced effect on human monocytes, with nanomolar concentrations (250–2000 ng/mL) resulting in rapid and irreversible depletion of cellular ATP to levels below 20% of controls within 60 min.[223] Exposure of human monocytes to lower doses (10–200 ng/mL) of hemolysin results in the release of large amounts of interleukin 1 beta within 1–2 h.[223] 50% of cellular ATP is observed to be depleted within 90 min when incubated with only 0.3–3.0 colony-forming units of *E. coli* per monocyte.[223] In neutrophils, hemolysin causes defects in membrane permeability, resulting in ATP efflux and propidium iodide influx.[243] These cellular membrane lesions do not appear to be repaired.[243] Human serum albumin, lipoproteins and IgG protect erythrocytes and platelets, but not neutrophils, from the effects of hemolysin.[243] The binding of hemolysin to neutrophils results in granule exocytosis and loss of cellular phagocytic killing ability.[243] This is expected to impair or destroy the typical immunoprotective properties of these cells.

HlyA has also been shown to play an important role in giving *E. coli* the ability to translocate across cellular monolayers.[237] In this case, translocation was associated with decreased epithelial integrity of colonic cell cultures, which exhibited increased ion permeability. Experimentation has demonstrated that hemolysin secreting *E. coli* strains were able to translocate, whereas non-hemolytic mutants were unable to do so, and did not compromise the epithelial barrier.[237] In human uroepithelial cell lines, incubation with α-hemolysin substantially reduced IL-6 secretion by LPS through degradation of β-catenin and NF-κB subunit RelA.[244] Co-culture of a normal human colon mucosal epithelial cell line with *hly*+ *E. coli* activated expression of HIF1α and the glucose transporter GLUT1, which often upregulated in many types of cancer as an indication of reprogrammed energy metabolism, and repressed expression of the tumor suppressor BIM,[245] suggesting undelaying mechanisms of tumor promotion.

7.3.4 α-Hemolysin toxicity in animal models

Much like the case of CNF, there has been relatively little cancer-specific research pertaining to animal models and hemolysin exposure. However, some relevant studies on hemolysin toxicity are discussed in this section. In a rat model, *Hly*+ strains provoked higher urinary shedding of leukocytes and erythrocytes, indicating an affect of hemolysin on inflammatory reactions.[246] Rat models have revealed that hemolysin mediates neutrophil toxicity, resulting in in vivo necrosis and lysis.[247] In the same study mentioned above that showed the importance of hemolysin in *E. coli* translocation ability in vitro, hemolysin was also demonstrated to induce focal leaks in native rat colon.[237] The rat colons exposed to hemolytic *E. coli* in this experiment were observed to have both absent nuclei and notable bacterial adherence to the epithelial surface in addition to decreased epithelial integrity.[237]

7.3.5 Clinical and epidemiological data to support carcinogenic potential in humans

The effect of hemolysin has also been examined using *E. coli* strains taken directly from human patients. α-Hemolysin synthesized by *E. coli* strains isolated from ileal mucosa of patients with Crohn's disease has been shown to adhere to differentiated intestinal cells, and may disrupt the intestinal barrier.[248] Hemolysin has also been shown to be associated with bacteremia in hospitalized cancer patients.[249] One study found that 26% of blood samples from cancer patients with bacteremia were positive for hemolysin-producing *E. coli*, while only 6% of fecal samples from healthy patients were positive for hemolysin-producing *E. coli*.[249]

In a very interesting outcome, *hly+ E. coli* has been implicated in adenomagenesis and colorectal cancer in females but not males.[245] *Hly+ E. coli* has been found to be relatively more prevalent in stools from females with adenoma and colorectal cancer, and to be correlated with poor survival in female but not male colorectal cancer patients.[245] In this experiment, *hly+ E. coli* was detected in 36.1% of female colorectal cancer patients, 39.3% of female adenoma patients as compared to just 6.7% of female healthy controls.[245] The association of *hly+ E. coli* with colorectal cancer was shown to be independent of both cancer stage as well as patient age.[245] It is hypothesized that the reason *hly+ E. coli* appears to be correlated to increased negative outcomes in female as opposed to males is due to immunological, hormonal or behavioral differences.[245] It is unlikely that genetic factors are causative of this discrepancy due to the fact that in vitro experimentation demonstrated both cytotoxic and carcinogenic effects on male derived cellular lines.[245]

8. *Salmonella enterica* avirulence protein A (AvrA)

8.1 Background of *Salmonella enterica* infection in humans

Salmonella enterica is a Gram-negative, facultative anaerobe, not a symbiotic commensal, but an intracellular pathogen to both humans and animals.

Salmonella infection poses a major public health concern worldwide as foodborne illness.[250] There has been also a concern that long-standing *Salmonella* infection may increase the risk of colorectal cancer in humans as increased risk of colorectal tumor has been reported in several populations worldwide.[251,252] The *Salmonella* genome contains a remarkably large number (up to 23) of pathogenicity islands (SPI),[253,254] synthesizing over 60 effectors.[255] We describe here Avirulence protein A (AvrA), one of the bacterial type three secretion system effectors as there have been recent significant findings to support its oncogenicity.

8.2 Microbial characteristics of AvrA

AvrA represents the least characterized effector from the SPI1 which is originally thought to be non-virulent.[256] AvrA is directly injected into host cytoplasm by type 3 secretion system (T3SS) with a filament like-needle that is encoded by the same SPI1 in a finely coordinated manner.[257,258] The AvrA gene is present in most *Salmonella enterica* isolates from humans and animals. This is particularly the case for non-typhoid *Salmonella* as its prevalence was reported to be 98–100% (514/523 and 185/185),[259,260] while it was absent in typhoid strains, which has been linked to gallbladder cancer. Natural AvrA genetic variants have been reported for some *Salmonella typhimurium* strains, which lacks an internal leucine residue (L140) and has a profound effect on AvrA function.[261,262] Among non-typhoidal strains, AvrA protein expression is known to vary markedly with clinical presentation. AvrA protein is not often produced by clinical isolates from systemic disease, but it is often detectable in those from limited enteritis.[263] Interestingly, whereas expressions of most effectors from SPI1 are controlled at the transcriptional level,[264] it is now clear that expression of AvrA is regulated in a post-transcriptional manner and that mRNA transcription takes place constitutively in all AvrA-positive strains,[265] as AvrA remains silent when cloned into *E. coli*.[266] In fact, a number of Proteobacteria have evolved global regulation systems of post-transcriptional gene expression for their various virulence factors to be responsive to changes in the environment and to flourish in specialized host niches. *Salmonella enterica* is equipped with the Csr (carbon storage regulator) system, consisting of a small RNA binding protein (CrsA) and non-coding RNAs (CsrB/CsrC).[267] CsrA binding to the 5′ untranslated and/or early coding regions of mRNAs alters mRNA translation, turnover and/or transcript elongation, leading to either decay or stabilization of specific mRNA targets, while CsrB/CsrC containing multiple CsrA binding sites bind and sequester CsrA, thereby antagonizing

CsrA activities.[267,268] Overexpression of either CsrA or CsrB shuts down AvrA expression, but that constitutional CsrA expression was required for AvrA expression.[265]

8.3 Chemical and molecular characteristics of AvrA

AvrA belong to a superfamily of the Yersinia outer protein J (YopJ) acetyltransferases, consisting of 287 amino acids.[262,269,270] The regulatory regions are located at flanking N- and C-terminal segments (residues 1–45 and 220–287, respectively) and the intervening residues (46–219) form the catalytic domain, which is structurally homologous to the CE clan peptidases.[262,270] In addition, AvrA possesses ubiquitin-like protein protease and deubiquitination activities. These enzymatic activities have been considered crucial for bacterial survival and virulence through post-translational modifications (PTMs) of host proteins to modulate the activity of key proteins on oncogenic and inflammatory pathways. Specific host targets for deubiquitination include IκBα and β-catenin and those for O (threonine)- and N^{ε}(lysine)-acetylation include MAPKK[271] and p53.[272,273] p53 acetylation has been shown to increase its stability and transcriptional activities.[274]

8.4 In vitro toxicity data of AvrA

AvrA transfection to HeLa cells led to inhibition of activation of NF-κB, a key pro-inflammatory transcription factor, and augmented epithelial apoptosis.[275] AvrA transfection to human 293 T cells induced repression of c-Jun N-terminal kinase and NF-κB signaling pathways[271] as acetylation of critical amino acid residues in MAPKK blocks MAPKK phosphorylation activities.[272] Colonization of AvrA-positive *Salmonella* with human colon cancer cells corroborated this finding showing inhibition of IκBα degradation.[276] The ubiquitination of β-catenin decreased in the presence of AvrA expression, indicating that the expression of the *Salmonella* AvrA effector stabilizes β-catenin by decreasing the ubiquitination of β-catenin.[276] Further colonization experiments with human colon cancer cell lines demonstrated modulation of several members of the Wnt pathway by AvrA expression, i.e., upregulation of Wnt11 and Wnt2 and downregulation of Wnt1,[277–279] which was also confirmed in vivo with murine models.

8.5 AvrA toxicity in animal models

In both transgenic *Drosophila* and murine models, AvrA was shown to potently inhibit c-Jun N-terminal kinase (JNK) and NF-κB signaling pathways and, in the mouse intestinal mucosa, infection with AvrA-positive *Salmonella* dampened the proapoptotic innate immune response to *Salmonella*, suggesting a survival strategy for intracellular pathogen, where the bacteria elicit transient inflammation but do not destroy epithelial cells.[271] In a murine short-term infection model, epithelial cell proliferation of the large intestine was increased in mice infected with AvrA-positive *Salmonella* than those infected with AvrA-negative *Salmonella*.[276,278] Further studies have revealed that infection with AvrA-expressing *Salmonella* increased the Wnt/β-catenin activity, intestinal stem cell population, and cell proliferation in the infected intestinal mucosa.[280] In their murine chronic infection model, Lu et al. found that AvrA persistently regulated β-catenin post-translational modifications, including phosphorylation and acetylation. Moreover, the upstream regulator Akt, transcription factors, T cell factors, nuclear β-catenin, and β-catenin target genes were enhanced in mice infected with *Salmonella*-expressing AvrA, suggesting its functional role in promoting intestinal renewal.[281] The subsequent studies using mouse chemical carcinogenesis models have demonstrated that colorectal tumor incidence indeed markedly increased (almost doubled) in the AvrA+ *Salmonella* infected mice, compared with mice without bacterial gavage or mice infected with AvrA− *Salmonella*.[282] While the Wnt1 protein level in the mouse intestine was decreased in the AvrA+ infected mice compared with those infected with an AvrA strain,[279] Wint11 expression was enhanced in animals infected with AvrA+ strains.[277] In the same animal model colonized with *Salmonella* AvrA-sufficient or AvrA-deficient bacterial strains, AvrA-expressing bacteria activated the STAT3 pathway, which is predicted to enhance proliferation and promote tumorigenesis, providing new insights regarding a STAT3-dependent mechanism by which the specific bacterial product AvrA enhances the development of infection-associated colon cancer.[283]

8.6 Clinical and epidemiological data to support carcinogenic potential in humans

There have been sparse data from epidemiological and clinical studies to directly support oncogenicity of AvrA in humans. However, Lu et al. repotted that *AvrA*-gene positive *Salmonella* was detectable in human feces and demonstrated that AvrA protein was immunohistochemically detectable in human colorectal mucosa and that its expression in tumor adjacent mucosa in colorectal cancer cases was significantly higher than in normal mucosa of non-cancer patients.[284] Consistent with the results from in vitro and animal studies, decreased Wint1 expression was observed in human colorectal cancer tissue.[279]

9. Concluding remarks

We have conducted an in-depth review on several emerging bacterial toxins that may be involved in pathogenesis of human orodigestive tract cancer. It was clear that these bacterial toxins have biological activities to induce several hallmarks of cancer.[285] Some toxins (CDT, colibactin and CNF) can directly interact with DNA or chromosomes leading to their breakdowns, causing mutations and genome instability, which is considered a crucial step for tumor initiation. Other cancer hallmarks commonly manipulated by bacterial toxins are cell proliferation, replication and death. While sustained proliferation and unlimited replication of epithelial cells are essential for tumor growth, enhanced cell apoptosis and senescence can lead to evasion of immune surveillance if they occur preferentially in hematologic cells, and to tumor stimulatory SASP[45] if they primarily involve epithelial and stromal cells. Manipulation of cell proliferation and death by bacterial toxins often are achieved by induction or degradation of oncogenes or tumor suppressors. The two most often modulated pathways were p53[286] and β-catenin/Wnt pathways,[287,288] while others, such as JAK-STAT, MAPK/ERK, Ras and Myc pathways are also exploited by PMT, Bft and AvrA as discussed above.

Wnt/β-catenin signaling pathway controls several crucial biological processes, such as cell motility, proliferation and the induction of differentiation. The activation of its canonical pathway is dependent on the binding of Wnt glycoprotein ligands to Frizzled (Fzd) receptors and its co-receptor, low-density lipoprotein receptor-related protein 5/6 (LRP5/6) on the plasma membrane, forming the ternary complex Wnt–Fzd–LRP5/6. Subsequent recruitment of a scaffolding protein disheveled (Dvl) leads to the inhibition of Axin-mediated β-catenin phosphorylation, and thus β-catenin stabilization, cytoplasmic accumulation, and nuclear translocation. Once in the nucleus, β-catenin activates Wnt-dependent gene expression.[154] Wnt/β-catenin has recently emerged as an important target of several virulence factors produced by bacteria. The mechanisms are diverse and include the repression of Wnt inhibitors' expression by the epigenetic modification of histones, blocking Wnt–Fzd ligand binding, manipulating β-catenin nuclear translocation, down- or upregulation of Wnt family members, post-translational modification of β-catenin, and inhibition of Axin-1 expression, which promotes β-catenin activity,[154,280,287] while others, such as HlyA directly degrades β-catenin through its protease activity and suppresses its activity.[244,288] Accordingly, outcomes of manipulations of the Wnt/β-catenin pathway by these bacterial toxins are diverse. Both *Salmonella* AvrA that deubiquitinates β-catenin to prevent its degradation and *Fusobacterium* FadA that binds E-cadherin to stimulate β-catenin lead to increased cell proliferation and tumor formation,[154,282,288] while *E. coli* HlyA results in immunosuppression.[244,288]

Many bacteria as well as oncoviruses, such as HPV and EBV, are known to repress p53 through direct cleavage or proteasome degradation.[286] However, as we discussed above, other bacterial toxins, such as CDT and colibactin, upregulate P53 due to their genotoxic effects, while AvrA introduces a post-translational modification by acetylation to stabilize and stimulate this pathway.[273] Resulting sustained apoptosis has been speculated to lead eventually to the development of SASP, which in turn promotes tumorigenesis.[45] An additional way to resist cell death has been recognized to be enhancing autophagy, which is considered a pro-survival signal in challenging cellular environment.[285] CDT exploits this strategy.

An emerging hallmark of cancer is reprograming of cellular metabolism and energetics from mitochondrial respiration to aerobic glycolysis, so-called the Warburg effect.[285] While this metabolic shift caused by exposure to LPS/endotoxin has been well documented,[289,290] the effects of the exotoxins reviewed here on cellular energetics have been less well delineated. Yet, there are ample laboratory data to support this possibility. CDT from *Campylobacter* and *Helicobacter* species and HlyA from E coli have been shown to damage mitochondria, which may lead to reduced respiratory function.[291–293] On the other hand, activation of key signaling molecules, such as HIF1α and GLUT1 by HlyA[245] and c-Myc by Bft,[181] on the glycolysis pathway has been noted. Tan et al. directly assessed glycolysis and mitochondria biogenesis in mouse melanoma cells exposed to *E. coli* CNF, reporting an increase in glycolysis and a decrease in mitochondrial biogenesis.[294] However, conflicting data exist, suggesting that explore to CNF enhances mitochondrial volume and function in other cell culture systems,[295,296] and thus the net effect on cellular energetics in human non-transformed cells is unknown.

Many bacterial toxins, such as CDT, CTLP, PMT, FadA, Bft, CFA and HlyA, also exert biological activates to degrade intercellular junction proteins, including tight junction ZO-1 and E-cadherin, as well as extracellular matrix, partly through activation of matrix metalloproteinases, which leads to increased mucosal permeability, decreased mucosal defense and disruption of cellular communications, and activates tumor invasion.[285] On the other hand, bacteria effectors, such as AvrA, stabilizes tight junction to accommodate their own survival in the host.[297,298]

However, by far, the cancer hallmark "tumor-promoting inflammation" is most often controlled by bacterial toxins. In fact, all but Fap2 reviewed here act on this pathway, through diverse mechanisms and with distinct outcomes. A common aspect of infection-related cancer is the induction of chronic inflammation through various mechanisms.[1,299] Whereas most bacteria toxins reviewed here and others induce inflammation and often produce tumor-promoting cytokines, such as IL-17 and IL-6,[187] some pathogenic bacteria have evolved to synthesize bacterial toxins, such as AvrA, HlyA and CTLP, with the ability to temper the inflammatory response to create a suitable niche for their survival and proliferation in the host, not killing the host or host cells they infected. Acute excessive inflammation is a short-term response that either clears the pathogenic bacteria from the host through recruitment of immunological cells, or results in fatal consequences (host death). Thus, it does not lead to the development cancer. Accordingly, anti-inflammatory properties are particularly important and common in many T3SS pathogens with numerous virulence factors that induce strong pro-inflammatory reactions, such as *Salmonella* and *E coli*.[272] The NF-κB network is often manipulated by bacterial toxins to both up- and down-regulate inflammatory responses.[300]

Table 1 summarizes the strength of evidence from human studies and mechanistic studies to support carcinogenicity of each selected toxin. Causal association has not been established for any of the toxins or bacteria reviewed here and the strength in human studies simply reflect consistency in statistical associations. It is important to note that the vast majority of in vitro studies with pathogens are short-term infections that induce a range of host cell responses. Besides, intestinal cells have short turnover rate of 3–4 days. Thus, the interpretation of transitory cellular signaling in the context of in vitro infection as tumorigenic may not be necessarily warranted. The tumor-promoting effects seen in animal models thus far have been delivered from genetically susceptible or chemically induced carcinogenesis models, and no transgenic animal models for the genes encoding these toxins have been tested. Overall, despite growing laboratory evidence to support oncogenic potential of selected bacterial toxins reviewed here,[3,301,302] evidence from human studies has been rather limited. This is partly because most microbiome research to date have focused on

Table 1 Summary of existing evidence to support carcinogenicity of selected bacterial toxins.

Toxins	Mechanistic (in vitro/animal)	Human (clinical/ epidemiological)	Target organs
CDT	a	b	Oral, enterohepatic, genital
CTLP	c	a	Oro-digestive tract/organ
PMT	b	c	Ororespiratory tract
FadA	a	c	Oral, colorectum
Fap2	c	c	Oral, colorectum
Bft	a	a	Colorectum
Colibactin	a	a	Colorectum
CNF	c	b	Colorectum
α-Hemolysin	c	b	Colorectum
AvrA	a	b	Colorectum

[a]Strong mechanisms with animal model/association with cancer in multiple studies.
[b]Strong mechanisms only/association with cancer from single reports.
[c]Suggestive mechanisms only/presence in target organs or in circulation.

global community composition based on 16S rRNA gene sequencing. More targeted research will certainly help advance establishing the bacterial etiology of orodigestive tract cancer, which includes studies testing various human tissue samples, including not only tumors, but also surrounding mucosa, precursor lesions and healthy mucosa, as well as fecal, oral rinse/plaque and pre-diagnostic blood samples to directly detect bacterial toxins and their molecular signatures and to assess antibody responses. In addition, we need more physiologically relevant experimental animal models, which can reflect chronic infection in vivo in humans as well as take bacterial-bacterial interactions among microbiome into consideration. Such effort may lead to the development of new probiotic, prebiotic or antibiotic approaches or immunotherapies to prevent cancer development and progression.

Acknowledgments

A.A.V. is a post-doctoral researcher with Healthy Urban Waters (HUW) supported by the Fred A. and Barbara M. Erb Family Foundation. A.A.V. was also supported by a National Institutes of Health (NIH) Post-doctoral fellowship from the NIH Common Fund and Office of Scientific Workforce Diversity under three linked awards RL5GM118981, TL4GM118983, 1UL1GM118982 administered by the National Institute of General Medical Sciences.

References

1. Kuper H, Adami H-O, Trichopoulos D. Infections as a major preventable cause of human cancer. *J Intern Med*. 2000;248(3):171–183.
2. Plummer M, de Martel C, Vignat J, Ferlay J, Bray F, Franceschi S. Global burden of cancers attributable to infections in 2012: a synthetic analysis. *Lancet Glob Health*. 2016;4(9):e609–e616.
3. van Elsland D, Neefjes J. Bacterial infections and cancer. *EMBO Rep*. 2018;19(11): e46632.
4. Backert S, Tegtmeyer N. Type IV secretion and signal transduction of Helicobacter pylori CagA through interactions with host cell receptors. *Toxins*. 2017;9(4):115.
5. Hatakeyama M. Helicobacter pylori CagA—a bacterial intruder conspiring gastric carcinogenesis. *Int J Cancer*. 2006;119(6):1217–1223.
6. Tegtmeyer N, Neddermann M, Asche CI, Backert S. Subversion of host kinases: a key network in cellular signaling hijacked by Helicobacter pylori CagA. *Mol Microbiol*. 2017;105:358–372.
7. El-Etr SH, Mueller A, Tompkins LS, Falkow S, Merrell DS. Phosphorylation-independent effects of CagA during interaction between Helicobacter pylori and T84 polarized monolayers. *J Infect Dis*. 2004;190(8):1516–1523.
8. Selbach M, Moese S, Meyer TF, Backert S. Functional analysis of the Helicobacter pylori cag pathogenicity island reveals both VirD4-CagA-dependent and VirD4-CagA-independent mechanisms. *Infect Immun*. 2002;70(2):665–671.
9. Ohnishi N, Yuasa H, Tanaka S, et al. Transgenic expression of *Helicobacter pylori* CagA induces gastrointestinal and hematopoietic neoplasms in mouse. *Proc Natl Acad Sci U S A*. 2008;105(3):1003–1008.
10. Qin J, Li R, Raes J, et al. A human gut microbial gene catalogue established by metagenomic sequencing. *Nature*. 2010;464:59.
11. Joossens M, Huys G, Cnockaert M, et al. Dysbiosis of the faecal microbiota in patients with Crohn's disease and their unaffected relatives. *Gut*. 2011;60(5):631–637.
12. Marchesi JR, Dutilh BE, Hall N, et al. Towards the human colorectal cancer microbiome. *PLoS One*. 2011;6(5):e20447.
13. Hajishengallis G, Lamont RJ. Breaking bad: manipulation of the host response by Porphyromonas gingivalis. *Eur J Immunol*. 2014;44(2):328–338.
14. Todar K. *Online Textbook of Bacteriology*. Madison, WI: University of Wisconsin; 2008–2012.
15. Moreira APB, Texeira TFS, Ferreira AB, do Carmo Gouveia Peluzio M, de Cássia Gonçalves Alfenas R. Influence of a high-fat diet on gut microbiota, intestinal permeability and metabolic endotoxaemia. *Br J Nutr*. 2012;108(5):801–809.
16. Vasquez AA, Ram JL, Qazazi MS, Sun J, Kato I. Oral microbiome: potential link to systemic diseases and oral cancer. In: Sun J, Dudeja PK, eds. *Mechanisms Underlying Host-Microbiome Interactions in Pathophysiology of Human Diseases*. Boston, MA: Springer US; 2018:195–246.
17. Trent MS, Stead CM, Tran AX, Hankins JV. Invited review: diversity of endotoxin and its impact on pathogenesis. *J Endotoxin Res*. 2006;12(4):205–223.
18. Lemichez E, Barbieri JT. General aspects and recent advances on bacterial protein toxins. *Cold Spring Harb Perspect Med*. 2013;3(2):a013573.
19. Henkel JS, Baldwin MR, Barbieri JT. Toxins from bacteria. *EXS*. 2010;100:1–29.
20. Veith PD, Glew MD, Gorasia DG, Reynolds EC. Type IX secretion: the generation of bacterial cell surface coatings involved in virulence, gliding motility and the degradation of complex biopolymers. *Mol Microbiol*. 2017;106(1):35–53.
21. Abby SS, Cury J, Guglielmini J, Néron B, Touchon M, Rocha EPC. Identification of protein secretion systems in bacterial genomes. *Sci Rep*. 2016;6:23080.

22. Guerra L, Cortes-Bratti X, Guidi R, Frisan T. The biology of the cytolethal distending toxins. *Toxins*. 2011;3(3):172.
23. Nesic D, Hsu Y, Stebbins CE. Assembly and function of a bacterial genotoxin. *Nature*. 2004;429(6990):429–433.
24. Hontz JS, Villar-Lecumberri MT, Potter BM, Yoder MD, Dreyfus LA, Laity JH. Differences in crystal and solution structures of the cytolethal distending toxin B subunit: relevance to nuclear translocation and functional activation. *J Biol Chem*. 2006;281(35):25365–25372.
25. Faïs T, Delmas J, Serres A, Bonnet R, Dalmasso G. Impact of CDT toxin on human diseases. *Toxins*. 2016;8(7):220.
26. Guerra L, Teter K, Lilley BN, et al. Cellular internalization of cytolethal distending toxin: a new end to a known pathway. *Cell Microbiol*. 2005;7(7):921–934.
27. Guerra L, Nemec KN, Massey S, et al. A novel mode of translocation for cytolethal distending toxin. *Biochim Biophys Acta (BBA)—Mol Cell Res*. 2009;1793(3):489–495.
28. Jinadasa RN, Bloom SE, Weiss RS, Duhamel GE. Cytolethal distending toxin: a conserved bacterial genotoxin that blocks cell cycle progression, leading to apoptosis of a broad range of mammalian cell lineages. *Microbiology*. 2011;157(7):1851–1875.
29. Shenker BJ, Dlakić M, Walker LP, et al. A novel mode of action for a microbial-derived immunotoxin: the cytolethal distending toxin subunit B exhibits phosphatidylinositol 3,4,5-triphosphate phosphatase activity. *J Immunol*. 2007;178(8):5099–5108.
30. Boesze-Battaglia K, Alexander D, Dlakić M, Shenker BJ. A journey of cytolethal distending toxins through cell membranes. *Front Cell Infect Microbiol*. 2016;6:81.
31. Pickett CL, Whitehouse CA. The cytolethal distending toxin family. *Trends Microbiol*. 1999;7(7):292–297.
32. Ahmed HJ, Svensson LA, Cope LD, et al. Prevalence of cdtABC genes encoding cytolethal distending toxin among Haemophilus ducreyi and Actinobacillus actinomycetemcomitans strains. *J Med Microbiol*. 2001;50(10):860–864.
33. Buc E, Dubois D, Sauvanet P, et al. High prevalence of mucosa-associated E. coli producing cyclomodulin and genotoxin in colon cancer. *PLoS One*. 2013;8(2):e56964-e.
34. Yamasaki S, Asakura M, Tsukamoto T, Faruque SM, Deb R, Ramamurthy T. Cytolethal distending toxin (cdt): genetic diversity, structure and role in diarrheal disease. *Toxin Rev*. 2006;25(1):61–88.
35. Lara-Tejero M, Galán JE. A bacterial toxin that controls cell cycle progression as a deoxyribonuclease I-like protein. *Science*. 2000;290(5490):354–357.
36. Nishikubo S, Ohara M, Ikura M, et al. Single nucleotide polymorphism in the cytolethal distending toxin B gene confers heterogeneity in the cytotoxicity of actinobacillus actinomycetemcomitans. *Infect Immun*. 2006;74(12):7014–7020.
37. AbuOun M, Manning G, Cawthraw SA, et al. Cytolethal distending toxin (CDT)-negative *Campylobacter jejuni* strains and anti-CDT neutralizing antibodies are induced during human infection but not during colonization in chickens. *Infect Immun*. 2005;73(5):3053–3062.
38. Haghjoo E, Galán JE. *Salmonella typhi* encodes a functional cytolethal distending toxin that is delivered into host cells by a bacterial-internalization pathway. *Proc Natl Acad Sci U S A*. 2004;101(13):4614–4619.
39. Spanò S, Ugalde JE, Galán JE. Delivery of a *Salmonella typhi* exotoxin from a host intracellular compartment. *Cell Host Microbe*. 2008;3(1):30–38.
40. Miller R, Wiedmann M. Dynamic duo—the salmonella cytolethal distending toxin combines ADP-ribosyltransferase and nuclease activities in a novel form of the cytolethal distending toxin. *Toxins*. 2016;8(5):121.
41. DiRienzo JM. Uptake and processing of the cytolethal distending toxin by mammalian cells. *Toxins*. 2014;6(11):3098.

42. Ueno Y, Ohara M, Kawamoto T, et al. Biogenesis of the *Actinobacillus actinomycetemcomitans* cytolethal distending toxin holotoxin. *Infect Immun.* 2006;74(6): 3480–3487.
43. O'Donoghue EJ, Krachler AM. Mechanisms of outer membrane vesicle entry into host cells. *Cell Microbiol.* 2016;18(11):1508–1517.
44. Guidi R, Levi L, Rouf SF, Puiac S, Rhen M, Frisan T. Salmonella enterica delivers its genotoxin through outer membrane vesicles secreted from infected cells. *Cell Microbiol.* 2013;15(12):2034–2050.
45. Coppé J-P, Desprez P-Y, Krtolica A, Campisi J. The senescence-associated secretory phenotype: the dark side of tumor suppression. *Annu Rev Pathol.* 2010;5:99–118.
46. Shenker BJ, McKay T, Datar S, Miller M, Chowhan R, Demuth D. *Actinobacillus actinomycetemcomitans* immunosuppressive protein is a member of the family of cytolethal distending toxins capable of causing a G2 arrest in human T cells. *J Immunol.* 1999;162(8):4773–4780.
47. Cortes-Bratti X, Karlsson C, Lagergård T, Thelestam M, Frisan T. The *Haemophilus ducreyi* cytolethal distending toxin induces cell cycle arrest and apoptosis via the DNA damage checkpoint pathways. *J Biol Chem.* 2001;276(7):5296–5302.
48. Ando-Suguimoto ES, da Silva MP, Kawamoto D, Chen C, DiRienzo JM, Mayer MPA. The cytolethal distending toxin of *Aggregatibacter actinomycetemcomitans* inhibits macrophage phagocytosis and subverts cytokine production. *Cytokine.* 2014;66(1):46–53.
49. Shenker BJ, Walker LP, Zekavat A, Dlakić M, Boesze-Battaglia K. Blockade of the PI-3K signalling pathway by the *Aggregatibacter actinomycetemcomitans* cytolethal distending toxin induces macrophages to synthesize and secrete pro-inflammatory cytokines. *Cell Microbiol.* 2014;16(9):1391–1404.
50. Akifusa S, Poole S, Lewthwaite J, Henderson B, Nair SP. Recombinant *Actinobacillus actinomycetemcomitans* cytolethal distending toxin proteins are required to interact to inhibit human cell cycle progression and to stimulate human leukocyte cytokine synthesis. *Infect Immun.* 2001;69(9):5925–5930.
51. Aragon V, Chao K, Dreyfus LA. Effect of cytolethal distending toxin on F-actin assembly and cell division in chinese hamster ovary cells. *Infect Immun.* 1997;65(9): 3774–3780.
52. Frisan T, Cortes-Bratti X, Chaves-Olarte E, Stenerlöw B, Thelestam M. The *Haemophilus ducreyi* cytolethal distending toxin induces DNA double-strand breaks and promotes ATM-dependent activation of RhoA. *Cell Microbiol.* 2003;5(10):695–707.
53. Seiwert N, Neitzel C, Stroh S, et al. AKT2 suppresses pro-survival autophagy triggered by DNA double-strand breaks in colorectal cancer cells. *Cell Death Dis.* 2017;8:e3019.
54. Damek-Poprawa M, Korostoff J, Gill R, DiRienzo JM. Cell junction remodeling in gingival tissue exposed to a microbial toxin. *J Dent Res.* 2013;92(6):518–523.
55. Blazkova H, Krejcikova K, Moudry P, Frisan T, Hodny Z, Bartek J. Bacterial intoxication evokes cellular senescence with persistent DNA damage and cytokine signalling. *J Cell Mol Med.* 2010;14(1–2):357–367.
56. Guidi R, Guerra L, Levi L, et al. Chronic exposure to the cytolethal distending toxins of Gram-negative bacteria promotes genomic instability and altered DNA damage response. *Cell Microbiol.* 2013;15(1):98–113.
57. Guerra L, Guidi R, Frisan T. Do bacterial genotoxins contribute to chronic inflammation, genomic instability and tumor progression? *FEBS J.* 2011;278(23): 4577–4588.
58. Graillot V, Dormoy I, Dupuy J, et al. Genotoxicity of cytolethal distending toxin (CDT) on isogenic human colorectal cell lines: potential promoting effects for colorectal carcinogenesis. *Front Cell Infect Microbiol.* 2016;6:34.

59. Frisan T, Nagy N, Chioureas D, Terol M, Grasso F, Masucci MG. A bacterial genotoxin causes virus reactivation and genomic instability in Epstein–Barr virus infected epithelial cells pointing to a role of co-infection in viral oncogenesis. *Int J Cancer.* 2019;144(1):98–109.
60. Ge Z, Schauer DB, Fox JG. In vivo virulence properties of bacterial cytolethal-distending toxin. *Cell Microbiol.* 2008;10(8):1599–1607.
61. Fox JG, Ge Z, Whary MT, Erdman SE, Horwitz BH. *Helicobacter hepaticus* infection in mice: models for understanding lower bowel inflammation and cancer. *Mucosal Immunol.* 2010;4:22.
62. García A, Ihrig MM, Fry RC, et al. Genetic susceptibility to chronic hepatitis is inherited codominantly in *Helicobacter hepaticus*-infected AB6F1 and B6AF1 hybrid male mice, and progression to hepatocellular carcinoma is linked to hepatic expression of lipogenic genes and immune function-associated networks. *Infect Immun.* 2008;76(5):1866–1876.
63. Diwan BA, Sipowicz M, Logsdon D, et al. Marked liver tumorigenesis by *Helicobacter hepaticus* requires perinatal exposure. *Environ Health Perspect.* 2008;116(10):1352–1356.
64. Ge Z, Rogers AB, Feng Y, et al. Bacterial cytolethal distending toxin promotes the development of dysplasia in a model of microbially induced hepatocarcinogenesis. *Cell Microbiol.* 2007;9(8):2070–2080.
65. Fox JG, Feng Y, Theve EJ, et al. Gut microbes define liver cancer risk in mice exposed to chemical and viral transgenic hepatocarcinogens. *Gut.* 2010;59(1):88–97.
66. Jain D, Prasad KN, Sinha S, Husain N. Differences in virulence attributes between cytolethal distending toxin positive and negative *Campylobacter jejuni* strains. *J Med Microbiol.* 2008;57(3):267–272.
67. Fox JG, Rogers AB, Whary MT, et al. Gastroenteritis in NF-κB-deficient mice is produced with wild-type *Camplyobacter jejuni* but not with *C. jejuni* lacking cytolethal distending toxin despite persistent colonization with both strains. *Infect Immun.* 2004;72(2):1116–1125.
68. Young VB, Knox KA, Pratt JS, et al. In vitro and in vivo characterization of *Helicobacter hepaticus* cytolethal distending toxin mutants. *Infect Immun.* 2004;72(5):2521–2527.
69. Shen Z, Feng Y, Rogers AB, et al. Cytolethal distending toxin promotes *Helicobacter cinaedi*-associated typhlocolitis in interleukin-10-deficient mice. *Infect Immun.* 2009; 77(6):2508–2516.
70. Shen Z, Feng Y, Muthupalani S, et al. Novel *Helicobacter species H. japonicum* isolated from laboratory mice from Japan induces typhlocolitis and lower bowel carcinoma in C57BL/129 IL10-/- mice. *Carcinogenesis.* 2016;37(12):1190–1198.
71. Ge Z, Feng Y, Ge L, Parry N, Muthupalani S, Fox JG. *Helicobacter hepaticus* cytolethal distending toxin promotes intestinal carcinogenesis in 129Rag2-deficient mice. *Cell Microbiol.* 2017;19(7):e12728.
72. Péré-Védrenne C, Prochazkova-Carlotti M, Rousseau B, et al. The cytolethal distending toxin subunit CdtB of *Helicobacter hepaticus* promotes senescence and endoreplication in xenograft mouse models of hepatic and intestinal cell lines. *Front Cell Infect Microbiol.* 2017;7:268.
73. He Z, Gharaibeh RZ, Newsome RC, et al. *Campylobacter jejuni* promotes colorectal tumorigenesis through the action of cytolethal distending toxin. *Gut.* 2019;68(2): 289–300.
74. Pimentel M, Morales W, Rezaie A, et al. Development and validation of a biomarker for diarrhea-predominant irritable bowel syndrome in human subjects. *PLoS One.* 2015;10(5):e0126438.
75. Rezaie A, Park SC, Morales W, et al. Assessment of anti-vinculin and anti-cytolethal distending toxin B antibodies in subtypes of irritable bowel syndrome. *Dig Dis Sci.* 2017;62(6):1480–1485.

76. Scanu T, Spaapen Robbert M, Bakker Jeroen M, et al. *Salmonella* manipulation of host signaling pathways provokes cellular transformation associated with gallbladder carcinoma. *Cell Host Microbe*. 2015;17(6):763–774.
77. Shimoyama T, Takahashi R, Abe D, Mizuki I, Endo T, Fukuda S. Serological analysis of *Helicobacter hepaticus* infection in patients with biliary and pancreatic diseases. *J Gastroenterol Hepatol*. 2010;25(s1):S86–S89.
78. Murakami K, Takahashi R, Ono M, et al. Serodiagnosis of Helicobacter hepaticus infection in patients with liver and gastrointestinal diseases: western blot analysis and ELISA using a highly specific monoclonal antibody for *H. hepaticus* antigen. *J Gastroenterol*. 2011;46(9):1120–1126.
79. Hamada T, Yokota K, Ayada K, et al. Detection of Helicobacter hepaticus in human bile samples of patients with biliary disease. *Helicobacter*. 2009;14(6):545–551.
80. Mäkinen PL, Mäkinen KK, Syed SA. Role of the chymotrypsin-like membrane-associated proteinase from *Treponema denticola* ATCC 35405 in inactivation of bioactive peptides. *Infect Immun*. 1995;63(9):3567–3575.
81. Ishihara K. Virulence factors of *Treponema denticola*. *Periodontol 2000*. 2010;54(1): 117–135.
82. Fenno JC. *Treponema denticola* interactions with host proteins. *J Oral Microbiol*. 2012;4. https://doi.org/10.3402/jom.v4i0.9929.
83. Goetting-Minesky MP, Godovikova V, Li JJ, et al. Conservation and revised annotation of the *Treponema denticola prcB-prcA-prtP* locus encoding the dentilisin (CTLP) protease complex. *Mol Oral Microbiol*. 2013;28(3):181–191.
84. Lee SY, Bian X-L, Wong GWK, Hannam PM, McBride BC, Fenno JC. Cleavage of *Treponema denticola* PrcA polypeptide to yield protease complex-associated proteins Prca1 and Prca2 is dependent on PrtP. *J Bacteriol*. 2002;184(14):3864–3870.
85. Godovikova V, Goetting-Minesky MP, Fenno JC. Composition and localization of *Treponema denticola* outer membrane complexes. *Infect Immun*. 2011;79(12):4868–4875.
86. Godovikova V, Wang H-T, Goetting-Minesky MP, et al. *Treponema denticola* PrcB is required for expression and activity of the PrcA-PrtP (Dentilisin) complex. *J Bacteriol*. 2010;192(13):3337–3344.
87. Heuner K, Bergmann I, Heckenbach K, Göbel UB. Proteolytic activity among various oral *Treponema* species and cloning of a prtP like gene of Treponema socranskii subsp. socranskii. *FEMS Microbiol Lett*. 2001;201(2):169–176.
88. Correia FF, Plummer AR, Ellen RP, et al. Two paralogous families of a two-gene subtilisin operon are widely distributed in oral treponemes. *J Bacteriol*. 2003;185(23): 6860–6869.
89. Veith PD, Glew MD, Gorasia DG, Chen D, O'Brien-Simpson NM, Reynolds EC. Localization of outer membrane proteins in *Treponema denticola* by quantitative proteome analyses of outer membrane vesicles and cellular fractions. *J Proteome Res*. 2019;18:1567–1581.
90. Fenno JC, Müller KH, McBride BC. Sequence analysis, expression, and binding activity of recombinant major outer sheath protein (Msp) of *Treponema denticola*. *J Bacteriol*. 1996;178(9):2489–2497.
91. Caimano MJ, Bourell KW, Bannister TD, Cox DL, Radolf JD. The *Treponema denticola* major sheath protein is predominantly periplasmic and has only limited surface exposure. *Infect Immun*. 1999;67(8):4072–4083.
92. Puthenveetil R, Kumar S, Caimano MJ, et al. The major outer sheath protein forms distinct conformers and multimeric complexes in the outer membrane and periplasm of *Treponema denticola*. *Sci Rep*. 2017;7(1):13260.
93. Fenno JC, Wong GWK, Hannam PM, McBride BC. Mutagenesis of outer membrane virulence determinants of the oral spirochete *Treponema denticola*. *FEMS Microbiol Lett*. 1998;163(2):209–215.

94. Uitto VJ, Pan YM, Leung WK, et al. Cytopathic effects of *Treponema denticola* chymotrypsin-like proteinase on migrating and stratified epithelial cells. *Infect Immun.* 1995;63(9):3401–3410.
95. Fenno JC, Hannam PM, Leung WK, Tamura M, Uitto V-J, McBride BC. Cytopathic effects of the major surface protein and the chymotrypsinlike protease of *Treponema denticola. Infect Immun.* 1998;66(5):1869–1877.
96. Grenier D, Uitto VJ, McBride BC. Cellular location of a *Treponema denticola* chymotrypsinlike protease and importance of the protease in migration through the basement membrane. *Infect Immun.* 1990;58(2):347–351.
97. Chi B, Qi M, Kuramitsu HK. Role of dentilisin in *Treponema denticola* epithelial cell layer penetration. *Res Microbiol.* 2003;154(9):637–643.
98. Ellen RP, Ko KS, Lo CM, Grove DA, Ishihara K. Insertional inactivation of the prtP gene of *Treponema denticola* confirms dentilisin's disruption of epithelial junctions. *J Mol Microbiol Biotechnol.* 2000;2(4):581–586.
99. Yamazaki T, Miyamoto M, Yamada S, Okuda K, Ishihara K. Surface protease of *Treponema denticola* hydrolyzes C3 and influences function of polymorphonuclear leukocytes. *Microbes Infect.* 2006;8(7):1758–1763.
100. AR J, Baek KJ, Shin JE, Choi Y. Mechanisms of IL-8 suppression by *Treponema denticola* in gingival epithelial cells. *Immunol Cell Biol.* 2014;92(2):139–147.
101. Li H, Lu S, Chen Y, et al. AKT2 phosphorylation of hexokinase 2 at T473 promotes tumorigenesis and metastasis in colon cancer cells via NF-κB, HIF1α, MMP2, and MMP9 upregulation. *Cell Signal.* 2019;58:99–110.
102. Pietruszewska W, Bojanowska-Pozniak K, Kobos J. Matrix metalloproteinases MMP1, MMP2, MMP9 and their tissue inhibitors TIMP1, TIMP2, TIMP3 in head and neck cancer: an immunohistochemical study. *Otolaryngol Pol.* 2016;70(3):32–43.
103. Kessenbrock K, Plaks V, Werb Z. Matrix metalloproteinases: regulators of the tumor microenvironment. *Cell.* 2010;141(1):52–67.
104. Miao D, Fenno JC, Timm JC, Joo NE, Kapila YL. The *Treponema denticola* chymotrypsin-like protease dentilisin induces matrix metalloproteinase-2-dependent fibronectin fragmentation in periodontal ligament cells. *Infect Immun.* 2011;79(2):806–811.
105. Ateia IM, Sutthiboonyapan P, Kamarajan P, et al. *Treponema denticola* increases MMP-2 expression and activation in the periodontium via reversible DNA and histone modifications. *Cell Microbiol.* 2018;20(4):e12815.
106. Leung WK, Wu Q, Hannam PM, McBride BC, Uitto V-J. *Treponema denticola* may stimulate both epithelial proliferation and apoptosis through MAP kinase signal pathways. *J Periodontal Res.* 2002;37(6):445–455.
107. Ishihara K, Kuramitsu HK, Miura T, Okuda K. Dentilisin activity affects the organization of the outer sheath of *Treponema denticola. J Bacteriol.* 1998;180(15):3837–3844.
108. Capone R, Wang HT, Ning Y, Sweier DG, Lopatin DE, Fenno JC. Human serum antibodies recognize *Treponema denticola* Msp and PrtP protease complex proteins. *Oral Microbiol Immunol.* 2008;23(2):165–169.
109. Dejonckheere E, Vandenbroucke RE, Libert C. Matrix metalloproteinase8 has a central role in inflammatory disorders and cancer progression. *Cytokine Growth Factor Rev.* 2011;22(2):73–81.
110. Narikiyo M, Tanabe C, Yamada Y, et al. Frequent and preferential infection of *Treponema denticola, Streptococcus mitis,* and *Streptococcus anginosus* in esophageal cancers. *Cancer Sci.* 2004;95(7):569–574.
111. Nieminen MT, Listyarifah D, Hagström J, et al. *Treponema denticola* chymotrypsin-like proteinase may contribute to orodigestive carcinogenesis through immunomodulation. *Br J Cancer.* 2017;118:428.

112. Listyarifah D, Nieminen MT, Mäkinen LK, et al. *Treponema denticola* chymotrypsin-like proteinase is present in early-stage mobile tongue squamous cell carcinoma and related to the clinicopathological features. *J Oral Pathol Med.* 2018;47(8):764–772.
113. Kylmä AK, Jouhi L, Listyarifah D, et al. *Treponema denticola* chymotrypsin-like protease as associated with HPV-negative oropharyngeal squamous cell carcinoma. *Br J Cancer.* 2018;119(1):89–95.
114. Yang Y, Cai Q, Shu X-O, et al. Prospective study of oral microbiome and colorectal cancer risk in low-income and African American populations. *Int J Cancer.* 2019;144(10):2381–2389.
115. Wilson BA, Ho M. Cellular and molecular action of the mitogenic protein-deamidating toxin from *Pasteurella multocida. FEBS J.* 2011;278(23):4616–4632.
116. Orth JHC, Aktories K. Molecular biology of *Pasteurella multocida* toxin. In: Aktories K, JHC O, Adler B, eds. *Pasteurella multocida: Molecular Biology, Toxins and Infection.* Berlin, Heidelberg: Springer Berlin Heidelberg; 2012:73–92.
117. Orth JHC, Aktories K. *Pasteurella multocida* toxin activates various heterotrimeric G proteins by deamidation. *Toxins.* 2010;2(2):205.
118. Lax AJ, Grigoriadis AE. *Pasteurella multocida* toxin: the mitogenic toxin that stimulates signalling cascades to regulate growth and differentiation. *Int J Med Microbiol.* 2001;291(4):261–268.
119. Higgins TE, Murphy AC, Staddon JM, Lax AJ, Rozengurt E. *Pasteurella multocida* toxin is a potent inducer of anchorage-independent cell growth. *Proc Natl Acad Sci U S A.* 1992;89(10):4240–4244.
120. Wilson BA, Ho M. *Pasteurella multocida* toxin interaction with host cells: entry and cellular effects. *Curr Top Microbiol Immunol.* 2012;361:93–111.
121. Wilson BA, Ho M. *Pasteurella multocida*: from zoonosis to cellular microbiology. *Clin Microbiol Rev.* 2013;26(3):631–655.
122. Pullinger GD, Bevir T, Lax AJ. The *Pasteurella multocida* toxin is encoded within a lysogenic bacteriophage. *Mol Microbiol.* 2004;51(1):255–269.
123. Clemons NC, Bannai Y, Haywood EE, et al. Cytosolic delivery of multidomain cargos by the N terminus of *Pasteurella multocida* toxin. *Infect Immun.* 2018;86(8):e00248–e00318.
124. Danielsen EM, Christiansen N, Danielsen EM. *Pasteurella multocida* toxin: targeting mast cell secretory granules during kiss-and-run secretion. *Tissue Cell.* 2016;48(1):1–9.
125. Hopkins AM, Walsh SV, Verkade P, Boquet P, Nusrat A. Constitutive activation of Rho proteins by CNF-1 influences tight junction structure and epithelial barrier function. *J Cell Sci.* 2003;116(Pt. 4):725–742.
126. Seo B, Choy EW, Maudsley S, Miller WE, Wilson BA, Luttrell LM. *Pasteurella multocida* toxin stimulates mitogen-activated protein kinase via Gq/11-dependent transactivation of the epidermal growth factor receptor. *J Biol Chem.* 2000;275(3):2239–2245.
127. Liu F, Yang X, Geng M, Huang M. Targeting ERK, an Achilles' Heel of the MAPK pathway, in cancer therapy. *Acta Pharm Sin B.* 2018;8(4):552–562.
128. Pópulo H, Lopes JM, Soares P. The mTOR signalling pathway in human cancer. *Int J Mol Sci.* 2012;13(2):1886–1918.
129. Oubrahim H, Wong A, Wilson BA, Chock PB. Mammalian target of rapamycin complex 1 (mTORC1) plays a role in *Pasteurella multocida* toxin (PMT)-induced protein synthesis and proliferation in Swiss 3T3 cells. *J Biol Chem.* 2013;288(4):2805–2815.
130. Ozgen N, Obreztchikova M, Guo J, et al. Protein kinase D Links Gq-coupled receptors to cAMP response element-binding protein (CREB)-Ser133 phosphorylation in the heart. *J Biol Chem.* 2008;283(25):17009–17019.

131. Preuß I, Hildebrand D, Orth JHC, Aktories K, Kubatzky KF. *Pasteurella multocida* toxin is a potent activator of anti-apoptotic signalling pathways. *Cell Microbiol.* 2010;12(8):1174–1185.
132. Kolenbrander PE, Palmer Jr RJ, Periasamy S, Jakubovics NS. Oral multispecies biofilm development and the key role of cell-cell distance. *Nat Rev Microbiol.* 2010;8(7):471–480.
133. Orth JHC, Aktories K, Kubatzky KF. Modulation of host cell gene expression through activation of STAT transcription factors by *Pasteurella multocida* toxin. *J Biol Chem.* 2007;282(5):3050–3057.
134. Hildebrand D, Walker P, Dalpke A, Heeg K, Kubatzky KF. *Pasteurella multocida* toxin-induced Pim-1 expression disrupts suppressor of cytokine signalling (SOCS)-1 activity. *Cell Microbiol.* 2010;12(12):1732–1745.
135. Kubatzky K, Kloos B, Hildebrand D. Signaling cascades of *Pasteurella multocida* toxin in immune evasion. *Toxins.* 2013;5(9):1664.
136. Bagley KC, Abdelwahab SF, Tuskan RG, Lewis GK. *Pasteurella multocida* toxin activates human monocyte-derived and murine bone marrow-derived dendritic cells in vitro but suppresses antibody production in vivo. *Infect Immun.* 2005;73(1):413–421.
137. Hildebrand D, Heeg K, Kubatzky K. *Pasteurella multocida* toxin manipulates T cell differentiation. *Front Microbiol.* 2015;6:1273.
138. Pani G, Galeotti T, Chiarugi P. Metastasis: cancer cell's escape from oxidative stress. *Cancer Metastasis Rev.* 2010;29(2):351–378.
139. Elias B, Boros G, Albert M, et al. Clinical and pathological effects of the dermonecrotic toxin of *Bordetella bronchiseptica* and *Pasteurella multocida* in specific-pathogen-free piglets. *Nihon Juigaku Zasshi.* 1990;52(4):677–688.
140. Balur MB, Koçak HE, Altınay S, Özdamar K, Taşkın Ü, Oktay MF. Is submucosal fat injection effective in atrophic rhinitis? An experimental animal study. *Eur Arch Otorhinolaryngol.* 2017;274(10):3637–3642.
141. Correa P, Haenszel W, Cuello C, et al. Gastric precancerous process in a high risk population: cross-sectional studies. *Cancer Res.* 1990;50(15):4731–4736.
142. Correa P, Haenszel W, Cuello C, et al. Gastric precancerous process in a high risk population: cohort follow-up. *Cancer Res.* 1990;50(15):4737–4740.
143. Choudat D, Le Goff C, Delemotte B, et al. Occupational exposure to animals and antibodies against *Pasteurella multocida. Br J Ind Med.* 1987;44(12):829–833.
144. Donnio PY, Le Goff C, Avril JL, Pouedras P, Gras-Rouzet S. *Pasteurella multocida:* oropharyngeal carriage and antibody response in breeders. *Vet Res.* 1994;25(1):8–15.
145. Avril JL, Donnio PY, Pouedras P. Selective medium for *Pasteurella multocida* and its use to detect oropharyngeal carriage in pig breeders. *J Clin Microbiol.* 1990;28(6): 1438–1440.
146. Donnio PY, Avril JL, Andre PM, Vaucel J. Dermonecrotic toxin production by strains of *Pasteurella multocida* isolated from man. *J Med Microbiol.* 1991;34(6):333–337.
147. Harris PJ, Osswald MB. *Pasteurella multocida* epiglottitis: a review and report of a new case with associated chronic lymphocytic leukemia. *Ear Nose Throat J.* 2010;89(12):E4–E7.
148. Véleza M, Casanas B, Greene JN. *Pasteurella multocida* infections in cancer patients. *Asian Biomed.* 2010;4(3):449.
149. Brennan CA, Garrett WS. *Fusobacterium nucleatum*—symbiont, opportunist and oncobacterium. *Nat Rev Microbiol.* 2019;17(3):156–166.
150. Han YW, Wang X. Mobile microbiome: oral bacteria in extra-oral infections and inflammation. *J Dent Res.* 2013;92(6):485–491.
151. Han YW, Ikegami A, Rajanna C, et al. Identification and characterization of a novel adhesin unique to oral fusobacteria. *J Bacteriol.* 2005;187(15):5330–5340.

152. Rubinstein MR, Wang X, Liu W, Hao Y, Cai G, Han YW. *Fusobacterium nucleatum* promotes colorectal carcinogenesis by modulating E-cadherin/β-catenin signaling via its FadA adhesin. *Cell Host Microbe*. 2013;14(2):195–206.
153. Nithianantham S, Xu M, Yamada M, Ikegami A, Shoham M, Han YW. Crystal structure of FadA adhesin from *Fusobacterium nucleatum* reveals a novel oligomerization motif, the leucine chain. *J Biol Chem*. 2009;284(6):3865–3872.
154. Rubinstein MR, Baik JE, Lagana SM, et al. *Fusobacterium nucleatum* promotes colorectal cancer by inducing Wnt/β-catenin modulator Annexin A1. *EMBO Rep*. 2019;20(4): e47638.
155. Yang Y, Weng W, Peng J, et al. *Fusobacterium nucleatum* increases proliferation of colorectal cancer cells and tumor development in mice by activating toll-like receptor 4 signaling to nuclear factor-κB, and up-regulating expression of microRNA-21. *Gastroenterology*. 2017;152(4):851–866.e24.
156. Butt J, Jenab M, Pawlita M, et al. Antibody responses to *Fusobacterium nucleatum* proteins in prediagnostic blood samples are not associated with risk of developing colorectal cancer. *Cancer Epidemiol Biomarkers Prev*. 2019;28(9):1552–1555.
157. Coppenhagen-Glazer S, Sol A, Abed J, et al. Fap2 of *Fusobacterium nucleatum* is a galactose-inhibitable adhesin involved in coaggregation, cell adhesion, and preterm birth. *Infect Immun*. 2015;83(3):1104–1113.
158. Kaplan CW, Lux R, Huynh T, Jewett A, Shi W, Haake SK. *Fusobacterium nucleatum* apoptosis-inducing outer membrane protein. *J Dent Res*. 2005;84(8):700–704.
159. Guevarra Jr LA, Afable ACF, Belza PJO, et al. Immunogenicity of a Fap2 peptide mimotope of *Fusobacterium nucleatum* and its potential use in the diagnosis of colorectal cancer. *Infect Agent Cancer*. 2018;13:11.
160. Jewett A, Hume WR, Le H, et al. Induction of apoptotic cell death in peripheral blood mononuclear and polymorphonuclear cells by an oral bacterium, *Fusobacterium nucleatum*. *Infect Immun*. 2000;68(4):1893–1898.
161. Kaplan CW, Ma X, Paranjpe A, et al. *Fusobacterium nucleatum* outer membrane proteins Fap2 and RadD induce cell death in human lymphocytes. *Infect Immun*. 2010;78(11):4773–4778.
162. Gur C, Ibrahim Y, Isaacson B, et al. Binding of the Fap2 protein of *Fusobacterium nucleatum* to human inhibitory receptor TIGIT protects tumors from immune cell attack. *Immunity*. 2015;42(2):344–355.
163. Abed J, Emgard JE, Zamir G, et al. Fap2 mediates *Fusobacterium nucleatum* colorectal adenocarcinoma enrichment by binding to tumor-expressed Gal-GalNAc. *Cell Host Microbe*. 2016;20(2):215–225.
164. Jones CG, Lawton JH, Shachak M. Organisms as ecosystem engineers. *Oikos*. 1994;69(3):373–386.
165. Yang KR, Mooney SM, Zarif JC, Coffey DS, Taichman RS, Pienta KJ. Niche inheritance: a cooperative pathway to enhance cancer cell fitness through ecosystem engineering. *J Cell Biochem*. 2014;115(9):1478–1485.
166. Boleij A, Hechenbleikner EM, Goodwin AC, et al. The *Bacteroides fragilis* toxin gene is prevalent in the colon mucosa of colorectal cancer patients. *Clin Infect Dis*. 2015;60(2):208–215.
167. Van Tassell RL, Lyerly DM, Wilkins TD. Purification and characterization of an enterotoxin from Bacteroides fragilis. *Infect Immun*. 1992;60(4):1343–1350.
168. Myers LL, Shoop DS, Stackhouse LL, et al. Isolation of enterotoxigenic *Bacteroides fragilis* from humans with diarrhea. *J Clin Microbiol*. 1987;25(12):2330–2333.
169. Sears CL, Geis AL, Housseau F. *Bacteroides fragilis* subverts mucosal biology: from symbiont to colon carcinogenesis. *J Clin Invest*. 2014;124(10):4166–4172.
170. Moore WE, Cato EP, Holdeman LV. Some current concepts in intestinal bacteriology. *Am J Clin Nutr*. 1978;31(10 suppl):S33–S42.

171. Wexler HM. *Bacteroides*: the good, the bad, and the nitty-gritty. *Clin Microbiol Rev*. 2007;20(4):593–621.
172. Simon GL, Gorbach SL. Intestinal flora in health and disease. *Gastroenterology*. 1984;86(1):174–193.
173. Moncrief JS, Obiso R, Barroso LA, et al. The enterotoxin of *Bacteroides fragilis* is a metalloprotease. *Infect Immun*. 1995;63(1):175–181.
174. Kling JJ, Wright RL, Moncrief JS, Wilkins TD. Cloning and characterization of the gene for the metalloprotease enterotoxin of *Bacteroides fragilis*. *FEMS Microbiol Lett*. 1997;146(2):279–284.
175. Wilson MM, Anderson DE, Bernstein HD. Analysis of the outer membrane proteome and secretome of *Bacteroides fragilis* reveals a multiplicity of secretion mechanisms. *PLoS One*. 2015;10(2):e0117732.
176. Coyne MJ, Roelofs KG, Comstock LE. Type VI secretion systems of human gut Bacteroidales segregate into three genetic architectures, two of which are contained on mobile genetic elements. *BMC Genomics*. 2016;17:58.
177. Chatzidaki-Livanis M, Geva-Zatorsky N, Comstock LE. Bacteroides fragilis type VI secretion systems use novel effector and immunity proteins to antagonize human gut Bacteroidales species. *Proc Natl Acad Sci U S A*. 2016;113(13):3627–3632.
178. Silverman JM, Agnello DM, Zheng H, et al. Haemolysin coregulated protein is an exported receptor and chaperone of type VI secretion substrates. *Mol Cell*. 2013; 51(5):584–593.
179. Obiso RJ, Azghani AO, Wilkins TD. The *Bacteroides fragilis* toxin fragilysin disrupts the paracellular barrier of epithelial cells. *Infect Immun*. 1997;65(4):1431–1439.
180. Popoff M. Concepts in bacterial virulence. In: Russell W, Herwald H, eds. *Contributions to Microbiology*. Switzerland: Reinhardt Druck, Basel, Karger; 2004:31–34.
181. Wu S, Morin PJ, Maouyo D, Sears CL. *Bacteroides fragilis* enterotoxin induces c-Myc expression and cellular proliferation. *Gastroenterology*. 2003;124(2):392–400.
182. Donelli G, Fabbri A, Fiorentini C. *Bacteroides fragilis* enterotoxin induces cytoskeletal changes and surface blebbing in HT-29 cells. *Infect Immun*. 1996;64(1):113–119.
183. Wu S, Rhee KJ, Zhang M, Franco A, Sears CL. *Bacteroides fragilis* toxin stimulates intestinal epithelial cell shedding and gamma-secretase-dependent E-cadherin cleavage. *J Cell Sci*. 2007;120(pt 11):1944–1952.
184. Goodwin AC, Shields CED, Wu S, et al. Polyamine catabolism contributes to enterotoxigenic *Bacteroides fragilis*-induced colon tumorigenesis. *Proc Natl Acad Sci U S A*. 2011;108(37):15354–15359.
185. Allen J, Hao S, Sears CL, Timp W. Epigenetic changes induced by *Bacteroides fragilis* toxin. *Infect Immun*. 2019;87(6):pii: e00447-18.
186. Housseau F, Sears CL. Enterotoxigenic *Bacteroides fragilis* (ETBF)-mediated colitis in Min (Apc+/−) mice: a human commensal-based murine model of colon carcinogenesis. *Cell Cycle*. 2010;9(1):3–5.
187. Wang L, Yi T, Kortylewski M, Pardoll DM, Zeng D, Yu H. IL-17 can promote tumor growth through an IL-6-Stat3 signaling pathway. *J Exp Med*. 2009;206(7): 1457–1464.
188. Zhu Q, Jin Z, Wu W, et al. Analysis of the intestinal lumen microbiota in an animal model of colorectal cancer. *PLoS One*. 2014;9(6):e90849.
189. Toprak NU, Yagci A, Gulluoglu BM, et al. A possible role of *Bacteroides fragilis* enterotoxin in the aetiology of colorectal cancer. *Clin Microbiol Infect*. 2006;12(8): 782–786.
190. Chen W, Liu F, Ling Z, Tong X, Xiang C. Human intestinal lumen and mucosa-associated microbiota in patients with colorectal cancer. *PLoS One*. 2012;7(6):e39743.

191. Dejea CM, Fathi P, Craig JM, et al. Patients with familial adenomatous polyposis harbor colonic biofilms containing tumorigenic bacteria. *Science*. 2018;359(6375): 592–597.
192. Faïs T, Delmas J, Barnich N, Bonnet R, Dalmasso G. Colibactin: more than a new bacterial toxin. *Toxins (Basel)*. 2018;10(4):151.
193. Nougayrède JP, Homburg S, Taieb F, et al. *Escherichia coli* induces DNA double-strand breaks in eukaryotic cells. *Science*. 2006;313(5788):848–851.
194. Healy AR, Vizcaino MI, Crawford JM, Herzon SB. Convergent and modular synthesis of candidate precolibactins. Structural revision of precolibactin A. *J Am Chem Soc*. 2016;138(16):5426–5432.
195. Xue M, Kim CS, Healy AR, et al. Structure elucidation of colibactin and its DNA cross-links. *Science*. 2019;365(6457):eaax2685.
196. Vizcaino MI, Crawford JM. The colibactin warhead crosslinks DNA. *Nat Chem*. 2015;7(5):411–417.
197. Wilson MR, Jiang Y, Villalta PW, et al. The human gut bacterial genotoxin colibactin alkylates DNA. *Science*. 2019;363(6428):pii: eaar7785.
198. Li Z-R, Li J, Cai W, et al. Macrocyclic colibactin induces DNA double-strand breaks via copper-mediated oxidative cleavage. *Nat Chem*. 2019;11(10):880–889.
199. Arthur JC, Perez-Chanona E, Mühlbauer M, et al. Intestinal inflammation targets cancer-inducing activity of the microbiota. *Science*. 2012;338(6103):120–123.
200. McCarthy AJ, Martin P, Cloup E, Stabler RA, Oswald E, Taylor PW. The genotoxin colibactin is a determinant of virulence in *Escherichia coli* K1 experimental neonatal systemic infection. *Infect Immun*. 2015;83(9):3704–3711.
201. Raisch J, Rolhion N, Dubois A, Darfeuille-Michaud A, Bringer MA. Intracellular colon cancer-associated *Escherichia coli* promote protumoral activities of human macrophages by inducing sustained COX-2 expression. *Lab Invest*. 2015;95(3):296–307.
202. Johnson JR, Johnston B, Kuskowski MA, Nougayrede JP, Oswald E. Molecular epidemiology and phylogenetic distribution of the *Escherichia coli* pks genomic island. *J Clin Microbiol*. 2008;46(12):3906–3911.
203. Mousa JJ, Yang Y, Tomkovich S, et al. MATE transport of the *E. coli*-derived genotoxin colibactin. *Nat Microbiol*. 2016;1:15009.
204. Secher T, Samba-Louaka A, Oswald E, Nougayrède JP. *Escherichia coli* producing colibactin triggers premature and transmissible senescence in mammalian cells. *PLoS One*. 2013;8(10):e77157.
205. Cuevas-Ramos G, Petit CR, Marcq I, Boury M, Oswald E, Nougayrède JP. *Escherichia coli* induces DNA damage in vivo and triggers genomic instability in mammalian cells. *Proc Natl Acad Sci U S A*. 2010;107(25):11537–11542.
206. Marcq I, Martin P, Payros D, et al. The genotoxin colibactin exacerbates lymphopenia and decreases survival rate in mice infected with septicemic *Escherichia coli*. *J Infect Dis*. 2014;210(2):285–294.
207. Tomkovich S, Yang Y, Winglee K, et al. Locoregional effects of microbiota in a preclinical model of colon carcinogenesis. *Cancer Res*. 2017;77(10):2620–2632.
208. Payros D, Secher T, Boury M, et al. Maternally acquired genotoxic *Escherichia coli* alters offspring's intestinal homeostasis. *Gut Microbes*. 2014;5(3):313–325.
209. Cougnoux A, Dalmasso G, Martinez R, et al. Bacterial genotoxin colibactin promotes colon tumour growth by inducing a senescence-associated secretory phenotype. *Gut*. 2014;63(12):1932–1942.
210. Eklöf V, Löfgren-Burström A, Zingmark C, et al. Cancer-associated fecal microbial markers in colorectal cancer detection. *Int J Cancer*. 2017;141(12):2528–2536.
211. Nowrouzian FL, Oswald E. *Escherichia coli* strains with the capacity for long-term persistence in the bowel microbiota carry the potentially genotoxic pks island. *Microb Pathog*. 2012;53(3–4):180–182.

212. Knust Z, Schmidt G. Cytotoxic necrotizing factors (CNFs)—a growing toxin family. *Toxins (Basel)*. 2010;2(1):116–127.
213. Hoffmann C, Pop M, Leemhuis J, Schirmer J, Aktories K, Schmidt G. The *Yersinia pseudotuberculosis* cytotoxic necrotizing factor (CNFY) selectively activates RhoA. *J Biol Chem*. 2004;279(16):16026–16032.
214. Sugai M, Hatazaki K, Mogami A, et al. Cytotoxic necrotizing factor type 2 produced by pathogenic *Escherichia coli* deamidates a gln residue in the conserved G-3 domain of the rho family and preferentially inhibits the GTPase activity of RhoA and rac1. *Infect Immun*. 1999;67(12):6550–6557.
215. De Rycke J, Phan-Thanh L, Bernard S. Immunochemical identification and biological characterization of cytotoxic necrotizing factor from *Escherichia coli*. *J Clin Microbiol*. 1989;27(5):983–988.
216. Caprioli A, Falbo V, Roda LG, Ruggeri FM, Zona C. Partial purification and characterization of an *Escherichia coli* toxic factor that induces morphological cell alterations. *Infect Immun*. 1983;39(3):1300–1306.
217. Usein CR, Damian M, Tatu-Chitoiu D, et al. Prevalence of virulence genes in *Escherichia coli* strains isolated from Romanian adult urinary tract infection cases. *J Cell Mol Med*. 2001;5(3):303–310.
218. Falbo V, Pace T, Picci L, Pizzi E, Caprioli A. Isolation and nucleotide sequence of the gene encoding cytotoxic necrotizing factor 1 of *Escherichia coli*. *Infect Immun*. 1993; 61(11):4909–4914.
219. Yu H, Kim KS. Ferredoxin is involved in secretion of cytotoxic necrotizing factor 1 across the cytoplasmic membrane in *Escherichia coli* K1. *Infect Immun*. 2010; 78(2):838–844.
220. Kouokam JC, Wai SN, Fällman M, Dobrindt U, Hacker J, Uhlin BE. Active cytotoxic necrotizing factor 1 associated with outer membrane vesicles from uropathogenic *Escherichia coli*. *Infect Immun*. 2006;74(4):2022–2030.
221. Blumenthal B, Hoffmann C, Aktories K, Backert S, Schmidt G. The cytotoxic necrotizing factors from *Yersinia pseudotuberculosis* and from *Escherichia coli* bind to different cellular receptors but take the same route to the cytosol. *Infect Immun*. 2007;75(7):3344–3353.
222. Schmidt G, Sehr P, Wilm M, Selzer J, Mann M, Aktories K. Gln 63 of Rho is deamidated by *Escherichia coli* cytotoxic necrotizing factor-1. *Nature*. 1997;387(6634): 725–729.
223. Bhakdi S, Muhly M, Korom S, Schmidt G. Effects of *Escherichia coli* hemolysin on human monocytes. Cytocidal action and stimulation of interleukin 1 release. *J Clin Invest*. 1990;85(6):1746–1753.
224. Zhang Z, Aung KM, Uhlin BE, Wai SN. Reversible senescence of human colon cancer cells after blockage of mitosis/cytokinesis caused by the CNF1 cyclomodulin from *Escherichia coli*. *Sci Rep*. 2018;8(1):17780.
225. Thomas W, Ascott ZK, Harmey D, Slice LW, Rozengurt E, Lax AJ. Cytotoxic necrotizing factor from *Escherichia coli* induces RhoA-dependent expression of the cyclooxygenase-2 Gene. *Infect Immun*. 2001;69(11):6839–6845.
226. Boyer L, Travaglione S, Falzano L, et al. Rac GTPase instructs nuclear factor-kappaB activation by conveying the SCF complex and IkBalpha to the ruffling membranes. *Mol Biol Cell*. 2004;15(3):1124–1133.
227. Travaglione S, Fabbri A, Fiorentini C. The Rho-activating CNF1 toxin from pathogenic *E. coli*: a risk factor for human cancer development? *Infect Agent Cancer*. 2008;3:4.
228. Rippere-Lampe KE, Lang M, Ceri H, Olson M, Lockman HA, O'Brien AD. Cytotoxic necrotizing factor type 1-positive *Escherichia coli* causes increased inflammation and tissue damage to the prostate in a rat prostatitis model. *Infect Immun*. 2001;69(10):6515–6519.

229. Rippere-Lampe KE, O'Brien AD, Conran R, Lockman HA. Mutation of the gene encoding cytotoxic necrotizing factor type 1 (cnf(1)) attenuates the virulence of uropathogenic *Escherichia coli*. *Infect Immun.* 2001;69(6):3954–3964.
230. Guo Y, Zhang Z, Wei H, et al. Cytotoxic necrotizing factor 1 promotes prostate cancer progression through activating the Cdc42-PAK1 axis. *J Pathol.* 2017;243(2):208–219.
231. Raisch J, Buc E, Bonnet M, et al. Colon cancer-associated B2 *Escherichia coli* colonize gut mucosa and promote cell proliferation. *World J Gastroenterol.* 2014;20(21): 6560–6572.
232. Dubois D, Delmas J, Cady A, et al. Cyclomodulins in urosepsis strains of *Escherichia coli*. *J Clin Microbiol.* 2010;48(6):2122–2129.
233. Cortajarena AL, Goñi FM, Ostolaza H. Asp-863 is a key residue for calcium-dependent activity of *Escherichia coli* RTX toxin alpha-haemolysin. *FEBS Lett.* 2003;546(2–3): 271–275.
234. Jorgensen SE, Short EC, Kurtz HJ, Mussen HK, Wu GK. Studies on the origin of the alpha-haemolysin produced by *Escherichia coli*. *J Med Microbiol.* 1976;9(2):173–189.
235. LovelL R, Rees TA. A filterable haemolysin from *Escherichia coli*. *Nature.* 1960; 188:755–756.
236. Smith HW. The haemolysins of Escherichia coli. *J Pathol Bacteriol.* 1963;85:197–211.
237. Troeger H, Richter JF, Beutin L, et al. *Escherichia coli* alpha-haemolysin induces focal leaks in colonic epithelium: a novel mechanism of bacterial translocation. *Cell Microbiol.* 2007;9(10):2530–2540.
238. Goebel W, Hedgpeth J. Cloning and functional characterization of the plasmid-encoded hemolysin determinant of *Escherichia coli*. *J Bacteriol.* 1982;151(3):1290–1298.
239. Bhakdi S, Mackman N, Nicaud JM, Holland IB. *Escherichia coli* hemolysin may damage target cell membranes by generating transmembrane pores. *Infect Immun.* 1986;52(1):63–69.
240. Felmlee T, Pellett S, Welch RA. Nucleotide sequence of an *Escherichia coli* chromosomal hemolysin. *J Bacteriol.* 1985;163(1):94–105.
241. Gentschev I, Dietrich G, Goebel W. The *E. coli* alpha-hemolysin secretion system and its use in vaccine development. *Trends Microbiol.* 2002;10(1):39–45.
242. Balsalobre C, Silván JM, Berglund S, Mizunoe Y, Uhlin BE, Wai SN. Release of the type I secreted alpha-haemolysin via outer membrane vesicles from *Escherichia coli*. *Mol Microbiol.* 2006;59(1):99–112.
243. Bhakdi S, Greulich S, Muhly M, et al. Potent leukocidal action of *Escherichia coli* hemolysin mediated by permeabilization of target cell membranes. *J Exp Med.* 1989;169(3):737–754.
244. Dhakal BK, Mulvey MA. The UPEC pore-forming toxin α-hemolysin triggers proteolysis of host proteins to disrupt cell adhesion, inflammatory, and survival pathways. *Cell Host Microbe.* 2012;11(1):58–69.
245. Jin Y, Tang S, Li W, et al. Hemolytic *E. coli* promotes colonic tumorigenesis in females. *Cancer Res.* 2016;76(10):2891–2900.
246. Marre R, Hacker J, Henkel W, Goebel W. Contribution of cloned virulence factors from uropathogenic *Escherichia coli* strains to nephropathogenicity in an experimental rat pyelonephritis model. *Infect Immun.* 1986;54(3):761–767.
247. Russo TA, Davidson BA, Genagon SA, et al. *E. coli* virulence factor hemolysin induces neutrophil apoptosis and necrosis/lysis in vitro and necrosis/lysis and lung injury in a rat pneumonia model. *Am J Physiol Lung Cell Mol Physiol.* 2005;289(2):L207–L216.
248. Darfeuille-Michaud A, Neut C, Barnich N, et al. Presence of adherent *Escherichia coli* strains in ileal mucosa of patients with Crohn's disease. *Gastroenterology.* 1998; 115(6):1405–1413.
249. Hilali F, Ruimy R, Saulnier P, et al. Prevalence of virulence genes and clonality in *Escherichia coli* strains that cause bacteremia in cancer patients. *Infect Immun.* 2000;68(7):3983–3989.

250. Cianflone NFC. Salmonellosis and the GI tract: more than just peanut butter. *Curr Gastroenterol Rep*. 2008;10(4):424–431.
251. Kato I, Boleij A, Kortman GA, et al. Partial associations of dietary iron, smoking and intestinal bacteria with colorectal cancer risk. *Nutr Cancer*. 2013;65(2):169–177.
252. Mughini-Gras L, Schaapveld M, Kramers J, et al. Increased colon cancer risk after severe *Salmonella* infection. *PLoS One*. 2018;13(1):e0189721.
253. Hayward MR, AbuOun M, La Ragione RM, et al. SPI-23 of S. Derby: role in adherence and invasion of porcine tissues. *PLoS One*. 2014;9(9):e107857.
254. Sabbagh SC, Forest CG, Lepage C, Leclerc J-M, Daigle F. So similar, yet so different: uncovering distinctive features in the genomes of *Salmonella enterica* serovars Typhimurium and Typhi. *FEMS Microbiol Lett*. 2010;305(1):1–13.
255. Van Engelenburg SB, Palmer AE. Imaging type-III secretion reveals dynamics and spatial segregation of *Salmonella* effectors. *Nat Methods*. 2010;7(4):325–330.
256. Phoebe Lostroh C, Lee CA. The *Salmonella* pathogenicity island-1 type III secretion system. *Microbes Infect*. 2001;3(14–15):1281–1291.
257. Costa TRD, Felisberto-Rodrigues C, Meir A, et al. Secretion systems in Gram-negative bacteria: structural and mechanistic insights. *Nat Rev Microbiol*. 2015;13(6):343–359.
258. Gaytan MO, Martinez-Santos VI, Soto E, Gonzalez-Pedrajo B. Type three secretion system in attaching and effacing pathogens. *Front Cell Infect Microbiol*. 2016;6:129.
259. Huehn S, La Ragione RM, Anjum M, et al. Virulotyping and antimicrobial resistance typing of *Salmonella enterica* serovars relevant to human health in Europe. *Foodborne Pathog Dis*. 2009;7(5):523–535.
260. Prager R, Mirold S, Tietze E, et al. Prevalence and polymorphism of genes encoding translocated effector proteins among clinical isolates of *Salmonella enterica*. *Int J Med Microbiol*. 2000;290(7):605–617.
261. Du F, Galán JE. Selective inhibition of Type III secretion activated signaling by the *salmonella* effector AvrA. *PLoS Pathog*. 2009;5(9):e1000595.
262. Labriola JM, Zhou Y, Nagar B. Structural analysis of the bacterial effector AvrA identifies a critical helix involved in substrate recognition. *Biochemistry*. 2018;57(33):4985–4996.
263. Streckel W, Wolff A-C, Prager R, Tietze E, Tschäpe H. Expression profiles of effector proteins SopB, SopD1, SopE1, and AvrA differ with systemic, enteric, and epidemic strains of *Salmonella enterica*. *Mol Nutr Food Res*. 2004;48(7):496–503.
264. Ellermeier JR, Slauch JM. Adaptation to the host environment: regulation of the SPI1 type III secretion system in *Salmonella enterica* serovar Typhimurium. *Curr Opin Microbiol*. 2007;10(1):24–29.
265. Kerrinnes T, Zelas ZB-B, Streckel W, et al. CsrA and CsrB are required for the post-transcriptional control of the virulence-associated effector protein AvrA of *Salmonella enterica*. *Int J Med Microbiol*. 2009;299(5):333–341.
266. Ben-Barak Z, Streckel W, Yaron S, Cohen S, Prager R, Tschäpe H. The expression of the virulence-associated effector protein gene avrA is dependent on a *Salmonella enterica*-specific regulatory function. *Int J Med Microbiol*. 2006;296(1):25–38.
267. Timmermans J, Van Melderen L. Post-transcriptional global regulation by CsrA in bacteria. *Cell Mol Life Sci*. 2010;67(17):2897–2908.
268. Vakulskas CA, Potts AH, Babitzke P, Ahmer BMM, Romeo T. Regulation of bacterial virulence by Csr (Rsm) systems. *Microbiol Mol Biol Rev*. 2015;79(2):193–224.
269. Ma K-W, Ma W. YopJ family effectors promote bacterial infection through a unique acetyltransferase activity. *Microbiol Mol Biol Rev*. 2016;80(4):1011–1027.
270. Zhang Z-M, Ma K-W, Yuan S, et al. Structure of a pathogen effector reveals the enzymatic mechanism of a novel acetyltransferase family. *Nat Struct Mol Biol*. 2016;23:847.

271. Jones RM, Wu H, Wentworth C, Luo L, Collier-Hyams L, Neish AS. *Salmonella* AvrA coordinates suppression of host immune and apoptotic defenses via JNK pathway blockade. *Cell Host Microbe.* 2008;3(4):233–244.
272. Sun J. Pathogenic bacterial proteins and their anti-inflammatory effects in the eukaryotic host. *Anti-Inflammatory Anti-Allergy Agents Med Chem.* 2009;8(3):214–227.
273. Wu S, Ye Z, Liu X, et al. *Salmonella typhimurium* infection increases p53 acetylation in intestinal epithelial cells. *Am J Physiol Gastrointest Liver Physiol.* 2010;298(5): G784–G794.
274. Reed SM, Quelle DE. p53 acetylation: regulation and consequences. *Cancer.* 2015; 7(1):30–69.
275. Collier-Hyams LS, Zeng H, Sun J, et al. Cutting edge: *Salmonella* AvrA effector inhibits the key proinflammatory, anti-apoptotic NF-κB pathway. *J Immunol.* 2002;169(6): 2846–2850.
276. Ye Z, Petrof EO, Boone D, Claud EC, Sun J. *Salmonella* effector AvrA regulation of colonic epithelial cell inflammation by deubiquitination. *Am J Pathol.* 2007; 171(3):882–892.
277. Liu X, Wu S, Xia Y, et al. Wingless homolog Wnt11 suppresses bacterial invasion and inflammation in intestinal epithelial cells. *Am J Physiol Gastrointest Liver Physiol.* 2011;301(6):G992–G1003.
278. Liu X, Lu R, Wu S, et al. Wnt2 inhibits enteric bacterial-induced inflammation in intestinal epithelial cells. *Inflamm Bowel Dis.* 2012;18(3):418–429.
279. Wang J, Lu R, Fu X, et al. Novel regulatory roles of Wnt1 in infection-associated colorectal cancer. *Neoplasia.* 2018;20(5):499–509.
280. Liu X, Lu R, Wu S, Sun J. *Salmonella* regulation of intestinal stem cells through the Wnt/β-catenin pathway. *FEBS Lett.* 2010;584(5):911–916.
281. Lu R, Liu X, Wu S, et al. Consistent activation of the β-catenin pathway by *Salmonella* type-three secretion effector protein AvrA in chronically infected intestine. *Am J Physiol Gastrointest Liver Physiol.* 2012;303(10):G1113–G1125.
282. Lu R, Wu S, Zhang YG, et al. Enteric bacterial protein AvrA promotes colonic tumorigenesis and activates colonic beta-catenin signaling pathway. *Oncogenesis.* 2014;3:e105.
283. Lu R, Wu S, Zhang Y-G, et al. *Salmonella* protein AvrA activates the STAT3 signaling pathway in colon cancer. *Neoplasia.* 2016;18(5):307–316.
284. Lu R, Bosland M, Xia Y, Zhang Y-G, Kato I, Sun J. Presence of *Salmonella* AvrA in colorectal tumor and its precursor lesions in mouse intestine and human specimens. *Oncotarget.* 2017;8(33):55104–55115.
285. Hanahan D, Weinberg RA. Hallmarks of cancer: the next generation. *Cell.* 2011; 144(5):646–674.
286. Siegl C, Rudel T. Modulation of p53 during bacterial infections. *Nat Rev Microbiol.* 2015;13:741.
287. Silva-García O, Valdez-Alarcón JJ, Baizabal-Aguirre VM. Wnt/β-catenin signaling as a molecular target by pathogenic bacteria. *Front Immunol.* 2019;10:2135.
288. Rogan MR, Patterson LL, Wang JY, JW MB. Bacterial manipulation of Wnt signaling: a host-pathogen Tug-of-Wnt. *Front Immunol.* 2019;10:2390.
289. Escoll P, Buchrieser C. Metabolic reprogramming of host cells upon bacterial infection: why shift to a Warburg-like metabolism? *FEBS J.* 2018;285(12):2146–2160.
290. Ramond E, Jamet A, Coureuil M, Charbit A. Pivotal role of mitochondria in macrophage response to bacterial pathogens. *Front Immunol.* 2019;10:2461.
291. Canonico B, Campana R, Luchetti F, et al. *Campylobacter jejuni* cell lysates differently target mitochondria and lysosomes on HeLa cells. *Apoptosis.* 2014;19(8): 1225–1242.

292. Liyanage NPM, Manthey KC, Dassanayake RP, Kuszynski CA, Oakley GG, Duhamel GE. *Helicobacter hepaticus* cytolethal distending toxin causes cell death in intestinal epithelial cells via mitochondrial apoptotic pathway. *Helicobacter.* 2010;15(2): 98–107.
293. Bielaszewska M, Rüter C, Kunsmann L, et al. Enterohemorrhagic *Escherichia coli* hemolysin employs outer membrane vesicles to target mitochondria and cause endothelial and epithelial apoptosis. *PLoS Pathog.* 2013;9(12):e1003797.
294. Tan X, Xu A, Zhao T, et al. Simulated microgravity inhibits cell focal adhesions leading to reduced melanoma cell proliferation and metastasis via FAK/RhoA-regulated mTORC1 and AMPK pathways. *Sci Rep.* 2018;8(1):3769.
295. Travaglione S, Loizzo S, Rizza T, et al. Enhancement of mitochondrial ATP production by the *Escherichia coli* cytotoxic necrotizing factor 1. *FEBS J.* 2014;281(15): 3473–3488.
296. Fiorentini C, Matarrese P, Straface E, et al. Toxin-induced activation of Rho GTP-binding protein increases Bcl-2 expression and influences mitochondrial homeostasis. *Exp Cell Res.* 1998;242(1):341–350.
297. Lin Z, Zhang Y-G, Xia Y, Xu X, Jiao X, Sun J. *Salmonella enteritidis* effector AvrA stabilizes intestinal tight junctions via the JNK pathway. *J Biol Chem.* 2016;291(52): 26837–26849.
298. Liao AP, Petrof EO, Kuppireddi S, et al. *Salmonella* type III effector AvrA stabilizes cell tight junctions to inhibit inflammation in intestinal epithelial cells. *PLoS One.* 2008;3(6):e2369.
299. Merchant JL. Inflammation, atrophy, gastric cancer: connecting the molecular dots. *Gastroenterology.* 2005;129(3):1079–1082.
300. Gambhir S, Vyas D, Hollis M, Aekka A, Vyas A. Nuclear factor kappa B role in inflammation associated gastrointestinal malignancies. *World J Gastroenterol.* 2015;21(11): 3174–3183.
301. Gagnaire A, Nadel B, Raoult D, Neefjes J, Gorvel J-P. Collateral damage: insights into bacterial mechanisms that predispose host cells to cancer. *Nat Rev Microbiol.* 2017;15:109.
302. Rosadi F, Fiorentini C, Fabbri A. Bacterial protein toxins in human cancers. *Pathog Dis.* 2016;74(1):ftv105.

CHAPTER SIX

Mucosa microbiome of gastric lesions: Fungi and bacteria interactions

Zikai Wang, Rongrong Ren, Yunsheng Yang*
Department of Gastroenterology and Hepatology, Chinese PLA General Hospital, Beijing, China
*Corresponding author: e-mail address: sunny301ddc@126.com

Contents

Abstract

Many components of the gastric non-*Helicobacter pylori* microbiota have been identified recently thanks to advances in DNA sequencing techniques. Several lines of evidence support the hypothesis that the gastric microbiome is essential for gastric disorders such as gastric cancer. Microbial interactions impact the pathophysiology of various gastric disorders. This chapter provides an overview of recent findings regarding general gastric microbial community profiling, microbial interactions in the stomach, and microbial characteristics in various gastric disorders.

Progress in Molecular Biology and Translational Science, Volume 171
ISSN 1877-1173
https://doi.org/10.1016/bs.pmbts.2020.03.004

1. Introduction

Humans are colonized by symbiotic microorganisms that constitute an integrated and functional ecosystem that influences health as well as disease. Microbes associated with the human body include bacteria, archaea, fungi, and viruses. Humans are a type of super-organism composed of the human body itself and symbiotic microorganisms. Researching human diseases only from the perspective of the human body reveals only a partial view of a condition; the role of commensal microbiota in human beings must be considered as well. Microorganisms and their metabolites play important roles in human energy metabolism, nutrient absorption, immune function and other important physiological activities. In fact, a variety of human diseases are induced when commensalism between the host and the microorganisms is disrupted. Early microbiological studies relied on culture-based methods. However, the overall compositional and functional characteristics of microbiome in humans could not be investigated using culture-based methods because 99% of such microorganisms are uncultured.

Currently, metagenomics (combining the power of genomics, bioinformatics, and systems biology) provides new access to the complex microbial world. Recent microbiological studies are firmly supported technically by high-throughput sequencing techniques and bioinformatics analyses that provide fascinating insights into the microbiota present in the human gastrointestinal (GI) tract.[1] Nevertheless, the overall microbiome in stomach has rarely been examined,[2] and the characteristics of the gastric microbiome associated with respect to various gastric disorders has not been evaluated deeply. Furthermore, the human GI tract contains a variable fungal microbiome; however, the characteristics of the gastric fungal microbiome, and its specific role under gastric lesions has not yet been studied extensively. Microbial interactions are common, complex and species-dependent, and have dramatic effects on human beings[3]; however, the bacterial and fungal interactions in the stomach under various gastric disorders, and their potential pathogenic mechanisms, remain unclear. This chapter will review recent evidence for a microbiota concept in the stomach and provide an overview of recent findings regarding the general characteristics of gastric bacterial and fungal microbiome, microbial interactions in stomach, and the microbial characteristics under various gastric disorders.

2. General characteristics of gastric microbiome

2.1 Bacterial microbiome in the stomach

The stomach is a special area in the human GI microecosystem. The unique ecological environment and characteristic microbial community are characterized by gastric acid secretion, mucus thickness and peristalsis. Because the stomach connects the esophagus and oral cavity on the proximal side and the duodenum on the distal side, the microflora of the other parts of the human GI tract can enter the stomach. It was once believed that gastric acid could kill the bacteria entering the stomach and the gastric environment with the low pH value was not suitable for bacterial colonization; nevertheless, culture-based studies demonstrated that large numbers of acid-resistant bacterial strains exist in the stomach and are derived from the transient flora in the mouth and food, including *Streptococcus*, *Neisseria* and *Lactobacillus*, with content generally $<10^3$ colony-forming units (CFU)/mL.[4] Importantly, *Helicobacter pylori* was isolated from the stomach in 1984, thereby initiating a new era of digestive diseases research.

With the development of molecular biology and bacterial 16S rDNA identification techniques, the stomach flora was explored using new molecular biological methods. Early studies found gastric mucosal-associated microbes such as *Enterococcus*, *Pseudomonas*, *Staphylococcus* and *Stomatococcus*.[5] At present, the identification of gastric microflora has increased dramatically with the development of metagenomics and high-throughput sequencing technology. Many studies began to investigate the characteristics of gastric microbiome in individuals with various gastric disorders, mainly focused on the bacterial microbiome. A study using 16S rDNA sequencing analysis identified 128 kinds of phylotypes belonging to eight classes, and obtained 1056 non-*H. pylori* clones from the stomach of patients with gastric diseases.[6] Another study using a 16S rDNA high-throughput sequencing analysis of the gastric mucosa-associated microbiota in *H. pylori*-negative patients with gastritis obtained a total of 1223 non-*H. pylori* clones that could be classified into 133 kinds of phylotypes belonging to eight bacterial classes.[7] Although these two studies were conducted in populations within different regions and ethnic groups, the compositions of stomach microflora were very similar among those populations. These two studies identified approximately 130 phylotypes belonging to 7–8 classes, and 77.4% and 79.8% of clone fragments separately for each study were homogeneous. The two species with highest abundance, *Streptococcus* and *Prevotella*, were the same in these studies.

These are the earliest studies of the gastric microbiome based on high-throughput sequencing methods. Another study conducted pyrosequencing analyses of gastric mucosa-associated flora in 6 healthy subjects and obtained 262 phylotypes belonging to 13 classes, including some that had not been confirmed by other studies, including *Chlamydia* and *Cyanobacteria.*[8] Generally, the human stomach holds a core microbiome. The gastric microbiota composition is highly variable between individuals; however, recent studies detected five major phyla in stomach, including Firmicutes, Bacteroidites, Actinobacteria, Fusobacteria and Proteobacteria; the predominant genera in the stomach include *Prevotella*, *Streptococcus*, *Veillonella*, *Rothia* and *Haemophilus*, *etc.*[9–14] The characteristic stomach microbiota associated with various gastric disorders will be introduced in the followed section.

2.2 Fungal microbiome in the stomach

It is now evident that the commensal fungal microbiota is an important part of the human GI ecosystem. Early studies based on culture-dependent methods reported that fungal microflora could be detected in the GI tract of 70% of healthy adults.[15] Most of these are aerobes or facultative anaerobes, and that the number of fungal microbes in the human stomach is 0–10^2 CFU/mL.[15] Next-generation sequencing of the fungal 18S rRNA gene or internal transcribed spacer sequencing has allowed better characterization of the GI mycobiome.[3] However, only a few studies have been conducted on the fungal microbiota in the human GI tract using culture-dependent methods or high-throughput DNA sequencing analysis.

Candida albicans can grow well in highly acidic environments,[16] and some genotypes may increase the severity of gastric mucosal lesions.[17] *Candida*-associated gastric ulcer (GU) is characterized by the ulcer's delayed healing process and even perforation. As with *H. pylori*, *Candida* spp. may play an etiologic role in GU, as they can be detected in *H. pylori*-negative patients. Other views suggested that fungal microbes alone cannot play a pathogenic role as they colonize only the gastric epithelium, and the presence of fungal microbes in stomach is a secondary phenomenon of GU. *Candida* is the most frequently identified species, and the gastric mycosis can be caused by *Basidiobolus* spp., *Mucor* spp., *Cryptococcus* spp., and others. Furthermore, the fungal infection and colonization in stomach is common in patients with GC. *H. pylori* infection is the definitive risk factor for GC, so interactions between *Candida* sp. and *H. pylori* may play an etiologic role in the gastric carcinogenesis. One study of gastric juice from immunocompromised patients only found 19–81 fungal operational taxonomic units, *Candida* and *Phialemonium* were the only two genera found in all gastric fluid samples.[18] Generally, there is little knowledge about the distribution of the fungal microbiota and its function in stomach.[19] Further studies are needed to explain the diversity and abundance of the gastric fungal microbiome, and its potential pathogenic role in different gastric disorders.

2.3 Factors influencing the gastric microbiome

The gastric microbiota is affected by factors such as host genotype, age, race, dietary habits, beverages, tobacco use, lifestyle, geographic location, ethnicity, drugs treatment and medical interventions; physiological factors such as pH, anatomy, peristalsis, oxygen, nutrition, bile acids, the mucus layer and the mucosal inflammation also influence the gastric microbiome.[20–22] This section introduces several important influence factors of gastric microbiota.

When gastric microbial communities of different anatomical gastric positions are compared, some studies showed no significant difference in the alpha diversity of microbiota between the samples of gastric antrum and body mucosa, further indicating substantial similarity between these two anatomical sites in health and disease status.[14,23] Several studies showed that the bacterial community was similar between the gastric lesion surface and the non-lesion mucosa; however, other studies showed taxonomical and functional composition differences between non-malignant and tumor tissues.[14] Some studies suggested that the gastric mucosal-related microbiome of cardia and non-cardia GC were similar.[24] The GI microbiota distribution is spatially specific from the mucosa layer to the gastric juice. Compared with gastric mucosa-related microbiota, the gastric juice-associated microbiota is relatively unstable and is easily affected by diet and other factors.[18] One study thoroughly washed gastric mucosal biopsy samples, but found no change in the constitution of gastric mucosa-associated flora. In this manner, the gastric flora proved to be closely associated with gastric mucosa and could not be readily washed away.[7] This finding suggests that many non-*H. pylori* microbes could be resident flora in stomach, and not just transient flora from the oral cavity.[25]

Stomach surgery such as gastrectomy could influence the gastric microbiota. The diversity of stomach microbiota becomes more abundant after subtotal gastrectomy. *Ralstonia* and *Helicobacter* were the top two genera of discriminant abundance in the stomach of GC patients before surgery, while *Streptococcus* and *Prevotella* were the two most abundant genera after tumor excision. N-nitrosation genes were prevalent before surgery, whereas bile salt hydrolase, NO and N_2O reductase were prevalent subsequently.[26] In addition, bariatric surgeries such as Roux-en-Y gastric bypass and adjustable gastric banding were used in severe obesity to improve the metabolism and inflammation. Bariatric surgeries can improve the decreased gut microbial gene richness and restore obesity-associated intestinal dysbiosis; however, its influence on the gastric microbiome remains unclear.

Proton-pump inhibitors (PPIs) are widely used to treat acid-related gastric disorders and are recommended for *H. pylori* eradication. It is hypothesized that PPIs may affect the microbiota directly by targeting P-type ATPase enzymes of naturally occurring bacteria like *H. pylori*. PPIs may increase the number of gastric non-*H. pylori* microflora indirectly by suppressing gastric acidity. Several studies revealed very similar bacterial numbers in the gastric fluid microbiota between PPI users and PPI non-users. However, PPI administration induced a small but significant increase in the intersubject diversity of gastric microbiota. Another study showed that PPIs did not significantly influence gastric mucosal-associated microbial composition; however, the relative abundance of Streptococcaceae and *Streptococcus* increased significantly in PPI-treated patients with dyspepsia.[27] The influence of PPIs on the gastric microbiome requires further investigation.

Antibiotics influence the intestinal microbiome. However, the effect of antibiotics on gastric microbiome are unclear. Several animal studies show that antibiotics can reduce *Lactobacilli* spp. and promote fungal colonization of the gastric epithelium. Based on the human studies, treatment with cefoperazone induces the alteration of gastric microbiota, with the overgrowth of *Enterococci* and a significant reduction of *Lactobacilli*. Treatment for *H. pylori* infection include antibiotics, which may increase drug resistant strains in the GI tract and disturb the GI microbiome. The potential role of probiotics in gastric disorders and dysbiosis remains uncertain. The majority of studies focus on the probiotic effectiveness for the treatment of *H. pylori* infection. The Maastricht V Consensus stated that some probiotics may be effective for *H. pylori* eradication. Probiotics such as *Lactobacillus*, *Bifidobacterium* and *Saccharomyces boulardii* may raise the *H. pylori* eradication rate, and decrease the relapse rate, while reducing the GI side effects of antibiotic therapy such as nausea, vomiting, diarrhea, abdominal pain and taste disturbance.[28] In the future, studies should be performed to define the influence of probiotics on the gastric microbiome.

3. Microbial interactions in the stomach

3.1 *Helicobacter pylori* and other components of the gastric microbiome

H. pylori colonizes the stomachs of approximately 50% of individuals worldwide. It is the major pathogenic factor of chronic gastritis and peptic ulcer and is one of the risk factors for GC and mucosa-associated lymphoid tissue lymphoma. The interactions between *H. pylori* and non-*H. pylori* microbiota in the stomach indeed exist and may play an important role in gastric diseases. Stomach colonization was negative even after several weeks of oral implantation of *H. pylori* in specific pathogen-free animal models; however, it was usually positive in germ-free animal models, suggesting that other stomach flora affected the intragastric colonization of *H. pylori* and its activity in gastric disorders.[12] Studies based on animal models have found that some probiotics including *Lactobacillus*, *Bifidobacterium* and *Saccharomyces* exhibit antagonistic effects, and can prevent the adhesion, colonization and growth of *H. pylori* in the stomach. Some *Lactobacillus* species prevented gastric colonization of *H. pylori* in Mongolian gerbils, while *Eubacterium limosum* promoted the colonization of *H. pylori*.[29] Long-term intragastric colonization of *H. pylori* affected the distribution and number of flora in the stomach.[30] For example, the propagation of *Lactobacilli* was inhibited, while *Enterococcus*, *Staphylococcus*, *Bifidobacterium* and *Bacteroides* were rarely affected. Some *Lactobacillus* species were dominant strains in the stomach of Mongolian gerbils that were not affected by *H. pylori* infection; *L. gasseri* and *L. reuteri* existed in the stomach and suppressed the colonization and growth of *H. pylori*.[31] Indeed, the *in vivo* colonization of *H. pylori* was closely related with other stomach flora; however, most of these studies were based on animal models and were limited by their methods.[1,32] Human studies also confirmed that *H. pylori* interacted with other gastric flora. *H. pylori* infection is usually accompanied by colonization of non-*H. pylori* microbiota. Non-*H. pylori* microbiota can persist in the stomach as antigenic stimulators that enhance the immune response caused by *H. pylori* infection and that their co-infection could promote the development of atrophic gastritis. Culture analysis of gastric mucosa-associated microbiota showed that some *Lactobacillus* species competed with *H. pylori* for colonization resources. Whether non-*H. pylori* microbiota in stomach can inhibit or promote the intragastric colonization of *H. pylori* is completely understood.[1,32]

The influence of *H. pylori* on the gastric microbiota in various gastric disorders has been investigated using sequencing methods. The colonization of *H. pylori* in stomach could reduce gastric bacterial diversity, and eradication of *H. pylori* could restore the relative abundance of other bacterial microbes to levels similar to individuals without *H. pylori*.[9] Certainly, the contradictory findings suggested no effect of *H. pylori* on taxonomic diversity and richness of gastric microbiome; however, *H. pylori* could reduce the number of microbial interactions significantly.[6,14,33] Furthermore, one study showed that infection with *H pylori* strains possessing cytotoxin-associated gene A (CagA) could decrease the gastric bacterial diversity with reduced abundance of *Roseburia* and increased abundances of *Helicobacter* and *Haemophilus* genera; however, another study showed the presence of CagA gene had no significant influence on accompanying gastric microbial taxa.[34,35] The detection of *H. pylori* sequences in stomach using high-throughput sequencing methods does not necessarily indicate active infection. For example, some studies showed that the relative abundance of *H. pylori* of >1% was a positive infection of *H. pylori*, whereas <1% was considered as negative.[24] Attempts have been made to characterize the gastric microbiome with or without *H. pylori* infection; however, no consistent microbial characteristics have been obtained.

3.2 Bacterial and fungal interactions in the stomach

To date, major of studies about bacterial and fungal interactions have focused on interactions between *C. albicans* and several pathogenic bacteria in pulmonary and oral diseases or in various catheter- and ventilator-related infections.[36] Pathogenic bacteria in hospitals such as *Pseudomonas aeruginosa*, *Acinetobacter baumannii*, *Staphylococcus epidermidis*, and others, commonly co-inhabit with *C. albicans*.[36] Polymicrobial infections with *C. albicans* and these pathogens cause a range of infectious syndromes. Mixed species biofilms develop through co-aggregation between *C. albicans* and several pathogenic bacteria on the surface of exogenous medical devices and some susceptible sites of the human body; these are thought to be important for the development of infectious disorders.[37]

Only few data on the microbial interactions in the stomach are currently available. This section introduces the potential bacterial-fungal interactions (*e.g.*, fungal *C. albicans* and bacterial *H. pylori*) that are possibly associated with the gastric microecosystem. The characteristics of the potential bacterial-fungal interactions in the stomach were shown in Fig. 1. Several

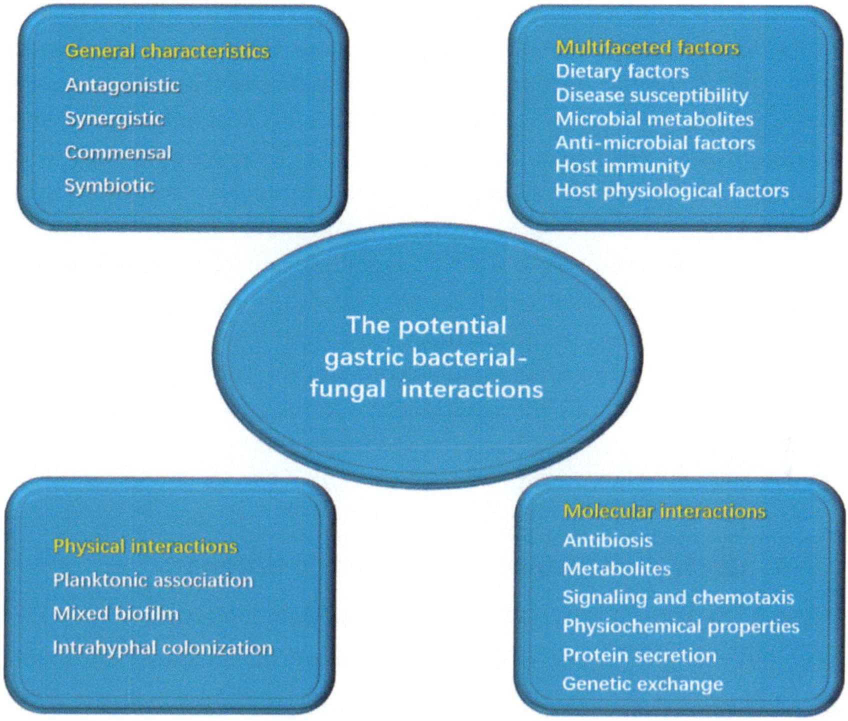

Fig. 1 Profiling of bacterial-fungal interactions in the stomach.

studies found the intracellular occurrence of *H. pylori* in *Candida* spp., and confirmed that *H. pylori* in the vacuole of *Candida* spp. could be viable and may express *H. pylori*-specific proteins such as VacA, urease and peroxiredoxin.[38,39] Accordingly, the vacuole of *Candida* may function as reservoirs and vehicles of *H. pylori*, possibly offering essential nutrients for *H. pylori* survival and multiplication, and protecting *H. pylori* against hostile gastric conditions. These interactions between *H. pylori* and *Candida* may play crucial roles in their gastric colonization. *C. albicans* and *Streptococcus* spp. can form a mixed biofilm with antimicrobial antagonism through the polysaccharides on the cell surface of *Streptococcus*.[40,41] In a simulated *in vivo* model, a synergistic effect between *C. albicans* and *Streptococci* increased the invasive ability of *C. albicans* and promoted biofilm formation by *Streptococcus* spp. in the oral and upper GI mucosa.[40] *Lactobacillus* spp. inhibited the growth and virulence of *C. albicans* by the production of hydrogen peroxide and organic acids, but could not eradicate *C. albicans*.[42,43] Various *Lactobacillus* strains have different abilities for inhibiting the growth of *C. albicans* in the GI tract.[44] The effects of *Lactobacillus* spp. on the host immune response may be involved in the modulation of fungal virulence.[45,46] In addition to these strain-specific interactions, interactions between fungal and bacterial microbiota as a whole have also been confirmed. For example, the germ-free mouse model exhibited an increased susceptibility to intestinal colonization of *C. albicans*.[47] Disturbances of the bacterial community in the GI tract promoted *C. albicans* colonization,[48,49] suggesting that the normal bacterial microbiota of the GI tract has an inhibitory effect against fungal colonization and invasion. It should be noted that conclusions drawn from *in vivo* and *in vitro* studies, and from mammalian and non-mammalian models, may not be entirely consistent, and could be due to the participation of the host immune system. Overall, the bacterial and fungal interactions in stomach are complex and undefined; more interactions remain to be investigated.

4. Gastric microbiome and gastric lesions

4.1 Stomach microbiome and gastric cancer

Gastric cancer (GC) is one of the most common malignancies worldwide. The pathogenesis of GC is a multistep process affected by multiple factors that remain largely unclear. It is generally believed that environmental, dietary, *H. pylori* infection and host genetic factors participate in the pathogenesis of GC. *H. pylori* has been recognized as a high microbial risk factor of initializing carcinogenesis; however, only 1–3% of individuals with *H. pylori* infection develop GC.[50] Studies have shown that *H. pylori* eradication does not completely prevent persistent inflammation of the gastric mucosa and GC development, suggesting that the pathogenesis of GC is not entirely due to *H. pylori*.[51,52] It has been shown that gastritis caused by *H. pylori* infection and the long-term use of PPI appears to inhibit gastric acid secretion, allowing successful colonization of new microbes, possibly contributing to malignant transformation. This hypothesis is further supported by animal studies showing that transgenic hypergastrinemic insulin-gastrin mice were more susceptible to *H. pylori*-induced GC than were germ-free mice.[53]

The overall knowledge regarding the role of gastric non-*H. pylori* microbes in the development of GC remains limited. Whether other gastric flora facilitate or inhibit the effect of *H. pylori* in GC development remains unknown; only a few studies have examined this issue. Recently, many studies investigated the characteristics of gastric microbiome of gastric tumorigenesis by using the high-throughput sequencing technology, and provided evidence that the dysbiosis of microbiota in stomach are associated with gastric tumorigenesis.[10,14,24,26,54–56] The microbial diversity and richness of gastric microbiome gradually reduces across histological stages of gastric tumorigenesis from gastritis, intestinal metaplasia, intraepithelial neoplasia to intestinal-type GC. One study showed that *Parvimonas micra, Dialister pneumosintes, Slackia exigua, Peptostreptococcus stomatis, Prevotella intermedia, Fusobacterium nucleatum, Prevotella oris* and *Catonella morbi* were significantly enriched in stomach of GC compared with those of other precancerous stages, and specifically, there was increased presence of oral cavity bacterial species such as *P. stomatis* and *D. pneumosintes*.[24] Several studies showed consistent increases in abundance of lactic acid bacteria in GC patients including *Streptococcus, Lactobacillus, Bifidobacterium* and *Lactococcus*. These lactic acid bacteria supply exogenous lactate as a fuel source for cancer cells that promote inflammation, angiogenesis, metastasis, epithelial-mesenchymal transition and immune evasion, and produce reactive oxygen species and N-nitroso compounds.[57] Nevertheless, no consistent component of the gastric microbiome associated with different histological stages of gastric tumorigenesis has been identified.

Recent studies used bioinformatics analysis methods to predict the metagenome functional contents based on the microbial community profile obtained from the 16S rRNA gene sequences. For example, several pathways such as N-nitrosation genes, purine metabolism and peptidoglycan biosynthesis were enriched in GC patients.[24,26,55] Studies of stomach microbiome mainly focused on gastric mucosa samples that are more influenced by the contamination with human host DNA. The real meaning of the metagenome functional prediction based on 16S rDNA sequencing is debatable, and the detailed functions of microbiome in gastric tumorigenesis needs to be further investigated. Several studies tried to build the microbial predictive model for identifying different histological stages of GC. For example, one study showed that oral microbes such as *P. stomatis, S. anginosus, P. micra, S. exigua* and *D. pneumosintes* could be used as biomarkers for GC.[24] Another study investigated microbial biomarkers of common tongue coatings in GC patients[25] and showed that bacterial *Capnocytophaga leadbetteri* and fungal *Ampelomyces* spp. were potential biomarkers of white-thin coating; fungal *Blastobotrys adeninivorans* was a potential biomarker of yellow-thick coating. However, the accuracy of predictive model as a potential microbial biomarker for GC requires further evaluation.

Fungal infection and colonization in the stomach is common in patients with GC. Interactions between *Candida* spp. and *H. pylori* may play an etiological role in gastric carcinogenesis. The carcinogenicity of fungi is mainly due to the long-term intake of exogenous mycotoxins in contaminated food. Mycotoxins are small-molecule secondary metabolites of fungi that possess diverse structures and can be classified into over 400 species, of which about 30 are pathogenic to human beings.[58] The toxic effects of mycotoxins associated with the digestive system include carcinogenicity. The detection rate of aflatoxins in food is high in those regions of China in which gastric and esophageal carcinomas are most common.[59] Sterigmatocystin is a late metabolite in the aflatoxin pathway, and is also carcinogenic and mutagenic. *In vitro* and animal studies confirmed that sterigmatocystin promoted metaplasia and may participate in the progression of GC in the presence of *H. pylori.*[60,61] Trichothecenes, represented by deoxynivalenol and nivalenol, are produced by *Fusarium* spp. The detection rate of deoxynivalenol and nivalenol in contaminated food in the Linxian and Cixian areas of China with a high incidence of GC is much higher, and feeding with contaminated grains in these areas induced benign and malignant tumors in mice.[62] Epidemiological, *in vivo*, and *in vitro* studies should be conducted to clarify the carcinogenicity of the mycotoxins.

4.2 Stomach microbiome and gastric ulcer

The pathogenic mechanisms of gastric ulcer (GU) involve *H. pylori* infection, the excessive gastric acid secretion and the use of non-steroidal anti-inflammatory drugs. Studies found alterations of the gastric microbiota among GU patients, and showed that some non-*H. pylori* microbes such as *Streptococci* were significantly associated with GU; however, whether other non-*H. pylori* microbiota participate in the occurrence of GU remains unclear. Recently, several studies based on the 16S rRNA sequencing method showed that Mycobacterium, Bacteroides, and Firmicutes were predominant in GU, and that *Prevotella*, *Neisseria*, and *Streptococcus* were identified in *H. pylori*-positive mucosal tissues.[6,63] *Candida*-associated GU is associated with delayed healing of ulcers and even with perforation. As with *H. pylori*, *Candida* spp. may play an etiologic role in GU, because they are detected in *H. pylori*-negative patients with GU.[64] Other investigators suggest that fungi alone cannot play a pathogenic role because they colonize only gastric epithelium,[65] and the presence of fungi in the stomach is a secondary phenomenon of GU.[66] One study showed that 66.7% of gastric disorder patients with *Candida* in the gastric mucosa were co-colonized with *H. pylori*.[65] Although causation *versus* secondary colonization was never examined in these studies, the co-existence of *C. albicans* with *H. pylori* may indicate synergism in pathogenesis of GU, and hypha formation by *Candida* could contribute to the mechanism of ulcer perforation.[65] Overall, the characteristics of gastric microbiome in GU patients have not been investigated deeply.

4.3 Stomach microbiome and the other gastric disorders

For chronic gastritis, increased abundance Firmicutes and decreased abundance Proteobacteria were observed in *H. pylori*-positive patients with antral gastritis compared with their abundances in *H. pylori*-negative subjects.[7] The reduction of gastric acid barrier in atrophic gastritis allows other microbial colonization in stomach. Generally, the microbial diversity and species richness vary among atrophic gastritis, healthy subjects and GC patients; a shift from *Prevotella* to *Streptococcus* in the most relevant bacterial genera was observed in atrophic gastritis.[24] One study showed more substantial alterations of gastric microbiota in patients with atrophic gastritis; however, there were relatively few alterations in patients receiving PPI therapy. Autoimmune atrophic gastritis resulted in a different, more diverse gastric microbial pattern than that of the stomach of patients with *H. pylori*-induced atrophic gastritis, which may be due to differences in acid secretion between these conditions or factors such as different immune profiles.[67]

Gastric polyps are focal elevated lesions within the gastric epithelium mucosa. The current limited systematic studies of gastric polyps focus mainly on the relationship between gastric polyps and cancer as well as the role of *H. pylori* in the pathogenesis. The possible pathogeneses related to the development of gastric polyps include hereditary factors, bile reflux, *H. pylori* infection, and others, while none have been directly proved. As such, the etiology and biological characteristics of gastric polyps and their long-term effects in humans remain unclear. Studies of gastric polyps are far less detailed than those of colonic polyps. Recently, a study showed that the stomach microbiota diversity in patients with gastric polyps was significantly lower than that of the healthy controls; and patients with fundic gland polyps and those with hyperplastic polyps had considerably similar gastric microbiota compositions.[23] These findings suggest that stomach dysbiosis may be closely related to the occurrence and development of gastric polyps.

Functional dyspepsia is a very prevalent disorder, characterized by postprandial distress and/or epigastric pain that is unexplained after routine clinical evaluation. To date, the definite pathophysiology of functional dyspepsia remains to be clarified, although it may involve gastric motility disturbances and visceral hypersensitivity. The role of gastric microbiome in the pathophysiology of functional dyspepsia should be considered. The analysis of gastric juice-related microbiota among functional dyspepsia patients showed different bacterial composition even at the phylum level compared with healthy subjects, with the characteristics of a Bacteroidetes > Proteobacteria abundance and the absence of Acidobacteria.[68] In the future, the role of gastric microbiome in functional GI disorders should be further investigated.

4.4 Microbial therapy for gastric disorders

Gastric dysbiosis in several gastric disorders has been confirmed using high-throughput sequencing techniques, and there is a compelling rationale for modulating the disturbed stomach microbiota among various gastric diseases. Currently, microbial therapies such as probiotics and fecal microbiota transplantation have been used in human diseases associated with intestinal dysbiosis. For example, fecal microbiota transplantation has been recommended as an alternative to standard therapy with antibiotics for *Clostridium difficile* infection. Nevertheless, the efficacy of microbial therapy in gastric disorders remains uncertain. At present, the majority of studies focus on the efficacy of probiotics supplementation in the treatment of *H. pylori* infection. The Maastricht V Consensus recommended that certain probiotics may have a beneficial effect on *H. pylori* eradication, and may be effective in reducing GI side effects caused by *H. pylori* eradication therapies.[69] Several probiotics such as *Lactobacillus*, *Bifidobacterium* and *Saccharomyces boulardii* maybe increase the *H. pylori* eradication rate, decrease the relapse rate, and reduce the GI side effects of antibiotic therapy such as nausea, vomiting, diarrhea, abdominal pain and taste disturbance.[69] More studies should be performed to evaluate the true role of probiotics on the treatment of *H. pylori* infection, and to define the influence of probiotics on the *H. pylori* and the other gastric microbiota. Other potential microbial therapy for gastric diseases should be further investigated; for example, the microbiota transplantation may be performed on the animal models for some gastric diseases.

5. Conclusion

The human stomach indeed harbors its own microbiota. The maintenance of gastric microbial homeostasis is essential for stomach health, and the bacterial and fungal microbiome has been implicated in various gastric disorders. Further studies are needed to explore: (1) the definite characteristics of bacterial and fungal microbiome in stomach under health conditions and various gastric disorders; (2) the potential pathogenic mechanisms of disturbed gastric microbiome in gastric diseases based on the well-designed functional analyses; (3) complex microbial interactions such as the relationship between *H. pylori* and other gastric microbiota and the host-microbial interactions in stomach; and (4) the potential diagnostic and therapeutic microbial biomarkers for gastric disorders such as gastric cancer and peptic ulcer. Metagenomics combined with human genome-wide association studies, as well as metabonomics, metaproteomics and culturomics, will contribute to our understanding of the role of gastric microbiome in the stomach.

Acknowledgments

This work was supported the National High Technology Research and Development Program of China (863 Program, ID: 2015AA02701).

Conflicts of interest

The authors declare no conflict of interest.

References

1. Wang ZK, Yang YS. Upper gastrointestinal microbiota and digestive diseases. *World J Gastroenterol: WJG*. 2013;19(10):1541–1550.
2. Human Microbiome Project C. Structure, function and diversity of the healthy human microbiome. *Nature*. 2012;486(7402):207–214.
3. Peleg AY, Hogan DA, Mylonakis E. Medically important bacterial-fungal interactions. *Nat Rev Microbiol*. 2010;8(5):340–349.
4. Bacteria in the stomach. *Lancet*. 1981;2(8252):906–907.
5. Monstein HJ, Tiveljung A, Kraft CH, Borch K, Jonasson J. Profiling of bacterial flora in gastric biopsies from patients with Helicobacter pylori-associated gastritis and histologically normal control individuals by temperature gradient gel electrophoresis and 16S rDNA sequence analysis. *J Med Microbiol*. 2000;49(9):817–822.
6. Bik EM, Eckburg PB, Gill SR, et al. Molecular analysis of the bacterial microbiota in the human stomach. *Proc Natl Acad Sci U S A*. 2006;103(3):732–737.
7. Li XX, Wong GL, To KF, et al. Bacterial microbiota profiling in gastritis without Helicobacter pylori infection or non-steroidal anti-inflammatory drug use. *PLoS One*. 2009;4(11): e7985.

8. Stearns JC, Lynch MD, Senadheera DB, et al. Bacterial biogeography of the human digestive tract. *Sci Rep*. 2011;1:170.
9. Li TH, Qin Y, Sham PC, Lau KS, Chu KM, Leung WK. Alterations in gastric microbiota after *H. Pylori* eradication and in different histological stages of gastric carcinogenesis. *Sci Rep*. 2017;7:44935.
10. Ramaswamy K, Correa M, Koshy A. Non-healing gastric ulcer associated with Candida infection. *Indian J Med Microbiol*. 2007;25(1):57–58.
11. Soeta N, Terashima M, Gotoh M, et al. An improved rapid quantitative detection and identification method for a wide range of fungi. *J Med Microbiol*. 2009;58(pt 8): 1037–1044.
12. Rickes S, Schultze U, Monkemuller K, Malfertheiner P. Walter Krienitz—his life and intuitive description of bacteria in the stomach. *Dtsch Med Wochenschr*. 2006;131(23):1341–1343.
13. Dias-Jacome E, Libanio D, Borges-Canha M, Galaghar A, Pimentel-Nunes P. Gastric microbiota and carcinogenesis: the role of non-Helicobacter pylori bacteria—a systematic review. *Rev Esp Enferm Dig*. 2016;108(9):530–540.
14. Stringer AM, Gibson RJ, Bowen JM, Keefe DM. Chemotherapy-induced modifications to gastrointestinal microflora: evidence and implications of change. *Curr Drug Metab*. 2009;10(1):79–83.
15. Schulze J, Sonnenborn U. Yeasts in the gut: from commensals to infectious agents. *Dtsch Arztebl Int*. 2009;106(51–52):837–842.
16. Zwolinska-Wcislo M, Budak A, Bogdal J, Trojanowska D, Stachura J. Fungal colonization of gastric mucosa and its clinical relevance. *Med Sci Monit*. 2001;7(5):982–988.
17. Gong YB, Zheng JL, Jin B, et al. Particular *Candida albicans* strains in the digestive tract of dyspeptic patients, identified by multilocus sequence typing. *PLoS One*. 2012;7(4): e35311.
18. von Rosenvinge EC, Song Y, White JR, Maddox C, Blanchard T, Fricke WF. Immune status, antibiotic medication and pH are associated with changes in the stomach fluid microbiota. *ISME J*. 2013;7(7):1354–1366.
19. Playford EG, Webster AC, Sorrell TC, Craig JC. Systematic review and meta-analysis of antifungal agents for preventing fungal infections in liver transplant recipients. *Eur J Clin Microbiol Infect Dis*. 2006;25(9):549–561.
20. Olszak T, An D, Zeissig S, et al. Microbial exposure during early life has persistent effects on natural killer T cell function. *Science*. 2012;336(6080):489–493.
21. Wu GD, Chen J, Hoffmann C, et al. Linking long-term dietary patterns with gut microbial enterotypes. *Science*. 2011;334(6052):105–108.
22. Flint HJ, Scott KP, Louis P, Duncan SH. The role of the gut microbiota in nutrition and health. *Nat Rev Gastroenterol Hepatol*. 2012;9(10):577–589.
23. Ren R, Wang Z, Sun H, et al. The gastric mucosal-associated microbiome in patients with gastric polyposis. *Sci Rep*. 2018;8(1):13817.
24. Coker OO, Dai Z, Nie Y, et al. Mucosal microbiome dysbiosis in gastric carcinogenesis. *Gut*. 2018;67(6):1024–1032.
25. Xu J, Xiang C, Zhang C, et al. Microbial biomarkers of common tongue coatings in patients with gastric cancer. *Microb Pathog*. 2019;127:97–105.
26. Clyne M, May FEB. The interaction of Helicobacter pylori with TFF1 and its role in mediating the tropism of the bacteria within the stomach. *Int J Mol Sci*. 2019;20(18):4400.
27. Paroni Sterbini F, Palladini A, Masucci L, et al. Effects of proton pump inhibitors on the gastric mucosa-associated microbiota in dyspeptic patients. *Appl Environ Microbiol*. 2016;82(22):6633–6644.

28. Bruno G, Rocco G, Zaccari P, Porowska B, Mascellino MT, Severi C. Helicobacter pylori infection and gastric dysbiosis: can probiotics administration be useful to treat this condition? *Can J Infect Dis Med Microbiol*. 2018;2018: 6237239.
29. Zaman C, Osaki T, Hanawa T, Yonezawa H, Kurata S, Kamiya S. Analysis of the microflora in the stomach of Mongolian gerbils infected with Helicobacter pylori. *J Gastroenterol Hepatol*. 2010;25(suppl 1):S11–S14.
30. Yin YN, Wang CL, Liu XW, et al. Gastric and duodenum microflora analysis after long-term Helicobacter pylori infection in Mongolian Gerbils. *Helicobacter*. 2011;16(5): 389–397.
31. Sun YQ, Monstein HJ, Nilsson LE, Petersson F, Borch K. Profiling and identification of eubacteria in the stomach of Mongolian gerbils with and without Helicobacter pylori infection. *Helicobacter*. 2003;8(2):149–157.
32. Alarcon T, Llorca L, Perez-Perez G. Impact of the microbiota and gastric disease development by Helicobacter pylori. *Curr Top Microbiol Immunol*. 2017;400:253–275.
33. Yang I, Woltemate S, Piazuelo MB, et al. Different gastric microbiota compositions in two human populations with high and low gastric cancer risk in Colombia. *Sci Rep*. 2016;6:18594.
34. Zhao Y, Gao X, Guo J, et al. Helicobacter pylori infection alters gastric and tongue coating microbial communities. *Helicobacter*. 2019;24(2): e12567.
35. Klymiuk I, Bilgilier C, Stadlmann A, et al. The human gastric microbiome is predicated upon infection with Helicobacter pylori. *Front Microbiol*. 2017;8:2508.
36. Shirtliff ME, Peters BM, Jabra-Rizk MA. Cross-kingdom interactions: *Candida albicans* and bacteria. *FEMS Microbiol Lett*. 2009;299(1):1–8.
37. Harriott MM, Noverr MC. Importance of Candida-bacterial polymicrobial biofilms in disease. *Trends Microbiol*. 2011;19(11):557–563.
38. Saniee P, Siavoshi F, Nikbakht Broujeni G, Khormali M, Sarrafnejad A, Malekzadeh R. Immunodetection of Helicobacter pylori-specific proteins in oral and gastric Candida yeasts. *Arch Iran Med*. 2013;16(11):624–630.
39. Siavoshi F, Salmanian AH, Akbari F, Malekzadeh R, Massarrat S. Detection of Helicobacter pylori-specific genes in the oral yeast. *Helicobacter*. 2005;10(4):318–322.
40. Diaz PI, Xie Z, Sobue T, et al. Synergistic interaction between *Candida albicans* and commensal oral streptococci in a novel in vitro mucosal model. *Infect Immun*. 2012;80(2):620–632.
41. Holmes AR, McNab R, Jenkinson HF. Candida albicans binding to the oral bacterium Streptococcus gordonii involves multiple adhesin-receptor interactions. *Infect Immun*. 1996;64(11):4680–4685.
42. Fitzsimmons N, Berry DR. Inhibition of *Candida albicans* by Lactobacillus acidophilus: evidence for the involvement of a peroxidase system. *Microbios*. 1994;80(323):125–133.
43. Noverr MC, Huffnagle GB. Regulation of *Candida albicans* morphogenesis by fatty acid metabolites. *Infect Immun*. 2004;72(11):6206–6210.
44. Wagner RD, Pierson C, Warner T, et al. Biotherapeutic effects of probiotic bacteria on candidiasis in immunodeficient mice. *Infect Immun*. 1997;65(10):4165–4172.
45. Martinez RC, Seney SL, Summers KL, Nomizo A, De Martinis EC, Reid G. Effect of *Lactobacillus rhamnosus* GR-1 and *Lactobacillus reuteri* RC-14 on the ability of *Candida albicans* to infect cells and induce inflammation. *Microbiol Immunol*. 2009;53(9):487–495.
46. Spear GT, Zariffard MR, Cohen MH, Sha BE. Vaginal IL-8 levels are positively associated with *Candida albicans* and inversely with lactobacilli in HIV-infected women. *J Reprod Immunol*. 2008;78(1):76–79.
47. Hummel RP, Oestreicher EJ, Maley MP, Macmillan BG. Inhibition of *Candida albicans* by *Escherichia coli* in vitro and in the germfree mouse. *J Surg Res*. 1973;15(1):53–58.

48. Rosenbach A, Dignard D, Pierce JV, Whiteway M, Kumamoto CA. Adaptations of *Candida albicans* for growth in the mammalian intestinal tract. *Eukaryot Cell.* 2010;9(7):1075–1086.
49. White SJ, Rosenbach A, Lephart P, et al. Self-regulation of *Candida albicans* population size during GI colonization. *PLoS Pathog.* 2007;3(12): e184.
50. Wroblewski LE, Peek Jr RM, Wilson KT. Helicobacter pylori and gastric cancer: factors that modulate disease risk. *Clin Microbiol Rev.* 2010;23(4):713–739.
51. Fukase K, Kato M, Kikuchi S, et al. Effect of eradication of Helicobacter pylori on incidence of metachronous gastric carcinoma after endoscopic resection of early gastric cancer: an open-label, randomised controlled trial. *Lancet.* 2008;372(9636):392–397.
52. Wong BC, Lam SK, Wong WM, et al. Helicobacter pylori eradication to prevent gastric cancer in a high-risk region of China: a randomized controlled trial. *JAMA.* 2004; 5291(2):187–194.
53. Lertpiriyapong K, Whary MT, Muthupalani S, et al. Gastric colonisation with a restricted commensal microbiota replicates the promotion of neoplastic lesions by diverse intestinal microbiota in the Helicobacter pylori INS-GAS mouse model of gastric carcinogenesis. *Gut.* 2014;63(1):54–63.
54. van de Veerdonk FL, Kullberg BJ, van der Meer JW, Gow NA, Netea MG. Host-microbe interactions: innate pattern recognition of fungal pathogens. *Curr Opin Microbiol.* 2008;11(4):305–312.
55. Ferreira RM, Pereira-Marques J, Pinto-Ribeiro I, et al. Gastric microbial community profiling reveals a dysbiotic cancer-associated microbiota. *Gut.* 2018;67(2):226–236.
56. Eun CS, Kim BK, Han DS, et al. Differences in gastric mucosal microbiota profiling in patients with chronic gastritis, intestinal metaplasia, and gastric cancer using pyrosequencing methods. *Helicobacter.* 2014;19(6):407–416.
57. Vinasco K, Mitchell HM, Kaakoush NO, Castano-Rodriguez N. Microbial carcinogenesis: lactic acid bacteria in gastric cancer. *Biochim Biophys Acta Rev Cancer.* 2019;1872(2):188309.
58. Bennett JW, Klich M. Mycotoxins. *Clin Microbiol Rev.* 2003;16(3):497–516.
59. Li Z, Cui J, Zhang X, Kang W. Aflatoxin G1 reduces the molecular expression of HLA-I, TAP-1 and LMP-2 of adult esophageal epithelial cells in vitro. *Toxicol Lett.* 2010;195(2–3):169–173.
60. Kusunoki M, Misumi J, Shimada T, et al. Long-term administration of the fungus toxin, sterigmatocystin, induces intestinal metaplasia and increases the proliferative activity of PCNA, p53, and MDM2 in the gastric mucosa of aged Mongolian gerbils. *Environ Health Prev Med.* 2011;16(4):224–231.
61. Misumi J. The mechanisms of gastric cancer development produced by the combination of Helicobacter pylori with Sterigmatocystin, a mycotoxin. *Nihon Rinsho.* 2004;62(7):1377–1386.
62. Hsia CC, Wu ZY, Li YS, Zhang F, Sun ZT. Nivalenol, a main Fusarium toxin in dietary foods from high-risk areas of cancer of esophagus and gastric cardia in China, induced benign and malignant tumors in mice. *Oncol Rep.* 2004;12(2):449–456.
63. Chen X, Xia C, Li Q, Jin L, Zheng L, Wu Z. Comparisons between bacterial communities in mucosa in patients with gastric antrum ulcer and a duodenal ulcer. *Front Cell Infect Microbiol.* 2018;8:126.
64. Sasaki K. Candida-associated gastric ulcer relapsing in a different position with a different appearance. *World J Gastroenterol: WJG.* 2012;18(32):4450–4453.
65. Karczewska E, Wojtas I, Sito E, et al. Assessment of co-existence of Helicobacter pylori and Candida fungi in diseases of the upper gastrointestinal tract. *J Physiol Pharmacol.* 2009;60(suppl 6):33–39.

66. Wu CS, Wu SS, Chen PC. A prospective study of fungal infection of gastric ulcers: clinical significance and correlation with medical treatment. *Gastrointest Endosc.* 1995;42(1):56–58.
67. Sokol H, Conway KL, Zhang M, et al. Card9 mediates intestinal epithelial cell restitution, T-helper 17 responses, and control of bacterial infection in mice. *Gastroenterology.* 2013;145(3):591–601.e3.
68. Standaert-Vitse A, Sendid B, Joossens M, et al. *Candida albicans* colonization and ASCA in familial Crohn's disease. *Am J Gastroenterol.* 2009;104(7):1745–1753.
69. Malfertheiner P, Megraud F, O'Morain CA, et al. Management of Helicobacter pylori infection-the Maastricht V/Florence Consensus Report. *Gut.* 2017;66(1):6–30.

CHAPTER SEVEN

Microbiota derived factors as drivers of type 1 diabetes

Bree J. Tillett, Emma E. Hamilton-Williams*
The University of Queensland Diamantina Institute, The University of Queensland, Woolloongabba, QLD, Australia
*Corresponding author: e-mail address: e.hamiltonwilliams@uq.edu.au

Contents

Abstract

Type 1 diabetes (T1D) is an autoimmune disease caused by complex interactions between host genetics and environmental factors, culminating in the T-cell mediated destruction of the insulin producing cells in the pancreas. The rapid increase in disease frequency over the past 50 years or more has been too rapid to attribute to genetics. Dysbiosis of the gut microbiota is currently being widely investigated as a major contributor to environmental change driving increased T1D onset. In this chapter, we discuss the major changes in gut microbiota composition and function linked to T1D risk as well as the potential origin of these changes including infant diet, antibiotic use and host genetics. We examine the interaction between inflammation and gut barrier function and the dysbiotic gut microbiota that have been linked to T1D.

Progress in Molecular Biology and Translational Science, Volume 171
ISSN 1877-1173
https://doi.org/10.1016/bs.pmbts.2020.04.008

1. Introduction

The establishment of a stable, healthy gut microbiota is fundamental to the development of an effective and tolerant immune system. Increasingly, in modern times this symbiosis is disturbed causing disease. An imbalance in the gut's bacteria—termed dysbiosis, is associated with a range of conditions, for example, metabolic diseases such as type 2 diabetes and obesity, cardio-vascular diseases such as atherosclerosis, neurological diseases such as Huntington's disease as well as autoimmune diseases such as type 1 diabetes (T1D). Like many of these disorders, T1D is not classically described as a gut-associated condition; however, there is growing evidence supporting a link between the gut microbiota and susceptibility, initiation and progression to T1D.

T1D is a chronic autoimmune disease characterized by the activation of islet-specific autoreactive T cells which initiate an immune response against the insulin-producing pancreatic islet β cells. Activated immune cells infiltrate the pancreatic islets causing β cell apoptosis. As inflammation progresses, insulin regulation is impaired leading to hyperglycemia and ultimately the cessation of endogenous insulin production. Increased genetic risk, most strongly from the human leukocyte antigen (HLA) complex but also from 100 or more non-HLA loci, predisposes to loss of immune tolerance and is a prerequisite for disease development.[1] The first sign of islet-specific autoimmunity is usually the production of islet specific autoantibodies (IAb) detectable in serum (seropositivity). Common autoantibodies found in T1D include antibodies targeting insulin, glutamic acid decarboxylase (GAD), protein tyrosine phosphatase (IA2) and zinc transporter 8.[2–5] Disease risk increases substantially as multiple autoantibodies develop during disease progression. Positivity for a single islet autoantibody in children with HLA-conferred genetic risk for T1D increased disease progression from 0.4% to 14.5% compared to those with no autoantibodies over a 10-year follow-up. This increased substantially to 69.7% in those who developed multiple autoantibodies.[6] Though no glycemic symptoms are observed at this point, the presence of two or more islet autoantibodies in now defined as Stage 1 of T1D.[7] Stage 2 is defined as when these individuals develop abnormal glucose tolerance but do not require insulin and Stage 3 is typical clinical diagnosis with clinical symptoms such as polyurea and weight loss.[7]

The time to progression through these the stages of T1D as well as the age of disease onset varies widely. Furthermore, since World War 2 the incidence of T1D has been steadily increasing.[8–10] Despite traditional classification as a genetic disorder, the steady increase in incidence and variable rates of disease progression point to the increasing importance of environmental factors influencing disease onset. In this scenario, exposure to external factors associated with modern lifestyles in at-risk individuals results in exposure to triggers that accelerate progression or remove protective factors. A wide range of environmental factors have been linked to risk of T1D progression including infant diet, probiotic use, viral infection, maternal obesity, maternal vitamin D levels, infant weight gain, Caesarean delivery, psychological stress and an altered gut microbiota (reviewed in Ref. 11). This wide range of environmental determinants, some of which are likely not yet known, indicate a complex multifactorial disease with multiple timepoints and disease stages that may be influenced by external factors.

2. Functional dysbiosis of the gut microbiota in type 1 diabetes

An intensive research effort over the past decade has investigated the hypothesis that individuals with T1D have an altered gut microbiota to healthy individuals and that these changes may be traced prior to disease onset. The presence of an altered gut microbiota before disease onset or even prior to seroconversion would lend support to the premise that dysbiosis acts as a "trigger" of disease, rather than a consequence of ongoing autoimmunity or metabolic disturbances. This hypothesis was first investigated in rodent models of T1D via direct manipulation of the gut microbiota via raising animals in germ-free conditions,[12–14] ablation and remodeling of the bacterial communities using antibiotics,[15–20] colonization with new flora[14,21–25] or manipulation of the microbiota with various diets.[26–30] The vast majority of these studies concluded that alterations in the gut microbiota led changes in disease susceptibility often accompanied by altered immune activation and intestinal immunity. Some of these manipulations were performed by introducing a new microbiota into the mother resulting in disease protection transmitted to her offspring. This demonstrated that very early colonization with altered microflora can have a profound impact on disease. Manipulations that protect from disease have been linked with reductions in inflammatory conditions in the gut and pancreatic draining lymph node.[13,15,17,22,26,29,30] In human studies, a wide variation in the specific taxa that are associated with T1D or autoimmunity has been observed. Some commonalities that have been observed include a reduction in bacterial diversity linked to disease progression.[31–33] Several studies have observed an increase in the relative abundance of Bacteroides and a decrease in short-chain fatty acid (SCFA) producing species, particularly butyrate.[32,34–38] However, not all studies have seen reduced diversity and have found either differences in other taxa or have not seen significant differences at all.[39–42] These differences can in part be explained by differences in the geographic location of the subjects,[43] sample handling and sequencing methods, low sample size and selection of the control group. However, it appears that the "signal" for dysbiosis linked to islet autoimmunity at least on the taxonomic level is relatively modest potentially because of the inter-individual variability in the gut microbiota is too great to overcome or the microbiota changes really are only subtle. An alternative explanation is that changes in the functional interactions between the bacterial community and the host are linked to the development of islet autoimmunity and that a wide variety of possible community profiles can lead to this functional dysbiosis. Understanding these interactions requires both a deeper understanding of the genetic potential of the communities present via deep metagenomic sequencing as well as other approaches to study the relationship between intestinal inflammation and the microbiota.

3. Reduced short-chain fatty acid production is associated with disease

Production of SCFA including acetate, butyrate and propionate via fermentation of dietary fiber is a major function of a healthy gut microbiota. SCFA provide an important energy source for intestinal enterocytes and help promote intestinal barrier function.[44] It is believed that reduced fiber intake associated with Western diet is linked to a decrease in SCFA producing species in the gut linked to a number of diseases.[45] SCFA stimulate mucin production and integrity of tight junctions between intestinal epithelial cells reducing undesired interactions between luminal antigens and the host immune system. Lastly, there is evidence to suggest that SCFA's can act as nucleotide-binding oligomerization domain-like receptors which are key modulators of inflammation and regulation of T-cells. Current theory suggests that loss of SCFA producing bacteria and subsequent gradual breakdown of immune tolerance and increased inflammation contributes to progression of autoimmunity and inflammatory disease.

In the largest T1D microbiome study to date, Vatanen et al. analyzed the stool microbiomes from The Environmental Determinants of Diabetes in the Young (TEDDY) study. They included over 750 subjects across six clinical centers and four countries collected longitudinally from 3 months of age until clinical endpoint which included subjects that developed islet autoimmunity ($n = 267$), clinical disease ($n = 101$) and matched controls ($n = 415$).[38] They observed that the T1D subjects had an increase in genes from the fermentation pathway (q=0.019) as did islet autoantibody+ subjects from Finland (q=0.049). The fermentation pathways that were reduced included pathways involved in butyrate and acetate production (cohort-wide T1D subjects, q=0.043 and q=0.14, respectively), as well as propionate production (German T1D subjects, q=0.019). While this study did not specifically report on the timing of the decreased capacity for fermentation in the group that progressed to T1D, their data suggests that reduced biosynthesis of SCFA may start in islet autoantibody+ subjects but is more pronounced following diagnosis.

In order to investigate very early functional changes in the gut microbiota, Endesfelder et al. compared the microbiomes at 6 months of age from 19 high-risk children who subsequently developed islet autoimmunity with 21 children that remained seronegative.[35] This study identified a transition in composition of the gut microbiota from a community with high abundance of *Bifidobacterium* which digest human milk oligosaccharides to communities containing specialist bacteria such as *Ruminococcus*, *Blautia*, and *Akkermansia* following introduction of other food groups. A third community present in the cohort was dominated by *Bacteroides* and these children tended toward consuming the most complex diets and this group was associated with increased risk of islet autoimmunity. They identified a reduction in butyrate production via co-fermentation of acetate in this subgroup dominated by *Bacteroides*. Both of the sub-groups associated with protection from islet autoimmunity had higher abundance of the mucin degrading bacteria *Akkermansia,* which is able to fully degrade mucin and also stimulates further mucin production by goblet cells. In agreement with this finding, *Akkermansia* was found to be significantly lower in abundance in TEDDY children that progressed to T1D by 16S rRNA compositional analysis.[42] In contrast, *Bacteroides* is a partial mucin degrader and can switch energy source to fermentation of carbohydrates obtained from the luminal contents resulting in H_2 production. This switch toward *Bacteroides* may then promote sulfate reducing bacteria and methanogens that can out compete acetogens for substrates such as H_2.[46] Additionally, H_2 production can also inhibit production of SCFA via negative feedback.[47] Hence, a complex network of cross-feeding occurs between different taxa which is dependent on diet. Thus, the establishment of a microbial community with robust SCFA producing potential may be linked to infant diet and established in very early life.

4. Infant diet and the establishment of the microbiota in at risk subjects

4.1 Breastfeeding

Dietary factors, particularly infant diet, have long been considered as potential risk factors for T1D. While early on the focus was on food antigens as potential antigenic triggers via molecular mimicry, know research has shifted to investigating dietary involvement in shaping the microbiota function, protection from infection, inflammation and immune development. Recently, several large-scale studies have tackled the issue of infant diet and T1D risk using either population-based birth cohorts or case-control studies of at-risk children with some discrepancies between studies. In 2017, Lund-Blix et al. investigated breastfeeding and T1D risk in two Scandinavian population-based cohorts totaling 155,392 children of whom 504 children developed T1D.[48] They found that children who were never breastfed had a twofold higher risk of developing T1D than those who were breastfed (Hazard ratio 2.29). However, the duration of breastfeeding and exclusivity of breastfeeding did not impact disease risk. In an earlier pooled-analysis of data from 43 previous studies (40 case-control and 3 cohort studies), data from 9874 individuals with T1D and a larger number of controls were assessed. The authors concluded that a weak protective effect of breastfeeding was found. Recently several cohort studies in high-risk children, including the Finnish Type 1 Diabetes Prediction and Prevention Nutrition Study (DIPP[49]), The Diabetes Autoimmunity Study in the Young (DAISY[50]) and TEDDY[51] have all concluded that breast-feeding did not independently impact risk of islet autoimmunity or T1D. Case-control studies like TEDDY have the advantage that far more detailed dietary information is collected and the cases and controls are carefully matched for confounders such as HLA. However, these studies are biased toward children with either a family history of T1D or high-risk HLA as well as other factors influencing subject selection (e.g., distance lived from study centers) that may bias the results. Furthermore, subject behavior may be modified by enrolment in such studies.

The beneficial health effects provided by components of breastmilk including human milk oligosaccharides has long been acknowledged.

Recently human breastmilk oligosaccharides (HMO) were fed to T1D susceptible NOD mice and found to delay disease onset and reduce insulitis.[52] HMO feeding to the mice was associated with an increase in abundance of taxa including *Akkermansia*, *Ruminococcus* and *Lachnospriaceae* as well as an increase in fecal SCFA concentration. Another study found feeding NOD mice human milk in drinking water greatly reduced diabetes onset.[53] Feeding human milk strongly and stably increased *Akkermansia* abundance while *Bifidobacterium* abundance was slightly increased. These data support a mechanistic link between breastfeeding and establishment of mucin degrading bacteria such as *Akkermansia* in the mature microbial community with the ability to cross-feed and support the growth of other SCFA producing taxa.

4.2 Introduction of solid foods

As well as breastfeeding, other infant feeding practices such as the type and timing of introduction of solid foods strongly influence the maturation of the gut microbiota and potentially T1D risk. Of note in these studies is that practices for introducing solid foods vary widely between cultures which likely contributes to the variability in findings. Both early (<4 months) and late introduction (≥6 months) of solid foods was associated with increased risk of islet autoimmunity in the US-based DAISY study.[50] In the German BABYDIET and BABYDIAB studies of high-risk children, early introduction (<3 months) of gluten-containing cereals increased risk of developing islet autoimmunity and T1D, but the numbers of subjects in this early introduction category were low.[54] The Norwegian MIDIA study followed 908 children with high-risk HLA genetics and found no associations with age of introduction of solid foods and disease risk.[55] Recently, Hakola et al. found in a population of 5915 Finnish children with high-risk HLA (DIPP cohort) that early introduction of solid foods (<3 months) increased the risk of developing multiple islet autoantibodies or T1D early (up to 3 years) but not in children that developed autoimmunity or T1D older than 3 years.[49] This divergence in risk patterns in children that develop disease earlier in life may be related to disease heterogeneity or different endotypes in T1D.[56] Children that develop disease earlier have differences in immune phenotype, genetic risk profile and clinical characteristics at onset.[57,58] Introduction of solid food had a significant association with the structure of the microbiota between 7 and 10 months of age, although breastfeeding has a more pronounced and durable association up to

14 months of age.[42] In the TEDDY metagenomic studies, functional changes in the genetic potential of the gut microbiota coincided with major dietary events such as cessation of breastfeeding (reduced L-lactate dehydrogenase) and introduction of solid foods (increased transketolase, linked to digestion of fiber).[38]

5. Antibiotic use and disease risk

Another potential trigger destabilizing SCFA producing species such as *Bifidobacterium* is antibiotic use. Vatanen et al. found in the TEDDY study that *bifidobacteria* were highly susceptible to antibiotics.[38] The abundance of *Bifidobacterium* members *B. bifidum*, *B. pseudocatenulatum*, *B. adolescentis*, *B. dentium* and *B. catenulatum* were significantly reduced following a course of antibiotics. In contrast, the abundance of *B. longum* and *B. breve* did not systematically decline following antibiotics, but rather these taxa were outcompeted by other community members following antibiotics. In animal studies, NOD mice given a pulsed therapeutic antibiotics (PAT) regimen to simulate treatment in children showed faster onset of compared to mice that had been given a continuous exposure to low does antibiotics confirming that regular "attacks" on the microbiota reduce stability and protective capacity.[15] However, in human studies there has been little evidence of an association between antibiotic use in either prenatally in the mother or during childhood with risk of islet autoimmunity or T1D.[59–61]

6. Pro-inflammatory microbial species increase in abundance prior to T1D progression

Several studies have found that a decrease in SCFA producing bacteria persists post seroconversion.[18,33,38] In progressors (individuals who progress from seroconversion to clinical diagnosis), there may be a further decline in the number of bacterial taxa present (alpha diversity) associated with an increase in abundance of pro-inflammatory species after seroconversion.[31,32] Kostic et al. examined the microbiomes from stool of 33 HLA-matched infants from birth until the age of 3.[31] This study found that in children that later progressed to T1D, there was a relative overabundance of *Blautia*, *Rikenelaceae* and the *Ruminococcus* and *Streptococcus* genera post seroconversion. These groups of bacteria contain "pathobionts" which are commensal bacteria which under certain conditions can switch to become pathogenic. In the children that progressed to T1D, there was a spike in relative abundance of *Ruminococcus gnavus* and *Streptococcus infantarius* at the same time as reduction in overall alpha diversity compared with both IAb+ non-progressors and children which remained autoantibody negative ($P < 0.025$). Conversely, the relative abundance of anti-inflammatory species such as *Lachnospiraceae* and *Villonellaceae* decreased with a complete absence of *Coprococcus eutactud* and *Dialliister invisus*. Seroconverters displayed an intermediate abundance of these groups compared with the non-seroconverters and the progressors. While a caveat of this study was the small number of children that progressed to clinical T1D, these data suggest a shift to a more pro-inflammatory microbiota is associated with disease state particularly the period after initiation of islet autoimmunity but before disease onset.

In addition to alterations in the taxa present before disease onset, Kostic et al. also found specific microbial metabolic pathways were associated with disease progression.[31] There was an increase in genes related to multiple sugar transport system and a decrease in genes involved in biosynthesis of a number of amino acids. This depicts a shift in metabolic function away from the synthesis of nutrients to passive transport of their required nutrients, which is common for auxotrophs which thrive in inflammatory environments. Direct measurement of stool metabolites found that the taxa that were overabundant in the T1D progressor group correlated with increased sphingomyelin and decreased lithocholic acid which both regulate intestinal immune function and inflammation. Serum metabolites, particularly triglycerides also correlated positively with *Blautia* and *Ruminococcus* members but negatively with *Veillonella* abundance. This suggests that these taxa may have systemic effects on host metabolism in support of a potential role in modifying disease risk.

7. Increased intestinal permeability and impaired barrier function precede the onset of T1D

The intestinal barrier is made up of the epithelial cell layer and the mucous layer and serves to shield the host from the luminal contents including commensal and pathogenic microbes. As well as providing a physical barrier, various epithelial cells secrete a range of host defense molecules including immunoglobulins and antimicrobial peptides which can directly attack bacteria.[62] Cell junctions including tight junctions, adheren junctions and desmosomes maintain the epithelial barrier while tight junctions regulate paracellular transport and intestinal permeability.[63] The gut microbiota can interact with the intestinal barrier and can strengthen and weaken its integrity via various mechanisms. For example, by degrading or stimulating mucous production, by stimulating enterocyte growth or modulating the tight junctions. In addition, the epithelial layer contains numerous immune cells which continually survey the environment, sampling luminal antigens, regulating inflammation and mounting immune responses to pathogens. Disturbances in intestinal permeability have been linked to inflammation and disease and are likely a major mechanism by which a dysbiotic microbiota can impact health.[44]

A reduction in gut barrier function as measured by uptake of sugars in patients with T1D has long been described. This impairment was found in subjects at high-risk of disease (first-degree relatives with islet autoantibodies) as well as in new-onset subjects and those with long-term disease.[64–67] Recently, longitudinal studies have allowed monitoring both intestinal permeability and microbial composition across time prior to disease diagnosis.[33] Harbison et al. studied children with either recent-onset T1D ($n = 29$), children with islet autoantibodies ($n = 18$) including 12 with multiple antibodies and 6 of these who progressed to T1D during the course of the study along with control children (autoantibody negative siblings and unrelated controls, $n = 41$). They observed that children with recent onset T1D had higher intestinal permeability than both siblings and unrelated controls. Furthermore, those multiple islet autoantibody+ children that progressed to T1D had higher intestinal permeability than those that did not progress. Interestingly, permeability dropped over time arguing against a direct relationship with hyperglycemia which may increase closer to disease onset. This rather suggested that intestinal dysfunction was occurring relatively early relative to diagnosis. Intriguingly, a drop in bacterial diversity occurred in the progressor sub-group closer to the time of disease diagnosis when intestinal permeability had lowered suggesting that intestinal permeability may occur prior to dysbiosis. A limitation of this study was the small numbers of children that progressed to T1D. However, performing such an invasive intestinal permeability test in larger longitudinal cohorts is unlikely to be feasible. Therefore, alternative assays to measure gut barrier function using non-invasive sampling (e.g., stool) would be advantageous to the field.

8. Intestinal inflammation and immune activation in T1D

Current theory proposes that inflammation driven by translocation of bacterial derived antigens or endotoxin across an impaired gut barrier is a key determinant driving progression from seroconversion to clinical onset. In support of this concept, Westerholm-Ormio et al. showed that small intestinal biopsies from pediatric patients with T1D displayed increased immune activation including increased HLA-II expression and IL-1α and IL-4 positive cells, even though they found no evidence of structural abnormality to the intestine.[68] Pellegrini et al. found increased infiltration of the gut mucosa with macrophage/monocyte lineage cells in subjects with T1D compared to those without.[69] This study also identified upregulation of cytokines (IL-1β and TNFα) and chemokine genes indicating a distinct inflammatory profile of the gut mucosa of patients with T1D which was associated with specific gut bacteria. Another study found that the percentage of regulatory T cells (Tregs) was decreased in the duodenum of T1D patients compared with healthy subjects and that dendritic cells isolated from the lamina propria had a reduced capacity to convert CD4+CD25-T cells into Tregs.[70] Thus, while overt immune infiltration and structural damage to the intestine is not seen in T1D, these studies indicate that immune activation is present suggesting an increased immune response to either commensal bacteria or gut derived pathogens is likely occurring.

By studying the host and bacterial proteins present in stool samples using metaproteomics, we investigated whether changes in proteins derived from the intestinal mucosa could be found in subjects with recent-onset T1D or islet autoimmunity that may suggest increased inflammation or altered barrier function.[71] We found the new-onset subjects had significantly increased abundance of galectin-3 and fibrillin-1 human derived proteins in their stool. Galectin-3 is expressed by a variety of immune cells and is commonly upregulated in a wide variety of autoimmune disorders, either in serum or on the target tissues where it is thought to drive inflammation.[72] Fibrillin-1 is an extracellular matrix protein which plays a central role in storage and regulation of the bioavailability of the anti-inflammatory cytokine TGF-beta and other growth factors.[73] Fibrillin-1 microfibrils are chemotactic for inflammatory cells such as macrophages.[74] These data support that inflammation is present within the intestine of recent-onset T1D subjects. In the islet autoantibody+ subjects, we observed reductions in proteins derived from the mucous layer, namely CLCA1 and IgA heavy chain. Furthermore, in a network analysis we found that proteins and taxa with an increased abundance in control subjects positively correlated with mucous layer proteins (e.g., mucin-2 and FCGBP) as well as the known health associated and butyrate producing species such as *Faecalibacterium prausnitzii*.[75] Together, these studies support the notion that the intestinal lamina propria contains increased immune activation which is likely related to altered barrier function. Whether this immune activation precedes an altered microbiota or an altered microbiota triggers intestinal inflammation or both is still under investigation.

9. Host genetics may predispose to intestinal inflammation in T1D

While environmental factors such as diet clearly are major drivers in shaping the composition of the gut microbiota, host-derived factors also play a role.[76,77] Genetic predisposition to T1D is strongly linked to altered immune function including the *HLA* loci (altered T cell repertoire), loci linked to Treg function and immune tolerance (*IL2RA*, *Ptpn22*, *Ptpn2*, *CTLA4*, *IL10* and others) and loci linked to innate immunity and viral response (*IFIH1*, *TYK2*, *DEXI*, *ERBB3*).[78–80] Studies in both mouse and human have linked genetic variants linked to T1D susceptibility to alterations in the gut microbiota. Using the NOD mouse model and genetic linkage in a human cohort, we demonstrated that T1D associated variants linked to regulatory T cell function (including *IL2* and *CTLA4*) modified the composition of the gut microbiota.[81] This effect could be replicated by experimentally altering IL2 availability. Mice expressing T1D linked susceptibility alleles had increased lymphocyte infiltration in the lamina propria, reduced intestinal Tregs and reduced goblet mucin production, suggesting a direct feedback between the T1D linked immune variants and intestinal host defense. IL-10, a regulatory cytokine produced by Tregs in the gut, has been shown to directly stimulate goblet cell mucin production.[82] Thus IL-10 may be the mechanistic link in this process. It is well established that both thymic-and non-thymic derived Tregs contribute to the colonization dynamics of the gut microbiota[83,84] and it therefore follows that genes that profoundly influence Treg function may then modify the community composition.

The strongest genetic association with T1D comes from the HLA region. Russell and co-workers used a Swedish general population study using stool samples from 403 children with matched HLA genotype data.[85] By stratifying the subjects according to T1D HLA risk, they found that the overall beta diversity in the cohort was significantly associated with HLA risk group along with a number of individual taxa. Another study looked at serum antibody responses toward a panel of commensal bacteria.[86] Anti-commensal antibody responses differed between subjects with recent-onset T1D and healthy controls with lower total Ig and IgA responses observed toward *A. colihominis*, *B. animalis*, *B. fragilis*, *B. vulgatus*, and *C. perfringens* in the T1D subjects. In a cohort of islet autoantibody+ subjects, they observed a difference in the anti-commensal antibody responses of children that later progressed to T1D in a strongly HLA dependent fashion. This suggests that intestinal immunity toward commensal bacteria is modified by HLA haplotype and this could at least in part underlie the strong influence of the HLA on T1D risk. This would then support a model whereby genetic susceptibility to T1D sets up an intestinal immune environment predisposed to immune activation from the intestinal microbiota, which then skews the types of bacteria which are able to colonize the gut (Fig. 1). This acts in

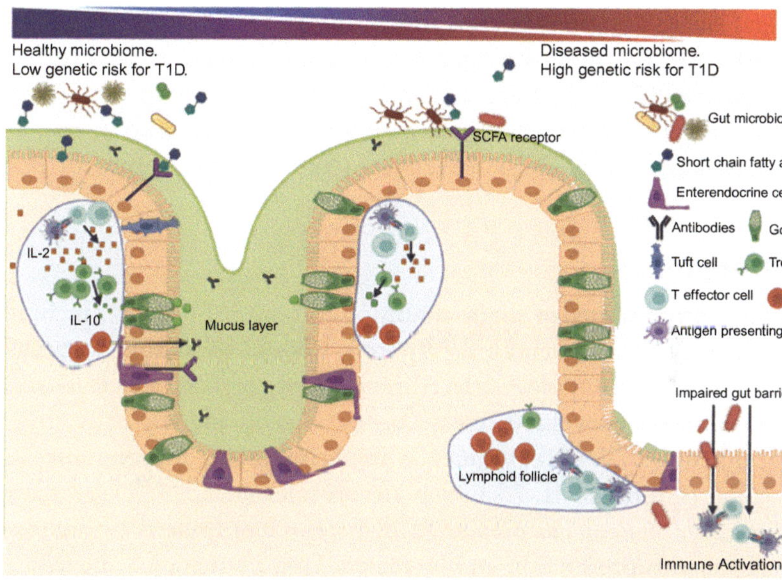

Fig. 1 Proposed interactions between host genetics, immune activation and the microbiome in T1D. Low genetic risk for T1D is linked to increased T cell production of IL-2 and IL-2 receptor signaling, which supports Treg numbers and function in the large intestine including IL-10 production. Increased IL-10 production by Tregs can act on the goblet cells to increase mucin production reducing interactions between the host antigen-presenting cells and bacteria. This tolerogenic environment together with dietary fiber supports a diverse microbiome including short chain fatty acid (SCFA) producing bacteria. High genetic risk for T1D is linked to lower IL-2 signaling, reducing immune tolerance and altering bacterial diversity. Under this scenario the gut barrier is weakened through reduced mucus production leaving the host exposed to inflammatory bacteria. B cell antibody responses to both commensal and non-commensal bacteria differ with high and low-risk HLA alleles.

concert with early colonization of the gut with a dysbiotic community that is then evolved by early infant diet and other environmental exposures that are still being elucidated.

10. Potential of microbiome targeted therapy for T1D

Considerable interest now exists in therapeutic strategies to enhance colonization of beneficial bacteria either in subjects with T1D to slow disease progression or in at-risk individuals with an aim to prevent T1D. The two main approaches being considered for T1D are probiotics and prebiotics. Probiotics are typically lab cultured commensal bacterial strains which are ingested, with the aim of colonization. Typically, probiotic strains are SCFA producers that originated from various fermented foods such as *lactobacillus* or *bifidobacterium* strains from dairy products. Prebiotics on the other hand, are the food for "good" gut bacteria such as resistant starches and digestible fibers. Prebiotics encourage population growth of beneficial bacteria already present, rather than directly introducing new strains. Pro and prebiotics have been effective in reducing symptoms of many diseases.[87]

10.1 Probiotics

In rodents, introduction of single strains of commensal bacteria, including traditional and 'next-generation' probiotic strains, were able to prevent disease onset. For example, early administration of *Clostridium butyricum* and *Lactobacillaceae* prevented disease onset by reducing the activation of pro-inflammatory IL-1β cytokines while stimulating indoleamine 2,3-dioxygenase (IDO) and IL-33 which are protolerogenic.[23,88] Oral gavage of *Akkermansia muciniphila* prevented diabetes accompanied by increased intestinal mucous production and increased Treg frequency.[22] *Lactobacillus johnsonii* and *Lactobacillis casei* have also been shown to have disease protective effects.[24,89]

Human use of probiotics was investigated as an environmental protective factor in the TEDDY study, which found that probiotic supplementation in the first month of life reduced the risk of islet autoimmunity, but only in children with high-risk HLA genotypes.[51] Administration of probiotics during this very early life period may aid in colonization of the probiotic strains, resulting in a more pronounced impact on disease progression. To date, few clinical trials have been completed investigating the potential for probiotics to mitigate disease risk or preserve residual beta cell function. A clinical study commenced in 2017 aims to assess the combined effects of

6-month supplementation with *Lactobacillus rhamnosus* and *Bifidobacterium lactis* on Beta-cell function in newly diagnosed children.[90] A number of registered clinical trials include supplementation with *Lactobacillus johnsonii* or various combined treatments of *Bifidobacterium* and *lactobacillus* either in at-risk or newly diagnosed cohorts. They will investigate parameters such as intestinal permeability, progression to islet autoimmunity or c-peptide levels and other immune parameters. These studies will all take considerable time to complete and many factors such as the timing of the intervention and patient stratification (e.g., HLA type) must be considered.

10.2 Prebiotics

A second approach, which is not restricted by colonization rates, is the use of prebiotics. Currently human trials have used prebiotics aim to encourage growth of SCFA producing bacteria or directly deliver SCFA to the gut. Ho et al. administered oligofructose-enriched inulin for 12 weeks to children with T1D and found higher c-peptide, reduced intestinal permeability and an increase in the relative abundance of commensal Bifidobacterium in the inulin treated group.[91] It is yet to be tested whether this improved gut integrity could be applied to at risk individuals, where it has been shown that the gut permeability is compromised and associated with disease risk, and hence delay disease. Oral supplementation with the short chain fatty acid butyrate (as opposed to a fiber supplement encouraging SCFA producing bacteria, but still a prebiotic), has recently been trialed.[92] In this study 4 g of sodium butyrate or a placebo was taken orally for 1 month with a washout period of 1 month. Despite changes in fecal levels of SCFA associated with the supplement, no changes in T cell phenotypes, glycemic indices or microbiome were observed in this study. This lack of effect may be related to the relatively short treatment period or the route which likely resulted in the SCFA delivery to the small intestine. In in mouse models, SCFA conjugated to resistant starch delivered the prebiotic to the fermentative bacteria of the colon and this was disease protective.[28] Clinical trials using this approach in patients are underway and will assess taxonomic and functional changes in the gut microbiome, stool proteins, immune phenotype, glycemic indices and SCFA levels. Other approaches under investigation include combining pro-and prebiotic's (termed synbiotics).[93] Future studies may also look to combine microbiome targeting therapies together with immunotherapy with a goal of making the intestinal immune niche more hospitable to beneficial bacteria while directly promoting growth of beneficial bacteria and immune tolerance.

11. Conclusion

Numerous studies have now demonstrated that functional changes in the gut microbiota occur before disease onset and these are associated with reduced intestinal barrier function and inflammation. However, the effect sizes observed are generally modest likely due to the large variation in the gut microbiota composition between individuals and between geographic regions with numerous confounders. Establishing causality of a link between the gut microbiota and T1D onset in humans still remains to be achieved. This will require a placebo-controlled intervention to formally establish that a link between long-term modulation of the gut microbiota, autoimmune phenotype and clinical features of T1D indeed exists. Further studies are required utilizing longitudinal cohorts and interventional studies to fully understand the role of the gut microbiota in T1D as well as its therapeutic potential. Ongoing questions in the field remain such as the role of the mother's microbiome, intestinal virome, and mycobiota as well as the interactions with the host.

Acknowledgments

E.E.H.-W. is funded by grants from the Juvenile Diabetes Research Foundation (3-SRA-2019-730-S-B), Diabetes Australia (Y18M1-HAME) and The Children's Hospital Foundation (50316). This work was carried out at the Translational Research Institute, Woolloongabba, QLD 4102, Australia. The Translational Research Institute is supported by a grant from the Australian Government.

References

1. Pociot F, Lernmark A. Genetic risk factors for type 1 diabetes. *Lancet*. 2016; 387(10035):2331–2339.
2. Palmer JP, Asplin CM, Clemons P, et al. Insulin antibodies in insulin-dependent diabetics before insulin treatment. *Science*. 1983;222(4630):1337–1339.
3. Baekkeskov S, Aanstoot HJ, Christgau S, et al. Identification of the 64K autoantigen in insulin-dependent diabetes as the GABA-synthesizing enzyme glutamic acid decarboxylase. *Nature*. 1990;347(6289):151–156.
4. Payton MA, Hawkes CJ, Christie MR. Relationship of the 37,000- and 40,000-M(r) tryptic fragments of islet antigens in insulin-dependent diabetes to the protein tyrosine phosphatase-like molecule IA-2 (ICA512). *J Clin Invest*. 1995;96(3):1506–1511.
5. Yu L, Boulware DC, Beam CA, et al. Zinc transporter-8 autoantibodies improve prediction of type 1 diabetes in relatives positive for the standard biochemical autoantibodies. *Diabetes Care*. 2012;35(6):1213–1218.
6. Ziegler AG, Rewers M, Simell O, et al. Seroconversion to multiple islet autoantibodies and risk of progression to diabetes in children. *JAMA*. 2013;309(23):2473–2479.

7. Insel RA, Dunne JL, Atkinson MA, et al. Staging presymptomatic type 1 diabetes: a scientific statement of JDRF, the Endocrine Society, and the American Diabetes Association. *Diabetes Care.* 2015;38(10):1964–1974.
8. Patterson CC, Dahlquist GG, Gyurus E, Green A, Soltesz G, EURODIAB Study Group. Incidence trends for childhood type 1 diabetes in Europe during 1989-2003 and predicted new cases 2005-20: a multicentre prospective registration study. *Lancet.* 2009;373(9680):2027–2033.
9. EURODIAB ACE Study Group. Variation and trends in incidence of childhood diabetes in Europe. *Lancet.* 2000;355(9207):873–876.
10. Haynes A, Bulsara MK, Jones TW, Davis EA. Incidence of childhood onset type 1 diabetes in Western Australia from 1985 to 2016: evidence for a plateau. *Pediatr Diabetes.* 2018;19:690–692.
11. Craig ME, Kim KW, Isaacs SR, et al. Early-life factors contributing to type 1 diabetes. *Diabetologia.* 2019;62(10):1823–1834.
12. Wen L, Ley RE, Volchkov PY, et al. Innate immunity and intestinal microbiota in the development of Type 1 diabetes. *Nature.* 2008;455(7216):1109–1113.
13. Alam C, Bittoun E, Bhagwat D, et al. Effects of a germ-free environment on gut immune regulation and diabetes progression in non-obese diabetic (NOD) mice. *Diabetologia.* 2011;54(6):1398–1406.
14. King C, Sarvetnick N. The incidence of type-1 diabetes in NOD mice is modulated by restricted flora not germ-free conditions. *PLoS One.* 2011;6(2):e17049.
15. Livanos AE, Greiner TU, Vangay P, et al. Antibiotic-mediated gut microbiome perturbation accelerates development of type 1 diabetes in mice. *Nat Microbiol.* 2016;1(11):16140.
16. Brown K, Godovannyi A, Ma C, et al. Prolonged antibiotic treatment induces a diabetogenic intestinal microbiome that accelerates diabetes in NOD mice. *ISME J.* 2016;10(2):321–332.
17. Candon S, Perez-Arroyo A, Marquet C, et al. Antibiotics in early life alter the gut microbiome and increase disease incidence in a spontaneous mouse model of autoimmune insulin-dependent diabetes. *PLoS One.* 2015;10(5):e0125448.
18. Hu Y, Peng J, Tai N, et al. Maternal antibiotic treatment protects offspring from diabetes development in nonobese diabetic mice by generation of tolerogenic APCs. *J Immunol.* 2015;195(9):4176–4184.
19. Tormo-Badia N, Hakansson A, Vasudevan K, Molin G, Ahrne S, Cilio CM. Antibiotic treatment of pregnant non-obese diabetic mice leads to altered gut microbiota and intestinal immunological changes in the offspring. *Scand J Immunol.* 2014;80(4): 250–260.
20. Zhang XS, Li J, Krautkramer KA, et al. Antibiotic-induced acceleration of type 1 diabetes alters maturation of innate intestinal immunity. *Elife.* 2018;7:e37816.
21. Markle JG, Frank DN, Mortin-Toth S, et al. Sex differences in the gut microbiome drive hormone-dependent regulation of autoimmunity. *Science.* 2013;339(6123):1084–1088.
22. Hanninen A, Toivonen R, Poysti S, et al. Akkermansia muciniphila induces gut microbiota remodelling and controls islet autoimmunity in NOD mice. *Gut.* 2018;67(8):1445–1453.
23. Jia L, Shan K, Pan LL, et al. Clostridium butyricum CGMCC0313.1 protects against autoimmune diabetes by modulating intestinal immune homeostasis and inducing pancreatic regulatory T cells. *Front Immunol.* 2017;8:1345.
24. Matsuzaki T, Nagata Y, Kado S, et al. Prevention of onset in an insulin-dependent diabetes mellitus model, NOD mice, by oral feeding of Lactobacillus casei. *APMIS.* 1997;105(8):643–649.
25. Valladares R, Sankar D, Li N, et al. Lactobacillus johnsonii N6.2 mitigates the development of type 1 diabetes in BB-DP rats. *PLoS One.* 2010;5(5):e10507.

26. Alam C, Valkonen S, Palagani V, Jalava J, Eerola E, Hanninen A. Inflammatory tendencies and overproduction of IL-17 in the colon of young NOD mice are counteracted with diet change. *Diabetes*. 2010;59(9):2237–2246.
27. Chen K, Chen H, Faas MM, et al. Specific inulin-type fructan fibers protect against autoimmune diabetes by modulating gut immunity, barrier function, and microbiota homeostasis. *Mol Nutr Food Res*. 2017;61(8):1601006.
28. Marino E, Richards JL, McLeod KH, et al. Gut microbial metabolites limit the frequency of autoimmune T cells and protect against type 1 diabetes. *Nat Immunol*. 2017;18(5):552–562.
29. Hansen CH, Krych L, Buschard K, et al. A maternal gluten-free diet reduces inflammation and diabetes incidence in the offspring of NOD mice. *Diabetes*. 2014;63(8): 2821–2832.
30. Emani R, Asghar MN, Toivonen R, et al. Casein hydrolysate diet controls intestinal T cell activation, free radical production and microbial colonisation in NOD mice. *Diabetologia*. 2013;56(8):1781–1791.
31. Kostic AD, Gevers D, Siljander H, et al. The dynamics of the human infant gut microbiome in development and in progression toward type 1 diabetes. *Cell Host Microbe*. 2015;17(2):260–273.
32. Giongo A, Gano KA, Crabb DB, et al. Toward defining the autoimmune microbiome for type 1 diabetes. *ISME J*. 2011;5(1):82–91.
33. Harbison JE, Roth-Schulze AJ, Giles LC, et al. Gut microbiome dysbiosis and increased intestinal permeability in children with islet autoimmunity and type 1 diabetes: a prospective cohort study. *Pediatr Diabetes*. 2019;20(5):574–583.
34. Mejia-Leon ME, Petrosino JF, Ajami NJ, Dominguez-Bello MG, de la Barca AM. Fecal microbiota imbalance in Mexican children with type 1 diabetes. *Sci Rep*. 2014;4:3814.
35. Endesfelder D, Engel M, Davis-Richardson AG, et al. Towards a functional hypothesis relating anti-islet cell autoimmunity to the dietary impact on microbial communities and butyrate production. *Microbiome*. 2016;4:17.
36. de Goffau MC, Luopajarvi K, Knip M, et al. Fecal microbiota composition differs between children with beta-cell autoimmunity and those without. *Diabetes*. 2013; 62(4):1238–1244.
37. de Groot PF, Belzer C, Aydin O, et al. Distinct fecal and oral microbiota composition in human type 1 diabetes, an observational study. *PLoS One*. 2017;12(12):e0188475.
38. Vatanen T, Franzosa EA, Schwager R, et al. The human gut microbiome in early-onset type 1 diabetes from the TEDDY study. *Nature*. 2018;562(7728):589–594.
39. Endesfelder D, Castell WZ, Ardissone A, et al. Compromised gut microbiota networks in children with anti-islet cell autoimmunity. *Diabetes*. 2014;63:2006–2014.
40. Stewart CJ, Nelson A, Campbell MD, et al. Gut microbiota of Type 1 diabetes patients with good glycaemic control and high physical fitness is similar to people without diabetes: an observational study. *Diabet Med*. 2017;34(1):127–134.
41. Alkanani AK, Hara N, Gottlieb PA, et al. Alterations in intestinal microbiota correlate with susceptibility to type 1 diabetes. *Diabetes*. 2015;64(10):3510–3520.
42. Stewart CJ, Ajami NJ, O'Brien JL, et al. Temporal development of the gut microbiome in early childhood from the TEDDY study. *Nature*. 2018;562(7728):583–588.
43. Kemppainen KM, Ardissone AN, Davis-Richardson AG, et al. Early childhood gut microbiomes show strong geographic differences among subjects at high risk for type 1 diabetes. *Diabetes Care*. 2015;38(2):329–332.
44. Hiippala K, Jouhten H, Ronkainen A, et al. The potential of Gut commensals in reinforcing intestinal barrier function and alleviating inflammation. *Nutrients*. 2018;10(8):E988.
45. Makki K, Deehan EC, Walter J, Backhed F. The impact of dietary fiber on gut microbiota in host health and disease. *Cell Host Microbe*. 2018;23(6):705–715.

46. Dar SA, Kleerebezem R, Stams AJ, Kuenen JG, Muyzer G. Competition and coexistence of sulfate-reducing bacteria, acetogens and methanogens in a lab-scale anaerobic bioreactor as affected by changing substrate to sulfate ratio. *Appl Microbiol Biotechnol.* 2008;78(6):1045–1055.
47. Chassard C, Lacroix C. Carbohydrates and the human gut microbiota. *Curr Opin Clin Nutr Metab Care.* 2013;16(4):453–460.
48. Lund-Blix NA, Dydensborg Sander S, Stordal K, et al. Infant feeding and risk of type 1 diabetes in two large Scandinavian birth cohorts. *Diabetes Care.* 2017;40(7): 920–927.
49. Hakola L, Takkinen HM, Niinisto S, et al. Infant feeding in relation to the risk of advanced islet autoimmunity and type 1 diabetes in children with increased genetic susceptibility: a Cohort Study. *Am J Epidemiol.* 2018;187(1):34–44.
50. Frederiksen B, Kroehl M, Lamb MM, et al. Infant exposures and development of type 1 diabetes mellitus: the Diabetes Autoimmunity Study in the Young (DAISY). *JAMA Pediatr.* 2013;167(9):808–815.
51. Uusitalo U, Liu X, Yang J, et al. Association of early exposure of probiotics and islet autoimmunity in the TEDDY Study. *JAMA Pediatr.* 2016;170(1):20–28.
52. Xiao L, Van't Land B, Engen PA, et al. Human milk oligosaccharides protect against the development of autoimmune diabetes in NOD-mice. *Sci Rep.* 2018;8(1):3829.
53. Sane F, Scuotto A, Pierrat V, Kacet N, Hober D, Romond MB. Diabetes progression and alterations in gut bacterial translocation: prevention by diet supplementation with human milk in NOD mice. *J Nutr Biochem.* 2018;62:108–122.
54. Chmiel R, Beyerlein A, Knopff A, Hummel S, Ziegler AG, Winkler C. Early infant feeding and risk of developing islet autoimmunity and type 1 diabetes. *Acta Diabetol.* 2015;52(3):621–624.
55. Lund-Blix NA, Stene LC, Rasmussen T, Torjesen PA, Andersen LF, Ronningen KS. Infant feeding in relation to islet autoimmunity and type 1 diabetes in genetically susceptible children: the MIDIA Study. *Diabetes Care.* 2015;38(2):257–263.
56. Battaglia M, Ahmed S, Anderson MS, et al. Introducing the endotype concept to address the challenge of disease heterogeneity in type 1 diabetes. *Diabetes Care.* 2020;43(1):5–12.
57. Leete P, Mallone R, Richardson SJ, Sosenko JM, Redondo MJ, Evans-Molina C. The effect of age on the progression and severity of type 1 diabetes: potential effects on disease mechanisms. *Curr Diab Rep.* 2018;18(11):115.
58. Howson JM, Cooper JD, Smyth DJ, et al. Evidence of gene-gene interaction and age-at-diagnosis effects in type 1 diabetes. *Diabetes.* 2012;61(11):3012–3017.
59. Haupt-Jorgensen M, Morgen CS, Jess T, et al. Maternal antibiotic use during pregnancy and type 1 diabetes in children—a National Prospective Cohort Study. *Diabetes Care.* 2018;41(12):e155–e157.
60. Mikkelsen KH, Knop FK, Vilsboll T, Frost M, Hallas J, Pottegard A. Use of antibiotics in childhood and risk of Type 1 diabetes: a population-based case-control study. *Diabet Med.* 2017;34(2):272–277.
61. Kemppainen KM, Vehik K, Lynch KF, et al. Association between early-life antibiotic use and the risk of islet or celiac disease autoimmunity. *JAMA Pediatr.* 2017;171(12):1217–1225.
62. Hooper LV, Macpherson AJ. Immune adaptations that maintain homeostasis with the intestinal microbiota. *Nat Rev Immunol.* 2010;10(3):159–169.
63. Ulluwishewa D, Anderson RC, McNabb WC, Moughan PJ, Wells JM, Roy NC. Regulation of tight junction permeability by intestinal bacteria and dietary components. *J Nutr.* 2011;141(5):769–776.
64. Bosi E, Molteni L, Radaelli MG, et al. Increased intestinal permeability precedes clinical onset of type 1 diabetes. *Diabetologia.* 2006;49(12):2824–2827.

65. Maffeis C, Martina A, Corradi M, et al. Association between intestinal permeability and faecal microbiota composition in Italian children with beta cell autoimmunity at risk for type 1 diabetes. *Diabetes Metab Res Rev*. 2016;32(7):700–709.
66. Secondulfo M, Iafusco D, Carratu R, et al. Ultrastructural mucosal alterations and increased intestinal permeability in non-celiac, type I diabetic patients. *Dig Liver Dis*. 2004;36(1):35–45.
67. Sapone A, de Magistris L, Pietzak M, et al. Zonulin upregulation is associated with increased gut permeability in subjects with type 1 diabetes and their relatives. *Diabetes*. 2006;55(5):1443–1449.
68. Westerholm-Ormio M, Vaarala O, Pihkala P, Ilonen J, Savilahti E. Immunologic activity in the small intestinal mucosa of pediatric patients with type 1 diabetes. *Diabetes*. 2003;52(9):2287–2295.
69. Pellegrini S, Sordi V, Bolla AM, et al. Duodenal mucosa of patients with type 1 diabetes shows distinctive inflammatory profile and microbiota. *J Clin Endocrinol Metab*. 2017;102(5):1468–1477.
70. Badami E, Sorini C, Coccia M, et al. Defective differentiation of regulatory FoxP3+ T cells by small-intestinal dendritic cells in patients with type 1 diabetes. *Diabetes*. 2011;60(8):2120–2124.
71. Gavin PG, Mullaney JA, Loo D, et al. Intestinal metaproteomics reveals host-microbiota interactions in subjects at risk for type 1 diabetes. *Diabetes Care*. 2018;41(10):2178–2186.
72. de Oliveira FL, Gatto M, Bassi N, et al. Galectin-3 in autoimmunity and autoimmune diseases. *Exp Biol Med (Maywood)*. 2015;240(8):1019–1028.
73. Zeyer KA, Reinhardt DP. Fibrillin-containing microfibrils are key signal relay stations for cell function. *J Cell Commun Signal*. 2015;9(4):309–325.
74. Guo G, Booms P, Halushka M, et al. Induction of macrophage chemotaxis by aortic extracts of the mgR Marfan mouse model and a GxxPG-containing fibrillin-1 fragment. *Circulation*. 2006;114(17):1855–1862.
75. Sokol H, Pigneur B, Watterlot L, et al. Faecalibacterium prausnitzii is an anti-inflammatory commensal bacterium identified by gut microbiota analysis of Crohn disease patients. *Proc Natl Acad Sci USA*. 2008;105(43):16731–16736.
76. Kurilshikov A, Wijmenga C, Fu J, Zhernakova A. Host genetics and gut microbiome: challenges and perspectives. *Trends Immunol*. 2017;38(9):633–647.
77. Rothschild D, Weissbrod O, Barkan E, et al. Environment dominates over host genetics in shaping human gut microbiota. *Nature*. 2018;555(7695):210–215.
78. Polychronakos C, Li Q. Understanding type 1 diabetes through genetics: advances and prospects. *Nat Rev Genet*. 2011;12(11):781–792.
79. Dos Santos RS, Marroqui L, Velayos T, et al. DEXI, a candidate gene for type 1 diabetes, modulates rat and human pancreatic beta cell inflammation via regulation of the type I IFN/STAT signalling pathway. *Diabetologia*. 2019;62(3):459–472.
80. Dendrou CA, Cortes A, Shipman L, et al. Resolving TYK2 locus genotype-to-phenotype differences in autoimmunity. *Sci Transl Med*. 2016;8(363):363ra149.
81. Mullaney JA, Stephens JE, Costello ME, et al. Type 1 diabetes susceptibility alleles are associated with distinct alterations in the gut microbiota. *Microbiome*. 2018;6:35.
82. Hasnain SZ, Tauro S, Das I, et al. IL-10 promotes production of intestinal mucus by suppressing protein misfolding and endoplasmic reticulum stress in goblet cells. *Gastroenterology*. 2013;144(2):357–368.e359.
83. Campbell C, Dikiy S, Bhattarai SK, et al. Extrathymically generated regulatory T cells establish a niche for intestinal border-dwelling bacteria and affect physiologic metabolite balance. *Immunity*. 2018;48(6):1245–1257.e1249.
84. Cebula A, Seweryn M, Rempala GA, et al. Thymus-derived regulatory T cells contribute to tolerance to commensal microbiota. *Nature*. 2013;497(7448):258–262.

85. Russell JT, Roesch LFW, Ordberg M, et al. Genetic risk for autoimmunity is associated with distinct changes in the human gut microbiome. *Nat Commun*. 2019;10(1):3621.
86. Paun A, Yau C, Meshkibaf S, et al. Association of HLA-dependent islet autoimmunity with systemic antibody responses to intestinal commensal bacteria in children. *Sci Immunol*. 2019;4(32):eaau8125.
87. Mishra SP, Wang S, Nagpal R, et al. Probiotics and prebiotics for the amelioration of type 1 diabetes: present and future perspectives. *Microorganisms*. 2019;7(3):67.
88. Dolpady J, Sorini C, Di Pietro C, et al. Oral probiotic VSL#3 prevents autoimmune diabetes by modulating microbiota and promoting indoleamine 2,3-dioxygenase-enriched tolerogenic intestinal environment. *J Diabetes Res*. 2016;2016:7569431.
89. Lau K, Benitez P, Ardissone A, et al. Inhibition of type 1 diabetes correlated to a Lactobacillus johnsonii N6.2-mediated Th17 bias. *J Immunol*. 2011;186(6):3538–3546.
90. Groele L, Szajewska H, Szypowska A. Effects of Lactobacillus rhamnosus GG and Bifidobacterium lactis Bb12 on beta-cell function in children with newly diagnosed type 1 diabetes: protocol of a randomised controlled trial. *BMJ Open*. 2017;7(10):e017178.
91. Ho J, Nicolucci AC, Virtanen H, et al. Effect of prebiotic on microbiota, intestinal permeability, and glycemic control in children with type 1 diabetes. *J Clin Endocrinol Metab*. 2019;104(10):4427–4440.
92. de Groot PF, Nikolic T, Imangaliyev S, et al. Oral butyrate does not affect innate immunity and islet autoimmunity in individuals with longstanding type 1 diabetes: a randomised controlled trial. *Diabetologia*. 2020;63(3):597–610.
93. Zare Javid A, Aminzadeh M, Haghighi-Zadeh MH, Jamalvandi M. The effects of synbiotic supplementation on glycemic status, lipid profile, and biomarkers of oxidative stress in type 1 diabetic patients. A placebo-controlled, double-blind, randomized clinical trial. *Diabetes Metab Syndr Obes*. 2020;13:607–617.

CHAPTER EIGHT

Diet, nutrients and the microbiome

Wendy J. Dahl*, Daniela Rivero Mendoza, Jason M. Lambert
Department of Food Science and Human Nutrition, University of Florida, Gainesville, FL, United States
*Corresponding author: e-mail address: wdahl@ufl.edu

Contents

Abstract

Although there is associative evidence linking fecal microbiome profile to health and disease, many studies have not considered the confounding effects of dietary intake. Consuming food provides fermentable substrate which sustains the microbial ecosystem that resides with most abundance in the colon. Western, Mediterranean and vegetarian dietary patterns have a role in modulating the gut microbiota, as do trending restrictive diets such the paleolithic and ketogenic. Altering the amount or ratio of carbohydrate, protein and fat, particularly at the extremes of intake, impacts the microbiome. Diets high in fermentable carbohydrates support the relative abundance of *Bifidobacterium*, *Prevotella*, *Ruminococcus*, *Dorea* and *Roseburia*, among others, capable of degrading polysaccharides, oligosaccharides and sugars. Conversely, very high fat diets increase bile-resistant organisms such as *Bilophila* and *Bacteroides*. Food form,

Progress in Molecular Biology and Translational Science, Volume 171
ISSN 1877-1173
https://doi.org/10.1016/bs.pmbts.2020.04.006

whole foods vs. ultra-processed, alters the provision of macronutrient substrate to the colon due to differing digestibility, and thereby may impact the microbiota and its metabolic activity. In addition, phytochemicals in plant-based foods have specific and possibly prebiotic effects on the microbiome. Further, food ingredients such as certain low-calorie sweeteners enhance *Bifidobacterium* spp. The weight of evidence to date suggests a high level of interindividual variability in the human microbiome vs. clearly defined, dietary-induced profiles. Healthful dietary patterns, emphasizing plant foods high in microbial-available carbohydrate, support favorable microbiome profiles active in saccharolytic fermentation. Future research into diet and microbiome should consider the balance of gut microbial-generated metabolites, an important link between microbiome profile and human health.

1. Introduction

With the advent of low-cost, 16S rRNA gene sequencing and open-access software for genomic classification,[1] research into the relationships between health, disease and fecal microbiome has flourished. Given the anatomical proximity, much of the research first focused on gastrointestinal disease and to date, fecal microbiome profile has been shown to be associated with irritable bowel syndrome,[2] inflammatory bowel disease,[3] and celiac disease,[4,5] among other gastrointestinal conditions and diseases.[6] Although considerable research has focused on the relationship between gut microbiota and risk of colorectal cancer, this association, specifically in obese individuals, has been recently questioned.[7] The gut microbiota and its metabolic activity has the potential to modulate inflammation[8] and intestinal permeability,[9] thus the relationships between fecal microbiome and diseases of inflammatory etiologies have also been investigated. Fecal microbiome profile has been associated with type 2 diabetes,[10] cardiovascular disease,[11] and non-alcoholic fatty liver disease (NAFLD).[12–14] Although a possible relationship between fecal microbiome and obesity has received considerable attention, the evidence suggests a minor difference, if any, between the microbiome profiles of normal weight and obese adults.[15] Similarly, there is little evidence for such a relationship in children.[16] Most recently, the possibility of the gut microbiota being a factor in the etiology of neurological diseases such as Parkinson's disease[17] and amyotrophic lateral sclerosis (ALS)[18] as well as psychiatric diseases, such as schizophrenia,[19] have been suggested. As sampling offers ease of access, the majority of published studies relating microbiome to human health report on fecal analysis, and thus for the purposes of this chapter, microbiome will be synonymous with fecal microbiome.

The relationship between diet, microbiota and health is not a recent concept, with published research dating back more than a century.[20] Studies predating the current molecular techniques for classification of intestinal microorganisms demonstrated that diet composition modulates intestinal gas,[20,21] and therefore bacterial metabolism, as well as the production of proteolytic metabolites such as phenols.[22] However, the vastness and considerable interindividual diversity of the gut microbial ecosystem, described by the collective genomes of the microorganisms isolated from fecal samples, have more recently been revealed and what characterizes a healthful microbiota profile remains elusive.[23] Similarly, dysbiosis, although considered to be an imbalance in the microbiota, has no single definition.[24,25] In addition, microbial diversity, thought to be a relatively simple metric to categorize health vs. disease risk, has not been shown to be a reliable indicator by meta-analyses.[26] Instead, compositional changes in species[26] and their metabolic functions[24] need to be examined to clearly elucidate the connections between diet, gut microbiota, and health.

Armor et al. emphasize that beyond microbiome profile, the functional activity of microbiota, specifically the microbial-generated metabolome, and its relationships to disease are not well understood and require investigation.[27] The benefits of short chain fatty acids, primarily sourced from the intestinal fermentation of microbial-available carbohydrates, have been explored. Butyrate, an essential fuel for the colonocytes, has roles in intestinal homeostasis and mitigation of inflammation.[28] Acetate and propionate, the two most plentiful short chain fatty acids produced during intestinal anaerobic fermentation of primarily sugars, have roles in energy homeostasis, among others.[29] With respect to proteolytic fermentation, less is known of the health effects of branched chain fatty acids, produced through fermentation of valine and leucine, whereas the microbial-generated products of sulfur-containing amino acids are pro-inflammatory and cytotoxic.[30] Phenolic metabolites, products of tyrosine catabolism, negatively impact the intestinal epithelial cells and gut integrity and once absorbed have deleterious systematic effects at high levels.[30] The many other metabolites resulting from amino acid fermentation as well as from microbial fat catabolism, with their various local and systematic effects, are reviewed elsewhere.[30] Beyond the products of macronutrient fermentation, there are perhaps innumerable microbial-generated metabolites resulting from the bioconversion of dietary-sourced substrates, the minor constituents of foods, that have the potential to impact human health—a comprehensive review of which is beyond the scope of this chapter.

Bacteroidetes and Firmicutes are the dominant phyla in the human microbiota, and their ratio has been suggested to be an indicator of health,[23] whereas Actinobacteria, Proteobacteria, Fusobacteria, and others are in lower abundance. Given the complexity of the human fecal microbiota, the search is on for indicator organisms, whether diagnostic or etiologic, and thus, potential therapeutic targets. Some probable candidates are *Bifidobacterium* spp., given the many probiotic organisms arising from this genus and their association with "robustness" of the gut microbiota as well as the support of diversity.[31] A potential candidate is *Faecalibacterium prausnitzii*, reduced in IBD,[3] major depression,[32] chronic kidney disease (CKD),[33] multiple sclerosis,[34] low-grade inflammation and cardiovascular disease,[8] as well as cancer.[35] In addition, this species is negatively associated with ageing[36] and enhanced with physical activity.[37] Evidence suggests that *Akkermansia muciniphila* is associated with health and may be anti-inflammatory.[38] It is also positively associated with physical activity.[37] Although suppressed in IBD,[3] *A. muciniphila* appears to be increased in multiple sclerosis,[34] thus its potential role in health is less certain. The relative representation of *Roseburia* spp. also may be beneficial as they are reduced in various disease states and are possibly anti-inflammatory.[39] Cross-continent sampling suggests that *Prevotella,* associated with plant-based diets, may be a biomarker of health.[40] However, it has also been suggested that certain *Prevotella* spp. may contribute to pathogenesis,[41] and thus, the contribution of *Prevotella* to health vs. disease needs clarification. In addition to these individual genera, there is a logical expectation of identifying additional microorganisms indicative of health or disease, as well as combinations of beneficial organisms due to the complex associations of cross-feeding that may result in net positive health effects.[42]

Although there is mounting evidence that the microbiota and its complex metabolic activity impacts human health, one of the major limitations of this body of research is that many studies have not examined dietary intake or longer-term dietary patterns. As the provision of fermentable dietary-sourced substrate is one of the primary determinants of the microbiota and its functional activity,[43,44] consideration of diet is critical to delineating microbiome relationships with health and disease. The purpose of this chapter is to summarize the research to date which has explored the relationship between dietary patterns, nutrients, food and the microbiome in primarily healthy individuals. As infant microbiota is less stable than that of adults and the effects of infant feeding are published elsewhere,[45] studies on infants are excluded from this review. In addition, a comprehensive review of the effects of dietary interventions on the microbiome of specific patient populations is beyond the scope of this review.

2. Macronutrients and microbiome

The Acceptable Macronutrient Distribution Ratios (AMDR), as percentages of total energy for protein (10–35%), carbohydrate (45–65%) and fat (20–35%), set by the former Institute of Medicine of the National Academies in North America, provides guidance as to macronutrient intake ranges that will provide adequate levels of essential nutrients while reducing the risk of chronic disease.[46] One of the outcomes that was not considered when these recommendations were set, was the potential for macronutrient intakes to modulate microbiota and thereby impact human health. In recent years, studies have explored the effects of manipulating macronutrients on the microbiome. Table 1 provides details of the macronutrient intervention trials discussed below.

2.1 Carbohydrate

Given that all dietary fiber, by definition, is indigestible by human saccharolytic enzymes, its fate is primarily fermentation in colon,[44] and if resistant to fermentation, as in the case of lignified plant material, fecal excretion.[47] Dietary fiber is a dietary component with significant impact on the microbiota and its metabolism.[48] Epidemiologic evidence suggests that populations consuming agrarian diets higher in plant sourced foods, and thus higher in fiber, may favor greater microbial diversity and higher relative abundance of *Prevotella* over *Bacteroides*, whereas diets composed of processed food high in fat, added sugar and low in fiber may enhance *Proteobacteria*.[49] However, microbial exposure, lifestyle, geography, and a host of other factors confound this associative evidence. Wu et al. postulated, based on the microbiomes of 98 individuals, that long-term dietary patterns higher in protein and animal fat are associated with *Bacteroides* and those higher in carbohydrates favor *Prevotella*, and that these "enterotypes" are relatively stable,[50] although others have since questioned this interpretation and conclude that clear separations are not supported by current data.[40]

Interventional trials provide evidence as to whether fiber alters the microbiome in healthy individuals. In a systematic review and meta-analysis of fiber trials, So et al. reported that provision of fiber generally results in a higher relative abundance of *Bifidobacterium* spp. and *Lactobacillus* spp. compared to lower fiber diets or placebo.[51] In these studies of healthy individuals, no differences were seen for alpha diversity or other predetermined bacteria of interest. A major limitation, however, is that most studies carried

Table 1 Characteristics of the intervention trials examining macronutrient composition and microbiome profile.

Author (year) country	Study design	Sample size (male/female)	Age (years ± SE)	Macronutrient ratio: % energy protein: carbohydrate:fat(diet characteristics)	Microbiome outcome
Beaumont (2017) France	2-week run-in period followed by a 3-week, 3-arm parallel intervention period (5-week total), randomized, double-blind	38 (13/25)	G1: 31.0 ± 2.0 G2: 28.0 ± 2.0 G3: 31.0 ± 2.0	Run-in period: 15:50:35 (individualized menus providing their habitual energy intake) G1: 15:54:31 (control—maltodextrin) G2: 34:36:30 (casein) G3: 32:38:30 (soy)	No significant differences between dietary treatments
Cotillard (2013) France	6-week ER high-protein period followed by a 6-week weight-maintenance period (12-week total)	49 (8/41)	NR	Period 1: 35:44:25 (1200 kcal/d for women and 1500 kcal/d for men with low glycemic index carbohydrates and soluble fibers) Period 2: ratio NR (20% increase in total energy intake)	Gene richness increased significantly in the low gene count group (but no in high gene count group) after the energy-restricted diet and weight-maintenance phase
Dahl (2016) United States	2-week 3-parallel arm intervention and 2-week control, separated by a 1-week washout and a final 1-week washout (6-week total), randomized, double-blinded crossover	57 (22/45)	G1: 20.3 ± 1.7 G2: 21.3 ± 2.7 G3: 20.8 ± 2.0	Ratios NR G1: (RS-4 soluble fiber with high viscosity) G2: (RS-4 soluble fiber with low viscosity) G3: (RS-4 insoluble fiber with low viscosity)	G2: Significant decrease in the proportion of Firmicutes and in the ratio of Firmicutes to Bacteroidetes, decreased Lachnospiraceae mainly due to a decrease in *Coprococcus*

David (2014) United States	4-day baseline period followed by 5-day intervention periods separated by 6-day washouts and a final 6-day washouts (26 days total), crossover	11 (6/5)	28.1 ± 1.1	PBD: 10:68:22 (rich in grains, legumes, fruits, and vegetables; ad libitum) ABD: 30:0.6:69.5 (meats, eggs, and cheeses; ad libitum)	ABD increased bile-tolerant microorganisms (*Alistipes*, *Bilophila*, and *Bacteroides*) and decreased Firmicutes that metabolize dietary plant polysaccharides (*Roseburia*, *Eubacterium rectale*, and *Ruminococcus bromii*)
Ford (2020) United States	2-week baseline period followed by 4, 2-week controlled high-protein intervention periods, separated by 2-week washouts (18-week total), placebo-controlled, randomized, double-blind, crossover	26 (0/26)	73.7 ± 5.6	T1: 30:46:27 (high protein alone) T2: 30:46:27 (high protein plus multistrain probiotic formulation) T3: 30:46:27 (high protein plus prebiotic) T4: 30:46:27 (high protein plus synbiotic) Commercially available foods were provided to participants to meet each participant's energy requirement. Washouts consisted of their usual diet	Representation of *Lactobacillus*, *Lactococcus*, and *Streptococcus* were increased during high-protein diets, whereas *Roseburia* and *Anaerostipes* were suppressed compared to washout periods
Lang (2018) United States	2-week baseline period followed by 3, 4-week intervention periods separated by 2-week washouts (18-week total), randomized, crossover	109 (NR)	NR (range:21–65)	LT1: 25:41:34 (nonmeat protein) LT2: 26:39:35 (lean cuts of red meat) LT3: 23:46:31 (white meat) HT1: 24:42:35 (nonmeat protein) HT2: 24:41:35 (lean cuts of red meat) HT3: 24:42:34 (white meat) Either low (7% E) or high (15% E) saturated fat groups. Remaining protein source consisted of eggs, dairy, and vegetable protein All fish, seafood, and processed meats were excluded from diets	No significant changes in microbiome community in ordination space or in the Firmicutes/Bacteroidetes ratios. Participant characteristics outweighed dietary interventions and drove microbiome community composition

Continued

Table 1 Characteristics of the intervention trials examining macronutrient composition and microbiome profile.—cont'd

Author (year) country	Study design	Sample size (male/ female)	Age (years ± SE)	Macronutrient ratio: % energy protein: carbohydrate:fat(diet characteristics)	Microbiome outcome
Martínez (2010) United States	2-week baseline period and 3-week 3-parallel arm interventions separated by 2-week washouts and a final 2-week washout (17-week total), double-blind, crossover	10 (5/5)	NR (range:23–38)	Ratios NR T1: NR (RS-2—Hi-Maize 260) T2: NR (RS4—Fibersym® RW) T3: NR (native wheat starch (Midsol 50)	RS-4 but not RS-2 increased Actinobacteria and Bacteroidetes and decreased Firmicutes. RS4 increased *Bifidobacterium adolescentis* and *Parabacteroides distasonis*, while RS2 increased proportions of *Ruminococcus bromii* and *Eubacterium rectale* when compared to RS4
Salonen (2014) Scotland (Secondary analysis of Walker 2011)	7-week weight-maintenance period followed by a 3-week ER high-protein period (10-week total), randomized, crossover	14 (14/0)	54.0 ± 4.0	Period 1: 13:52:35 (plus 27.7 g/d non-starch polysaccharides NSP for 7 day, and either RS-3 or high in NSP for 3 weeks each) Period 2: ratio NR (ER diet reduced in carbohydrate but high in protein)	Multiple Ruminococcaceae phylotypes increased on RS3, whereas mostly Lachnospiraceae phylotypes increased on NSP. *Bifidobacterium* were reduced on ER. RS3 decreased the diversity of the microbiota
Walker (2011) Scotland	7-week weight-maintenance period followed by a 3-week ER high protein period (10-week total), randomized, crossover	14 (14/0)	54.0 ± 4.0	Period 1: 13:52:35 (plus 27.7 g/d non-starch polysaccharides NSP for 7 day, and either RS-3 or high in NSP for 3 weeks each) Period 2: ratio NR (ER diet reduced in carbohydrate but high in protein)	Proportions (of relatives) of *Eubacterium rectale* and *Ruminococcus bromii* increased on the RS diet whereas *Collinsella aerofaciens* decreased on ER. *Ruminococci bromii* increased on RS3 compared to NSP. *Oscillibacter* increased on the RS3 and ER diets. Firmicutes bacteria increased on the RS3 and decreased on ER.

Wan (2018) China	7-day run-in feeding period, and a 6-month, 3-parallel arm intervention period, randomized, observer blind	217 (103/114)	LFD: 23.3 ± 3.4 MFD: 23.6 ± 4.0 HFD: 23.4 ± 4.1	LFD: 14:66:20 MFD: 14:56:30 HFD: 14:46:40 Isoenergetic diets. Soybean oil, white rice, and white wheat flour intake from total energy vary in the 3 diets.	LFD increased abundance of *Blautia* and *Faecalibacterium*, whereas HFD was associated with increased *Alistipes*, *Bacteroides* and decreased *Faecalibacterium*.
Wu (2012), United States	10-day 2-arm parallel intervention, randomized	10 (NR)	NR	G1: ratio NR (high-fat/low-fiber diet) G2: ratio NR (low-fat/high-fiber diet)	Changes in microbiome composition were detectable within 24 h of initiating intervention but the magnitude of the changes was not sufficient to switch individuals between the enterotype, *Prevotella* vs. *Bacteroides*.

SE, Standard error; G, Group; ER, Energy restricted; NR, Not reported; RS, Resistant starch; NSP, Non-starch polysaccharides; T, Treatment; L, Low; H, High; E, Total energy; PBD, Plant-based diet; ABD, Animal-based diet; LFD, Low-fat diet; MFD, Moderate-fat diet; HFD, High-fat diet.

out and thus reviewed, tested isolated prebiotic fibers, known to be somewhat selectively utilized,[52] and some studies were not designed to measure comprehensive microbiome profile changes but only genera of interest, specifically *Bifidobacterium*. The sub-analysis of prebiotic fibers, including fructans and galacto-oligosaccharides, showed a consistent increase in *Bifidobacterium* spp. and *Lactobacillus* spp.; thus the prebiotic studies drove the overall response to fiber in this analysis. In reviews of patient populations, provision of prebiotic fiber to hospital patients[53] and in some studies of IBD patients[54] also have shown to increase *Bifidobacterium* spp.,[53] with and without health benefits. In a recent cross-sectional study, Lin et al. examined the dietary fiber intake of 151 adults, estimated from food frequency questionnaires, and the relationship to microbiome.[55] Higher fiber intake from mixed sources was associated with a higher abundance of *Lachnospira* and *Faecalibacterium* and lower abundance of *Actinomyces*, *Odoribacter* and *Oscillospira*. At a species level, abundance of *Faecalibacterium prausnitzii* was enhanced with higher fiber intake, whereas *Eubacterium dolichum* and *Bacteroides uniformis* were lower. More research is needed to determine the microbiome changes to controlled dietary interventions testing diverse sources of complex dietary fibers. It is important to note that fiber, given its various structures of constituent sugars and food sources, may not have a uniform or consistent impact on the microbiome. In addition, the microbiome response to an increase in fiber also may differ among individuals given their very unique microbiome profiles.[56]

Resistant starch, starch that is not digested in the small intestine, may provide as much or more carbohydrate substrate to the microbiota than fiber.[57] The interventional trials examining the effects of resistant starch on microbiota have been reviewed; however, the limited number of studies have primarily focused on short chain fatty acid concentrations with few assessing comprehensive microbiome profile.[58] Martinez et al. examined the effects of resistant starch types 2 (RS2) and 4 (RS4) on fecal microbiota composition.[59] With RS4 provision, phylum-level changes were noted, specifically increased Actinobacteria and Bacteriodetes, and a decrease in Firmicutes were seen. Species level changes were seen with RS4, specifically increased proportions of *Bifidobacterium adolescentis* and *Parabacteroides distasonis*. Abundance of *Ruminococcus bromii* and *Eubacterium rectale* were higher with RS2 compared to RS4. Similarly, feeding young adults a resistant potato starch (RS4) reduced Firmicutes.[60] Salonen et al. reported that the provision of type 3 resistant starch (RS3) decreased the diversity of the microbiota.[61] *Oscillospira guillermondii*, *Ruminococcus bromii*, *Sporobacter termitis*, *Clostridium leptum*, *Clostridium cellulosi* and certain *Alistipes* spp. were increased while decreased Lachnospiraceae, a family noted for utilizing plant polysaccharides, was seen. Resistant starch, whether it is physically inaccessible, granular, retrograded, or chemically modified, may have differing effects on the microbiome and the response to resistant starch intake also may depend on the individual's baseline microbiome profile. In addition, high intakes of a specific resistant starch may have selective effects vs. supporting a diversity of saccharolytic bacteria.

Sugars vary in their digestibility, therefore depending on the source and digestive characteristics of the individual, sugar intake may impact the microbiome. For example, adult lactase persistence is uncommon in the majority of the global population, thus dietary lactose primarily escapes digestion[62] and is subsequently fermented. Li et al. compared 4-weeks of milk consumption in lactose maldigesters to digesters and assessed differences in microbiota.[63] In lactose maldigesters but not digesters, milk intake increased Actinobacteria, *Bifidobacterium*, *Anaerostipe*, and *Blautia*, but decreased *Megamonas*. Diversity and richness of microbiota were unaffected. Similarly, unabsorbed sugar alcohols, naturally occurring and added to foods as low calorie sweeteners, provide substrate for intestinal fermentation. Ruiz-Ojeda et al. in a review of sweeteners and microbiota, reported that sugar alcohol intake, specifically isomaltose, maltitol, lactitol, and xylitol increases abundance of *Bifidobacterium*.[64] High intakes of fructose will result in significant malabsorption,[65] and thus, would also be expected to modulate the microbiome in specific ways. In an effort to decrease intakes of sugars, many foods are sweetened using sugar substitutes. Non-nutritive sweeteners are approved food additives and exhibit high intensity sweetness 30–20,000 times that of sucrose. They include, aspartame, saccharin, acesulfame potassium, sucralose, advantame, neotame, stevia glycosides, and cyclamate among others. Given their high intensity sweetness, most are consumed in very low quantities and are also absorbed in the small intestine,[64] and thus would not be expected to exert any direct effects on the microbiome. As expected, there is very little evidence to suggest that non-nutritive sweeteners modulate human gut microbiota.[64]

2.2 Fat

Studies to test the effects of non-digestible carbohydrates such as fiber and resistant starch supplementation on microbiome can be designed without manipulating the provision of protein, fat and digestible carbohydrates, and without a major impact on energy intake. However, when the effects of differing intakes of protein, fat and digestible carbohydrate are tested, the ratios of the macronutrient contributions to energy or the energy level of the diet needs to change and thus, will confound the findings. Few studies have examined how modulating macronutrient profiles using controlled feeding trials affects the microbiome, two examining effects of very high fat diets on microbiome. David et al. evaluated a short-term diet in a trial of nine subjects providing an animal-sourced diet composed of 69.5% of energy from fat, 30% from protein, and essentially no carbohydrate compared to a very high fiber, vegetarian diet.[66] They showed an increased abundance of bile-tolerant bacteria including *Alistipes*, *Bilophia*, and *Bacteroides*, attributed to the high fat provision, and reduced abundance levels of *Roseburia*, *Eubacterium*, and *Ruminococcus*, due to the lack of fermentable carbohydrate substrate provided in the animal-sourced diet compared to the vegetarian diet. The animal-sourced experimental diet contained no fiber and a macronutrient composition not reflecting current US dietary patterns[67] or recommendations,[68] and which may pose risk if consumed in the long term.[69] Wu et al. provided a high-fat/low-fiber and low-fat/high-fiber diet to 10 subjects and found that the diets did not influence individual's enterotype, *Prevotella* vs. *Bacteroides*.[50] A recent meta-analysis examined the effect of high-fat diets and on the microbiome, based on the data from rodent models and these two high fat diets studies, and conclusions were made regarding the effect of fat on the microbiome, specifically, that high fat feeding increases the Firmicutes:Bacteroidetes ratio.[70] However, given that the analysis included data from only 29 human fecal samples, and considering the profound differences in diet composition between complex human diets and experimental chow—particularly the extreme fat levels used in some animal experiments, digestibility, differences in inter- and intra-individual variation in microbiome, the significance of the findings to human health is questionable.

In addition to research exploring very high fat diets, studies have examined the effects of moderate changes to fat intake and omega-3 supplementation on modulation of microbiome. Wan et al. in a trial of 217 young adults, examined the effects of a lower-fat diet (20% fat as energy), moderate-fat diet (30% fat), and higher-fat diet (40% fat) on microbiota over a 6-month period. The lower-fat diet increased alpha diversity, and relative abundance of *Blautia* and *Faecalibacterium*.[71] The higher-fat diet increased *Alistipes* and *Bacteriodes* and decreased *Faecalibacterium*. Watson et al. demonstrated decreased *Coprococcus* and *Faecalibacterium* and increased abundance of *Bifidobacterium*, *Oscillospira*, *Roseburia*, and *Lachnospira* with 4 g daily supplementation of eicosapentaenoic acid and docosahexaenoic acid.[72] In a cross-sectional of women, Menni et al. reported that estimated omega-3 intake and serum levels were associated with higher microbiome diversity and abundance of the Lachnospiraceae family, an effect independent of fiber intake.[73]

2.3 Protein

The effects of protein on the microbiome have been investigated in randomized controlled trials. Beaumont et al. in a crossover study of increased protein (casein and soy protein) and decreased starchy food intake in overweight adults showed no significant impact on microbiota for the subjects as a whole, but a high level of interindividual variability in response to the high protein diet was noted.[74] Ford et al. examined the effect of increasing protein intake to 29% while maintaining the AMDR in healthy older women.[75] The diet provided 15.5 g of fiber per day, a level similar to their subjects' baseline intake. Relative abundance of *Lactobacillus*, *Lactococcus*, and *Streptococcus* were enhanced with the high protein diet, whereas butyrate producers, *Roseburia* and *Anaerostipes*, were suppressed. Lang et al. provided a controlled diet to adults having a wide range of weight status, randomizing participants to high and low saturated fat groups, and to red meat, white meat, and non-meat protein source groups.[76] Interindividual differences explained the microbiome differences to a greater extent than did diet and the effects of fat concealed the small effects of dietary protein source.[76] The effects of high protein diets in the context of energy restriction also have been examined.[61,77,78] In a study of obese men, a high protein, reduced-carbohydrate, energy-restricted diet decreased *Bifidobacterium*.[61] A study of obese and overweight adults showed increased bacterial richness with a high protein, energy-restricted diet, but suggested that the change may have been due to the increased fiber provision.[77] In addition, Walker et al. examined the effects of a high protein, energy-restricted diet on the microbiota of overweight and reported decreased *Eubacterium rectale* and *Collinsella aerofaciens* in response to the diet.[78] Nonetheless, in these energy-restricted studies, the isolated effect of protein on the microbiota is difficult to discern from the impact of limiting energy intake.

3. Dietary patterns

Reduced risk of chronic disease is associated with healthful dietary patterns such as the Mediterranean diet[79] and plant-based diets.[80]

Considering the complexities of dietary intake and given that individuals consume foods rather than isolated nutrients, examining the impact of dietary patterns on microbiome may provide insight as to the mechanisms of health benefit and risk associated with various dietary patterns.

3.1 Western diet

The Western-style dietary pattern is characterized by high intakes of red meat, saturated fats, processed grains, added sugars, and low intakes of fiber. It is low in fruits, vegetables, legumes, nuts, seeds, and whole grains, foods that are encouraged by American dietary guidance[68] and elsewhere globally.[81] Many studies have used a Western diet as a control or comparator although few studies have thoroughly explored the microbiome profile of populations consuming the Western diet compared to non-Western diets. Recently, Tett et al. examined by meta-analyses a diverse "population structure" of a *Prevotella copri* complex present in the microbiome of non-Westernized populations and showed that this species, noted for carbohydrate utilization, is underrepresented in Western populations.[82] The high consumption of ultra-processed vs. whole foods is thought to contribute to the microbiome of individuals consuming the Western diet.[83] In a recent and relatively large trial of community-dwelling men, consuming a Western dietary pattern, was associated with higher abundance of a number of species including *Alistipes*, *Anaerotruncus*, *Collinsella*, *Coprobacillus*, *Desulfovibrio*, *Dorea*, *Eubacterium*, and *Ruminococcus*, whereas a more prudent diet was associated with higher relative abundance of *Faecalibacterium* as well as *Cetobacterium, Clostridium, Lachnospira, Paraprevotella, and Veillonella*.[84] Associations were significant for beta diversity but not alpha diversity. Deciphering the potential individual or collective benefits or risks of these species' differences requires further exploration.

3.2 Mediterranean diet

The Mediterranean diet is an eating pattern that has been traditionally consumed in countries bordering the Mediterranean Sea, such as Spain, Greece, and Italy. Although each of the Mediterranean countries has unique food traditions, consuming plant-sourced foods such as fruits, vegetables, whole grains, and legumes; few dairy foods, olive oil as the main source of fat, and moderate amounts of fish, poultry, and wine are consumed.[85] Recent studies have shown that adherence to the Mediterranean dietary pattern significantly impacts gut microbiome and metabolome. De Filippis et al. examined

153 individuals who habitually consumed an omnivore, vegetarian or vegan diet and some with high adherence to the Mediterranean diet.[86] Although there were no clear differences in microbiota composition between the three dietary patterns, they found an association with *Prevotella* and *Lachnospira*, genera known to degrade fiber, with diet quality as assessed by the Healthy Food Diversity index. Alpha diversity was not significantly different between the three dietary patterns. As follow-up, Meslier et al. conducted a Mediterranean diet interventional trial in overweight and obese individuals and demonstrated an increase in *Faecalibacterium prausnitzii*, a butyrate producer, and a decrease in *Ruminococcus gnavus*, suggested to be pro-inflammatory in response to the dietary intervention.[87]

Tosti et al. have proposed that gut microbial-generated metabolites may well be one of the mediating factors regarding the link between the Mediterranean diet and reduced risk of chronic disease.[79] Of note, lower adherence to the Mediterranean diet was associated with higher urinary trimethylamine-*N*-oxide (TMAO) TMAO excretion, a risk for cardiovascular disease.[86] Trimethylamine (TMA) is produced by the microbial bioconversion of such substrates as dietary carnitine and choline.[88] As dietary carnitine and choline are in higher concentrations in omnivore diets, particularly from red meat and eggs, the effect of diet on taxa capable of TMA generation has been suggested as a potential mechanism linking animal-sourced food intake to inflammatory disease risk. TMA, once absorbed is oxidized in the liver to TMAO, a compound associated with cardiovascular disease,[89] dementia,[90] and hypertension,[91] and is a potential marker of insulin resistance.[92] However, whether TMAO is causative or simply a metabolic marker has recently been questioned.[93]

3.3 Vegetarian and vegan diets

Vegetarian diets differ in their composition, with vegan diets being devoid of all food of animal origin. Vegetarian diets are widely considered to be health-enhancing, in part due to their generally higher diet quality.[94] Given that vegetarian diets are comprised of primarily plant-based foods, they provide higher levels of fiber and are generally lower in protein than omnivore diets, a balance that favors saccharolytic vs. proteolytic fermentation.[95] Trefflich et al. examined, by systematic review, the associations of vegan or vegetarian diets with microbiota composition and found no associations in the microbiome between consumers of vegetarian and vegan diets compared to omnivores, and concluded that high inter-individual variability precluded a description of a characteristic vegetarian or vegan microbiome profile.[96] In one such study, Losasso et al. in their cross-sectional analysis note that microbiome of vegetarian, vegans and omnivores may differ in abundance of Bacteroidetes; however, conclusions related to gut microbiota composition cannot be made based solely on dietary pattern.[97] Recently, Barrett et al. compared women consuming an omnivore or vegetarian diet during early pregnancy.[98] They found a slight reduction in beta diversity but not alpha diversity for those consuming a vegetarian diet and also, reduced abundance of *Collinsella* and *Holdemania* but higher *Roseburia* and *Lachnospiraceae*.[98] In summary, there does not seem to be a distinctive microbiome profile as a result of consuming a vegetarian diet, which may in part be due to the dietary variation of the vegetarian diets, and thus microbiome profile alone may not be the mechanism by which this dietary pattern exerts health benefits. However, the functional metabolism of the microbiota is also impacted by diet type, whether vegan, vegetarian or omnivore. The abundance of microbial genes and proteins that are responsible for various activities such as hydrolysis of polysaccharides and proteins and synthesis of vitamins is higher in vegans and vegetarians.[99]

4. Restrictive diets

4.1 Very low carbohydrate ketogenic diets

The ketogenic diet promotes high fat intake, moderate protein intake, and very low carbohydrate intake, driving the body into ketosis and reliance on fat for energy.[100] Foods that are commonly consumed on the ketogenic diet are meats, poultry, fish, shellfish, eggs, cheese, green leafy vegetables, nuts, oil, butter, mayonnaise, and cream. Additionally, non-nutritive sweeteners may be used to replace added sugar in the diet. Foods that contain higher amounts of carbohydrates, including grains, fruits, root vegetables, low-fat dairy, and legumes, are excluded. Therapeutic ketogenic diets are notably low in fiber.[101] A study in children for the treatment of intractable epilepsy examined fecal microbiota at baseline and after a ketogenic diet intervention and showed decreased abundance of Firmicutes and increased abundance of Bacteroidetes. Interestingly, they found that in the non-responsive group (no reduction in seizures), *Clostridiales*, *Ruminococcaceae*, *Rikenellaceae*, *Lachnospiraceae*, and *Alistipes* were enriched.[102] Lindefeldt et al. reported on 12 children prescribed a ketogenic diet for therapy-resistant epilepsy and found no change in alpha diversity with the diet prescription. Relative abundance of *Bifidobacterium*, *Eubacterium rectale* and *Dialister* decreased with the ketogenic diet.[103] Murtaza et al. examined elite athletes following a low carbohydrate, high fat (ketogenic) diet and found increased *Bacteroides* and *Dorea*, and decreased *Faecalibacterium*.[104] Swidsinski et al. reported on multiple sclerosis patients using fluorescence *in situ* hybridization (FISH) methodology only vs. gene sequencing and noted a short-term decrease in diversity that recovered and exceeded baseline after 6 months on the diet.[105] In a study testing a modified Mediterranean-Ketogenic diet compared to the American Heart Association diet in individuals with mild cognitive decline and normal cognitive function, Nagpal et al. reported that after 6 weeks, no difference in microbiome diversity was seen.[106] The modified Mediterranean-Ketogenic diet increased the abundance of Enterobacteriaceae, Akkermansia, Slackia, Christensenellaceae and Erysipelotriaceae. In addition, the genera *Bifidobacterium* and *Lachnobacterium* were reduced with the modified Mediterranean-Ketogenic diet. Interestingly, the modified Mediterranean-Ketogenic diet was associated with decreased Alzheimer's disease biomarkers in the cerebral spinal fluid.

4.2 Paleolithic diet

Although specific diet recommendations vary, the popular paleolithic diet purports to mimic a pre-agrarian diet of more than 10,000 year ago. Proponents of this diet generally recommend eating grass-fed meat, fish/seafood, fresh fruits and vegetables, eggs, nuts and seeds, and healthful oils sources such as olive oil, avocadoes, coconuts, macadamias, and walnuts. The consumption of cereal grains, legumes (including peanuts), dairy, refined sugar, potatoes, processed food, refined vegetables and salt is eliminated. Barone et al. compared the microbiomes of adults consuming a modern paleolithic diet to a Mediterranean diet and previously published data on hunters-gatherers.[107] The paleolithic diet supported diversity similar to that of the hunters-gatherers, and similar to the response seen in the David et al. study feeding a very high fat diet intervention,[66] as discussed above, bile-tolerant organisms were in high abundance. Of note, there was not one paleolithic diet. Pre-agrarian diets were adapted to food availability, and most were dissimilar to the current paleolithic diet composition,[108] and thus one could speculate that the microbiome profiles of paleolithic populations differed greatly by geography and diet.

5. Future research considerations

5.1 Dietary assessment

Diet is one of the key determinants of microbiome variation[56] and thus may be an important link to dietary patterns and health. Research designed to examine the microbiome profile and its relationship with disease have often neglected to consider food intake, the primary source of differential substrate for the microbiota.[43,109] Although some studies have considered estimated nutrient intake from food frequencies, 24-h dietary recalls, or food records, these methods are well known for their limitations,[110] particularly portion size estimation.[111] In addition, assessments of dietary intake primarily utilize nutrient analysis programs to approximate nutrient intakes based on nutrient databases such as the US Department of Agriculture's FoodData Central.[112] The contribution of various foods to the amount of substrate received by the microbiota may be far more variable than the nutrient content suggests. For example, the database provides an analysis of protein content from a representative sample of a food, although it is the digestibility of the protein, specifically the amount of protein (peptides and amino acids) that transits the small intestine to the colon, that may most impact the microbiome. In addition, nutrient databases may not provide estimates of phytochemicals, such as polyphenol content, that modulate the microbiome.[113] Specifically, it may be the malabsorbed nutrients as well as the non-nutrient food components that may have the most profound effect on the microbiota and its metabolism, and subsequently, microbiome profile. In addition, food additives, those intended for functional benefits but also those that are added for food function, are not commonly considered. Very few studies have undertaken controlled diet studies—providing food of known composition—and if so, they have not completed a compositional analysis of additional potential microbiome-altering components such as polyphenols. In addition, such studies, given the high cost and subject burden, are often short term.[66,75] Future research into the associations of diet and microbiome should utilize reliable and valid dietary assessment techniques as well as undertaking controlled feeding trials.

5.2 Intestinal transit and microbiota

Assessment of fecal microbiomes has limitations given that microbiome profiles,[114] and metabolite concentrations such as short chain fatty acids,[115] differ along the intestinal tract. Limitations to examination of the diet-microbiota relationship also include the confounding effect of intestinal transit time.[116] For example, distal colonic transit time has been shown to affect alpha-diversity and microbial metabolism.[117] Typical oral to anal transit time in healthy adults is 24–72 h. Transit time significantly less than 24 h may result in diarrhea-like, fast transit stools. Notably, with fast transit, fermentation of undigested foods stuffs may be limited and thus will affect the microbiota profile detected in the stools. In contrast, slow transit time, suggestive of constipation, allows for a significant time period for fermentation—this lengthy period of fermentation may result in complete fermentation of carbohydrate substrate and allowing ample time for facultative anaerobes to target other less efficient energy sources, such as protein. Stool form, as a proxy for intestinal transit time, explains a significant proportion of the inter-individual variation in microbiome profile,[116] thus studies exploring the relationship between diet and microbiome need to consider the confounding effects of subject transit time on outcomes.

5.3 Dietary patterns and foods vs. prebiotics

As described above, macronutrient composition as well as overall dietary pattern modulate the microbiome profile. Little research has explored the effect of micronutrient intake on the microbiome. Given the bioavailability of micronutrients and low relative intakes compared to macronutrients, it might be hypothesized that few directly impact the microbiota, with the possible exception of iron when supplemented.[118] Potential impacts of other minerals, such as calcium supplementation, on the microbiome may be of interest. Considerable research is exploring the effects of various phytochemicals on the microbiome, such as polyphenols.[119–121] The epidemiological evidence supports associations between dietary patterns and reduced chronic disease risk vs. supplementation. As we consume foods vs. nutrients in isolation, exploration of the beneficial effects of whole foods and the food groups that compose dietary patterns may be the most appropriate targets for future exploration of the microbiome and health, the findings of which can be directly translated to dietary recommendations focused on health. The research history of nutrition is fraught with missteps examining nutrients instead of dietary patterns, the most obvious example being the research into antioxidant vitamin supplementation vs. a diet rich in fruits and vegetables and cancer risk.[122,123] To date, research into dietary modification of the microbiome has seemed to follow a similar pattern to traditional nutrient-focused research, with many studies being focused on the original prebiotic compounds, short-chain, fermentable carbohydrates (e.g., oligofructose, galacto-oligosaccharides), that were purported to specifically enhance particular bacteria considered to be health enhancing. However, the health benefits of enhancing the relative abundance of these microorganisms, particularly as related to disease risk reduction, are unclear as most studies have not explored direct health benefits. Recent research has demonstrated that microbiome composition is dependent on how many days of diet and specific food choices vs. nutrient composition.[56] Further, as recently pointed out by Gurry et al. controlled testing of the effects of specific nutrients/substances is wrought with challenges as a standardized diet will not lead to a standard microbiome as required for such testing.[124] Instead, it may be more strategic to examine dietary patterns and whole foods vs. specific food components. To date, some studies have explored the effects of various food groups on the microbiome. For example, pulses, also known as grain legumes, have been shown to suppress proteolytic bacteria while enhancing butyrate producers[125]; whole grains increase *Bifidobacterium* and *Lactobacillus* while reducing *Escherichia coli* and *Clostridia* spp.[126]; and nuts contain many microbial-available constituents that have the potential to favorably modulate the microbiome and perhaps more so, the metabolome.[127]

6. Conclusions

Due to the high level of interindividual variability in the human microbiome, clearly defined, microbiome profiles induced by specific dietary patterns or nutrient intakes have not been realized. Healthful dietary patterns, emphasizing plant foods high in microbial-available carbohydrate, support favorable microbiome profiles with higher abundance of species capable of carbohydrate fermentation. Research exploring the effects of whole foods, particularly plant-sourced foods, should be pursued to understand the mechanisms connecting diet patterns to the modulation of the microbiota. Future research into diet and microbiome should consider the balance of gut microbial-generated metabolites, an important link between microbiome profile and human health.

References

1. Bolyen E, Rideout JR, Dillon MR, et al. Reproducible, interactive, scalable and extensible microbiome data science using QIIME 2. *Nat Biotechnol.* 2019;37(8):852–857.
2. Simpson CA, Mu A, Haslam N, Schwartz OS, Simmons JG. Feeling down? A systematic review of the gut microbiota in anxiety/depression and irritable bowel syndrome. *J Affect Disord.* 2020;266:429–446.

3. Pittayanon R, Lau JT, Leontiadis GI, et al. Differences in gut microbiota in patients with vs without inflammatory bowel diseases: a systematic review. *Gastroenterology*. 2020;158(4):930–946.e931.
4. Valitutti F, Cucchiara S, Fasano A. Celiac disease and the microbiome. *Nutrients*. 2019;11(10):2403.
5. Krishnareddy S. The microbiome in celiac disease. *Gastroenterol Clin North Am*. 2019;48(1):115–126.
6. Gorkiewicz G, Moschen A. Gut microbiome: a new player in gastrointestinal disease. *Virchows Arch*. 2018;472(1):159–172.
7. Greathouse KL, White JR, Padgett RN, et al. Gut microbiome meta-analysis reveals dysbiosis is independent of body mass index in predicting risk of obesity-associated CRC. *BMJ Open Gastroenterol*. 2019;6(1):e000247.
8. van den Munckhof ICL, Kurilshikov A, Ter Horst R, et al. Role of gut microbiota in chronic low-grade inflammation as potential driver for atherosclerotic cardiovascular disease: a systematic review of human studies. *Obes Rev*. 2018;19(12):1719–1734.
9. Fasano A. All disease begins in the (leaky) gut: role of zonulin-mediated gut permeability in the pathogenesis of some chronic inflammatory diseases. *F1000Res*. 2020;9:69.
10. Qin J, Li Y, Cai Z, et al. A metagenome-wide association study of gut microbiota in type 2 diabetes. *Nature*. 2012;490(7418):55–60.
11. Jie Z, Xia H, Zhong SL, et al. The gut microbiome in atherosclerotic cardiovascular disease. *Nat Commun*. 2017;8(1):845.
12. Kolodziejczyk AA, Zheng D, Shibolet O, Elinav E. The role of the microbiome in NAFLD and NASH. *EMBO Mol Med*. 2019;11(2):e9302.
13. Aron-Wisnewsky J, Vigliotti C, Witjes J, et al. Gut microbiota and human NAFLD: disentangling microbial signatures from metabolic disorders. *Nat Rev Gastroenterol Hepatol*. 2020. https://doi.org/10.1038/s41575-020-0269-9.
14. Mohammadi Z, Poustchi H, Motamed-Gorji N, et al. Fecal microbiota in non-alcoholic fatty liver disease and non-alcoholic steatohepatitis: a systematic review. *Arch Iran Med*. 2020;23(1):44–52.
15. Sze MA, Schloss PD. Looking for a Signal in the Noise: revisiting obesity and the microbiome. *MBio*. 2016;7(4):e01018-16.
16. Indiani C, Rizzardi KF, Castelo PM, Ferraz LFC, Darrieux M, Parisotto TM. Childhood obesity and Firmicutes/Bacteroidetes ratio in the gut microbiota: a systematic review. *Child Obes*. 2018;14(8):501–509.
17. Lin CH, Chen CC, Chiang HL, et al. Altered gut microbiota and inflammatory cytokine responses in patients with Parkinson's disease. *J Neuroinflammation*. 2019; 16(1):129.
18. Wright ML, Fournier C, Houser MC, Tansey M, Glass J, Hertzberg VS. Potential role of the gut microbiome in ALS: a systematic review. *Biol Res Nurs*. 2018;20(5):513–521.
19. Nguyen TT, Kosciolek T, Eyler LT, Knight R, Jeste DV. Overview and systematic review of studies of microbiome in schizophrenia and bipolar disorder. *J Psychiatr Res*. 2018;99:50–61.
20. Penfold WJ. Variability in the gas-forming power of intestinal bacteria. *Proc R Soc Med*. 1911;4(Pathol Sect):97–109.
21. Steggerda FR. Gastrointestinal gas following food consumption. *Ann N Y Acad Sci*. 1968;150(1):57–66.
22. Cummings JH, Hill MJ, Bone ES, Branch WJ, Jenkins DJ. The effect of meat protein and dietary fiber on colonic function and metabolism. II. Bacterial metabolites in feces and urine. *Am J Clin Nutr*. 1979;32(10):2094–2101.
23. Lloyd-Price J, Abu-Ali G, Huttenhower C. The healthy human microbiome. *Genome Med*. 2016;8(1):51.

24. Kriss M, Hazleton KZ, Nusbacher NM, Martin CG, Lozupone CA. Low diversity gut microbiota dysbiosis: drivers, functional implications and recovery. *Curr Opin Microbiol.* 2018;44:34–40.
25. Gagliardi A, Totino V, Cacciotti F, et al. Rebuilding the gut microbiota ecosystem. *Int J Environ Res Public Health.* 2018;15(8):e1679.
26. Ma ZS, Li L, Gotelli NJ. Diversity-disease relationships and shared species analyses for human microbiome-associated diseases. *ISME J.* 2019;13(8):1911–1919.
27. Armour CR, Nayfach S, Pollard KS, Sharpton TJ. A metagenomic meta-analysis reveals functional signatures of health and disease in the human gut microbiome. *mSystems.* 2019;4(4):e00332-18.
28. Liu H, Wang J, He T, et al. Butyrate: a double-edged sword for health? *Adv Nutr.* 2018;9(1):21–29.
29. Hu J, Lin S, Zheng B, Cheung PCK. Short-chain fatty acids in control of energy metabolism. *Crit Rev Food Sci Nutr.* 2018;58(8):1243–1249.
30. Oliphant K, Allen-Vercoe E. Macronutrient metabolism by the human gut microbiome: major fermentation by-products and their impact on host health. *Microbiome.* 2019;7(1):91.
31. Feng Y, Duan Y, Xu Z, et al. An examination of data from the American Gut Project reveals that the dominance of the genus Bifidobacterium is associated with the diversity and robustness of the gut microbiota. *Microbiology.* 2019;8(12):e939.
32. Cheung SG, Goldenthal AR, Uhlemann AC, Mann JJ, Miller JM, Sublette ME. Systematic review of gut microbiota and major depression. *Front Psych.* 2019;10:34.
33. Jiang S, Xie S, Lv D, et al. A reduction in the butyrate producing species Roseburia spp. and Faecalibacterium prausnitzii is associated with chronic kidney disease progression. *Antonie Van Leeuwenhoek.* 2016;109(10):1389–1396.
34. Mirza A, Forbes JD, Zhu F, et al. The multiple sclerosis gut microbiota: a systematic review. *Mult Scler Relat Disord.* 2020;37:101427.
35. Ferreira-Halder CV, Faria AVS, Andrade SS. Action and function of Faecalibacterium prausnitzii in health and disease. *Best Pract Res Clin Gastroenterol.* 2017;31(6):643–648.
36. Zhu Q, Huang L, Zhu J, Lu J, Qin H. Analysis of gut microbiota in long-lived older adults and their relatives: a gradual change with ageing. *Mech Ageing Dev.* 2018;178:1–8.
37. Bressa C, Bailen-Andrino M, Perez-Santiago J, et al. Differences in gut microbiota profile between women with active lifestyle and sedentary women. *PLoS One.* 2017;12(2): e0171352.
38. Derrien M, Belzer C, de Vos WM. Akkermansia muciniphila and its role in regulating host functions. *Microb Pathog.* 2017;106:171–181.
39. Tamanai-Shacoori Z, Smida I, Bousarghin L, et al. Roseburia spp.: a marker of health? *Future Microbiol.* 2017;12:157–170.
40. Gorvitovskaia A, Holmes SP, Huse SM. Interpreting Prevotella and Bacteroides as biomarkers of diet and lifestyle. *Microbiome.* 2016;4:15.
41. Precup G, Vodnar DC. Gut Prevotella as a possible biomarker of diet and its eubiotic versus dysbiotic roles: a comprehensive literature review. *Br J Nutr.* 2019;122(2): 131–140.
42. Rios-Covian D, Gueimonde M, Duncan SH, Flint HJ, de los Reyes-Gavilan CG. Enhanced butyrate formation by cross-feeding between Faecalibacterium prausnitzii and Bifidobacterium adolescentis. *FEMS Microbiol Lett.* 2015;362(21):fnv176.
43. Cummings JH, Englyst HN. Fermentation in the human large intestine and the available substrates. *Am J Clin Nutr.* 1987;45(Suppl 5):1243–1255.
44. Macfarlane GT, Macfarlane S. Bacteria, colonic fermentation, and gastrointestinal health. *J AOAC Int.* 2012;95(1):50–60.
45. Ho NT, Li F, Lee-Sarwar KA, et al. Meta-analysis of effects of exclusive breastfeeding on infant gut microbiota across populations. *Nat Commun.* 2018;9(1):4169.

46. Institute of Medicine. *Dietary Reference Intakes for Energy, Carbohydrate, Fiber, Fat, Fatty Acids, Cholesterol, Protein, and Amino Acids*. Washington, DC: The National Academies Press; 2005.
47. Nyman M, Asp NG, Cummings J, Wiggins H. Fermentation of dietary fibre in the intestinal tract: comparison between man and rat. *Br J Nutr*. 1986;55(3):487–496.
48. Holscher HD. Dietary fiber and prebiotics and the gastrointestinal microbiota. *Gut Microbes*. 2017;8(2):172–184.
49. Simpson HL, Campbell BJ. Review article: dietary fibre-microbiota interactions. *Aliment Pharmacol Ther*. 2015;42(2):158–179.
50. Wu GD, Chen J, Hoffmann C, et al. Linking long-term dietary patterns with gut microbial enterotypes. *Science*. 2011;334(6052):105–108.
51. So D, Whelan K, Rossi M, et al. Dietary fiber intervention on gut microbiota composition in healthy adults: a systematic review and meta-analysis. *Am J Clin Nutr*. 2018;107(6):965–983.
52. Vandeputte D, Falony G, Vieira-Silva S, et al. Prebiotic inulin-type fructans induce specific changes in the human gut microbiota. *Gut*. 2017;66:1968–1974.
53. Kamarul Zaman M, Chin KF, Rai V, Majid HA. Fiber and prebiotic supplementation in enteral nutrition: a systematic review and meta-analysis. *World J Gastroenterol*. 2015;21(17):5372–5381.
54. Wedlake L, Slack N, Andreyev HJ, Whelan K. Fiber in the treatment and maintenance of inflammatory bowel disease: a systematic review of randomized controlled trials. *Inflamm Bowel Dis*. 2014;20(3):576–586.
55. Lin D, Peters BA, Friedlander C, et al. Association of dietary fibre intake and gut microbiota in adults. *Br J Nutr*. 2018;120(9):1014–1022.
56. Johnson AJ, Vangay P, Al-Ghalith GA, et al. Daily sampling reveals personalized diet-microbiome associations in humans. *Cell Host Microbe*. 2019;25(6):789–802.e785.
57. Elia M, Cummings JH. Physiological aspects of energy metabolism and gastrointestinal effects of carbohydrates. *Eur J Clin Nutr*. 2007;61(Suppl 1):S40–S74.
58. Yang X, Darko KO, Huang Y, et al. Resistant starch regulates gut microbiota: structure, biochemistry and cell signalling. *Cell Physiol Biochem*. 2017;42(1):306–318.
59. Martinez I, Kim J, Duffy PR, Schlegel VL, Walter J. Resistant starches types 2 and 4 have differential effects on the composition of the fecal microbiota in human subjects. *PLoS One*. 2010;5(11):e15046.
60. Dahl W, Ford A, Ukhanova M, et al. Resistant potato starches (type 4 RS) exhibit varying effects on laxation with and without phylum level changes in microbiota: a randomised trial in young adults. *J Funct Foods*. 2016,23:1–11.
61. Salonen A, Lahti L, Salojarvi J, et al. Impact of diet and individual variation on intestinal microbiota composition and fermentation products in obese men. *ISME J*. 2014;8(11):2218–2230.
62. Fassio F, Facioni MS, Guagnini F. Lactose maldigestion, malabsorption, and intolerance: a comprehensive review with a focus on current management and future perspectives. *Nutrients*. 2018;10(11):e1599.
63. Li X, Yin J, Zhu Y, et al. Effects of whole milk supplementation on gut microbiota and cardiometabolic biomarkers in subjects with and without lactose malabsorption. *Nutrients*. 2018;10(10):e1403.
64. Ruiz-Ojeda FJ, Plaza-Diaz J, Saez-Lara MJ, Gil A. Effects of sweeteners on the gut microbiota: a review of experimental studies and clinical trials. *Adv Nutr*. 2019;10(suppl 1):S31–s48.
65. Jang C, Hui S, Lu W, et al. The small intestine converts dietary fructose into glucose and organic acids. *Cell Metab*. 2018;27(2):351–361.e353.
66. David LA, Maurice CF, Carmody RN, et al. Diet rapidly and reproducibly alters the human gut microbiome. *Nature*. 2014;505(7484):559–563.

67. Rehm CD, Penalvo JL, Afshin A, Mozaffarian D. Dietary intake among US adults, 1999-2012. *JAMA*. 2016;315(23):2542–2553.
68. US Department of Health and Human Services, US Department of Agriculture. *2015–2020 Dietary Guidelines for Americans*. 8th ed. 2015. http://health.gov/dietaryguidelines/2015/guidelines/. Published. Accessed.
69. Kosinski C, Jornayvaz FR. Effects of ketogenic diets on cardiovascular risk factors: evidence from animal and human studies. *Nutrients*. 2017;9(5):e517.
70. Bisanz JE, Upadhyay V, Turnbaugh JA, Ly K, Turnbaugh PJ. Meta-analysis reveals reproducible gut microbiome alterations in response to a high-fat diet. *Cell Host Microbe*. 2019;26(2):265–272.e264.
71. Wan Y, Wang F, Yuan J, et al. Effects of dietary fat on gut microbiota and faecal metabolites, and their relationship with cardiometabolic risk factors: a 6-month randomised controlled-feeding trial. *Gut*. 2019;68(8):1417–1429.
72. Watson H, Mitra S, Croden FC, et al. A randomised trial of the effect of omega-3 polyunsaturated fatty acid supplements on the human intestinal microbiota. *Gut*. 2018;67(11):1974–1983.
73. Menni C, Zierer J, Pallister T, et al. Omega-3 fatty acids correlate with gut microbiome diversity and production of N-carbamylglutamate in middle aged and elderly women. *Sci Rep*. 2017;7(1):11079.
74. Beaumont M, Portune KJ, Steuer N, et al. Quantity and source of dietary protein influence metabolite production by gut microbiota and rectal mucosa gene expression: a randomized, parallel, double-blind trial in overweight humans. *Am J Clin Nutr*. 2017;106(4):1005–1019.
75. Ford AL, Nagulesapillai V, Piano A, et al. Microbiota stability and gastrointestinal tolerance in response to a high protein diet with and without a prebiotic, probiotic and synbiotic: a randomized, double-blind, placebo-controlled trial in older women. *J Acad Nutr Diet*. 2020;120(4):500–516.e10.
76. Lang JM, Pan C, Cantor RM, et al. Impact of individual traits, saturated fat, and protein source on the gut microbiome. *MBio*. 2018;9(6):e01604–18.
77. Cotillard A, Kennedy SP, Kong LC, et al. Dietary intervention impact on gut microbial gene richness. *Nature*. 2013;500:585.
78. Walker AW, Ince J, Duncan SH, et al. Dominant and diet-responsive groups of bacteria within the human colonic microbiota. *ISME J*. 2010;5:220.
79. Tosti V, Bertozzi B, Fontana L. Health benefits of the Mediterranean diet: Metabolic and molecular mechanisms. *J Gerontol A Biol Sci Med Sci*. 2018;73(3):318–326.
80. Kahleova H, Levin S, Barnard N. Cardio-metabolic benefits of plant-based diets. *Nutrients*. 2017;9(8):e848.
81. Food and Agriculture Organization. *Food-based Dietary Guidelines*. Food and Agriculture Organization; 2020. *http://www.fao.org/nutrition/nutrition-education/food-dietary-guidelines/en/*. Published. Accessed March 28, 2020.
82. Tett A, Huang KD, Asnicar F, et al. The Prevotella copri complex comprises four distinct clades underrepresented in Westernized populations. *Cell Host Microbe*. 2019;26(5):666–679.e667.
83. Zinocker MK, Lindseth IA. The Western diet-microbiome-host interaction and its Role in Metabolic Disease. *Nutrients*. 2018;10(3):365.
84. Shikany JM, Demmer RT, Johnson AJ, et al. Association of dietary patterns with the gut microbiota in older, community-dwelling men. *Am J Clin Nutr*. 2019;110(4):1003–1014.
85. Widmer RJ, Flammer AJ, Lerman LO, Lerman A. The Mediterranean diet, its components, and cardiovascular disease. *Am J Med*. 2015;128(3):229–238.
86. De Filippis F, Pellegrini N, Vannini L, et al. High-level adherence to a Mediterranean diet beneficially impacts the gut microbiota and associated metabolome. *Gut*. 2016;65(11):1812–1821.

87. Meslier V, Laiola M, Roager HM, et al. Mediterranean diet intervention in overweight and obese subjects lowers plasma cholesterol and causes changes in the gut microbiome and metabolome independently of energy intake. *Gut*. 2020. pii: gutjnl-2019-320438.
88. Koeth RA, Lam-Galvez BR, Kirsop J, et al. L-carnitine in omnivorous diets induces an atherogenic gut microbial pathway in humans. *J Clin Invest*. 2019;129(1): 373–387.
89. Qi J, You T, Li J, et al. Circulating trimethylamine N-oxide and the risk of cardiovascular diseases: a systematic review and meta-analysis of 11 prospective cohort studies. *J Cell Mol Med*. 2018;22(1):185–194.
90. Vogt NM, Romano KA, Darst BF, et al. The gut microbiota-derived metabolite trimethylamine N-oxide is elevated in Alzheimer's disease. *Alzheimers Res Ther*. 2018;10(1):124.
91. Ge X, Zheng L, Zhuang R, et al. The gut microbial metabolite trimethylamine N-oxide and hypertension risk: a systematic review and dose-response meta-analysis. *Adv Nutr*. 2020;11(1):66–76.
92. Oellgaard J, Winther SA, Hansen TS, Rossing P, von Scholten BJ. Trimethylamine N-oxide (TMAO) as a new potential therapeutic target for insulin resistance and cancer. *Curr Pharm Des*. 2017;23(25):3699–3712.
93. DiNicolantonio JJ, McCarty M, OKeefe J. Association of moderately elevated trimethylamine N-oxide with cardiovascular risk: is TMAO serving as a marker for hepatic insulin resistance. *Open Heart*. 2019;6(1):e000890.
94. Parker HW, Vadiveloo MK. Diet quality of vegetarian diets compared with non-vegetarian diets: a systematic review. *Nutr Rev*. 2019;77(3):144–160.
95. Patel KP, Luo FJ, Plummer NS, Hostetter TH, Meyer TW. The production of p-cresol sulfate and indoxyl sulfate in vegetarians versus omnivores. *Clin J Am Soc Nephrol*. 2012;7(6):982–988.
96. Trefflich I, Jabakhanji A, Menzel J, et al. Is a vegan or a vegetarian diet associated with the microbiota composition in the gut? Results of a new cross-sectional study and systematic review. *Crit Rev Food Sci Nutr*. 2019;1–15.
97. Losasso C, Eckert EM, Mastrorilli E, et al. Assessing the influence of vegan, vegetarian and omnivore oriented westernized dietary styles on human gut microbiota: a cross sectional study. *Front Microbiol*. 2018;9:317.
98. Barrett HL, Gomez-Arango LF, Wilkinson SA, et al. A vegetarian diet is a major determinant of gut microbiota composition in early pregnancy. *Nutrients*. 2018; 10(7):890.
99. De Angelis M, Ferrocino I, Calabrese FM, et al. Diet influences the functions of the human intestinal microbiome. *Sci Rep*. 2020;10(1):4247.
100. Walczyk T, Wick JY. The ketogenic diet: making a comeback. *Consult Pharm*. 2017;32(7):388–396.
101. Dahl WJ, Niebergall EJ, Owen RJ. Implications of fiber inadequacy in the ketogenic diet: a case study. *ICAN*. 2011;3(5):288–290.
102. Zhang Y, Zhou S, Zhou Y, Yu L, Zhang L, Wang Y. Altered gut microbiome composition in children with refractory epilepsy after ketogenic diet. *Epilepsy Res*. 2018;145:163–168.
103. Lindefeldt M, Eng A, Darban H, et al. The ketogenic diet influences taxonomic and functional composition of the gut microbiota in children with severe epilepsy. *NPJ Biofilms Microbiomes*. 2019;5:5.
104. Murtaza N, Burke LM, Vlahovich N, et al. The effects of dietary pattern during intensified training on stool microbiota of elite race walkers. *Nutrients*. 2019;11(2):261.
105. Swidsinski A, Dorffel Y, Loening-Baucke V, et al. Reduced mass and diversity of the colonic microbiome in patients with multiple sclerosis and their improvement with ketogenic diet. *Front Microbiol*. 2017;8:1141.

106. Nagpal R, Neth BJ, Wang S, Craft S, Yadav H. Modified Mediterranean-ketogenic diet modulates gut microbiome and short-chain fatty acids in association with Alzheimer's disease markers in subjects with mild cognitive impairment. *EBioMedicine*. 2019;47:529–542.
107. Barone M, Turroni S, Rampelli S, et al. Gut microbiome response to a modern Paleolithic diet in a Western lifestyle context. *PLoS One*. 2019;14(8):e0220619.
108. Gibbons A. The evolution of diet. In: *National Geographic*. 2013. https://www.nationalgeographic.com/foodfeatures/evolution-of-diet/. Published Accessed.
109. Macfarlane GT, Macfarlane S. Fermentation in the human large intestine: its physiologic consequences and the potential contribution of prebiotics. *J Clin Gastroenterol*. 2011;45(Suppl):S120–S127.
110. Burrows TL, Ho YY, Rollo ME, Collins CE. Validity of dietary assessment methods when compared to the method of doubly labeled water: a systematic review in adults. *Front Endocrinol (Lausanne)*. 2019;10:850.
111. Amoutzopoulos B, Page P, Roberts C, et al. Portion size estimation in dietary assessment: a systematic review of existing tools, their strengths and limitations. *Nutr Rev*. 2020. pii: nuz107. https://doi.org/10.1093/nutrit/nuz107.
112. US Department of Agriculture. FoodData Central Agricultural Research Service. fdc.nal.usda.gov; 2019. Published. Accessed March 28, 2020.
113. Vetrani C, Maukonen J, Bozzetto L, et al. Diets naturally rich in polyphenols and/or long-chain n-3 polyunsaturated fatty acids differently affect microbiota composition in high-cardiometabolic-risk individuals. *Acta Diabetol*. 2020;https://doi.org/10.1007/s00592-020-01494-9.
114. Kashiwagi S, Naito Y, Inoue R, et al. Mucosa-associated microbiota in the gastrointestinal tract of healthy japanese subjects. *Digestion*. 2020. 101(2):107–120.
115. Cummings JH, Pomare EW, Branch WJ, Naylor CP, Macfarlane GT. Short chain fatty acids in human large intestine, portal, hepatic and venous blood. *Gut*. 1987; 28(10):1221–1227.
116. Vandeputte D, Falony G, Vieira-Silva S, Tito RY, Joossens M, Raes J. Stool consistency is strongly associated with gut microbiota richness and composition, enterotypes and bacterial growth rates. *Gut*. 2016;65(1):57–62.
117. Muller M, Hermes GDA, Canfora EE, et al. Distal colonic transit is linked to gut microbiota diversity and microbial fermentation in humans with slow colonic transit. *Am J Physiol Gastrointest Liver Physiol*. 2019;318:G361–G369.
118. Ghanchi A, James PT, Cerami C. Guts, germs, and iron: a systematic review on iron supplementation, iron fortification, and diarrhea in children aged 4–59 months. *Curr Dev Nutr*. 2019;3(3):nzz005.
119. Fernandez-Navarro T, Salazar N, Gutierrez-Diaz I, et al. Bioactive compounds from regular diet and faecal microbial metabolites. *Eur J Nutr*. 2018;57(2):487–497.
120. Chaplin A, Carpene C, Mercader J. Resveratrol, metabolic syndrome, and gut microbiota. *Nutrients*. 2018;10(11):1651.
121. Queipo-Ortuno MI, Boto-Ordonez M, Murri M, et al. Influence of red wine polyphenols and ethanol on the gut microbiota ecology and biochemical biomarkers. *Am J Clin Nutr*. 2012;95(6):1323–1334.
122. Vieira AR, Abar L, Vingeliene S, et al. Fruits, vegetables and lung cancer risk: a systematic review and meta-analysis. *Ann Oncol*. 2016;27(1):81–96.
123. Alsharairi NA. The effects of dietary supplements on asthma and lung cancer risk in smokers and non-smokers: a review of the literature. *Nutrients*. 2019;11(4):e725.
124. Gurry T, Gibbons SM, Nguyen LTT, et al. Predictability and persistence of prebiotic dietary supplementation in a healthy human cohort. *Sci Rep*. 2018; 8(1):12699.

125. Dahl WJ, Alvarez MM. Whole pulses and pulse fiber: modulating gastrointestinal function and the microbiome. In: *Health Benefits of Pulses*. Springer; 2019:91–108.
126. Tangestani H, Emamat H, Ghalandari H, et al. Whole grains, dietary fibers and the human gut microbiota: a systematic review of existing literature. *Recent Pat Food Nutr Agric*. 2020. https://doi.org/10.2174/2212798411666200316152252.
127. Lamuel-Raventos RM, Onge MS. Prebiotic nut compounds and human microbiota. *Crit Rev Food Sci Nutr*. 2017;57(14):3154–3163.

CHAPTER NINE

Interactions of probiotics and prebiotics with the gut microbiota

Rafael Ballan[a,b], Carolina Battistini[a,b], Douglas Xavier-Santos[a,b,c], Susana Marta Isay Saad[a,b,*]

[a]Department of Pharmaceutical and Biochemical Technology, School of Pharmaceutical Sciences, University of São Paulo, São Paulo, SP, Brazil
[b]Food Research Center, University of São Paulo, São Paulo, SP, Brazil
[c]Department of Civil, Chemical and Environmental Engineering, University of Genoa, Pole of Chemical Engineering, Genoa, Italy
[*]Corresponding author: e-mail address: susaad@usp.br

Contents

Abstract

The gut microbiota (GM) composition varies among individuals and is influenced by intrinsic (genetics, age) and extrinsic (environment, diet, lifestyle) factors. An imbalance or dysbiosis is directly associated with the development of several illnesses, due to the

Progress in Molecular Biology and Translational Science, Volume 171
ISSN 1877-1173
https://doi.org/10.1016/bs.pmbts.2020.03.008

potential increase in intestinal permeability leading to a systemic inflammation triggered by higher levels of circulating lipopolysaccharides and changes in the immune response caused by an overgrowth of a specific genus or of pathogens. These mechanisms may increase symptoms in gastrointestinal disorders or reduce glucose tolerance in metabolic diseases. Diet also has a significant impact on GM, and functional foods, namely prebiotics and probiotics, are a novel approach to reestablish the indigenous microbiota. Prebiotics, like inulin and polyphenols, are selectively utilized by GM, releasing short-chain fatty acids (SCFA) and other metabolites which may reduce the intestinal lumen pH, inhibit growth of pathogens, and enhance mineral and vitamin bioavailability. Probiotic microorganism may increase the microbial diversity of GM and improve the integrity of the intestinal barrier, leading to an improvement of baseline and pathologic inflammation. In this chapter, we will discuss the potential roles of prebiotics and probiotics in health and diseases throughout an individual's lifetime and proposed mechanisms of action.

1. Introduction

The term "microbiota" refers to the entire population of microorganisms in a given location. The largest population of microorganisms is the microbiota of the gut, which is comprised of more than 2000 bacterial species and is significantly influenced by intrinsic (genetics, age) and extrinsic (body mass index, smoking, physical activity, diet) factors, mainly by diet, which regulates microbial activity and gene expression (Fig. 1).[1–3] Studies have shown that several diseases are associated with an imbalance of the bacterial composition of the gut, which is referred to as a dysbiosis. Dietary interventions may be effective in restoring it to a healthier state.[4]

In this sense, functional foods, which in addition to their nutritional properties must confer a defined benefit to the consumer's health, seem to be interesting alternatives for dietary interventions targeting at improving the microbiota shape. The most known functional foods or ingredients are polyphenol, polyunsaturated fatty acids, prebiotics, and probiotics.[5,6]

Probiotics are defined as "live microorganisms that, when administered in adequate amounts, confer a health benefit on the host."[7] While the definition of probiotics has been debated, an expert consensus from 2013 has agreed on this description.[8] The major mechanisms of action of probiotics on human health are multifold and include competition with pathogens for nutrients and adhesion sites, production of bacteriocin, vitamins, and short-chain fatty acids (SCFA), immunomodulation, improvement of the intestinal barrier, and production of neurotransmitters.[9]

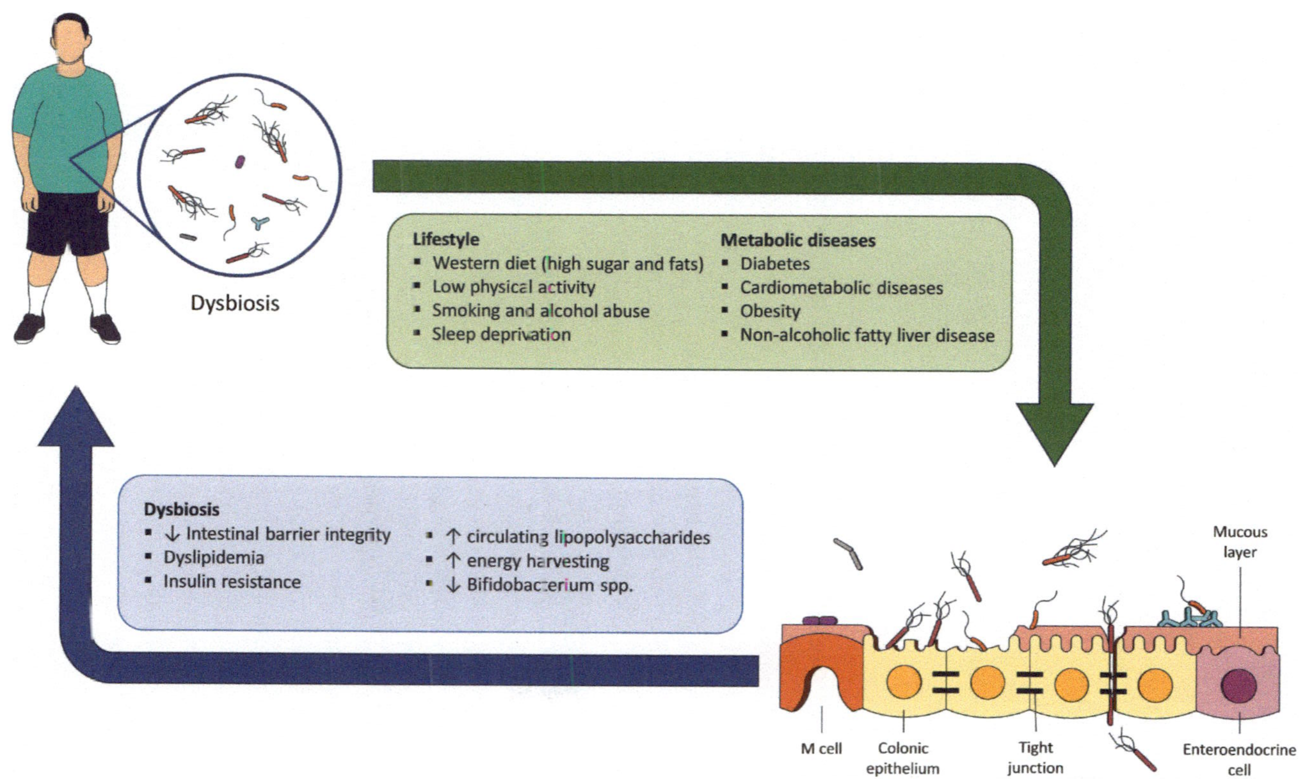

Fig. 1 Factors that influence the gut microbiota.

In the last years, studies related to probiotics and their effects on human health have increased exponentially. The vehicles or food matrices used to deliver probiotics affect the viability of the microorganisms and, therefore, have different impacts on the host microbiota. In addition, prebiotics and bioactive food compounds may interact with the commensal microorganisms that inhabit the human gut.

Several clinical trials have been conducted in order to evaluate the effect of probiotics, primarily those containing bacteria from the *Lactobacillus* and *Bifidobacterium* genera, on different food matrices and their impacts on the human health, leading to heterogeneous results. In this chapter, we will delve into different bioactive and probiotic foods and their effects on the human health.

2. An overview into the human microbiota

The human microbiota consists of a complex ecosystem that holds approximately 10^{14} bacteria. Much has been discussed about the factors that may contribute to its shape in the prenatal period, since the mother's oral microbiota has recently been identified as being similar to that found in the placenta, comprising Firmicutes, Tenericutes, Proteobacteria, Bacteroides, and Fusobacteria phyla. One hypothesis is that these bacteria are transported from the oral cavity to the fetus through the bloodstream.[10]

Outside of any in utero exposures, the first contact of the newborn with various microorganisms occurs during birth, which is influenced by the mode of delivery.[11] In the vaginal delivery, the skin and mucous membranes of the baby are in contact with the local microbiota, which leads to colonization by *Lactobacillus* spp., reflecting the mother's vaginal microbiota. In contrast, when delivery occurs by cesarean section, the newborn's mouth, intestines, and skin are colonized by microorganisms present in the birth and skin environments, such as *Staphylococcus* and *Propionibacteria*.[12,13]

During the first months of life, the gut microbiota adapts to the conditions of the surrounding environment and are influenced by anaerobic conditions, nutrient availability, and microbial interactions during community succession.[13] Cesarean-born infants often have fewer important maternally-transmitted microorganisms, such as Bacteroides and Bifidobacteria. Infants delivered via c-section continue to exhibit a lower diversity and reduced Th1 response during the first two years of life, although this difference in colonization between delivery modes is gradually reduced over time.[12,13]

Breastfeeding is the other major contributor to early microbial colonization of infants. Breast milk may contain up to 10^7 bacterial cells in 800 mL, mainly from the *Lactobacillus*, *Streptococcus*, and *Staphylococcus* genera.[14] In addition, it contains oligosaccharides that stimulate the selective growth of *Bifidobacterium* and *Lactobacillus* spp. and through fermentation in the gut, SCFA are released, reducing the colonic pH. Initially, this reduces the diversity of the infant's intestinal microbial community, which is likely due to acidification of the intestinal pH. Meanwhile, acidification acts as a defense against pathogens that cannot survive under these conditions.[15] Furthermore, the presence of immunoglobulin A, defensins, and lactoferrin in human milk provide other benefits to the infant's health,[16] and exclusive breastfeeding in the first months of life has been associated with a reduction in several childhood diseases, such as obesity, infections, and atopic diseases.[17–19]

After food introduction and, consequently, increased nutrient pool ingestion, the infant's intestinal microbiota resembles an adult's microbiota. It is generally reached in the third year of life and will be influenced together with the environment in which the child is inserted. A diverse microbiota enhances the vitamins and amino acids biosynthesis and the ability to metabolize carbohydrates.[13,20]

In adulthood, the microbiota is composed mainly of Firmicutes and Bacteroidetes, whereas members of other phyla, like Actinobacteria, Proteobacteria, and Verrucomicrobia are present in lower populations. The gut microbiota plays a role in various functions of the body, such as energy storage, intestinal barrier integrity, immune system development, including tolerogenicity, and neurotransmitter production. The host's microbiota is dynamic and, until now, it is difficult to establish how different factors, such as lifestyle, illnesses, geographical location, and age, may impact its stability.[21,22]

Presently, there is a collective effort to identify what would be a healthy microbiota, also known as a "core microbiota." However, due to the highly heterogenicity among individuals, it is easier to compare with the same individual along his lifetime. In fact, a core microbiome exists at the level of metabolic functions.[23] Throughout life, the gut microbiota is resilient, and transient or punctual situations do not seem to significantly affect its composition, such as adherence to a specific diet, use of antibiotics, and exposure to different environments.[24]

Several clinical trials have demonstrated that a number of diseases are related to changes in the relative abundance of a phylum or genus. For example, obesity or high caloric diet are associated with a greater relative abundance of Firmicutes over Bacteroidetes. This also occurs in populations that eat a typical western diet, which is rich in processed foods, carbohydrates, and sugars.[25,26] Other metabolic diseases, such as diabetes, cardiometabolic diseases, and low-grade inflammation associated with obesity are inversely associated with *Akkermansia muciniphila* abundance.[27]

In conclusion, mechanisms involved in shifts of the gut microbiota are still not well understood. However, dietary interventions with functional foods, like probiotics, are being studied extensively and have been proposed as a way of ameliorating various diseases and as a potential adjunctive therapy or preventative strategy, in spite of the fact that they are not able to promote permanent changes in the host's microbiota.

3. Functional foods

It is well known that functional foods may be good alternatives in preventing or ameliorating the *status* of certain diseases. Moreover, they can be used in combination with traditional therapies as adjuvants and, unlike antibiotics, will not have a detrimental effect on beneficial populations of bacteria. The major functional foods or ingredients explored are prebiotics, probiotics, the combinations of both in synbiotic foods, in addition to dietary fibers and bioactive compounds.

3.1 Prebiotics

A prebiotic is defined as "a substrate that is selectively utilized by host microorganisms conferring a health benefit."[5] The potential health benefits related to prebiotics include gut microbiota modulation along with beneficial microbial metabolites release (e.g., SCFA and tryptophan). However, these effects should be verified in the target host (animal or human).[5,28,29] Several prebiotics are commercially available, like fructo-oligosaccharides (FOS), inulin, β-galactooligosaccharides (GOS), lactulose, isomalto-oligosaccharides (IMO), and resistant starches (i.e., the fraction of starch that resists to digestion in the small intestine).[30]

Innumerous studies explored the potential health benefits of fiber consumption, with or without prebiotic effect. The main pathway associated with prebiotics involves selective fermentation by beneficial microorganism present in the gut, including *Lactobacillus* and *Bifidobacterium*, which produce acetate and lactate and in turn may stimulate production of butyrate by other bacteria. Moreover, the production of SCFA bestows diverse health benefits like improvement of mineral absorption and barrier function.[5,28,29]

In addition, prebiotics may be used to enhance fermentation processes or as encapsulating material. According to Oliveira et al., inulin, one of the most studied prebiotic compounds, may increase the acidification rate of milk by co-cultures of probiotic and starter strains.[31] Meanwhile, Santos et al. also reported higher resistance to GIT stress simulated in vitro by the probiotic strain *Lactobacillus acidophilus* La-5 microencapsulated with inulin when compared to free cells. Moreover, when incorporated in a food matrix, the survival rate was further increased.[32] Similarly, Rosolen et al. stated that the combination of whey and inulin as a coating material for *Lactococcus lactis* R7 promoted resistance to heat treatment and GIT stress simulated in vitro.[33]

3.2 Polyphenols

Polyphenols are a large group of compounds that are well recognized for their anti-inflammatory and antioxidant properties and their potential health benefits on cardiometabolic diseases and brain function. The most common phenolic compounds are flavonoids (e.g., flavonols, isoflavones, and anthocyanins), phenolic acids (e.g., ellagic and caffeic acids), lignans, and stilbenes (e.g., resveratrol), which can be consumed in black and green tea, red wine, apples, blueberries, dark chocolate, onions, almonds, soy, pomegranates, and coffee, among others.[34–36]

After ingestion, some polyphenols are absorbed in the small intestines, while a significant amount can reach the large intestines intact where they can be metabolized by the gut microbiota and release active metabolites, like phenolic acids. Moreover, studies have shown that some compounds may stimulate beneficial bacteria, like *Lactobacillus* spp., *Bifidobacterium* spp., *Akkermansia muciniphila*, and *Faecalibacterium prausnitzii*. Conversely, other compounds may inhibit pathogens, like *Helicobacter pylori*, *Staphylococcus aureus*, and *Listeria monocytogenes*.[34–36]

Hence, there is evidence that polyphenols have potential as prebiotics and an innovative option for gut microbiota modulation. Furthermore, fermentation by lactic acid bacteria may be an alternative to improve bioavailability of these compounds, which are generally low in unprocessed foods.[36–38]

3.3 Omega-3

Consumption of omega-3 polyunsaturated fatty acids has been associated with a reduction in cardiovascular diseases and improvement of cognition, depression, behavioral disorders, and brain function. Besides gut microbiota modulation, anti-inflammatory effects, and increased SCFA release, showing direct influence in the gut-brain axis.[39–42]

The major bioactive forms of omega-3 are eicosapentaenoic acid (EPA) and docosahexaenoic acid (DHA), which can be found in some fatty fishes or in nutraceutical supplements. Their precursors are α-linolenic acid (ALA), contained in nuts, flaxseed, and canola and soy oil.[39,40]

Studies have reported the potential of omega-3 fatty acids to modulate the gut microbiota and to act as prebiotic candidates. According to Prossomariti and colleagues, the consumption of EPA was able to improve the mucosal inflammation in ulcerative colitis (UC) patients, while decreasing *Clostridium* spp. in the feces in a short-term intervention.[41] In another study, a high omega-3 diet increased significantly *Bifidobacterium* and *Lactobacillus* population in mice throughout life. Meanwhile, it influenced positively brain function and behavior of the animals.[42]

3.4 Probiotics

Potential health benefits associated with probiotic consumption are widely known. Nonetheless, several factors should be taken into consideration when developing a food product or supplement with claims of health benefits. The microorganisms should be well recognized as GRAS (Generally Recognized as Safe), claims of health benefits may only be associated with select strains, not all genuses or species, or with the target population in that specific clinical trial (e.g., infants, chronic disease patients, elderly, etc.). In addition, a recommended daily dose of colony-forming unit (CFU) to achieve health benefits should be indicated and the microorganisms should be viable during the shelf-life of the products.[7,8,43]

The most common probiotics are species from the *Bifidobacterium* and *Lactobacillus* genera. However, other microorganisms may be classified as probiotics, such as *Akkermansia muciniphila*, *Faecalibacterium prausnitzii*, *Streptococcus thermophilus*, *Lactococcus lactis*, and *Saccharomyces bourlardii*.[8,44,45] Moreover, each strain has different multiplication and survival rates after passage through the gastrointestinal tract (GIT) depending on the food matrix (e.g., milk, soymilk, and fruit juice), pH, incorporated oxygen (e.g., stirred-yogurt), temperature of storage (e.g., room temperature, refrigerated or frozen), presence of prebiotics or any other food ingredient, and microencapsulation.[45–49]

In addition to health effects, probiotics can improve food matrices by reducing the level of undesirable compounds, like stachyose and raffinose in soymilk, or by producing compounds of interest, like vitamins.[46,47,50] Moreover, starter strains, such as *Streptococcus thermophilus*, are usually combined with probiotic cultures to reduce fermentation times, which when co-cultured with *Bifidobacterium* strains prevent off-flavors due to acetic acid release by species from this genus.[31,47,51]

Prebiotic compounds may be used in combination with probiotics, resulting in synbiotic foods, which have synergistic effects including reduction of fermentation time or enhancement of the survival of probiotics during passage through the GIT.[31,44] However, any changes in matrices may affect the behavior of strains differently. Bedani et al. reported that *Lactobacillus acidophilus* La-5 and *Bifidobacterium animalis* Bb-12 in fermented soymilk had higher survival rates after GIT stress simulated in vitro when compared to their fresh cultures, without any food matrix.[52] On the other hand, the addition of tropical fruit pulps in probiotic fermented soymilk showed the opposite effect, despite an improvement in the consumers' acceptance of the product.[53]

Besides prebiotics or other ingredients, emerging food technologies, such as microencapsulation and ohmic heating, can be applied to improve fermentation processes or to increase the resistance to GIT stress.[31–33,47,51] Nevertheless, it has recently been suggested that microbial metabolites alone are enough to confer health benefits and viable cells in the gut are not necessarily essential. This concept might completely change the directions of probiotic food and supplements research.[47]

In conclusion, there are several factors that influence probiotic stability in different matrices and their survival through the GIT, and consequently, their health benefits. Nevertheless, there is a lot of heterogeneity in the intervention studies reported, like participants' age, sex, body mass index, metabolic diseases, daily dose, duration of the study, different prebiotic and/or probiotic cultures administered, and different food matrices. Therefore, it is difficult to compare outcomes and identify one unique answer to fit all cases and design of a tailored treatment is likely a more biologically plausible approach to probiotic use.

4. Effects on healthy individuals

A limited number of clinical trials have studied the impact of probiotics or bioactive food consumption on the gut microbiota, especially in healthy individuals. Several studies evaluated the fecal recovery of probiotic strains in order to verify if the administered microorganisms survived through the GIT, in addition to checking the participants' adherence to the dietary intervention. Indeed, molecular biology techniques, like gene sequencing or real-time PCR quantification, are helpful in understanding the actual effects of probiotics on the gut microbiota. However, they are costly and are not always available to all researchers.

In a randomized double-blind placebo-controlled trial, Rampelli et al. administered a probiotic biscuit daily containing *Bifidobacterium longum* Bar33 and *Lactobacillus helveticus* Bar13, 10^9 CFUs per serving versus placebo to 32 elderly participants living in Italy. Fecal samples were collected at baseline (T0) and after 30 days (T30). When compared to healthy adults, elderly subjects had higher levels of pathobionts, such as *Clostridium difficile*, *C. perfringens*, *Bacillus cereus*, *Campylobacter*, and Enterobacteriaceae, whose influence on the host occurs indirectly via stimulation of the immune system. After treatment, the placebo group microbiota remained the same while the intervention group had decreased opportunistic pathogens. Furthermore, increased content of health-promoting bacteria such as *Akkermansia muciniphila*, *B. longum*, and Lactobacillaceae was observed in the treatment group. These results suggest that dietary intervention with probiotic food may improve age-related dysbiosis, promoting a transitory healthier gut microbiota.[54]

Fermented milk containing *Lactobacillus casei* strain *Shirota* (LCS) is known to promote immune system function in healthy individuals. Mechanisms, such as an increase in natural Killer (NK) cell function, activation markers on circulating T cells, and increased concentrations of IgA1 and IgA2 in the oral mucosa, are well described.[55,56] Nevertheless, little is known about the effects of its consumption on the intestinal microbiota. In a study conducted in 25 healthy Chinese adults, subjects consumed daily 100 mL of a probiotic drink containing 10^8 CFU/mL of viable LCS (Yakult). Fecal recovery showed that the strain survived the GIT and promoted the growth of beneficial bacteria, especially *Bifidobacterium* and *Anaerostipes*, an efficient lactate fermenter, suggesting a synergistic effect, once *Bifidobacterium* is a lactate producer. Furthermore, population of butyrate-producing bacteria, such as *Roseburia intestinalis* and *Clostridium*, declined. The strain demonstrated poor ability to persist in the GIT during follow-up (intervention, day 21: 10^7 CFU/g of feces; endpoint, day 42: 10^4 CFU/g of feces) and decreased ability to produce SCFA. This suggests that, despite modifying the microbiota composition of adults, regular consumption is necessary.[57,58]

Interestingly, Klein et al. conducted a cross-over study evaluating the impact of the consumption of a yogurt (300 g/day) containing probiotic strains *Lactobacillus acidophilus* 74-2 (9.3×10^8 CFU/g) and *Bifidobacterium lactis* 420 (3.0×10^6 CFU/g) on the host microbiota. After the run-in period (3 weeks), subjects consumed the probiotic or control yogurt for 5 weeks and then they have switched groups for another 5 weeks. Probiotics were significantly increased in fecal samples, mainly *B. lactis* 420 counts, but this was transient. The levels of cholesterol, LDL-c and HDL-c, and SCFA did not change, nor did other specific immune parameters.[59] It is noteworthy that adults were healthy, there was no washout period, and the probiotic doses administered were probably not sufficient to promote health benefits.[60]

The effects of the consumption of probiotic strains on healthy adults were recently reviewed by Khalesi et al. and found that current literature supports the use of probiotics to improve defection frequency, stool consistency, bowel movement, immune system responses, and vaginal lactobacilli concentration.[61] It seems that the use of probiotics in improving GI discomfort is most effective in healthy individuals.

The most well-known properties of prebiotics in healthy individuals is their ability to relieve intestinal discomfort or increase absorption of minerals. Improvement in stool consistency, defecation frequency, regulation of motility by SCFA-signaling hormones, flatulence and bloating, and increased calcium solubility by pH reduction are also described frequently.[62–64]

Supplementation of prebiotic GOS at 5 g/day increased calcium absorption, together with an increase in fecal bifidobacteria in young girls (9–13 years of age), suggesting that bifidobacteria may mediate this mechanism.[62] Similarly, the use of a mixture of inulin-type fructans (8 g/day) has been shown to be effective in increasing calcium absorption in adolescents and may be an interesting approach for improving or preventing bone mass loss.[65]

The effect of β-fructans, such as oligofructose and fructo-oligosaccharides, has been evaluated in healthy and ill subjects. A recent meta-analysis of human studies that tested β-fructans as a single supplementary ingredient and applied in a product or in the form of dietary supplementation demonstrated that short-chain and not long-chain β-fructans supplementation (degree of polymerization <10) contributed to an increased stool frequency (0.36 defecation ± 0.06 per day; $P < 0.001$). In addition, β-fructans improved stool consistency and stool wet weight. The effects promoted by β-fructans appear to improve these features up to a dose of 18 g/day, with no substantial effects above that. Furthermore, bifidobacteria growth stimulation appears to be dose-related, between 2.5 and 10 g/day, and doses above this only lead to marginal increases in stool frequency. Thus, different types of prebiotics might be useful for those with chronic constipation and may promote well-being by regulating bowel function.[66]

There are divergent opinions about studies designed to investigate the ability of probiotics to colonize the gut. Many studies fail to perform a follow-up of study participants' microbiota. However, since lifestyle is one of the factors shaping the autochthonous microbiota and probiotics have poor adherence to the intestinal mucosa, it is to be expected that the consumption of probiotic foods on a daily basis is recommended to produce the benefits observed in clinical trials.

5. Effects on various diseases

5.1 Application on early life conditions and diseases

Gastrointestinal disturbances are common in the early years of life. Therefore, many clinical trials are dedicated to preventing and treating diseases that affect this population, including acute and antibiotic-associated diarrhea, infantile colic, prevention of *Clostridium difficile*-associated diarrhea, necrotizing enterocolitis, *H. pylori* infection and neonatal sepsis.[67]

It is common for infants not to receive exclusive breastmilk and to minimally receive its benefits.[68] Thus, some infant formulas have been supplemented with prebiotics, such as GOS and FOS, aiming to prevent diseases. Enrichment of infant formulas with prebiotics has been effective in reducing respiratory tract infections to comparable levels observed breast fed children.[69,70]

5.1.1 Acute gastroenteritis, antibiotic associated diarrhea, Clostridium difficile*-associated diarrhea*

The most well-established condition for probiotic supplementation in infants is acute diarrhea. The use of probiotics is strongly recommended by the European Society of Pediatric Infectious Diseases (ESPGHAN) guidelines as an adjunct with rehydration for reducing the length of hospitalization (1.12 days, 95% confidence interval [CI] −1.16 to 0.38) and disease symptoms. In addition, the best-established strains with high quality clinical trials conducted are *L. rhamnosus* GG and *S. boulardii* CNCM I-745.[71]

Antibiotic-associated diarrhea (AAD) is a major global health concern in children, especially in Asia-Pacific where antibiotics are overused.[72,73] The main causative agent is *C. difficile* and its complications are severe.

According to current guidelines for the Asia-Pacific region, *L. rhamnosus* GG and *S. boulardii* CNCM I-745 may be considered in pediatric populations, since they are effective in the prevention of AAD and *C. difficile* infections.[74] In contrast, the ESPGHAN group strongly recommends the use of *L. rhamnosus* GG or *S. boulardii* for AAD prevention. The recommendation to use *S. boulardii* to prevent *C. difficile* associated diarrhea is not as strong and includes the caveat that physicians should be cautious and include evaluation of risk factors for these conditions before prescribing use of these probiotic strains.[75]

5.1.2 Necrotizing enterocolitis (NEC)

The incidence of NEC in preterm infants is high, mainly in those with very low birth weight (VLBW) <1500 g, increasing the interest in evaluating probiotic effects that improve intestinal barrier function, prevent bacterial translocation, promote mucosal IgA production, and inhibit potential pathogens.[76,77]

Several meta-analyses show that the combination of *B. infantis* + *Str. thermophilus* + *B. bifidus* and *L. acidophilus* + *B. infantis*, were the most effective in reducing risk of NEC and mortality in VLBW infants, but not for nosocomial sepsis.[78] Thus, Latin-American (LATAM) experts' consensus grades 1a (Systematic Review with homogeneity of RCTs) recommend use of the following probiotic strains: *B. brevis*, *L. rhamnosus* GG, *L. acidophilus*, and *L. reuteri* DSM 17938 and mixtures of *Bifidobacterium* and *Streptococcus*.[67]

Although several studies have demonstrated the efficacy of probiotics from the *Lactobacillus* and *Bifidobacterium* species in reducing the incidence of NEC and NEC-associated mortality, recently conducted high quality RCTs have not been able to demonstrate that BBG-001 strain *B. brevis* is effective in reducing the risk of NEC. However, the latest expert consensus for the Asia-Pacific region states that probiotics may be considered to prevent NEC due to evidence of their effectiveness in reducing risk and mortality in high-risk populations.[74]

5.1.3 Infant colic

Colic is perceived as a condition that requires intervention in infants. There is usually an increase in gas production and changes in bowel motility, resulting in abdominal pain. The *L. reuteri* DSM 17938 strain was effective and safe in many clinical trials for the prevention of colic episodes, reducing infant fussiness, regurgitation, and constipation. Therefore, Latin-American experts classify the recommendation of this strain as grade 1a (prevention) and 1b (treatment), while for Asia-Pacific population the evidence is considered weak, but the use of *L. reuteri* DSM 17938 may be considered.[67,74]

5.1.4 Inflammatory bowel disease

The role of probiotics in inflammatory bowel diseases is mainly targeted toward UC and Crohn's disease (CD). Probiotics are thought to modulate the host mucosal immune response, resulting in anti-inflammatory effects in IBD patients. Particularly for UC, guidelines suggest that commercial probiotic VSL#3, a multi-strain product including Bifidobacteria, Lactobacilli and Streptococcal species, is effective in inducing and maintaining remission of mild to moderate UC.[79] Another strain well studied is *E. coli Nissle*, for which the guidelines indicate that it may be as effective as standard mesalazine treatment in maintaining remission in UC.[74]

Crohn's disease management seems to be less effective, although several studies have shown changes in the intestinal microbiota to be relevant in this condition. One possible reason is that CD may be sub-classified, due to the genetic component present.[80] Current guidelines do not support effective probiotic therapy in maintaining remission of CD. The last expert panel conducted by Cameron et al. for the Asian-Pacific region concluded that the evidence is weak to indicate the use of probiotics in IBD in children, and did not recommend it overall.[74] In contrast, LATAM expert consensus recommended the use of VSL#3 only for UC (Evidence 1b—Individual RCT, with narrow Confidence Interval).[67,81]

5.1.5 Helicobacter pylori infection

With the increase in antibiotic resistance by pathogenic bacteria, interest for probiotics has grown with the objective of using them as a primary and adjuvant therapy. Some *Lactobacillus* species studied were unable to eradicate *H. pylori*, even in combination with drugs, such as antibiotics and proton pump inhibitors and probiotics only moderately improved eradication rates.[82,83] The recommendation for their use in the Asia-Pacific region is classified as "very low quality," but stronger evidence exists for the use of *S. boulardii*. There are currently no recommendations for probiotic use for *H. pylori* eradication in other geographic regions.[67,74,75]

It is noticeable that the evaluation of probiotic effects in the first years of life is directed toward health outcomes and not necessarily their effects on the intestinal microbiota. There are many RCTs conducted in infants, since they are a high-risk population with high mortality rates, garnering interest to improve the health status of this vulnerable population.

Additionally, specific guidelines have been developed for different regions, recognizing that each bacterial strain has unique properties that are affected by geography.

5.2 Adults

5.2.1 Metabolic diseases

With increasing prevalence of obesity and metabolic diseases such as diabetes mellitus type 2 (T2DM) in developed and emerging countries, the contributing factors for these diseases have been widely studied.[84] The GM plays a key role in energy homeostasis and low-grade inflammation, and its unbalance is associated with the development of the diseases mentioned above, a fact that has aroused the attention of researchers. Some bacterial species have been associated with obesity, such as *Lactobacillus* and *Bifidobacterium* species, Prevotellaceae and *Blautia coccoides*, as well as a high ratio between Firmicutes and Bacteroidetes.[85] Interventions that restore bacterial homeostasis in the gut may represent an important target for the use of probiotics and prebiotics and several studies have been conducted with the purpose of testing this approach. Such studies and respective results are discussed below.

In one study conducted by Stenman and colleagues, the strain *B. animalis* ssp. *lactis* 420 (B420) was administered to 225 healthy overweight or obese volunteers during a six-month period. The studied population was divided into four groups—probiotic only (B420, 10^{10} CFU/day), prebiotic Litesse® Ultra polydextrose alone (LU, 12 g/day), probiotic + prebiotic (B420: 10^{10} CFU/day + LU: 12 g/day), and placebo. In this study, B420 with or without LU was found to be effective in reducing body fat mass, waist circumference, energy intake, and body weight compared to placebo. This finding suggests a synergistic effect between the probiotic and the prebiotic and a major role for the chosen strain, as the prebiotic group did not show any difference when compared to the placebo. Results also demonstrated that expression of zonulin, a potential marker of intestinal permeability, was significantly correlated with changes in subjects' truncal fat mass ($r = 0.349$, $P < 0.0001$) in the B420 + LU group. Thus, the authors hypothesized that serum zonulin may be involved with gut microbiota modulation and intestinal integrity.[86]

Afterwards, Hibbert and colleagues aimed to explore possible changes in the microbiota and metabolites in the same population. The authors reported that *Lactobacillus* and *Akkermansia* were more abundant in the B420 group at the end of the intervention. *Methanobrevibacter*, Christensenellaceae, and *Akkermansia* were increased in B420 + LU, and LU alone groups, across the intervention period and had a negative correlation with the waist-area body fat mass at the end of the intervention.[87]

Akkermansia muciniphila, a mucin degrading bacteria, has gained attention since its abundance is known to be inversely associated with obesity, diabetes, cardiometabolic diseases, and low-grade inflammation.[88,89] Animal studies have shown that treatment with *A. muciniphila* reduces obesity and associated comorbidities, such as insulin resistance and hepatic steatosis.[90,91] Due to difficulties cultivating this bacteria, its effects after heat treatment were evaluated in animal studies and demonstrated an even greater impact on adiposity, glucose tolerance, and insulin resistance after *A. muciniphila* was heat-treated (HT). Thus, Depommier et al. developed a synthetic medium and conducted a clinical trial on overweight/obese humans as a proof-of-concept exploratory study.[92]

The study was conducted with 40 volunteers, and 32 completed the 3-month protocol with daily administration of *A. muciniphila* at a concentration of 10^{10} CFUs/day. Subjects who received the HT *A. muciniphila* had improved insulin sensitivity (+28.62 ± 7.02%, $P = 0.002$), reduced insulinemia (−34.08 ± 7.12%, $P = 0.006$), and decreased total plasma cholesterol levels (−8.68 ± 2.38%, $P = 0.02$), when compared with the placebo group. Together, a small reduction in body weight (−2.27 ± 0.92 kg, $P = 0.091$) was observed, compared with the placebo group. Fat mass (−1.37 ± 0.82 kg, $P = 0.092$) and hip circumference (−2.63 ± 1.14 cm, $P = 0.091$) were reduced compared to baseline. As this was the first human study with *A. muciniphila* administration, tolerability and safety were among the primary outcomes, while metabolic effects were the secondary. These interesting results will definitely arouse interest from other researchers to conduct other RCTs that may potentially replicate these results and validate beneficial effects on other metabolic disorders.[92] Despite the results obtained, supplementation with either HT or living cells of *A. muciniphila* did not affect the overall structure of the gut microbiome. In fact, it only increased the quantity of *A. muciniphila* recovered in the feces by 1.7 and 2.6 log, respectively, in the HT and living cells groups.[92]

In recent years, several clinical trials evaluating the effects of probiotics on obesity have been done, with positive results in weight reduction,[93–96] waist circumference,[93,97–100] and fat mass,[93,96,99,101] and mainly *Lactobacillus* and *Bifidobacterium* species were applied, besides mixtures of probiotic strains and synbiotics. Altogether, these studies demonstrated that the use of probiotics, especially if combined with a dietary intervention, were able to assist in weight loss and fat mass reduction in obesity, probably by modulating the intestinal microbiota.

The effects of probiotic and prebiotic food have also been investigated in other metabolic diseases such as T2DM and non-alcoholic fatty liver disease (NAFLD). Studies using the *Lactobacillus* La-5 and the *Bifidobacterium* Bb-12 strains as an intervention in individuals with T2DM showed positive results for fasting glucose[102] and hemoglobin fraction A1c (HbA1c),[103] as well as improved lipid profile.[102–104]

In a recent meta-analysis of RCTs, conducted by Zheng et al., the effects of different probiotics and synbiotics on inflammatory biomarkers and oxidative stress in diabetic individuals were evaluated. Despite the methodological limitations emphasized by the authors, the consumption of probiotics and synbiotics was able to reduce high-sensitivity C-reactive protein (standardized mean difference [SMD] = − 0.38; 95% confidence interval [CI] levels: −0.51, −0.24; $P=0.000$) and malondialdehyde (SMD = −0.61; 95% CI: −0.89, −0.32; $P=0.000$) levels, together with a total increase in the antioxidant capacity (SMD = 0.31; 95% CI: 0.09, 0.52; $P=0.006$), in nitric oxide (SMD, 0.62; 95% CI, 0.25–0.99; $P=0.001$), and glutathione (SMD = 0.41; 95% CI: 0.26, 0.55, $P=0.000$) levels, in diabetic patients compared to those in the placebo group. The authors highlighted the heterogeneity between strains and intervention time, as well as the low amount of available data. Nevertheless, the probiotic and synbiotic approach demonstrated efficacy in improving the evaluated biomarkers.[105]

Verifying the effects of probiotics on altered parameters in individuals with T2DM has been a challenge, due to the lack of literature, inappropriate experimental designs, and high heterogeneity in the administration of probiotics. In a meta-analysis that included only RCTs conducted with diabetic subjects or with associated risk factors, Sun et al. evaluated the following outcomes: fasting blood glucose, glycated hemoglobin (HbA1c), insulin, or HOMA-IR (Homeostatic Model Assessment-Insulin Resistance). The results of the 11 included studies showed that probiotics were able to reduce blood glucose (mean: −0.50 mmol/L; range: 0.09–1.29 mmol/L), HbA1c (mean: −0.48%; range: −0.3 to 1.21%), whereas they were not able to decrease insulin and HOMA-IR significantly.[106]

NAFLD is characterized by the accumulation of fat in the liver (over 5% of liver weight).[107] This classification involves simple steatosis to non-alcoholic steatohepatitis (NASH) and may lead to severe conditions such as fibrosis and cirrhosis.[108] Because obesity, inflammation, insulin resistance, and dyslipidemia are risk factors for the disease, researchers have been studying the potential effects of probiotics and prebiotics on the disease.[109]

Fatty liver improvement was reported by Ahn et al. in a study conducted with 34 obese and NAFLD individuals, using a mixture of *Lactobacillus* and

Bifidobacterium. Using 16S rRNA microbiome sequencing gene, an increase in *Ruminococcaceae-2*, *Lachnospiraceae-2*, *Coprococcus*, *Lachnospiraceae-1*, *Ruminococcus*, and *Dorea* was observed.[110] Oligofructose supplementation (8 g/day for 12 weeks followed by 16 g/day for 24 weeks) in adults with liver-biopsy-confirmed NASH was also able to modulate the intestinal microbiota, increasing *Bifidobacterium* and reducing *Clostridium* clusters XI and I, together with ameliorating liver steatosis histologically-confirmed in NASH patients, compared to placebo.[111]

There are studies supporting the use of probiotics and prebiotics in NAFLD. A recent review targeting the effect of synbiotics on NAFLD found evidence to suggest that synbiotics can reduce inflammation, insulin resistance, and anthropometric parameters in NAFLD patients. However, the effects on dyslipidemia and oxidative stress are inconsistent. Additional RCTs in larger study cohorts including more accurate methods for the evaluation of the severity of diseases, such as liver enzymes, are chief to improve our understanding of probiotic usage in NAFLD.[112]

5.2.2 Inflammatory bowel disease (IBD)

The two main forms of IBD are UC and CD. In these conditions, the intestinal immune system is imbalanced, leading to non-specific chronic inflammation of the intestinal tract.[113]

Studies assessing the intestinal microbiota of individuals with IBD found a dysbiosis that could contribute toward an intestinal barrier dysfunction. In CD it is common to observe a reduction in the complexity of the Firmicutes phyla, specifically a reduced abundance of *F. prausnitzii*, as well as an increase in enteropathogenic *E. coli*.[114,115] In UC, the intestinal microbiota is characterized by a high ratio of *B. fragilis*/*F. prausnitzii* and a low abundance of butyrate-producing bacteria.[116]

There is a lot of controversy in the literature about how effective and safe the use of probiotics and prebiotics in patients with IBD might be, due to the intestinal mucosal sensitivity and a potential worsening of the disease.

In a recent meta-analysis, Astó et al. proposed to evaluate the effects of prebiotics, probiotics, and synbiotics on UC. The meta-analysis included 12 placebo-controlled trials, involving a total of 886 individuals. At first, high heterogeneity between studies was found ($I^2 = 71\%$), but after a sub-analysis using only the UCDAI (Ulcerative Colitis Disease Activity Index) and DAI (Disease Activity Index) indexes, FDA-recommended scales, the heterogeneity reduced dramatically ($I^2 = 29\%$). In this case, a statistically significant difference between the probiotic and the placebo groups was observed, for which relative risk (RR) was 1.55 (CI 95% 1.13–2.15,

P-value = 0.007), and the OR (odds ratio) was 2.12 (CI 95% 1.36–3.31, P-value = 0.000, I^2 = 12%).[117,118] When a sub-analysis was performed for probiotics containing bifidobacteria, studies using this genus showed efficacy in inducing remission in patients with active UC (RR = 1.73 (CI 95% 1.23–2.43, P-value = 0.002, I^2 = 35%)) and OR = 2.50 (CI 95% = 1.33–4.70, P-value = 0.005, I^2 = 44%). For maintaining remission, no statistically significant differences were observed. The authors concluded that probiotics are a promising approach to achieve remission in patients with UC, especially bifidobacteria-containing products, but the results depend on the scale used in the studies (UCDAI and DAI).[117] The greater effect obtained with bifidobacteria-containing products seems to be related to a higher SCFA production in the intestinal lumen, which could reduce inflammation and maintain the gut barrier function.[119]

Although the administration of probiotics and prebiotics in CD appears to be a promising intervention, there is no current recommendation for how to use them in this disease. In fact, the European Crohn's and Colitis Organization recently published a review of complementary medicine in IBD. The organization maintained their position that evidence to support use of probiotics and prebiotics in CD for induction of remission, maintenance of remission, or even prevention of relapse in CD patients following surgically induced remission, is inconclusive. However, the organization recognizes that the use of *E. coli Nissle* 1917 or a multi-strain probiotic containing a combination of lactic acid bacteria, *Streptococcus* and *Bifidobacterium* may be effective in inducing and maintaining remission in UC.[120]

Here we discuss the main applications of probiotics and bioactive foods in various diseases explored more recently. With the new definition of prebiotics, it is likely that more RCTs, including phenolic compounds and omega-3 fatty acids, will be performed in the near future, as their beneficial health properties are recognized and well-studied. There are other applications for probiotics and bioactive foods which were not emphasized here but are worth mentioning including the use of galactooligosaccharides (GOS) and fructo-oligosaccharides (FOS) to reduce the incidence of atopic dermatitis[121] and prebiotic-supplemented formula to reduce the prevalence of allergies in the first years of life.[122]

6. Proposed mechanisms of action

There are several pathways associated with potential health benefits of prebiotic and probiotic consumption. Fig. 2 summarizes some of them that will be discussed hereafter.

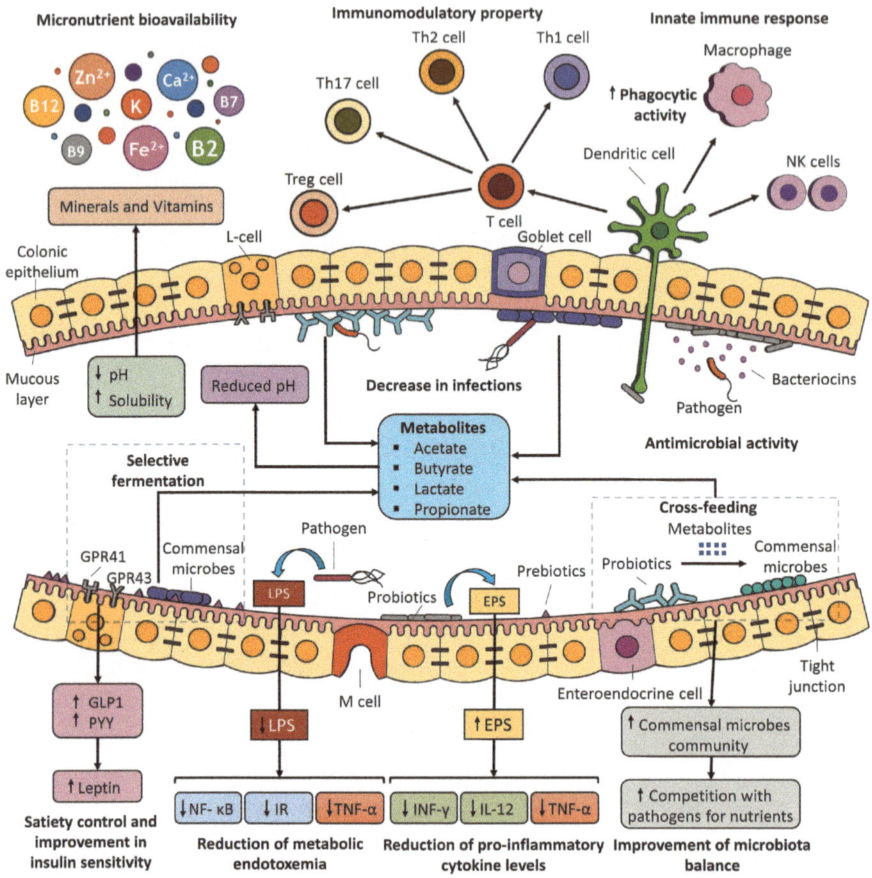

Fig. 2 Main mechanisms of action of prebiotics and probiotics. Prebiotics are selectively utilized by the commensal microbiota, releasing metabolites like short chain fatty acids (SCFA) and organic acids, reducing the lumen pH and thus increasing the absorption of minerals and inhibiting the growth of pathogens. Metabolic products from probiotics may stimulate butyrate producer bacteria by cross-feeding mechanism. Probiotics may also increase the phagocytic activity and modulate the production of immunoglobulins, improving the immune response, promoting the microbiota modulation by competition for nutrients and adhesion sites, in addition to bacteriocin release, reducing the pro-inflammatory response, and enhancing the barrier functions. L-cell, enteroendocrine L-cell; Th1 cell, type 1T helper cell; Th2 cell, type 2T helper cell; Th17 cell, type 17T helper cell; Treg cell, regulatory T cell; NK cells, natural killer cell; GPR 41, G protein-coupled 41; GPR 43, G protein-coupled 43; GLP1, glucagon like peptide 1; PYY, peptide YY; LPS, lipopolysaccharide; NF-κB, nuclear factor-κB; IR, insulin resistance; TNF-α, tumor necrosis factor-α; EPS, exopolysaccharide; INF-γ, interferon-γ; IL-12, interleukin 12; M cell, microfold cell.

6.1 Immunomodulation of the host by humoral response and cell-mediated functions

The ability to increase phagocytosis, natural killer cell function, and dendritic cell activity has been attributed to some probiotic strains.[123–125] Probiotics interact with gut associated lymphoid tissue (GALT), composed mainly of Peyer's patches and lymphoid cells distributed along the gut mucosa, where antigen-presenting cells (APC), including macrophages and dendritic cells, first interact with antigens and initiate an immune response. Bacteria have a conserved structure known as microbial-associated molecular patterns (MAMPs), which are recognized by pattern-recognition-receptors (PRPs), such as Toll-like receptors, Nod-like receptors, and C-type lectin receptors. Recognition of MAMP by these receptors precipitates the maturation of APCs to define the type of immune response that will ensue—regulatory (Treg) or effector (Th1, Th2 or Th17). This immunomodulatory property of probiotic strains is the main target for allergies and inflammatory bowel disease (IBD).[126,127]

Together, probiotics demonstrate the ability to upregulate adaptive immune responses, improving vaccine response and defense against pathogens through modulation of IgA and IgM production.[128,129] There are many cell-surface components in probiotics regulating pro-inflammatory cytokine production, such as the exopolysaccharides from *Bifidobacterium breve*. These metabolites are able to reduce the expression of interferon-γ, TNF-α (tumor necrosis factor-α), and IL-12, and impair the persistence of the pathogen *Citrobacter rodentium*. *Lactobacillus rhamnosus* strain GG is capable of reducing cytokine-induced epithelial cell apoptosis and protecting against experimental colitis, due to activation of the epidermal growth factor receptor pathway and secretion of protein p40.[130,131] Bacterial strains in other probiotics express different cell-surface architecture, like fimbriae, flagella, secreted proteins, and cell wall associated polysaccharides.[123]

6.2 Reduced lumen pH by production of organic acids

The quantitatively and metabolically most important short chain fatty acids (SCFA) are acetate, butyrate, and propionate, occurring as end products of the human colon fermentation process.[132] The production of SCFA as primary end products of the carbohydrate metabolism is well known and described everywhere. Probiotic species from the *Bifidobacterium* and *Lactobacillus* genera produce lactic and acetic organic acids, which reduce the local pH, inhibiting the growth of pathogenic microorganisms and provide a higher bioavailability of vitamins and minerals.[119,133] Although *Lactobacillus* and *Bifidobacterium* do not produce the main SCFA, butyrate, they promote the growth of commensal bacteria that do so, such as *Faecalibacterium prausnitzii* and *Roseburia intestinalis* by means of a cross-feeding mechanism.[134,135] It has been shown that SCFA can signal via cell-surface G protein-coupled receptors (GPCRs), such as GPR41 and GPR43, pathways related to the improvement in insulin sensitivity, regulation of energy intake through secretion of the hormones GLP-1 and PYY and satiety by increased levels of leptin. Increasing the levels of these organic acids in the intestinal lumen seems to be a promising approach against metabolic and intestinal diseases.[135,136]

6.3 Interactions with host gut microbiota

Probiotics may interact with the host microbiota in different ways: competition with pathogens for nutrients and epithelium adhesion, antagonism through the production of antimicrobial substances, cross-feeding other commensal bacteria, and inhibition of bacterial toxin production.[137,138] *Lactobacillus* spp. may produce antibacterial peptides like bacteriocins, including class II and III bacteriocins, which can be active in different mucosa inhibiting replication of pathogens.[139,140] The production of organic acids due to a saccharolytic property contributes to an acid environment that hinders the growth of pathogens.[141] As mentioned above, *F. prausnitzi* is able to use the *Bifidobacterium* fermentation end product acetate as substrate, characterizing the cross-feeding mechanism and consequently may improve the microbiota balance.[119] A probiotic *B. clausii* strain showed surprising results, inhibiting the cytotoxic effect of *Clostridium difficile* and *Bacillus cereus* through the secretion of an alkaline protease, elucidating a novel way to oppose enterotoxinogenic pathogens.[138]

6.4 Improvement in barrier function

The phenomenon known as "leaky gut" has been explored in the application of probiotics as a means of restoring intestinal barrier integrity in various diseases. Especially in metabolic diseases such as diabetes *mellitus* and obesity, probiotics have been associated with increased intestinal permeability, promoting the translocation of high amounts of LPS endotoxin from the cell wall of Gram-negative bacteria to the systemic circulation, causing what is called metabolic endotoxemia. When LPS is recognized by TLR-4 receptors in the intestinal mucosa, it promotes a pro-inflammatory response, leading to the activation of NF-κB and production of cytokines such as TNF-α, which results in increased insulin resistance due to an IRS-1 phosphorylation in serine.[142,143] Interestingly, some scientists have attempted to clarify this mechanism and showed controversial results. *Bifidobacterium lactis* and the association of *Lactobacillus rhamnosus* 19070-2 and *L. reuteri* DSM 12246 have been shown to be effective in improving the integrity of the intestinal barrier. On the other hand, *L. rhamnosus* GG (LGG) and *L. plantarum* 299v did not affect or aggravate the intestinal integrity.[144–147] These results illustrate how the mechanism of each strain must be elucidated, as well as the pathophysiology of the disease in order to effectively use probiotics.

6.5 Production of enzymes and vitamins

Production of bile salt hydrolase (BSH) and β-galactosidase enzymes by some probiotic bacteria may improve cholesterol levels and lactose digestion. A large proportion of the world's population suffers from undesirable symptoms when consuming milk or dairy (lactose-containing products) in their diet. Several strains have β-galactosidase activity and in clinical trials have demonstrated symptom relief in individuals with lactose intolerance and malabsorption.[148,149] In one clinical trial, *Lactobacillus reuteri* NCIMB 30242 reduced cholesterol levels and increased BSH activity, showing that probiotic strains with high levels of BSH may be useful in the management of chronic diseases.[150] Besides, other microorganisms, such as yogurt starter cultures (i.e., *Streptococcus thermophilus* and *Lactobacillus bulgaricus*), may have a similar effect on lactose intolerance and malabsorption due to their capacity to produce β-galactosidase.[151–153]

It is well known and established that several food-related lactic acid bacteria (LAB) and commensal bacteria have the ability to produce B-group and K vitamins.[154,155] Some species of *Bifidobacterium* like *Bifidobacterium bifidum* and *Bifidobacterium longum* subsp. *infantis* can produce high levels of folate via de novo synthesis, and therefore, could help in the recovery of the nutritional status. Vitamins like riboflavin, cobalamin, niacin, and pyridoxine (the last two on a smaller scale) may also be produced by LAB, representing an interesting way to bio-enrich food products.[156] Since there are studies linking chronic fatigue syndrome with changes in the microbiota and the B-group vitamins are important cofactors for several enzymes that participate in the ATP generation in the citric acid cycle, these probiotics may provide an alternative to improve the fatigue status in these diseases.[157]

With the increased application of genome sequencing techniques allowing the selection and understanding of the mechanisms by which strains produce vitamins, it is likely that enrichment of fermented foods with B-producing vitamins will be stimulated, even within public health policies, in order to prevent nutritional deficiencies.[155]

6.6 Production of neurochemicals

Bidirectional interactions between the gut and the brain, referred to as the gut-brain-axis, have gained attention due to large amounts of neurochemicals produced directly or indirectly by the gut, which are now known to be modulated by the host microbiota.[158] Such neurochemicals include serotonin, dopamine, acetylcholine, and gamma-aminobutyric acid. Interactions between the gut microbiota and brain are complex and not yet well understood, but a dysbiosis is associated with neurological disorders such as anxiety, depression, and degenerative processes.[159] The use of probiotics has been shown to relieve part of the symptoms related to these diseases, such as, improving global scores and reducing cortisol levels.[160] In the next future, with studies advancing in this area, safer and more assertive approaches should emerge and benefit people under these conditions.

6.7 Microbiota modulation by prebiotics

Given that prebiotics stimulate selective growth of indigenous gut bacteria and that the gut is a complex ecosystem, the ingestion of different prebiotics may result in an increase of particular species depending on the individual metabotype (i.e., the metabolic responses of a group of individuals).[161,162] Just like probiotics, different prebiotics can modulate the gut microbiota through different ways, with effects primarily on the immune response, pathogen-defense mechanisms, bowel function and stool consistency, satiety-related hormone production, along with other secondary effects on metabolism. Although prebiotics are primarily known as carbohydrate-based, recent research has shown that bioactive compounds present in plants may also have the ability to shape the gut microbiota. Phenolic compounds and polyunsaturated fatty acids are two critical ones.

Reduction of Th2-type immune response seems to be the main mechanism related to the prebiotics' ability to modulate the immune function. Several clinical trials performed on infants have shown a reduction in both the incidence and the prevalence of allergic diseases such as atopic dermatitis.[163] In vitro studies have demonstrated that the inhibition of pathogens occurs in the same manner as in the case of probiotics. So, the higher production of organic acids due to the increased proportion of beneficial bacteria leads to a reduction in the luminal pH, helping to maintain a stable microbiota in which commensal bacteria will reduce the availability nutrients to pathogens.[164]

Moreover, increasing SCFA concentrations due to prebiotic fermentation may result in changes in the intestinal motility and stool consistency, especially in infants.[63,165] However, it is important to emphasize that each SCFA may lead to different effects. As an example, higher propionate levels are related to delayed motility due to secretion of peptide YY (PYY), whereas higher butyrate levels increase motility, as shown in animal experiments.[64] Also, SCFA is capable of interacting with fatty acid receptors, GPR41 and GPR43, signaling the production of the anorexigenic hormones PYY and GLP-1, therefore suppressing appetite and improving insulin sensitivity. In addition, the administration of a synbiotic, containing L. *plantarum* ATCC 202195 + 150 mg of FOS and GOS alone, were able to prevent sepsis and decrease the intestinal permeability, showing that just as probiotics, prebiotics may have an influence on the expression of tight junction proteins, even though this mechanism has yet to be elucidated.[166,167] Together, these mechanisms suggest that the ingestion of prebiotics may be an adjuvant therapy in the prevention or management of infectious and inflammatory diseases.

Bioactive compounds have most recently been studied for their prebiotic-like effects on the gut microbiota. Among them, polyphenols and polyunsaturated fatty acids are the most frequently studied. Ingested polyphenols are metabolized by the microbiota, which may increase or decrease their bioavailability depending on the compound.[168] This biotransformation is highly dependent on the gut ecology and is heterogeneous inter-individual. Individuals can be stratified for their ability to produce specific polyphenol-derived gut microbiota metabolites (metabotypes), for example, is the low ability to produce equol from soy isoflavones in western populations (25–30%) compared to the eastern population (50%), who has the regular habit of consuming soy in their diet.[169–171] Thus, only the individuals producing higher amounts of equol benefit from the cardiometabolic effects of soy isoflavone consumption.[172]

The bioactive compounds of black and green tea, such as epigallocatechin, epicatechin, and catechin, are known potent inhibitors of pathogens including *Listeria monocytogenes*, *Salmonella typhimurium* DT104, and *Helicobacter pylori*. Other polyphenols, such as grape polyphenols and anthocyanidins, promote an increase in beneficial bacteria including the genera *Bifidobacterium* and *Lactobacillus*, as well as *F. prausnitzii* and *Akkermansia muciniphila*.[173,174] These bioactive compounds might also be able to modify the relative abundances of specific phyla, reducing the ratio of Firmicutes to Bacteroidetes.[175] Although promising, some of these changes have yet to be tested in humans.

Omega-3 fatty acids also seem to impact the intestinal microbiota, although only a limited number of studies demonstrate effects on the microbial composition. An increase in the genus *Bifidobacterium* and a reduction in enterobacteria appear to promote a state of eubiosis, leading to increased SCFA production and suppression of metabolic endotoxemia, which results in the improvement of inflammation.[39]

7. Future perspectives

Throughout this chapter, we have discussed the various effects of probiotics and prebiotics and their implications on the human health (Fig. 3).

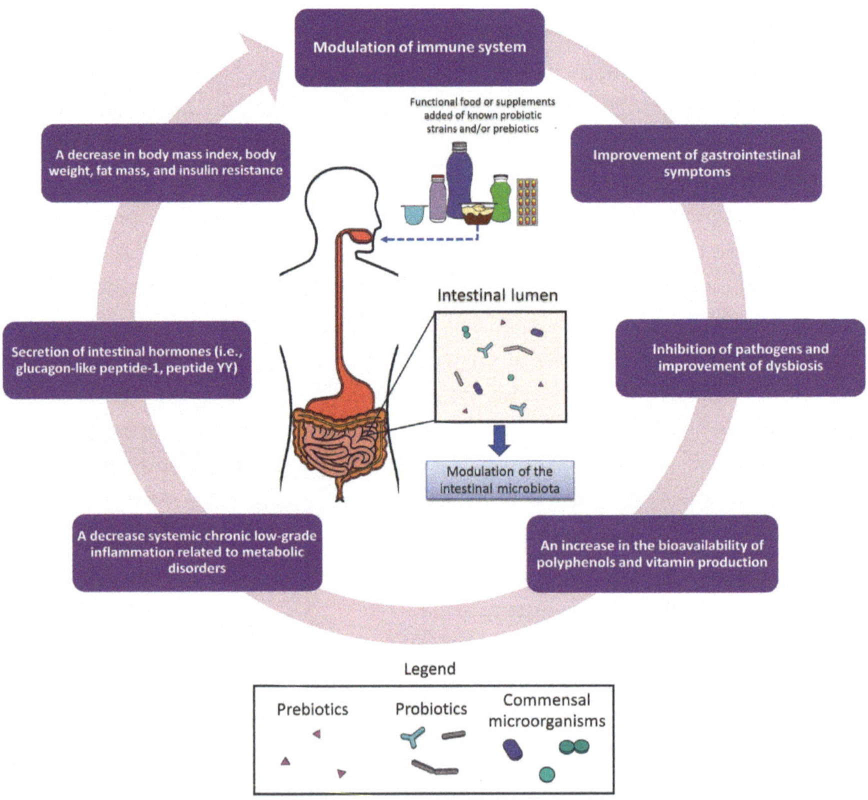

Fig. 3 Major health effects resulting from the consumption of prebiotics and probiotics.

However, it is noteworthy to mention that results in human studies are still heterogeneous and may not be applied universally to diseases, bacterial strains, and individuals from different geographic regions.

The key understanding when we talk about probiotics is that their effects are strain specific and are usually studied in a particular pathophysiological condition. Herewith, we have the differences among microbiotas of individuals, which are influenced by the type of birth, geographical location, individual genetics, lifestyle, and diet, besides other factors.

The availability of advanced genome sequencing techniques, as well as the large number of data available in libraries such as KEGG (Kyoto Encyclopedia of Genes and Genomes) and GenBank, are providing a better understand of the human microbiota, together with bioinformatics, which facilitates the processing of these large datasets.

As appropriate strains for certain health conditions are established, it will be possible to assign health claims and to make safe and effective recommendations for health professionals who are qualified to prescribe them. Effects observed with the use of bioactive and probiotic foods tend to be more significant in individuals with impaired health statuses, usually with an intestinal dysbiosis, allowing the bioactive/probiotic food to restore a health microbiota or at least improve this imbalance in microbes.

A smart use of food matrices, food technology, and different pharmaceutical forms should contribute by enabling adequate viability of probiotic strains and regular consumption, ensuring that the end consumer enjoys the benefit provided by these foods/supplements.

Lastly, the mechanism of action of each strain must be established in in vitro, in vivo, and proof-of-concept assays. With a good understanding of the microbiota of each individual or population, predictive models and algorithms may help in establishing which groups of individuals may benefit from the use of probiotics and prebiotics. Conducting better designed, multi-center clinical trials in larger populations with relevant and objective clinical outcomes could make the use of these supplements and functional foods a routine therapy and preventive health strategy for gastrointestinal well-being.

References

1. Wang J, Thingholm LB, Skiecevičie J, et al. Genome-wide association analysis identifies variation in vitamin D receptor and other host factors influencing the gut microbiota. *Nat Genet*. 2016;48(11):1396–1406.
2. Almeida A, Mitchell AL, Boland M, et al. A new genomic blueprint of the human gut microbiota. *Nature*. 2019;568(7753):499–504.

3. David LA, Maurice CF, Carmody RN, et al. Diet rapidly and reproducibly alters the human gut microbiome. *Nature*. 2014;505(7484):559–563.
4. Salvucci E. The human-microbiome superorganism and its modulation to restore health. *Int J Food Sci Nutr*. 2019;70(7):781–795.
5. Gibson GR, Hutkins R, Sanders ME, et al. Expert consensus document: The International Scientific Association for Probiotics and Prebiotics (ISAPP) consensus statement on the definition and scope of prebiotics. *Nat Rev Gastroenterol Hepatol*. 2017;14(8):491–502.
6. Canadian Food Inspection Agency. Nutrient Functional Claims. Available: https://www.inspection.gc.ca/food/requirements-and-guidance/labelling/industry/health-claims/eng/1392834838383/1392834887794?chap=8; 2019. Accessed 2 December 2019.
7. FAO/WHO. Guidelines for the evaluation of probiotics in food. Food and Agriculture Organization of the United Nations and World Health Organization Working Group Report. 2002.
8. Hill C, Guarner F, Reid G, et al. The International Scientific Association for Probiotics and Prebiotics consensus statement on the scope and appropriate use of the term probiotic. *Nat Rev Gastroenterol Hepatol*. 2014;11:506–514.
9. Sánchez B, Delgado S, Blanco-Míguez A, et al. Probiotics, gut microbiota, and their influence on host health and disease. *Mol Nutr Food Res*. 2017;61(1):1600240.
10. Aagaard K, Ma J, Antony KM, et al. The placenta harbors a unique microbiome. *Sci Transl Med*. 2014;6(237), 237ra65.
11. Dominguez-Bello MG, Costello EK, Contreras M, et al. Delivery mode shapes the acquisition and structure of the initial microbiota across multiple body habitats in newborns. *Proc Natl Acad Sci*. 2010;107(26):11971–11975.
12. Jakobsson HE, Abrahamsson TR, Jenmalm MC, et al. Decreased gut microbiota diversity, delayed Bacteroidetes colonisation and reduced Th1 responses in infants delivered by caesarean section. *Gut*. 2014;63(4):559–566.
13. Bäckhed F, Roswall J, Peng Y, et al. Dynamics and stabilization of the human gut microbiome during the first year of life. *Cell Host Microbe*. 2015;17(5):690–703.
14. Soto A, Martín V, Jiménez E, et al. Lactobacilli and bifidobacteria in human breast milk: influence of antibiotherapy and other host and clinical factors. *J Pediatr Gastroenterol Nutr*. 2014;59(1):78–88.
15. Yoshioka H, Iseki K, Fujita K. Development and differences of intestinal flora in the neonatal period in breast-fed and bottle-fed infants. *Pediatrics*. 1983;72(3):317–321.
16. Lönnerdal B. Nutritional and physiologic significance of human milk proteins. *Am J Clin Nutr*. 2003;77(6):1537S–1543S.
17. WHO Collaborative Study Team on the Role of Breastfeeding on the Prevention of Infant Mortality. Effect of breastfeeding on infant and child mortality due to infectious diseases in less developed countries: a pooled analysis. *Lancet (London, England)*. 2000; 355(9202):451–455.
18. Weng SF, Redsell SA, Swift JA, et al. Systematic review and meta-analyses of risk factors for childhood overweight identifiable during infancy. *Arch Dis Child*. 2012;97(12): 1019–1026.
19. Greer FR, Sicherer SH, Burks AE. The effects of early nutritional interventions on the development of atopic disease in infants and children: the role of maternal dietary restriction, breastfeeding, hydrolyzed formulas, and timing of introduction of allergenic complementary foods. *Pediatrics*. 2019;143(4), e20190281.
20. Yatsunenko T, Rey FE, Manary MJ, et al. Human gut microbiome viewed across age and geography. *Nature*. 2012;486(7402):222–227.
21. De Filippo C, Cavalieri D, Di Paola M, et al. Impact of diet in shaping gut microbiota revealed by a comparative study in children from Europe and rural Africa. *Proc Natl Acad Sci U S A*. 2010;107(33):14691–14696.

22. Bäckhed F. Programming of host metabolism by the gut microbiota. *Ann Nutr Metab.* 2011;58(s2):44–52.
23. Turnbaugh PJ, Hamady M, Yatsunenko T, et al. A core gut microbiome in obese and lean twins. *Nature.* 2009;457(7228):480–484.
24. Rajilić-Stojanović M, Heilig HGHJ, Tims S, et al. Long-term monitoring of the human intestinal microbiota composition. *Environ Microbiol.* 2013;15(4):1146–1159.
25. Bäckhed F, Manchester JK, Semenkovich CF, et al. Mechanisms underlying the resistance to diet-induced obesity in germ-free mice. *Proc Natl Acad Sci U S A.* 2007;104(3):979–984.
26. Jumpertz R, Le DS, Turnbaugh PJ, et al. Energy-balance studies reveal associations between gut microbes, caloric load, and nutrient absorption in humans. *Am J Clin Nutr.* 2011;94(1):58–65.
27. Cani PD, de Vos WM. Next-generation beneficial microbes: the case of *Akkermansia muciniphila. Front Microbiol.* 2017;8:1765.
28. Roager HM, Licht TR. Microbial tryptophan catabolites in health and disease. *Nat Commun.* 2018;9(1):1–10.
29. Sanders ME, Merenstein DJ, Reid G, et al. Probiotics and prebiotics in intestinal health and disease: from biology to the clinic. *Nat Rev Gastroenterol Hepatol.* 2019;16:605–616.
30. Yan YL, Hu Y, Gänzle MG. Prebiotics, FODMAPs and dietary fiber—conflicting concepts in development of functional food products? *Curr Opin Food Sci.* 2018; 20:30–37.
31. Oliveira RPS, Perego P, Converti A, et al. Growth and acidification performance of probiotics in pure culture and co-culture with *Streptococcus thermophilus*: the effect of inulin. *LWT—Food Sci Technol.* 2009;42(5):1015–1021.
32. Xavier dos Santos D, Casazza AA, Aliakbarian B, et al. Improved probiotic survival to *in vitro* gastrointestinal stress in a mousse containing *Lactobacillus acidophilus* La-5 microencapsulated with inulin by spray drying. *LWT.* 2019;99:404–410.
33. Rosolen MD, Bordini FW, de Oliveira PD, et al. Symbiotic microencapsulation of *Lactococcus lactis* subsp. *lactis* R7 using whey and inulin by spray drying. *LWT* 2019;115:108411.
34. Fraga CG, Croft KD, Kennedy DO, et al. The effects of polyphenols and other bioactives on human health. *Food Funct.* 2019;10(2):514–528.
35. Marhuenda-Muñoz M, Laveriano-Santos EP, Tresserra-Rimbau A, et al. Microbial phenolic metabolites: Which molecules actually have an effect on human health? *Nutrients.* 2019;11(11):1–15.
36. Singh AK, Cabral C, Kumar R, et al. Beneficial effects of dietary polyphenols on gut microbiota and strategies to improve delivery efficiency. *Nutrients* 2019;11(9):2216.
37. Manach C, Williamson G, Morand C, et al. Bioavailability and bioefficacy of polyphenols in humans. I. Review of 97 bioavailability studies. *Am J Clin Nutr.* 2005;81(1 Suppl):230–242.
38. Septembre-Malaterre A, Remize F, Poucheret P. Fruits and vegetables, as a source of nutritional compounds and phytochemicals: changes in bioactive compounds during lactic fermentation. *Food Res Int.* 2018;104:86–99.
39. Costantini L, Molinari R, Farinon B, et al. Impact of omega-3 fatty acids on the gut microbiota. *Int J Mol Sci.* 2017;18(12):2645.
40. NIH. Omega-3 Fatty Acids. Available at: https://ods.od.nih.gov/factsheets/Omega3FattyAcidsandHealth-HealthProfessional/; 2005. Accessed 1 December 2019.
41. Prossomariti A, Scaioli E, Piazzi G, et al. Short-term treatment with eicosapentaenoic acid improves inflammation and affects colonic differentiation markers and microbiota in patients with ulcerative colitis. *Sci Rep.* 2017;7(1):1–10.
42. Robertson RC, Seira Oriach C, Murphy K, et al. Omega-3 polyunsaturated fatty acids critically regulate behaviour and gut microbiota development in adolescence and adulthood. *Brain Behav Immun.* 2017;59:21–37.

43. Canadian Food Inspection Agency. Probiotic Claims. Available at: https://inspection.gc.ca/food/requirements-and-guidance/labelling/industry/health-claims/eng/1392834838383/1392834887794?chap=10; 2019. Accessed 30 November 2019.
44. Markowiak P, Ślizewska K. Effects of probiotics, prebiotics, and synbiotics on human health. *Nutrients*. 2017;9(9):1021.
45. Homayoni RA, Vaghef ME, Alipoor B, et al. The comparison of food and supplement as probiotic delivery vehicles. *Crit Rev Food Sci Nutr.* 2016;56(6):896–909.
46. Battistini C, Gullón B, Ichimura ES, et al. Development and characterization of an innovative synbiotic fermented beverage based on vegetable soybean. *Brazilian J Microbiol.* 2018;49(2):303–309.
47. Champagne CP, Gomes da Cruz A, Daga M. Strategies to improve the functionality of probiotics in supplements and foods. *Curr Opin Food Sci.* 2018;22:160–166.
48. Ranadheera RDCS, Baines SK, Adams MC. Importance of food in probiotic efficacy. *Food Res Int.* 2010;43(1):1–7.
49. Shori AB. Influence of food matrix on the viability of probiotic bacteria: A review based on dairy and non-dairy beverages. *Food Biosci.* 2016;13:1–8.
50. Albuquerque MAC, Bedani R, LeBlanc JG, et al. Passion fruit by-product and fructooligosaccharides stimulate the growth and folate production by starter and probiotic cultures in fermented soymilk. *Int J Food Microbiol.* 2017;261:35–41.
51. Tripathi MK, Giri SK. Probiotic functional foods: Survival of probiotics during processing and storage. *J Funct Foods*. 2014;9(1):225–241.
52. Bedani R, Rossi EA, Saad SMI. Impact of inulin and okara on *Lactobacillus acidophilus* La-5 and *Bifidobacterium animalis* Bb-12 viability in a fermented soy product and probiotic survival under *in vitro* simulated gastrointestinal conditions. *Food Microbiol.* 2013;34(2):382–389.
53. Bedani R, Vieira ADS, Rossi EA, et al. Tropical fruit pulps decreased probiotic survival to invitro gastrointestinal stress in synbiotic soy yoghurt with okara during storage. *LWT—Food Sci Technol.* 2014;55(2):436–443.
54. Rampelli S, Candela M, Severgnini M, et al. A probiotics-containing biscuit modulates the intestinal microbiota in the elderly. *J Nutr Heal Aging.* 2013;17(2):166–172.
55. Nagao F, Nakayama M, Muto T, et al. Effects of a fermented milk drink containing *Lactobacillus casei* strain Shirota on the immune system in healthy human subjects. *Biosci Biotechnol Biochem.* 2000;64(12):2706–2708.
56. Harbige LS, Pinto E, Allgrove J, et al. Immune response of healthy adults to the ingested probiotic *Lactobacillus casei* Shirota. *Scand J Immunol.* 2016;84(6):353–364.
57. Wang R, Chen S, Jin J, et al. Survival of *Lactobacillus casei* strain Shirota in the intestines of healthy Chinese adults. *Microbiol Immunol.* 2015;59(5):268–276.
58. Dong H, Rowland I, Yaqoob P. Comparative effects of six probiotic strains on immune function in vitro. *Br J Nutr.* 2012;108(3):459–470.
59. Klein A, Friedrich U, Vogelsang H, Jahreis G. *Lactobacillus acidophilus* 74-2 and *Bifidobacterium animalis* subsp *lactis* DGCC 420 modulate unspecific cellular immune response in healthy adults. *Eur J Clin Nutr.* 2008;62(5):584–593.
60. Ouwehand AC. A review of dose-responses of probiotics in human studies. *Benef Microbes.* 2017;8(2):143–151.
61. Khalesi S, Bellissimo N, Vandelanotte C, et al. A review of probiotic supplementation in healthy adults: Helpful or hype? *Eur J Clin Nutr.* 2019;73(1):24–37.
62. Whisner CM, Martin BR, Schoterman MHC, et al. Galacto-oligosaccharides increase calcium absorption and gut bifidobacteria in young girls: A double-blind cross-over trial. *Br J Nutr.* 2013;110(7):1292–1303.
63. Kanauchi O, Andoh A, Mitsuyama K. Effects of the modulation of microbiota on the gastrointestinal immune system and bowel function. *J Agric Food Chem.* 2013;61(42):9977–9983.

64. Hurst NR, Kendig DM, Murthy KS, et al. The short chain fatty acids, butyrate and propionate, have differential effects on the motility of the guinea pig colon. *Neurogastroenterol Motil.* 2014;26(11):1586–1596.
65. Abrams SA, Griffin IJ, Hawthorne KM. Young adolescents who respond to an inulin-type fructan substantially increase total absorbed calcium and daily calcium accretion to the skeleton. *J Nutr.* 2007;137(11):2524S–2526S.
66. De Vries J, Le Bourgot C, Calame W, Respondek F. Effects of β-fructans fiber on bowel function: a systematic review and meta-analysis. *Nutrients.* 2019;11(1):91.
67. Cruchet S, Furnes R, Maruy A, et al. The use of probiotics in pediatric gastroenterology: a review of the literature and recommendations by Latin-American experts. *Pediatr Drugs.* 2015;17(3):199–216.
68. Victoria CG, Bahl R, Barros AJ, et al. Breastfeeding in the 21st century: epidemiology, mechanisms, and lifelong effect. *The Lancet.* 2016;387(10017):475–490.
69. Ivakhnenko O, Nyankovskyy SL. Effect of the specific infant formula mixture of oligosaccharides on local immunity and development of allergic and infectious disease in young children: randomized study. *Pediatra Polska.* 2013;88(5):398–404.
70. Shahramian I, Kalvandi G, Javaherizadeh H, et al. The effects of prebiotic supplementation on weight gain, diarrhoea, constipation, fever and respiratory tract infections in the first year of life. *J Paediatr Child Health.* 2018;54(8):875–880.
71. Guarino A, Ashkenazi S, Gendrel D, et al. European Society for Pediatric Gastroenterology, Hepatology, and Nutrition/European Society for Pediatric Infectious Diseases. Evidence-based guidelines for the management of acute gastroenteritis in children in Europe. *J Pediatr Gastroenterol Nutr.* 2014;59(1):132–152.
72. Khamsarn S, Nampoonsak Y, Busamaro S, et al. Epidemiology of antibiotic use and antimicrobial resistance in selected communities in Thailand. *J Med Assoc Thai.* 2016;99:270–275.
73. At Thobari J, Satria CD, Ridora Y, et al. Antimicrobial use in an Indonesian community cohort 0-18 months of age. *PLoS One.* 2019;14(8), e0219097.
74. Cameron D, Hock QS, Kadim M, et al. Probiotics for gastrointestinal disorders: proposed recommendations for children of the Asia-Pacific region. *World J Gastroenterol.* 2017;23(45):7952–7964.
75. Szajewska H, Canani RB, Guarino A, et al. Probiotics for the prevention of antibiotic-associated diarrhea in children. *J Pediatr Gastroenterol Nutr.* 2016;62(3):495–506.
76. Deshpande G, Rao S, Patole S. Probiotics for prevention of necrotising enterocolitis in preterm neonates with very low birthweight: A systematic review of randomised controlled trials. *Lancet.* 2007;369(9573):1614–1620.
77. Martin CR, Walker WA. Probiotics: role in pathophysiology and prevention in necrotizing enterocolitis. *Semin Perinatol.* 2008;32(2):127–137.
78. AlFaleh K, Anabrees J. Probiotics for prevention of necrotizing enterocolitis in preterm infants. *Evidence-Based Child Heal A Cochrane Rev J.* 2014;9(3):584–671.
79. Guarner F, Khan AG, Garisch J, et al. World gastroenterology organisation global guidelines. *J Clin Gastroenterol.* 2012;46(6):468–481.
80. Weiser M, Simon JM, Kochar B, et al. Molecular classification of Crohn's disease reveals two clinically relevant subtypes. *Gut.* 2018;67(1):36–42.
81. Miele E, Pascarella F, Giannetti E, et al. Effect of a probiotic preparation (VSL#3) on induction and maintenance of remission in children with ulcerative colitis. *Am J Gastroenterol.* 2009;104(2):437–443.
82. Boonyaritichaikij S, Kuwabara K, Nagano J, et al. Long-term Administration of probiotics to asymptomatic pre-school children for either the eradication or the prevention of helicobacter pylori infection. *Helicobacter.* 2009;14(3):202–207.
83. Szajewska H, Albrecht P, Topczewska-Cabanek A. Randomized, double-blind, placebo-controlled trial: effect of *Lactobacillus* GG supplementation on *Helicobacter pylori*

eradication rates and side effects during treatment in children. *J Pediatr Gastroenterol Nutr*. 2009;48(4):431–436.
84. Engin A. The definition and prevalence of obesity and metabolic syndrome. *Adv Exp Med Biol*. 2017;960:1–17.
85. Álvarez-Mercado A, Navarro-Oliveros M, Robles-Sánchez C, et al. Microbial population changes and their relationship with human health and disease. *Microorganisms*. 2019;7(3):68.
86. Stenman LK, Lehtinen MJ, Meland N, et al. Probiotic with or without fiber controls body fat mass, associated with serum zonulin, in overweight and obese adults-randomized controlled trial. *EBioMedicine*. 2016;13:190–200.
87. Hibberd AA, Yde CC, Ziegler ML, et al. Probiotic or synbiotic alters the gut microbiota and metabolism in a randomised controlled trial of weight management in overweight adults. *Benef Microbes*. 2019;10(2):121–135.
88. Dao MC, Everard A, Aron-Wisnewsky J, et al. *Akkermansia muciniphila* and improved metabolic health during a dietary intervention in obesity: relationship with gut microbiome richness and ecology. *Gut*. 2016;65(3):426–436.
89. Li J, Zhao F, Wang Y, et al. Gut microbiota dysbiosis contributes to the development of hypertension. *Microbiome*. 2017;5(1):14.
90. Plovier H, Everard A, Druart C, et al. A purified membrane protein from *Akkermansia muciniphila* or the pasteurized bacterium improves metabolism in obese and diabetic mice. *Nat Med*. 2017;23(1):107–113.
91. Everard A, Plovier H, Rastelli M, et al. Intestinal epithelial N-acylphosphatidylethanolamine phospholipase D links dietary fat to metabolic adaptations in obesity and steatosis. *Nat Commun*. 2019;10(1).
92. Depommier C, Everard A, Druart C, et al. Supplementation with *Akkermansia muciniphila* in overweight and obese human volunteers: a proof-of-concept exploratory study. *Nat Med*. 2019;25(7):1096–1103.
93. Jung S, Lee YJ, Kim M, et al. Supplementation with two probiotic strains, *Lactobacillus curvatus* HY7601 and *Lactobacillus plantarum* KY1032, reduced body adiposity and Lp-PLA2 activity in overweight subjects. *J Funct Foods*. 2015;19:744–752.
94. Kadooka Y, Sato M, Imaizumi K, et al. Regulation of abdominal adiposity by probiotics (*Lactobacillus gasseri* SBT2055) in adults with obese tendencies in a randomized controlled trial. *Eur J Clin Nutr*. 2010;64(6):636–643.
95. Sanchez M, Darimont C, Panahi S, et al. Effects of a diet-based weight-reducing program with probiotic supplementation on satiety efficiency, eating behaviour traits, and psychosocial behaviours in obese individuals. *Nutrients*. 2017;9(3):284.
96. Szulińska M, Łoniewski I, van Hemert S, et al. Dose-dependent effects of multispecies probiotic supplementation on the lipopolysaccharide (LPS) level and cardiometabolic profile in obese postmenopausal women: a 12-week randomized clinical trial. *Nutrients*. 2018;10(6):773.
97. Gomes AC, de Sousa RGM, Botelho PB, et al. The additional effects of a probiotic mix on abdominal adiposity and antioxidant status: a double-blind, randomized trial. *Obesity*. 2017;25(1):30–38.
98. Higashikawa F, Noda M, Awaya T, et al. Antiobesity effect of *Pediococcus pentosaceus* LP28 on overweight subjects: a randomized, double-blind, placebo-controlled clinical trial. *Eur J Clin Nutr*. 2016;70(5):582–587.
99. Kim J, Yun JM, Kim MK, et al. *Lactobacillus gasseri* BNR17 supplementation reduces the visceral fat accumulation and waist circumference in obese adults: a randomized, double-blind placebo-controlled trial. *J Med Food*. 2018;21(5):454–461.
100. Pedret A, Valls RM, Calderón-Pérez L, et al. Effects of daily consumption of the probiotic *Bifidobacterium animalis* subsp. *lactis* CECT 8145 on anthropometric adiposity biomarkers in abdominally obese subjects: a randomized controlled trial. *Int J Obes (Lond)*. 2019;43(9):1863–1868.

101. Minami J, Iwabuchi N, Tanaka M, et al. Effects of *Bifidobacterium breve* B-3 on body fat reductions in pre-obese adults: a randomized, double-blind, placebo-controlled trial. *Biosci Microbiota, Food Health*. 2018;37(3):67–75.
102. Ejtahed HS, Mohtadi-Nia J, Homayouni-Rad A, et al. Effect of probiotic yogurt containing *Lactobacillus acidophilus* and *Bifidobacterium lactis* on lipid profile in individuals with type 2 diabetes *mellitus*. *J Dairy Sci*. 2011;94(7):3288–3294.
103. Tonucci LB, Olbrich dos Santos KM, Licursi de Oliveira L, et al. Clinical application of probiotics in type 2 diabetes *mellitus*: a randomized, double-blind, placebo-controlled study. *Clin Nutr*. 2017;36(1):85–92.
104. Mohamadshahi M, Veissi M, Haidari F, et al. Effects of probiotic yogurt consumption on lipid profile in type 2 diabetic patients: a randomized controlled clinical trial. *J Res Med Sci*. 2014;19(6):531–536.
105. Zheng HJ, Guo J, Jia Q, et al. The effect of probiotic and synbiotic supplementation on biomarkers of inflammation and oxidative stress in diabetic patients: a systematic review and meta-analysis of randomized controlled trials. *Pharmacol Res*. 2019; 142:303–313.
106. Sun J, Buys NJ. Glucose- and glycaemic factor-lowering effects of probiotics on diabetes: a meta-analysis of randomised placebo-controlled trials. *Br J Nutr*. 2016;115(7):1167–1177.
107. Chalasani N, Younossi Z, Lavine JE, et al. The diagnosis and management of nonalcoholic fatty liver disease: practice Guideline by the American Association for the Study of Liver Diseases, American College of Gastroenterology, and the American Gastroenterological Association. *Hepatology*. 2012;55(6):2005–2023.
108. Farrell GC, Larter CZ. Nonalcoholic fatty liver disease: from steatosis to cirrhosis. *Hepatology*. 2006;43(S1):S99–S112.
109. Bedogni G, Miglioli L, Masutti F, et al. Prevalence of and risk factors for nonalcoholic fatty liver disease: the dionysos nutrition and liver study. *Hepatology*. 2005; 42(1):44–52.
110. Ahn SB, Jun DW, Kim E, et al. Change of microbiota in patients with improved fatty liver and obesity. *J Hepatol*. 2018;68:S838–S839.
111. Bomhof MR, Parnell JA, Ramay HR, et al. Histological improvement of nonalcoholic steatohepatitis with a prebiotic: a pilot clinical trial. *Eur J Nutr*. 2019;58(4): 1735–1745.
112. Sangouni AA, Ghavamzadeh S. A review of synbiotic efficacy in non-alcoholic fatty liver disease as a therapeutic approach. *Diabetes Metab Syndr Clin Res Rev*. 2019;13(5): 2917–2922.
113. Baumgart DC, Sandborn WJ. Inflammatory bowel disease: clinical aspects and established and evolving therapies. *Lancet*. 2007;369(9573):1641–1657.
114. Miquel S, Martín R, Rossi O, et al. *Faecalibacterium prausnitzii* and human intestinal health. *Curr Opin Microbiol*. 2013;16(3):255–261.
115. Chassaing B, Darfeuillemichaud A. The commensal microbiota and enteropathogens in the pathogenesis of inflammatory bowel diseases. *Gastroenterology*. 2011;140(6): 1720–1728.e3.
116. Sitkin S, Pokrotnieks J. Clinical potential of anti-inflammatory effects of *Faecalibacterium prausnitzii* and butyrate in inflammatory bowel disease. *Inflamm Bowel Dis*. 2019;25(4): e40–e41.
117. Astó E, Méndez I, Audivert S, et al. The efficacy of probiotics, prebiotic inulin-type fructans, and synbiotics in human ulcerative colitis: a systematic review and meta-analysis. *Nutrients*. 2019;11(2):293.
118. US Department of Health and Human Services Food and Drug Administration. Ulcerative Colitis: Clinical Trial Endpoints Guidance for Industry. Available at: http://www.fda.gov/Drugs/GuidanceComplianceRegulatoryInformation/Guidances/default.htm; 2016. Accessed 12 December 2019.

119. Ríos-Covián D, Ruas-Madiedo P, Margolles A, et al. Intestinal short chain fatty acids and their link with diet and human health. *Front Microbiol.* 2016;7:185.
120. Torres J, Ellul P, Langhorst J, et al. European Crohn's and colitis organisation topical review on complementary medicine and psychotherapy in inflammatory bowel disease. *J Crohns Colitis.* 2019;13(6):673–685D.
121. Moro G, Arslanoglu S, Stahl B, et al. A mixture of prebiotic oligosaccharides reduces the incidence of atopic dermatitis during the first six months of age. *Arch Dis Child.* 2006;91(10):814–819.
122. Arslanoglu S, Moro GE, Boehm G, et al. Early neutral prebiotic oligosaccharide supplementation reduces the incidence of some allergic manifestations in the first 5 years of life. *J Biol Regul Homeost Agents.* 2012;26(3 Suppl):49–59.
123. Lebeer S, Vanderleyden J, De Keersmaecker SCJ. Host interactions of probiotic bacterial surface molecules: comparison with commensals and pathogens. *Nat Rev Microbiol.* 2010;8(3):171–184.
124. You J, Dong H, Mann ER, et al. Probiotic modulation of dendritic cell function is influenced by ageing. *Immunobiology.* 2014;219(2):138–148.
125. Qiu Y, Jiang Z, Hu S, et al. *Lactobacillus plantarum* enhanced IL-22 production in natural killer (NK) cells that protect the integrity of intestinal epithelial cell barrier damaged by enterotoxigenic Escherichia coli. *Int J Mol Sci.* 2017;18(11):2409.
126. Matsuoka K, Kanai T. The gut microbiota and inflammatory bowel disease. *Semin Immunopathol.* 2015;37(1):47–55.
127. Huang YJ, Marsland BJ, Bunyavanich S, et al. The microbiome in allergic disease: current understanding and future opportunities—2017 PRACTALL document of the American Academy of Allergy, Asthma & Immunology and the European Academy of Allergy and Clinical Immunology. *J Allergy Clin Immunol.* 2017;139(4):1099–1110.
128. Vitetta L, Saltzman E, Thomsen M, et al. Adjuvant probiotics and the intestinal microbiome: enhancing vaccines and immunotherapy outcomes. *Vaccine.* 2017;5(4):50.
129. Childs CE, Röytiö H, Alhoniemi E, et al. Xylo-oligosaccharides alone or in synbiotic combination with *Bifidobacterium animalis* subsp. *lactis* induce bifidogenesis and modulate markers of immune function in healthy adults: a double-blind, placebo-controlled, randomised, factorial cross-over study. *Br J Nutr.* 2014;111(11):1945–1956.
130. Yan F, Cao H, Cover TL, et al. Colon-specific delivery of a probiotic-derived soluble protein ameliorates intestinal inflammation in mice through an EGFR-dependent mechanism. *J Clin Invest.* 2011;121(6):2242–2253.
131. Fanning S, Hall LJ, Cronin M, et al. Bifidobacterial surface-exopolysaccharide facilitates commensal-host interaction through immune modulation and pathogen protection. *Proc Natl Acad Sci U S A.* 2012;109(6):2108–2113.
132. Rivière A, Selak M, Lantin D, et al. Bifidobacteria and butyrate-producing colon bacteria: importance and strategies for their stimulation in the human gut. *Front Microbiol.* 2016;7:979.
133. Macfarlane GT, Macfarlane S. Bacteria, colonic fermentation, and gastrointestinal health. *J AOAC Int.* 2012;95(1):50–60.
134. Chassard C, Bernalier-Donadille A. H_2 and acetate transfers during xylan fermentation between a butyrate-producing xylanolytic species and hydrogenotrophic microorganisms from the human gut. *FEMS Microbiol Lett.* 2006;254(1):116–122.
135. Venegas DP, De La Fuente MK, Landskron G, et al. Short-chain fatty acids (SCFAs) mediated gut epithelial and immune regulation and its relevance for inflammatory bowel diseases. *Front Immunol.* 2019;10:277.
136. Canfora EE, Jocken JW, Blaak EE. Short-chain fatty acids in control of body weight and insulin sensitivity. *Nat Rev Endocrinol.* 2015;11(10):577–591.
137. Van Baarlen P, Wells JM, Kleerebezem M. Regulation of intestinal homeostasis and immunity with probiotic lactobacilli. *Trends Immunol.* 2013;34(5):208–215.

138. Ripert G, Racedo SM, Elie AM, et al. Secreted compounds of the probiotic Bacillus clausii strain O/C inhibit the cytotoxic effects induced by *Clostridium difficile* and *Bacillus cereus* toxins. *Antimicrob Agents Chemother.* 2016;60(6):3445–3454.
139. Oelschlaeger TA. Mechanisms of probiotic actions—a review. *Int J Med Microbiol.* 2010;300(1):57–62.
140. Iseppi R, Messi P, Camellini S, et al. Bacteriocin activity of Lactobacillus brevis and Lactobacillus paracasei ssp. paracasei. *J Med Microbiol.* 2019;68(9):1359–1366.
141. Amaretti A, Di Nunzio M, Pompei A, et al. Antioxidant properties of potentially probiotic bacteria: *In vitro* and *in vivo* activities. *Appl Microbiol Biotechnol.* 2013;97(2): 809–817.
142. Kumar H, Kawai T, Akira S. Toll-like receptors and innate immunity. *Biochem Biophys Res Commun.* 2009;388(4):621–625.
143. Bron PA, Kleerebezem M, Brummer RJ, et al. Can probiotics modulate human disease by impacting intestinal barrier function? *Br J Nutr.* 2017;117(1):93–107.
144. Stratiki Z, Costalos C, Sevastiadou S, et al. The effect of a bifidobacter supplemented bovine milk on intestinal permeability of preterm infants. *Early Hum Dev.* 2007;83(9): 575–579.
145. Rosenfeldt V, Benfeldt E, Valerius NH, et al. Effect of probiotics on gastrointestinal symptoms and small intestinal permeability in children with atopic dermatitis. *J Pediatr.* 2004;145(5):612–616.
146. Gotteland M, Cruchet S, Verbeke S. Effect of lactobacillus ingestion on the gastrointestinal mucosal barrier alterations induced by indometacin in humans. *Aliment Pharmacol Ther.* 2001;15(1):11–17.
147. McNaught CE, Woodcock NP, Anderson ADG, et al. A prospective randomised trial of probiotics in critically ill patients. *Clin Nutr.* 2005;24(2):211–219.
148. Pakdaman MN, Udani JK, Molina JP, et al. The effects of the DDS-1 strain of *Lactobacillus* on symptomatic relief for lactose intolerance-a randomized, double-blind, placebo-controlled, crossover clinical trial. *Nutr J.* 2016;15:56.
149. Almeida CC, Lorena SLS, Pavan CR, et al. Beneficial effects of long-term consumption of a probiotic combination of *Lactobacillus casei* Shirota and *Bifidobacterium breve* Yakult may persist after suspension of therapy in lactose intolerant patients. *Nutr Clin Pract.* 2012;27(2):247–251.
150. Martoni CJ, Labbe A, Ganopolsky JG, et al. Changes in bile acids, FGF-19 and sterol absorption in response to bile salt hydrolase active *L. reuteri* NCIMB 30242. *Gut Microbes.* 2015;6(1):57–65.
151. Besseling-Van Der Vaart I, Heath MD, Guagnini F, et al. *In vitro* evidence for efficacy in food intolerance for the multispecies probiotic formulation Ecologic® Tolerance (Syngut™). *Benef Microbes.* 2016;7(1):111–118.
152. de Vrese M, Stegelmann A, Richter B, et al. Probiotics—compensation for lactase insufficiency. *Am J Clin Nutr.* 2001;73(2):421s–429s.
153. EFSA Panel on Dietetic Products NAA (NDA). Scientific Opinion on the substantiation of health claims related to live yoghurt cultures and improved lactose digestion (ID 1143, 2976) pursuant to Article 13(1) of Regulation (EC) No 1924/2006. *EFSA J.* 2010;8(10):1763.
154. Albuquerque MAC, Bedani R, Vieira ADS, et al. Supplementation with fruit and okara soybean by-products and amaranth flour increases the folate production by starter and probiotic cultures. *Int J Food Microbiol.* 2016;236:26–32.
155. LeBlanc JG, Chain F, Martín R, et al. Beneficial effects on host energy metabolism of short-chain fatty acids and vitamins produced by commensal and probiotic bacteria. *Microb Cell Fact.* 2017;16(1):79.
156. LeBlanc JG, Milani C, de Giori GS, et al. Bacteria as vitamin suppliers to their host: A gut microbiota perspective. *Curr Opin Biotechnol.* 2013;24(2):160–168.

157. Giloteaux L, Goodrich JK, Walters WA, et al. Reduced diversity and altered composition of the gut microbiome in individuals with myalgic encephalomyelitis/chronic fatigue syndrome. *Microbiome*. 2016;4:30.
158. O'Mahony SM, Clarke G, Borre YE, et al. Serotonin, tryptophan metabolism and the brain-gut-microbiome axis. *Behav Brain Res*. 2015;277:32–48.
159. Kim N, Choi H-J. Mind-altering with the gut: Modulation of the gut-brain axis with probiotics Young-Joon Oh. *Artic J Microbiol*. 2018;56(3):172–182.
160. Messaoudi M, Violle N, Bisson JF, et al. Beneficial psychological effects of a probiotic formulation (*Lactobacillus helveticus* R0052 and *Bifidobacterium longum* R0175) in healthy human volunteers. *Gut Microbes*. 2011;2(4):256–261.
161. Chung WSF, Walker AW, Louis P, et al. Modulation of the human gut microbiota by dietary fibres occurs at the species level. *BMC Biol*. 2016;14(1):3.
162. Tomás-Barberán FA, Selma MV, Espín JC. Interactions of gut microbiota with dietary polyphenols and consequences to human health. *Curr Opin Clin Nutr Metab Care*. 2016;19(6):471–476.
163. Pascal M, Perez-Gordo M, Caballero T, et al. Microbiome and allergic diseases. *Front Immunol*. 2018;9(JUL):1584.
164. Fooks LJ, Gibson GR. *In vitro* investigations of the effect of probiotics and prebiotics on selected human intestinal pathogens. *FEMS Microbiol Ecol*. 2006;39(1):67–75.
165. Vandenplas Y, Zakharova I, Dmitrieva Y. Oligosaccharides in infant formula: More evidence to validate the role of prebiotics. *Br J Nutr*. 2015;113(9):1339–1344.
166. Panigrahi P, Parida S, Nanda NC, et al. A randomized synbiotic trial to prevent sepsis among infants in rural India. *Nature*. 2017;548(7668):407–412.
167. Krumbeck JA, Rasmussen HE, Hutkins RW, et al. Probiotic *Bifidobacterium* strains and galactooligosaccharides improve intestinal barrier function in obese adults but show no synergism when used together as synbiotics. *Microbiome*. 2018;6(1):121.
168. Espín JC, González-Sarrías A, Tomás-Barberán FA. The gut microbiota: A key factor in the therapeutic effects of (poly)phenols. *Biochem Pharmacol*. 2017;139:82–93.
169. Atkinson C, Frankenfeld CL, Lampe JW. Gut bacterial metabolism of the soy Isoflavone Daidzein: exploring the relevance to human health. *Exp Biol Med*. 2005; 230(3):155–170.
170. Frankenfeld CL. O-Desmethylangolensin: the importance of Equol's lesser known cousin to human health. *Adv Nutr*. 2011;2(4):317–324.
171. Messina M, Nagata C, Wu AH. Estimated Asian adult soy protein and isoflavone intakes. *Nutr Cancer*. 2006;55(1):1–12.
172. Frankenfeld CL. Cardiometabolic risk and gut microbial phytoestrogen metabolite phenotypes. *Mol Nutr Food Res*. 2017;61(1):1500900.
173. Jin JS, Touyama M, Hisada T, et al. Effects of green tea consumption on human fecal microbiota with special reference to *Bifidobacterium* species. *Microbiol Immunol*. 2012; 56(11):729–739.
174. Gwiazdowska D, Juś K, Jasnowska-Małecka J, et al. The impact of polyphenols on *Bifidobacterium* growth. *Acta Biochim Pol*. 2015;62(4):895–901.
175. Roopchand DE, Carmody RN, Kuhn P, et al. Dietary polyphenols promote growth of the gut bacterium *Akkermansia muciniphila* and attenuate high-fat diet-induced metabolic syndrome. *Diabetes*. 2015;64:2847–2858.

CHAPTER TEN

Microbiome and cancer treatment: Are we ready to apply in clinics?

Stephen L. Chan[a,b,*]

[a]State Key Laboratory of Translational Oncology, Department of Clinical Oncology, Sir YK Pao Centre for Cancer, Institute of Digestive Disease, The Chinese University of Hong Kong, Hong Kong Special Administrative Region, China
[b]Hand in Hand Cancer Foundation Limited, Hong Kong Special Administrative Region, China
*Corresponding author: e-mail address: chanlam_stephen@cuhk.edu.hk

Contents

Abstract

Cancer treatment has been evolving in recent decades from surgery, conventional chemotherapy and radiation therapy to targeted therapies and more recently immunotherapies. Despite significant improvement in the efficacy of treatment with the discovery of novel therapies targeting particular cancer-related gene and proteins and more recently the immune system-modulating biologics, still only patients with specific subtypes of cancer benefit from those targeted therapies and there is room for further improvement of survival outcomes. As failure of cancer treatment is not uncommon in clinical practice, a lot of biomarker studies have been carried out with an aim to identify factors contributing to disease relapse and treatment failure. Gut microbiome is one of the research areas which warrants further investigation of its impact on cancer treatment as microbiota has long been proven to profoundly shape mammalian immunity. As there is increasing evidence showing a strong association between gut microbiota and clinical outcomes of immunotherapy, modulation of intestinal micro-ecological system may be a possible strategy to help improve therapeutic impact of immunotherapy in oncology practice.

Progress in Molecular Biology and Translational Science, Volume 171
ISSN 1877-1173
https://doi.org/10.1016/bs.pmbts.2020.04.004

1. Introduction

Cancer is the second leading cause of death worldwide secondary to cardiovascular disease.[1] According to the GLOBOCAN statistics, there were 18.1 million new cancer cases and about 9.6 million death related to cancer in 2018, and the cancer incidence and mortality are anticipated to reach 29.5 million and 16.4 million, respectively.[2] Despite the increasing disease burden, the rapid evolution of molecular medicine in the past decade have significantly driven paradigm shift of cancer treatment from conventional surgery, traditional chemotherapy and radiation therapy to molecular, biological and more recently immunotherapies.

While immunotherapy has become one of the treatment strategies in fighting against cancer through modulating the immune system of patients to induce anti-tumor effect, it is evident that some patients may experience resistance to the treatment.[3] More efforts are therefore required to investigate the underlying drug resistance mechanism as well as identify patients more suitable to be treated by immunotherapy, so as to improve the treatment efficacy.

Cancer precision medicine or personalized medicine is regarded as using therapeutics that are expected to confer benefit to a subset of patients whose cancer displays specific molecular features (most commonly genomic changes and gene or protein expression patterns).[4] With the increasing understanding of cancer at the molecular level, clinicians can yield more diagnostic and prognostic information of the disease so that treatment could be tailored for a particular group of patients instead of using "one-size-fits-all" approach such as traditional chemotherapy and radiation therapy in treating cancer patients.

In addition to molecular biomarkers of tumor that determine the efficacy of cancer treatment, intestinal microbiota, which play a role in developing innate and acquired immunity of individual, are now considered able to influence the peripheral immune system of patients and subsequently the clinical response to immunotherapy.[5–8] However, little is known about the specific composition of gut microbiome that can induce greater impact to the treatment efficacy and the underlying mechanism to reverse primary or acquired resistance to immunotherapy by modifying the intestinal microecological system.

2. Emerging development of immunotherapy

Use of therapy targeting particular genes and proteins are common in oncology practice, but most of the targeted therapies are merely specific to particular subtypes of cancer. Being currently regarded as the "fifth pillar" of cancer therapy following surgery, chemotherapy, radiation therapy and targeted therapy, immunotherapy has demonstrated the possibility to modulate the immune system by blocking the inhibitory action of T-cells such as programmed death receptor-1 (PD-1) and cytotoxic T lymphocyte antigen-4 (CTLA-4) to recognize and kill cancer cells.[9] The immense potential of immunotherapy was clearly revealed in a recent pooled survival analysis of more than 1800 metastatic melanoma patients who received the first anti-CTLA-4 agent ipilimumab. After a follow up of up to 10 years, the survival rate were more than 20% regardless of history of treatments before the use of ipilmumab,[10] while it is noteworthy that the historical 10-year survival rate was less than 10% before the emergence of this immune checkpoint inhibitor (ICI). Nowadays, a number of immune checkpoint inhibitors has gained regulatory approval for treatment of different types of cancers including melanoma, non-small cell lung cancer, renal cancer, lymphoma and colorectal cancer.[9,11]

Nevertheless, immunotherapy also has its imitation in treatment effectiveness. Generally, check-point inhibitor is effective in approximately 20% of patients. Taking pembrolizumab as an example, following the promising results of a phase II open-label study on advanced gastric cancer patients treated with pembrolizumab as third-line or more therapy in particular among those whose tumors with high level of PD-L1 expression,[12] further phase III randomized study of comparing pembrolizumab with paclitaxel as second-line therapy did not reveal significant improvement in overall survival among advanced gastric cancer patients with PD-L1-expressing tumors.[13] Although the report indicated the clinically meaningful 12-month survival of 40% emerging from the use of pembrolizumab, there were still 60% of patients who did not benefit from the use of immunotherapy. Selection of right patients for right immunotherapy remains an issue for oncologists.

3. Impact of microbiome on cancer treatment

Gut microbiome has been considered influential to health and disease such as cancer,[14,15] and it has long been recognized that gut microorganisms affect drug metabolism and toxicity.[16–18] Nowadays, there are mounting evidence that gut microbiota can modulate the patient's response to chemotherapy through immune interactions, xenometabolism and altered community structure.[19,20] Cyclophosphamide, a commonly used chemotherapeutic agent, was found to alter the gut microbiota which in

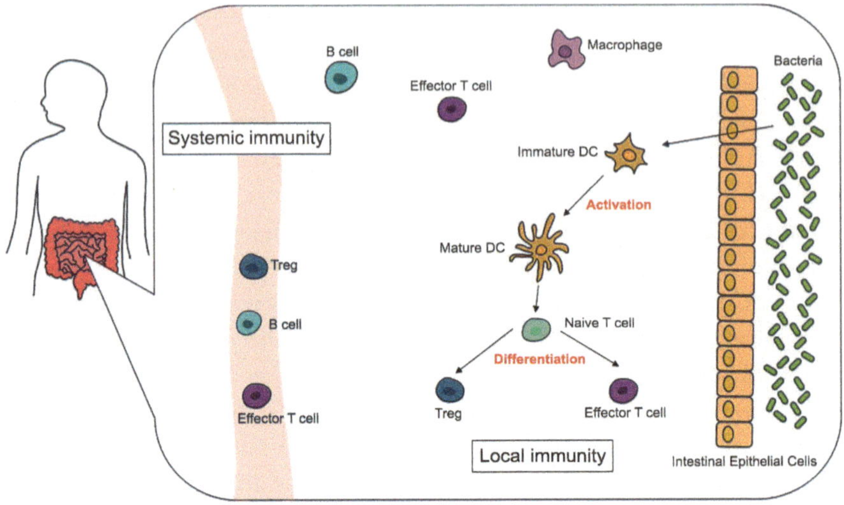

Fig. 1 Interaction of microbiome and immune cells in the bowel. Bacteria could interact directly or indirectly with immune cells leading to maturation of dendritic cells (DC) which induce differentiation of naive T cells to both effector T cells and regulatory T cells (Treg). T cells could impact on both local and systemic immunity.

turns help shape the anticancer immune response.[21] As cyclophosphamide often induces injury to the intestinal barrier which results in an increased trafficking of commensal bacterial to the secondary gut-associated lymphoid and activation of antitumor immune responses (Fig. 1).

In recent years, several studies have also demonstrated the influence of the gut microbiome on the response to ICIs[6–8,22,23] According to a retrospective study of 360 patients with advanced renal cell carcinoma (RCC) and non-small-cell lung cancer (NSCLC) treated with the single use of anti-PD1 or anti-PD-L1 monoclonal antibody or combination, patients who received antibiotics within 30 days of starting immunotherapy had significantly reduced survival benefit from ICI.[24] It was speculated that the gut dysbiosis, a perturbation of the microbial community, induced by antibiotics might decrease the gut microbiota diversity, shift the microbiota composition, and probably eradicate the most immunogenic bacteria to unleash the therapeutic potential of immunotherapy.

4. Discovery from translational studies and potential clinical applications

Basic science has become an important part of clinical science to translate laboratory findings into clinical applications. Several metagenomic studies on cancer patients' gut bacteria have provided important clues to increase our understanding on the interaction between the bacteria and anti-tumor immunity. One of the studies by Routy et al.[23] explored the association between fecal bacterial composition and response to ICI in RCC and NSCLC patients. The commensal most significantly associated with favorable tumor response to ICI and better survival outcomes was *Akkermansia muciniphila*, and both the bacteria species and diversity were also found different between responders and non-responders. Another metagenomic study by Gopalakrishnan et al.[22] further explored the casual relationship between favorable microbiota and response to ICI. Through investigating the immune profiling of germ-free mice having been transplanted with stool from responding patients (R-FMT) and non-responding patients (NR-FMT) to anti-PD1 therapy, the study observed a higher density of $CD8^+$ T-cell in tumor and increased number of $CD45^+$ immune and $CD8^+$ T cells in the gut of the R-FMT mice than those of NR-FMT mice. It seems favorable microbiota from responding patients is associated with enhanced systemic and local immune response. In addition to fecal microbiota transplantation, Routy et al.[23] also treated the transplanted mice with PD-1 monoclonal antibody and examined changes of immune cells in the tumor. Interestingly, an increased formation of intratumoral granulomas and the increased $CD4^+$/$Foxp3^+$ ratios in tumors from animals co-treated with anti-PD1 agent and *A. muciniphila* were observed. Favorable microbes could probably enhance helper T cell function with the treatment of PD-1 blockade. According to the currently available evidence, favorable gut microbiota is considered able to enhance antigen presentation as "pseudovaccination," enhance helper T cells functions and activate cytotoxic T cells.

Despite the encouraging results of the above-mentioned studies, it is important to explore the feasibility of clinically using favorable microbiota to maximize the therapeutic effect of immunotherapy so as to benefit a wider group of patients. Indeed, with more extensive research to understand the profile of favorable gut bacteria, it may help clinicians identify and select patients more suitable to be treated by immunotherapy, and predict the treatment response and prognosis. For non-responders, bacteriotherapy may be prescribed to improve response or even overcome resistance to treatment. Should fecal microbiota transplantation be considered, it is critical to identify an optimal donor with particular target bacteria for transplantation. To date, several bacteria such as *Bifidobacteria spp.*[7,25], *A. muciniphilia*[23], and *Bacteroides spp.*[26] were reported effective in improving anti-tumor immunity and controlling tumor growth *in vivo* (Table 1). Therefore, not only would a

Table 1 Potential gut microbiome to modulate immune checkpoint inhibitors.

Bacteria	Immunotherapy	Effect	References
Bifidobacterium	Anti-PD-1	Augmented T cell response	7
Akkermansia	Anti-PD-1	Increased recruitment of $CCR9^+CXCR3^+CD4^+$ T lymphocytes	23
Bifidobacterium	Anti-PD-L1	Induced tumor-specific T-cell response and increased accumulation of $CD8^+$ T cells within the tumor	25
Bacteroidales	CTLA-4 blockade	Decreased activation of splenic effector $CD4^+$ T cells and tumor-infiltrating lymphocytes	26

specific test be required to help identify an appropriate donor, selection of right bacteria is essential as bacteria may also play a role in inflammation-induced carcinogenesis. Nevertheless, modifying gut microbiota is a possible approach to restore patient's response to immunotherapy.

5. Way forward

Gut microbiome is becoming more important in the realm of cancer immunotherapy. Waves of clinical trials have been initiated to examine the clinical outcomes of modulating gut microbiota in the context of cancer therapy.[14] However, it remains challenging for bacteriotherapy be adopted into clinical practice at this stage as little is known about its mechanism of action on modulating the immunity during cancer treatment.

To date, there are several strategies that can modulate intestinal micro-ecological system such as fecal microbiota transplantation (FMT), administration of probiotics, alterations in diet or lifestyle, *etc.*,[14] but these are mostly applied to non-cancer patients. It is also noteworthy that not only the microbioal community varies among healthy individuals, the intestinal microbiota could have been modified by chemotherapy as well as prophylactic antibiotic when treating cancer patients. All these should be taken into consideration should bacteriotherapy be considered as an adjunct therapy to improve clinical outcomes of immunotherapy.

Several other questions need to be addressed, including but not limited to which species, signature or metabolites being the key player for immune modulation, whether reversal of resistant patients to responding patients be possible, how different between gastrointestinal (GI) and non-GI malignancies in terms of therapeutic outcomes, which particular modality of bacteriotherapy—FMT or probiotic therapy—more suitable for cancer patients. Further investigation is therefore required to determine the impact of gut microbiome in cancer treatment.

References

1. World Health Organization. Cancer. Available at http://www.who.int/news-room/fact-sheets/detail/cancer. Accessed January 18, 2020.
2. International Agency for Research on Cancer. Global Cancer Observatory. Available at http://gco.iarc.fr/. Accessed January 18, 2020.
3. O'Donnell JS, Long GV, Scolyer RA, et al. Resistance to PD1/PDL1 checkpoint inhibition. *Cancer Treat Rev*. 2017;52:71–81.
4. Yates LR, Seoane J, Le Tourneau C, et al. The European society for medical oncology (ESMO) precision medicine glossary. *Ann Oncol*. 2018;29:30–35.
5. Chaput N, Lepage P, Coutzac C, et al. Baseline gut microbiota predicts clinical response and colitis in metastatic melanoma patients treated with ipilimumab. *Ann Oncol*. 2017;28:1368–1379.
6. Chaput N, Lepage P, Coutzac C, et al. Baseline gut microbiota predicts clinical response and colitis in metastatic melanoma patients treated with ipilimumab. *Ann Oncol*. 2019;30:2012.
7. Matson V, Fessler J, Bao R, et al. The commensal microbiome is associated with anti-PD-1 efficacy in metastatic melanoma patients. *Science*. 2018;359:104–108.
8. Frankel AE, Coughlin LA, Kim J, et al. Metagenomic shotgun sequencing and unbiased metabolomic profiling identify specific human gut microbiota and metabolites associated with immune checkpoint therapy efficacy in melanoma patients. *Neoplasia* 2017;19:848–855.
9. Zaidi N, Jaffee EM. Immunotherapy transforms cancer treatment. *J Clin Invest*. 2019;129:46–47.
10. Schadendorf D, Hodi FS, Robert C, et al. Pooled analysis of long-term survival data from phase II and phase III trials of ipilimumab in unresectable or metastatic melanoma. *J Clin Oncol*. 2015;33:1889–1894.
11. Oiseth SJ, Aziz MS. Cancer immunotherapy: a brief review of the history, possibilities, and challenges ahead. *J Cancer Metastasis Treat*. 2017;3:250–261.
12. Fuchs CS, Doi T, Jang RW, et al. Safety and efficacy of pembrolizumab monotherapy in patients with previously treated advanced gastric and gastroesophageal junction cancer: Phase 2 clinical KEYNOTE-059 trial. *JAMA Oncol*. 2018;4:e180013.
13. Shitara K, Özgüroğlu M, Bang YJ, et al. Pembrolizumab versus paclitaxel for previously treated, advanced gastric or gastro-oesophageal junction cancer (KEYNOTE-061): a randomised, open-label, controlled, phase 3 trial. *Lancet*. 2018;392:123–133.
14. Helmink BA, Khan MAW, Hermann A, et al. The microbiome, cancer, and cancer therapy. *Nat Med*. 2019;25:377–388.
15. de Vos WM, de Vos EA. Role of the intestinal microbiome in health and disease: from correlation to causation. *Nutr Rev*. 2012;70(Suppl 1):S45–S56.
16. Scheline RR. Drug metabolism by intestinal microorganisms. *J Pharm Sci*. 1968;57:2021–2037.
17. Lindenbaum J, Tse-Eng D, Butler Jr VP, et al. Urinary excretion of reduced metabolites of digoxin. *Am J Med*. 1981;71:67–74.

18. Lin XB, Farhangfar A, Valcheva R, et al. The role of intestinal microbiota in development of irinotecan toxicity and in toxicity reduction through dietary fibres in rats. *PLoS One*. 2014;9:e83644.
19. Pouncey AL, Scott AJ, Alexander JL, et al. Gut microbiota, chemotherapy and the host: the influence of the gut microbiota on cancer treatment. *Ecancermedicalscience*. 2018;12:868.
20. Alexander JL, Wilson ID, Teare J, et al. Gut microbiota modulation of chemotherapy efficacy and toxicity. *Nat Rev Gastroenterol Hepatol*. 2017;14:356–365.
21. Viaud S, Saccheri F, Mignot G, et al. The intestinal microbiota modulates the anticancer immune effects of cyclophosphamide. *Science*. 2013;342:971–976.
22. Gopalakrishnan V, Spencer CN, Nezi L, et al. Gut microbiome modulates response to anti-PD-1 immunotherapy in melanoma patients. *Science*. 2018;359:97–103.
23. Routy B, Le Chatelier E, Derosa L, et al. Gut microbiome influences efficacy of PD-1-based immunotherapy against epithelial tumors. *Science*. 2018;359:91–97.
24. Derosa L, Hellmann MD, Spaziano M, et al. Negative association of antibiotics on clinical activity of immune checkpoint inhibitors in patients with advanced renal cell and non-small-cell lung cancer. *Ann Oncol*. 2018;29:1437–1444.
25. Sivan A, Corrales L, Hubert N, et al. Commensal Bifidobacterium promotes antitumor immunity and facilitates anti-PD-L1 efficacy. *Science*. 2015;350:1084–1089.
26. Vétizou M, Pitt JM, Daillère R, et al. Anticancer immunotherapy by CTLA-4 blockade relies on the gut microbiota. *Science*. 2015;350:1079–1084.

CHAPTER ELEVEN

Correlation and association analyses in microbiome study integrating multiomics in health and disease

Yinglin Xia*
Department of Medicine, University of Illinois at Chicago, Chicago, IL, United States
*Corresponding author: e-mail address: yxia@uic.edu

Contents

Progress in Molecular Biology and Translational Science, Volume 171
ISSN 1877-1173
https://doi.org/10.1016/bs.pmbts.2020.04.003

Abstract

Correlation and association analyses are one of the most widely used statistical methods in research fields, including microbiome and integrative multiomics studies. Correlation and association have two implications: dependence and co-occurrence. Microbiome data are structured as phylogenetic tree and have several unique characteristics, including high dimensionality, compositionality, sparsity with excess zeros, and heterogeneity.

These unique characteristics cause several statistical issues when analyzing microbiome data and integrating multiomics data, such as large p and small n, dependency, overdispersion, and zero-inflation. In microbiome research, on the one hand, classic correlation and association methods are still applied in real studies and used for the development of new methods; on the other hand, new methods have been developed to target statistical issues arising from unique characteristics of microbiome data. Here, we first provide a comprehensive view of classic and newly developed univariate correlation and association-based methods. We discuss the appropriateness and limitations of using classic methods and demonstrate how the newly developed methods mitigate the issues of microbiome data. Second, we emphasize that concepts of correlation and association analyses have been shifted by introducing network analysis, microbe-metabolite interactions, functional analysis, etc. Third, we introduce multivariate correlation and association-based methods, which are organized by the categories of exploratory, interpretive, and discriminatory analyses and classification methods. Fourth, we focus on the hypothesis testing of univariate and multivariate regression-based association methods, including alpha and beta diversities-based, count-based, and relative abundance (or compositional)-based association analyses. We demonstrate the characteristics and limitations of each approaches. Fifth, we introduce two specific microbiome-based methods: phylogenetic tree-based association analysis and testing for survival outcomes. Sixth, we provide an overall view of longitudinal methods in analysis of microbiome and omics data, which cover standard, static, regression-based time series methods, principal trend analysis, and newly developed univariate overdispersed and zero-inflated as well as multivariate distance/kernel-based longitudinal models. Finally, we comment on current association analysis and future direction of association analysis in microbiome and multiomics studies.

"Omics" refers to any type of biological data collected using the high-throughput techniques, such as next-generation sequencing, genome-sequencing. More than 30 -omics are introduced in literature. Among them, metagenomics, transcriptomics, proteomics, and metabolomics are the most studied omics. Many other specialized omics, such as lipidomics, fluxomics (metabolic flux analysis), toxicogenomics, nutrigenomics, and foodomics, are subcategories of the primary omics.[1,2,538]

In microbiome research, correlation and association analyses are essential. These analyses are conducted not only among taxa in microbiome data and between taxa and environmental factors, but also between microbiome and other omics. For example, such analyses are often conducted between microbiome and metabolism. The metabolite-microbe correlation and association analyses are seen in the microbiome research field. Various statistical models for analyzing the association among microbiome, host, and environmental factors have been developed.

The overall objectives of this chapter are to: (1) describe commonly used correlation and association methods and their implications; and (2) review and introduce correlation and association methods in microbiome and other omics studies, such as metabolomics and metagenomics. We highlight that current correlation and association analyses have moved beyond the context of microbiome to interomics by showing examples from the fields of microbiome and multiomics.

1. Microbiome is dynamically associated with environment, host factors, and other omics—Framework of association analysis in microbiome research

Microbiome research basically focuses on three factors: Environment, Microbiome, and Host (Fig. 1). Typically, there are mainly two themes in the microbiome studies: (1) to functionally characterize the relationship between microbiome features and host factors (e.g., biological, genetic), environmental factors (e.g., clinical or experimental conditions); and (2) to identify potential biological and environmental factors that are associated with microbiome composition. The goal of these studies is to understand mechanisms of host genetic and environmental factors that shape microbiome and microbiome affects on host's physiological properties. Insights

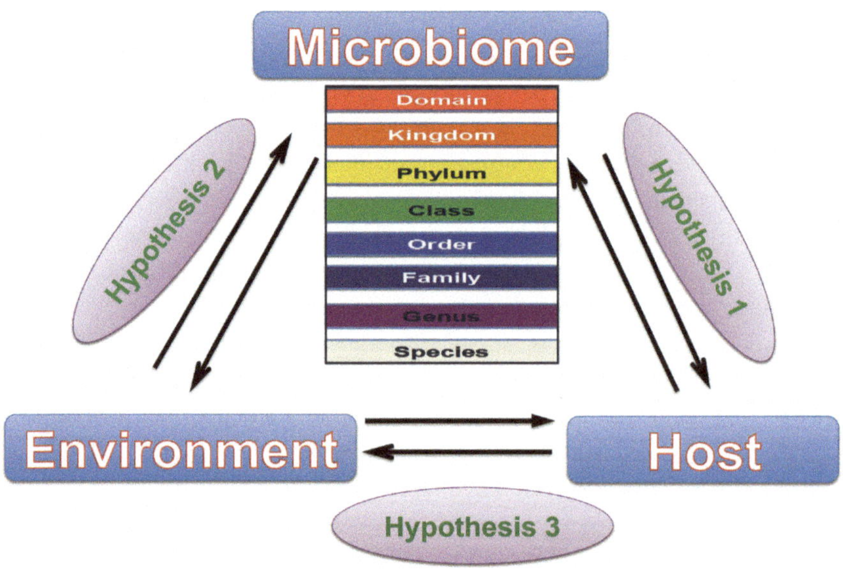

Fig. 1 Dynamic interactions among environment, microbiome, and host for the research hypotheses in microbiome studies.

gained from the studies potentially contribute to the development of therapeutic strategies in modulating the composition of microbiome in human diseases.[3,4]

The association among environmental factors, microbiome, and host factors is dynamic. The dynamic association can be formulated under a statistical framework of association analysis. We depict the framework in Fig. 2.

What is an association in microbiome and other omics? It is challenge to define association and measure association in the fields of microbiome and omics. However, we generally describe that association analysis in microbiome and omics studies is dynamic. First, it is because of a pair of taxa association within microbiome (associations of individual taxa). Second, microbiome is associated with various external factors (host and environmental factors). Third, microbiome is intrinsically dynamic because microbiome has two intrinsic factors: microbiome is evolutionary and temporal (e.g., microbiome's state follows maturation during life span and varies based on host health and disease).

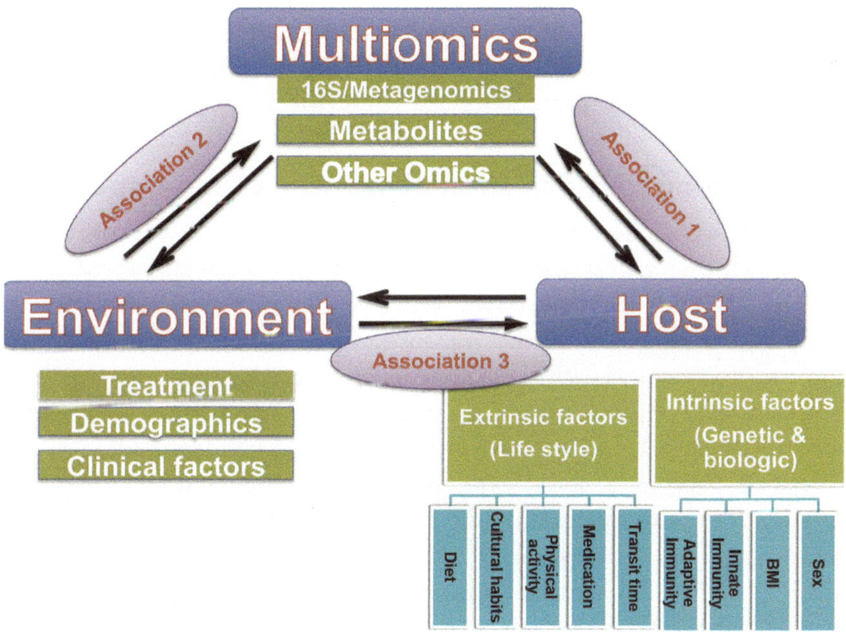

Fig. 2 A framework of correlation and association analyses in microbiome and integrative multiomics studies. Microbiome compositions could be associated with host factors (e.g., biologic and genetic factors). Microbiome compositions could also be associated with environmental factors or covariates including clinical or experimental conditions.

Currently, many microbiome and omics studies are still using the classic correlation and association methods. Some newly developed correlation and association methods are also based on these classic correlation and association methods. To understand the dynamic and complicated system of microbiome and its functions, we should review and treat associations of microbiome data as different traditional approaches in both concepts and methodologies.

2. Dependence and co-occurrence—Implications of correlation and association

As a measure of bivariate association, the correlation coefficient may be looked as from more than dozen of ways, for example, as the function of means, the standardized covariance, the ratio of two means, the ratio of two variances, the slope of a line, the cosine of an angle, and the tangent to an ellipse.[5] Basically, association can be defined by two ways: dependence and co-occurrence. In other words, in concept, there exist two kinds of association. The two concepts of association link together and share some common characteristics especially when more variables are involved, such as in association or correlation network analysis.

2.1 Measuring statistical strength of association—Dependence

In statistics, association implies "dependence." Statisticians are more interested in identifying both statistically significant and potentially novel association or dependence that could lead to interesting new insights and hypotheses. Regardless of the measures used, association analysis often requires a suitable measure to evaluate the dependences between variables.[6] The association of pair of variables is often analyzed via a correlation analysis. The pair of variables either were given or need to be identified, such as in correlation network analysis, gene expression analysis, and taxa abundance analysis. "It is this conception of correlation between two occurrences embracing all relationship from absolute independence to complete dependence"[7] (p. 157). Due to their linkage, the terms *association* and *correlation* are often used interchangeably. However, in a stricter sense, *correlation* refers to linear correlation. Actually *correlation* measures the strength of an association. *Association* refers to any relationship between a pair of variables.

In classic statistics, *dependence* or *association is defined as* any statistical relationship, whether causal or not, between two or more random variables or bivariate data, such as exposures and diseases. The method used to determine the strength of an association depends on the characteristics of the data for each variable. Although correlation measures the strength of an association, it does not measure the significance or importance of the association. A separate analysis of significant association in the correlation analysis can be conducted using a statistical test (e.g., a *t*-test) to compare the difference between the observed correlation coefficient and the expected correlation coefficient under the null hypothesis.

For correlation analysis, a zero value of coefficient suggests that the variables are statistically independent. From measuring and significant testing the association between two categorical variables, chi-square test is the standard method. For categorical variables, other measures of association include relative risk and odds ratio. Both relative risk and odds ratio measure the strength of an association, but do not directly test for statistical significance. The significance of association is suggested by their confidence intervals. Other statistical methods for measures of association include the point-biserial correlation coefficient, for one variable with an interval/ratio scale and the second variable with dichotomous outcomes; and serial correlation (also known as autocorrelation), using the Durbin-Watson test to perform significance of correlations.

2.2 Mining association rule—Co-occurrence

Researchers in data mining community focus on designing efficient algorithms to generate association rules from large datasets, and those from the machine learning study have explored the applications of association rule mining to classification.

2.2.1 Association implies "co-occurrence"

In data mining, the central task of association analysis is to discover sets of binary variables that co-occur together frequently in a given database.[6,8,9] The "association rules," a general data mining method, was developed in 1966 by Petr Hájek et al.[10,11] In 1993, Agrawal et al. introduced association rules for discovering regularities between products in large-scale market basket transaction data in market basket analysis.[8] Since then, association rule mining (ARM) has emerged as an important methodology and has been employed today in many application areas including bioinformatics and microbiome study.

2.2.2 Support and confidence

The strength of an association rule can be measured in terms of its support and confidence. Support is an indication of how often the pair of variables (i.e., Taxa or OTUs) appear in the dataset, while confidence is an indication of how frequently the rule has been found to be true; in other words, confidence indicates how frequently Y appears in selected pair of variables (i.e., Taxa or OTUs) that contain X. Mathematically, let $P(X)$ be the probability of appearance of X in S and let $P(Y \mid X)$ be the conditional probability of appearance of Y, given X appears. For a pair of variables X, $support(X)$ is defined as the fraction of sampling $Si \in S$ such that $X \subseteq Si$. Thus, $P(X)$ can be estimated as $P(X) = support(X)$. In terms of probabilistic metric, the support of a rule $X \rightarrow Y$ is defined as $support(X \rightarrow Y) = P(X \cup Y)$. Confidence measures reliability of an association rule $X \rightarrow Y$, defined as $confidence(X \rightarrow Y) = support(X \cup Y)/support(X)$. In terms of probability, confidence can be used to estimate $P(Y|X)$: $P(Y \mid X) = P(X \cup Y)/P(X) = confidence(X \rightarrow Y)$.

2.3 Applications of association rule mining

2.3.1 ARM first gained popularity for market-basket analysis

Based on Agrawal et al.,[8,12] the problem of association rule mining can be formulated this way: let $I = [i_1, i_2, \ldots, i_m]$ be a set of m binary attributes called items. Let $D = [T_1, T_2, \ldots, T_n]$ be a dataset of n transactions (also called the database). Where each transaction $T \subset D$ has a unique transaction identifier, called its TID and contains a subset of the items in I, i.e., $T \subseteq I$. Let X and Y be two itemsets s.t. $X \subset I$, $Y \subset I$, and $X \cap Y = \emptyset$. An association rule is defined as an implication of the form $X \rightarrow Y$, where X is called the antecedent (left-hand side or LHS) and Y is called the consequent (right-hand side or RHS) of the rule, respectively.

The transaction data can be represented in a binary format: a contingency table, where each row corresponds to a transaction and each column corresponds to an item. An item can be treated as a binary variable where if the item is present in a transaction, then its value is 1 and otherwise 0. Thus, the jth column $= 1$ indicates that item i_j presents and otherwise the jth column $= 0$, suggesting that item i_j is not present. Using itemset representation is an efficient way to record data because a specific item is not given in most transactions, so most columns equal to zero.

2.3.2 The application of ARM in the biomedicine and healthcare data

ARM has also been widely applied to the biomedicine and healthcare data, including being utilized to identify new and interesting patterns in surveillance data,[9] to discover significant association rules in a high-dimensional medical dataset,[13] to identify associations between patient's medications, laboratory results, and clinical problems using electronic health record

(EHR) data in biomedical informatics,[12,14] and to mine statistically significant and novel clinical associations from using an electronic medical record dataset of diagnoses and medications: the depressive disorders class as the consequent.[14] Abar et al. introduce an association rule mining in application of EMR dataset.[15] We briefly describe as below.

Let I be union of all medications and diagnoses for patients, then a set $E=[i_1, \ldots, i_k] \subseteq I$ is called a k-items of clinical itemset. Let a patient visit transaction T = (pid, vid, I) be defined over I, where vid is the patient visit ID, pid is the patient ID, and $I \subseteq I$ is the itemset corresponding to the current visit vid. The set of all visit transactions in a given database is denoted as the visit database V. An association is a rule of the form E ➔ Y, where E and Y are itemsets and $E \cap Y = \varnothing$.

2.3.3 Association rule mining in microbiome

Depending on the bioinformatics tools used to generate the high-throughput data, and the statistical packages prefer to analyze the data, microbiome data are typically structured as a sample-by-taxa (or OTUs) or taxa (or OTUs)-by-sample contingency table, while a meta-data contingency table consists of numeric and categorical attribute values.[16] In the context of a sample-by-taxa (or OTUs) contingency table, rows of the table consist of sample IDs, and columns consist of taxa or OTUs.

In terms of microbiome data, we can present the problem of association rule mining this way.

Consider a large dataset consisting of many sets of variables (i.e., Taxa or OTUs). Let X, Y be disjoint pair of variables (i.e., Taxa or OTUs) in a given dataset, i.e., $X \cap Y = \varnothing$, $V=[v_1, v_2, \ldots, v_m]$ be a set of m columns of Taxa (or OTUs), and $S=[s_1, s_2, \ldots, s_n]$ be a set of rows containing sampling (or chosen samples) called dataset, then an association rule is an implication expression of the form from antecedent X (left-hand side or LHS) ➔ consequent Y (right-hand side or RHS). We can treat each sampling or sequencing as a data processing action.

The association rules (ARs) suggest a strong relationship that exists between two pairs of variables (i.e., Taxa or OTUs), and minimum thresholds on support and confidence should be met.[8] Because very low support may occur simply by chance and the more frequent pair of variables (i.e., Taxa or OTUs) appear together, the higher reliability of the inference of association. Thus, support defines a desirable property of association rules: the efficient discovery, while confidence provides an estimate of the conditional probability of Y given X. Given frequent pair of variables mining, the goal of association rules with support and confidence is to determine which pair of variables (i.e., Taxa or OTUs) in sampling S naturally occur together within the dataset V. The pair of variables that occur together are called co-occurring groups of pairs. Thus support and confidence together measure the co-occurrence of X (left-hand side or LHS) and Y (right-hand side or RHS).

Based on the support-confidence framework, association rule mining is to eliminate uninteresting patterns. The drawback of support is: many potentially interesting patterns involving low support pair of variables (i.e., taxa or OTUs) might be eliminated by the support threshold, resulting in the rare item (in microbiome case, taxon or OTU) problem.[17,18] For example, by implementing support measure, taxa or OTUs that occur very infrequently in the dataset are excluded, although they would still produce interesting and potentially valuable association rules. The confidence measure also has pitfalls because it ignores the support of the sets of taxa (or OTUs) in the rule consequent[18] and it is sensitive to the frequency of the consequent taxa (or OTUs) in the dataset. These pitfalls result in the problem: consequents with higher support will automatically produce higher confidence values, even if no association exists between the pair of taxa (or OTUs).[17] Thus, an association detected by high confidence value can sometimes be misleading.

Taken together, support can be used to represent the significance of an association pattern and confidence can be used to measure the accuracy of a given rule. However, existing association rules under the support-confidence framework have limitations.

2.4 Correlation and association do not mean causality

2.4.1 Correlation does not mean causation

When two variables have been shown to be correlated, we indeed tempt to infer a cause-and-effect relationship between them. However, there are at least three reasons that correlation does not detect a cause-and-effect relationship: (1) The nature of correlation constraints only detects the linear relationship between variables; cause-and-effect relationship could be either linear or nonlinear. (2) A correlation suggests a symmetrical relationship or exchangeable roles played by the pair of variables, which indicates that the pair of variables could be either cause or effect. Correlation is just such a measure because correlation merely establishes the equal validness of X and Y, i.e., X could cause Y and Y also could cause X. In correlation the roles of the two variables X and Y can be interchanged. When the measure remains the same value, while the roles of the variables are interchanged, then such a measure tells us nothing about causality.[19] (3) When the two variables are identified as correlated, this may not be causal at all but rather be due to a latent or underlying variable, i.e., these two variables both depend on a third variable (often called a confound). Spurious correlations could happen. The "spurious correlation" could be due to the use of ratios.[20] Consequently, correlation does not indicate a direct causal relation, establishing that a correlation between two variables is not a sufficient condition and also not a necessary condition to establish a causal relationship (in either direction). Thus, we cannot use correlation to infer a causal relationship between the variables. Correlation indeed indicates the potential existence of causal relations between the variables. Actually, it was reviewed as a weaker form of causation.[21]

2.4.2 Association does not mean causation

Association signifies "dependence" or "co-occurrence." However, these two significations do not suggest a cause-and-effect relationship, because association in terms of dependence does not mean causation and association in terms of co-occurrence does not mean causation either.

We just reviewed the association's meaning of co-occurrence from ARs. ARs have been known to manifest when there is a causal relationship and also been used as starting points to arrive at potential causal relations.[19] However, ARs essentially do not indicate causation. Instead, it suggests a strong co-occurrence relationship between variables in the antecedent and consequent of the rule. To achieve causation, additional retrospective analyses and/or prospective experiments, such as randomized control trials, are required to remove confounding factors.[22] The results from association analysis should be interpreted with caution.

These two significations do not suggest a cause-and-effect relationship also due to confounders, spurious or false association could happen in the case of either "dependence" or "co-occurrence," which currently challenges metagenome-wide association studies.[23,24] Thus, it is not easy to translate associations into causal links.

The first step of discovering statistically sound association is to reduce the risk of finding any spurious correlations or spurious associations to a user-specified significance level. Second, a statistical hypothesis testing should be conducted because "any interpretation of the meaning of an association is necessarily hypothetical, and the number of possible alternative hypotheses is in general considerable"[25] (p. 42).

2.4.3 Establishing causation using sophisticated models

To establish causality, merely identifying association is not sufficient. The more sophisticated models are required. In 1843, the British philosopher John Stuart Mill discussed the phenomenon of cause, effect, and causation. Based on him, a causal relationship exists if three prerequisites are met[26]: first, the cause must temporally precede the effect. Second, the cause was related to the effect. Third, no plausible alternative explanation for the effect other than the cause, i.e., any plausible alternative explanations for a relationship between two variables must be ruled out. Actually, the concept of the cause temporally preceding the effect steps from English philosopher John Locke's *An essay concerning human understanding* (original work published in 1690).[27] In this classic book, he said: "That which produces any simple or complex idea, we denote by the general name *cause,* and that which is produced, *effect*" (1975, p. 324). The message from philosophy is: correlation or association is a necessary but not sufficient condition for causation. A counterfactual model is considered as a better way to understand what an effect is. This insight was thought going back at least to the 18th-century Scottish enlightenment philosopher David Hume[28] (p. 556).

Because various measures of association are still under development, selecting an appropriate measure of association is very important in analyzing association of microbiome data. The properties of a measure determine whether or not this measure is suitable for a specific application. Five important properties of a measure introduced by Tan et al.[6] include: symmetry under variable permutation, row/column scaling invariance, antisymmetry under row/column permutation, inversion invariance, and null invariance.

Several general guidelines for selecting an appropriate measure have been proposed, although their perspectives are different. For example, Liebetrau's[19] and Berry et al.'s[29] guidelines are based on the data nature and source and the intended use of the measure. Reynolds[29a] considered whether a measure is symmetric or asymmetric, and whether the interpretation is sensitive to confounding influences to Khamis[12] simply based on whether each type of the measured variables (i.e., level or scale of measurement) is continuous, discrete ordinal, or discrete nominal. Although these guidelines or advices are helpful for a microbiome researcher to choose appropriate measures of association, the proposed guidelines from the perspective of traditional measures of association or association analyses cannot be directly applied into microbiome data and other omics data.

We believe the following six points should be considered when selecting a measure:

1. *Whether it is used to measure association or goodness-of-fit testing.* Measures based on chi-squared statistic are effective for determining departures from independence. However, a test statistic generally makes a poor measure of association[30] (p. 97). For example, for measuring the significance of the association, the chi-square statistic is an excellent choice, but it is a common mistake to use the value of chi-squared statistic itself as the measure of association.
2. *Whether the measure treats the variables symmetrically or asymmetrically and whether the purposes of measure are to establish correlation, to measure agreement, or make predictions* (i.e., *to establish a cause-effect relationship*). Any symmetric measure can be used to measure correlation, while asymmetric measures should be used for prediction.[19] If we do not know or do not care which variable depends on another, then use a symmetric measure. In contract, if we know or want to test one variable causes another, it is usually to predict values of the dependent variable from independent variable or the causal variables.[1] In this case, the goal in an asymmetric measure is to use one variable to improve the predictability of another.[19] Among the measures, Goodman-Kruskal lambda, Goodman-Kruskal tau, Somer's d, Gini index, mutual information, and Jaccard measure are examples of the asymmetric measures, while the three correlation coefficients, phi coefficient, cosine, and odds ratio treat variables symmetrically.
3. *Whether the types of data are nominal, continuous, or ordinal.* Six combinations of measure of association can be organized: continuous-continuous, continuous-ordinal, continuous-nominal, ordinal-ordinal, ordinal-nominal, and nominal-nominal.
4. *Whether the measure is sensitive to marginal totals or confounding factors.*
5. *How different types of association deal with extreme values.*
6. *How to use correct variance estimates to make inference.*

3. Unique characteristics and statistical issues of microbiome and omics data

3.1 Structure of microbiome and omics data

Microbiome and other omics datasets are structured as a matrix of positive read numbers each representing either a measured value for variable i (feature) (e.g., microbial taxa abundance, gene expression, metabolite concentration, mRNA, or protein level) in sample j (object or site), or a ratio of two measured x_i values between two samples. They are usually generated either from common outputs of high-throughput sequencing, mass spectrometry, NMR-based metabolomics, and single sample microarrays (the former dataset structure) or from two-color microarrays or high-throughput quantitative PCR (the latter dataset structure).[31] The matrix is generally called feature table with each set of variables presenting a type of features (called feature type).

Omics data are often vaguely categorized into three main types: components data, interactions data, and data for functional states analyses.[32,33] Specifically, as reviewed by Xia et al.,[16] the taxa or OTU count data are high dimensional, discrete, sparse, and often have many zeros. These characteristics are unique not only because each characteristic is uniquely presented in microbiome data, but also mainly because they are interweaved or linked to each other (i.e., they are not independent or not excluded each other): each is the cause of others. Thus, clearly these mixed unique characteristic effects pose big challenges for integrating omics datasets and conducting data analysis.

3.2 High dimensionality—Large p and small n problem

Regardless of the experimental platforms that are used to generate the feature table or data type of omics, the feature table is high dimensional with the number of features (or explanatory variables) much higher than the number of samples.[34,35] It is a real challenge to integrate and analyze the high-dimensional data. First, due to the high-dimensional nature of microbiome and omics data, the notations and meanings of variables have been changed. In classic statistics, X is usually used to present the predictor variables, and Y for the response variables. In statistical integration of multiomics data studies, X could be presented for a set of predictor variables with one feature type and Y for a set of response variables with another feature type. For example, we consider two data matrices, X ($n \times p$) and Y ($n \times q$), where n represent the same individuals or samples, and p and q represent different sets of omics variables. Second, the high-dimensional microbiome and omics data with the number of taxa (OTUs) larger than the sample size, resulting in the large p and small n problem. In terms of data matrix, p (taxa or OTUs) refers to the number of columns, n (samples or data observations) refers to the number of rows, and then the large p and small n problem means that small n samples contain large p taxa. Graphically, there are n samples in a p dimensional space. It implies that the data are underdetermined ($p > n$) due to a singular covariance matrix. Statistically modeling high-dimensionality data is very challenge because without additional assumptions about the interaction of taxa or OTUs, any meaningful statistical inference about the interaction is inviable. The two not-excluded challenges are solving the large p and small n problem and using variable section to reduce the dimensions.[16]

3.3 Compositionality—Not independent problem

Compositionality is one of the unique characteristics of microbiome data. In his 1986 seminar work,[36] Aitchison described four characteristics of a compositional dataset: (1) each row presents a replicate, a single experimental or observational unit; (2) each column presents a specific ingredient or part of each composition; (3) each entry is nonnegative; and (4) the sum of the entries in each row is 1, or equivalently 100%. In summary, the main reason that drives a dataset to have compositionality is due to the constant-sum constraint.

Microbiome dataset has compositional characteristics. First, datasets generated from high-throughput sequencing are predefined or constrained to some constants, resulting in the total values of the data meaningless. Omics datasets including RNA sequencing (RNA-seq), 16S rRNA gene fragments sequencing, chromatin immunoprecipitation sequencing (ChIP-seq), metagenomic analysis, and selective growth experiments are composed of sequencing read counts mapped to a large number of features (e.g., OTUs, genes, species, or any taxonomic levels) in each sample. The observed number of reads (sequencing depth) is determined by the capacity of the machine (the sequencing platform used) and the number of samples that are multiplexed in the run.[37] The total reads reported from the high-throughput sequencing methods are large but finite.

Second, the ways of sample preparation and DNA/RNA extraction process make data carry only relative information in the measurements of omics.[38] For example, RNA sequencing starts with extraction of a fixed weight or volume tissue, DNA/RNA samples. In the end, a finite number of sequence fragment reads are obtained from a fixed volumes of total RNA. Thus, to reduce experimental biases due to sampling depth, OTU count data are normalized: each OTU count is divided by the total sum of counts in the sample (total library sizes), which results in microbiome data being compositional.

All these make microbiome dataset have compositional data structure with four compositional characteristics[39] (p. 347) as described in seminar work.[36] In summary, microbiome data essentially are compositional.[37,38,40–42]

Analyzing compositional data using standard data analysis techniques is problematic. First, using standard methods to analyze compositional data violates the assumptions of standard statistical tests. Standard methods (e.g., correlation analysis) rely on the assumption of the Euclidean geometry in real space,[43] while compositional data represent the special properties of the sample space, the simplex.[36] The violation of assumptions that the differences between parts are linear or additive makes most standard statistical methods and tests invalid[39] (p. 334).

Second, specifically applying correlation analysis to compositional data is misleading due to the problem of "spurious correlation."[20] Compositional feature of microbiome data challenges the appropriateness of Pearson's and Spearman's correlation analyses of microbiome data: the dependence of each pair components of compositional data violates the assumption that the paired data are random selected (independent) by linear (i.e., for Pearson) or rank (i.e., for Spearman) correlation; in other words, the components of compositional data are not linear or monotonic.[39] The challenges highlight the fact that the classic correlation analysis methods were originally designed for absolute values and could lead to spurious correlations for relative compositional data.[44]

Third, using standard graphical tools (e.g., scatter plot, QQ plot) for presentation of compositional data will be distorted.

Fourth, multivariate parametric modeling of compositional data violates the assumption of multivariate parametric analysis because the compositions are not multivariate normally distributed. For example, using MANOVA (or ANOVA) and multivariate (or univariate) linear regression to test hypotheses on the response is meaningless due to dependence of the mixture.

In summary, the compositionality effects of the data may introduce false-positive taxa-taxa or taxa-covariate associations,[40,45,46] precluding traditional statistical methods including correlation analysis for the detection of OTU-OTU relationships because it can result in spurious correlation.

3.4 Sparsity with excess zeros—Overdispersed and zero-inflated problems

For taxa count data, either taxonomy reads or OTU counts from amplicon sequencing experiments in microbiome studies or differential expression data from RNA-sequencing experiments, sparsity is seen as the absence of many taxa across samples and zeros are generated in most experiments. Thus, microbiome taxa abundance, especially the taxa abundance at lower taxonomic levels or OTU counts, often has many zeros and right skewed.[4,16,47] The reasons that many zeros exist in microbiome data may be due to structure itself and sampling (e.g., biological and technical variability); thus, both structural zeros and sampling zeros often appear in microbiome data.[39,48]

Zero and small values are one of main sources of sparsity. Sparsity is also due to the fact: the library sizes of DNA or RNA sequencing are widely different. The sparse microbiome data with many zeros drive OTU (taxon) count proportions to vary more than expected under the posed common multinomial regression, such as Poisson model, resulting in overdispersed and zero-inflated problems.

Sparsity is a central challenge in the analysis of 16S rRNA-sequence data.[49] The issues of sparsity with many zeros are a central topic in analysis of microbiome data. Modeling these kinds of data poses numerous challenges for traditional statistical tools. We described some critical challenges of modeling sparse data with many zeros in Chapter "What are microbiome data?" by Xia et al.[16] In summary, critical challenges (1) preclude parametric models to make accurate estimates of variance for meaningful inference and even make such estimates essentially impossible on samples that consist mostly of zeros. (2) When the taxa are sparse with many zeros, both distributions from taxa or OTUs abundance and the taxa or OTUs occurrence probability are skewed, which results in zero inflation. Thus, the taxa abundance with excess zeros cannot be correctly analyzed by any standard parametric model, such as normal, binomial, Poisson, negative-binomial, and beta distributions. (3) It also makes nonparametric methods invalid. Nonparametric methods are based on ranks or medians, thus generally insensitive or more "robust" to outliers and avoid making variance estimates that can be skewed by sparse samples. In the cases of many taxa having many zeros and few available samples, performing inference on the low-abundance taxa using the nonparametric methods is underpowered.

3.5 Heterogeneity—Challenging data integration, modeling, and meta-analysis

There are several reasons that drive microbiome data to be heterogeneous.

3.5.1 Sparsity of microbial taxa in microbiome compositional datasets

The abundances of microbial taxa are often sparse with an increasing number of zeros at lower taxonomic levels, and a right-skewed distribution with a long tail and only positive values.[50] In other words, it is often that a large set of microbiome taxa exists in whole dataset, and especially in low levels of taxa (e.g., genus and species), but a given sample is often dominated by only a few taxa with high abundance, while most other taxa having zeros or very small read counts. Thus, it is rare that any given taxa present in all samples and those present in a small proportion of samples prevail.[51] For example, some bacterial species are very abundant in some individuals, but completely absent in others.[52]

3.5.2 Technical variation

(i) *The batch effects are one important source of heterogeneity in microbiome data.* In metagenomic studies, experiments have large variations and batch effects. Numerous studies have reported batch effects in their experiments.[53–56] Recently, Wang and LêCao reviewed the methods of managing batch effects in microbiome data.[57]

(ii) *DNA extraction methods are another important source of heterogeneity in microbiome data.* The heterogeneity of microbiome studies is also due to the fact that the data used for analysis are generated using different protocols. As recently reviewed by Costea et al.[58] when analyzing data generated using different protocols, the results are often different.[59–64] To characterize how heterogeneous human fecal sample processing impacts the estimated results of microbiome, Costea et al.[58] tested 21 representative DNA extraction protocols on the same fecal samples and compared their differences due to library preparation and sample storage. This study and other studies such as Sinha et al.[65] showed that DNA extraction methods had the largest effect on the analysis outcome.

(iii) *Other important sources of technical variation are sample collection, storage, and bioinformatic analyses in microbiome data.* As recently reviewed by Schmidt et al.,[66] sample collection and storage[67–69] and bioinformatic analyses[70] are the important sources of technical variation.

Technical variation in microbiome analysis should be minimized to confidently assess the contributions of microbiota to human health and disease[58]; however, the heterogeneous microbiome data have posed technological challenges for integrating microbiome datasets, statistical modeling, comparisons of studies, and meta-analyses. Thus due to technical variation, most microbiome studies are limited in their interpretability.

3.5.3 Within-subject or temporal variation

Microbiome is dynamic, which varies temporally and particularly before the onset of more severe clinical symptoms. For example, in a recent longitudinal multiomic study, all longitudinal microbial measurement types showed significant variation within 2 weeks. Such within-subject variation and other potential sources of within-subject variation (e.g., transit time) pose higher challenge on sampling protocol to account for this kind of variation.[71]

In summary, heterogeneous microbiome data and studies pose big challenges on microbiome data integration, modeling, and meta-analysis. (*1*) *Data integration and integrative analyses are difficult.* Due to heterogeneity of sample collection and quality, integrating microbiome data and studies to construct datasets of samples for analysis is challenge. For example, it is not always possible to exactly match the same biosample across datasets instead of matching sample sets across datasets using nearby samples.[71] (*2*) *Statistical comparisons, modeling, and meta-analysis are difficult.* Due to heterogeneity, similar microbiome studies are often reported with inconsistent effects or conflicting results, which challenges statistical comparisons, modeling, and meta-analysis. Thus, currently the effective studies and methods or models for statistical comparisons, modeling, and meta-analysis are limited. For example, meta-analysis is designed to reduce study bias, ensure robust results, increase statistical power, and improve overall biological understanding of a study effect such as a clinical trial on similar experimental conditions or treatments,[72] such as IBD and obesity. However, current meta-analysis tools for microbiome data are rare, including web-based statistical tool "MicrobiomeAnalyst"[73] and R package "metamicrobiomeR."[74]

The reasons that hinder the development of statistical methods or models for effectively analyzing and conducting meta-analysis of microbiome data are that the proposing methods must be able to account for the heterogeneity of microbiome data, such as batch-effect correction between-study variation,[75–77] different patient cohorts and experimental designs,[75] and zero-inflated effect.[74]

In summary, the unique structure and characteristics of microbiome data pose big challenges on data integration and statistical analysis:

- **(i)** The large p and small n and sparse problems severely reduce the power for inferencing taxon-taxon or OTU-OTU association analyses, and require additional assumptions for accurate inference.
- **(ii)** The compositionality effects of the data may introduce false-positive taxon-taxon or taxa-covariate associations,[40,45,46] precluding traditional correlation analysis for the detection of taxon-taxon or OTU-OTU relationships because (1) the dependence of each pair components of compositional data violates the assumption that the paired data are random selected (independent) by linear or rank

correlation analysis; (2) the two components of compositional data are not linear or monotonic.[39] Thus, using Pearson's product-moment and Spearman's rank-order correlation methods to analyze microbiome data is not appropriate. The challenges highlight the fact that the classic correlation analysis methods were originally designed for absolute values and could lead to spurious correlations for relative compositional data.[44]

(iii) The heterogeneity effects of the data not only make difficult to integrate data, conduct comparisons of studies and meta-analysis, but also make interpretability difficult. For example, it was showed that in a microbiome study, only a very small fraction of interindividual gut microbiome variation can be explained at an estimated small effect size of 0.1–0.15.[66] It was also showed in a multiomic gut microbiome study of IBD that interindividual variation accounted for the majority of variance for all measurement types; however, even relatively large effects only explained a smaller proportion of variation.[71]

Traditional correlation and association methods are still in use for microbiome and other omics data. Thus, in following section, we introduce some classic correlation and association methods. Because microbiome data analysis is often challenged by various statistical properties of microbiome data (e.g., compositionality, high dimensionality, sparsity, and heterogeneity), we should consider microbiome data properties and the capabilities of these methods used, when using these classic correlation and association methods to analyze microbiome data. The assessment of methods also should be based on their capabilities (e.g., model assumption, parametric or nonparametric methods, robust or sensitive to outliers) and validation by researches.

4. Classic univariate correlation and association-based methods—Detecting correlation and association within and between omics datasets

More than 20 measures for analyzing the association between a pair of variables were originally developed in various fields such as statistics, social science, machine learning, and data mining.[6] We can divide these measures into two categories: symmetric and asymmetric measures. Generally symmetric measures are used for evaluating the variables within a dataset, whereas asymmetric measures are more suitable for analyzing association rules.[18]

This section reviews the basic terminology used in association analysis and describes and clarifies concepts that are relevant to association analysis.

We exchangeably use taxon-taxon interactions, among taxa interactions or microbe-microbe interaction. We review the most commonly used and practically useful measures in statistics and those from relevant fields. We select some symmetric measures and discuss their properties and comment on its appropriateness in application of microbiome data. In omics and specifically in microbiome studies, correlation analysis began with adopting the classic methods and processed to develop specific methods to target the features of genomic and microbiome data. Classic correlation analyses are still used. Among them, Pearson's product-moment and Spearman's rank-order correlations are most frequently used.

4.1 Commonly used measures of correlation

There are various measures for analyzing relationships between pairs of binary variables. Among them, correlation analysis is one of most often used statistical-based techniques. In 1885, Sir Francis Galton first defined the term "regression" and completed the theory of bivariate normal correlation.[78,79] Commonly used measures of correlation including Pearson's product-moment correlation coefficient and Spearman's rank-order correlation coefficient were still used in microbiome and metabolomic analyses.

4.1.1 Pearson's product-moment correlation coefficient

The product-moment correlation coefficient (also called Pearson's correlation coefficient) was named after Pearson because he described its properties in his famous 1896 paper.[80] Pearson's correlation coefficient is obtained by dividing the covariance of the two variables by the product of their standard deviations. Due to properties of Pearson's correlation coefficient closely related to the linear prediction problem, it is a measure of linear correlation. It is appropriate for measuring the strength of the linear relationship between two continuous variables of X and Y. It is the most familiar measure of dependence between two variables. The values of range lie in $[-1, +1]$. When its value is zero, then there is less of a relationship (closer to uncorrelated). The closer the coefficient is to either -1 or $+1$, the stronger the correlation between the variables. The value of correlation coefficient -1 indicates a perfect decreasing (inverse) linear relationship (anticorrelation), while $+1$ in the case of a perfect direct (increasing) linear relationship (correlation). If the variables X and Y are independent, then the value of Pearson's correlation coefficient is zero, but the converse is not true because the correlation coefficient detects only linear dependencies between two variables. Pearson's correlation coefficient close to zero can occur in all these three conditions: (1) if the two variables X and Y are independent, in this case, knowledge of the value of variable X does not increase the ability to predict the value of variable Y, or vice versa; (2) the variables X and Y are highly variable; or (3) the variables X and Y have a nonlinear relationship. By using Pearson product-moment correlation method, the two detection variables (e.g., X and Y) should meet the following assumptions: (1) both variables should be normally distributed, (2) the paired (X, Y) data are a random sample of quantitative data, (3) the relationship between each of the variables is a straight line, and (4) the data are normally distributed about the regression line.[39] Thus, we need to keep in mind that Pearson's correlation coefficient does not require normal distribution to run; however, it is sensitive to both outliers and data distribution.[81,82] Examples of using Pearson's product-moment correlation in microbiome and metabolism include: detecting the linear relationships between individual metabolites and identifying relationships between metabolite-OTU pairs,[83] determining the correlations between metabolites and bacteria,[84] and analyzing fecal microbiota and metabolites.[85]

4.1.2 Spearman's rank-order correlation coefficient

Spearman's rank correlation coefficient was first studied by Spearman.[86] This coefficient measures the strength of a monotonic association between two continuous variables of X and Y measured on an ordinal or ranked scale. Similar to Pearson's correlation coefficient, a coefficient of zero means "no association" between the variables and a value of -1 or $+1$ means "perfect inverse agreement" or "perfect agreement," respectively. Spearman's rank-order correlation is a nonparametric version of the Pearson's product-moment method.[87] Their main difference is that Pearson's correlation assesses linear relationships, whereas Spearman's correlation assesses monotonic relationships. A monotonic relationship occurs when the value of X variable increases, and the value of the Y variable either increases or decreases (whether linear or not). Spearman's rank correlation also assumes that the sample of paired (X, Y) data have been randomly selected; however, unlike the Pearson's product-moment correlation, it does not require that both variables should be normally distributed.[39] Thus, if one or both variables of interest have extreme values (sometimes called outliers), Spearman's rank correlation coefficient is a more appropriate measure of a linear relationship. Due to this property, it was considered as a different type of association measure instead of an alternative to Pearson's coefficient.[88,89]

You et al.[90] evaluated that Spearman's correlation is appropriate in detecting metabolite-microbe correlation, but Pearson's correlation is not because the complicated interomic relationships might be undetected by Pearson's correlation. Typically, the correlation analysis is conducted either between the relative abundance of taxa (OTUs, bacterial genera) or between them and the normalized metabolite concentrations. In different studies, the significant threshold of correlation coefficient was set differently, such as with P values ≤ 0.05, and correlation coefficient ≥ 0.70 denoting a strong correlation, ≥ 0.50 denoting a moderate correlation,[84,91] while other study considered both Spearman's correlation coefficient >0.60 and P value <0.01 as a valid co-occurrence event.[92] Because Bonferroni correction method is conserved, its implementation might eliminate many strong correlations; therefore, in practice, q values were often used for each Spearman's metabolite-OTU correlation for correcting multiple comparisons. The threshold of $q < 0.2$ or $q < 0.1$ was used to denote significance.[93] For example, Spearman's rank-order correlation was used to detect the monotonic relationships between individual metabolites and to identify relationships between metabolite-OTU pairs.[83] It has been often seen in detecting taxonomy and metabolite profiles,[91,93] examining associations between KEGG pathways and taxa.[94]

In summary, Spearman's rank correlation coefficient does not require the monotonic relationship to be represented by a linear relationship. Spearman's rank correlation coefficient is less sensitive to nonnormality in distributions and is more robust to outliers; thus, it is more appropriate to be used to assess nonlinear associations.

4.1.3 Kendall's rank correlation coefficient

Both Pearson's product-moment and Spearman's rank-order correlation coefficients are appropriate measures of two continuous variables. For continuous-ordinal variables or two ordinal variables, an appropriate measure of the strength of association is Kendall's rank correlation coefficient tau-b. Kendall tau rank correlation was developed by Maurice Kendall in 1938 and named after him.[95] Kendall's tau was originally defined for continuous bivariate variables, and the tau-a statistic tests the strength of association of the cross tabulations measured on an ordinal or rank scale. Tau-a does not make any adjustment for ties. However, for ordinal data, the ties must be taken into account. Unlike tau-a, the tau-b statistic makes adjustments for ties. Kendall discussed whether tau-a or tau-b is more appropriate for ranking data.[96,97]

The values of Kendall's rank correlation coefficient tau-b range also lie in $[-1, +1]$. The value of -1 indicates 100% negative association or perfect inversion, +1 for 100% positive association or perfect agreement, and a value of zero indicates the absence of association. Kendall's rank correlation method is used to measure the ordinal association between two measured variables by a nonparametric tau test. Similar to Spearman's rank correlation, Kendall's correlation also measures rank: the similarity of the ordering of the data when ranked by each of the variables.[95] In the case of the ordinal variable Y having a large number of levels, the Spearman's rank correlation coefficient is also suitable and their performances are generally comparable and the Spearman's rank correlation coefficient being somewhat better for large sample sizes.[98] Since the measure tau-b and its estimator cannot attain the extreme values +1 or −1 for nonsquare tables,[19] following Kendall, a measure related to tau-b was proposed by Stuart in 1953,[99] called tau-c (also called Kendall-Stuart tau-c or Stuart-Kendall tau-c). Tau-c is more suitable than tau-b for the analysis of data based on nonsquare (i.e., rectangular) contingency tables.[99,100] However, either due to its dependence on the dimensions of the table, makes it difficult to interpret,[19] or due to its analog to a normal version of product-moment, tau-c unlikely yields a useful interpretation,[101] even the appropriateness of such adjustment may be arguable.[102] Kendall's tau rank coefficient has been employed in microbiome association analysis and has been shown a slight power edge over Pearson's method.[103] One example of using Kendall's correlation was to test the association of smoking variables with taxa relative abundances.[104]

4.1.4 Matthews correlation coefficient (MCC)

MCC coefficient was introduced in machine learning by Brian W. Matthews in 1975 to assess the performance of protein secondary structure prediction.[105] It is a measure of the quality of binary (two-class) classifications or a method of measuring the agreement (correlation) between prediction and observation. MCC is defined in terms of True Positive (TP), True Negative (TN), False Positive (FP) and False Negative (FN) and uses the correlation between prediction and observation for estimating the quality of a given prediction.[106] Thus, the MCC measure is in essence a Pearson's correlation coefficient between the observed and predicted binary classifications. MCC can be interpreted as representing the correlation between the observed and expected classifications. The values of range lie in [−1, +1]. A value of +1 indicates a perfect agreement between prediction and observation, −1 indicates total disagreement between prediction and observation, while 0 suggesting a prediction no better than random. As the phi coefficient is analogous to Pearson's product-moment correlation coefficient for continuous variables, MCC is also related to Pearson's product-moment correlation coefficient.

The advantages of MCC include: (1) MCC is related to the chi-square statistic for a 2×2 contingency table. However, unlike the chi-square statistic, the measure of correlation makes MCC a more suitable measure of association. (2) MCC is considered as one of appropriate or best measures describing the contingency or confusion matrix of true and false positives and negatives.[7,107,108] (3) This measure is suitable for imbalanced data; i.e., it can be used for the classes even having very different sizes.[109] (4) MCC is widely used in bioinformatics as a performance metric. As an objective approach to assess the quality of the OTU assignments, the strength of the MCC is that it can assess any set of OTU assignments without relying on an external reference only requiring a distance matrix and a specific threshold, so MCC can synthesize the relationship between OTU assignments and the distances between sequences.[110] MCC is also used to target data imbalance. Data imbalance occurs when the sample size in the data classes is unevenly distributed. Such situation is encountered in bioinformatic analysis of microbiome data. In order to handle imbalanced data, Boughorbel et al.[109] developed a new Bayes classifier, called MCC classifier based on the MCC metric. For example, MCC was utilized to assess clustering quality between the observed and predicted values of the clustering scheme of OTU with TPs, TNs, FPs, and FNs.[21,111]

The commonly used correlation methods including Pearson, Spearman, and Kendall correlations do not take account for compositionality of microbiome data.

4.2 Commonly used measures of association

For two discrete nominal variables with just two levels for each variable, the data can be presented as a 2×2 contingency table. Several commonly used measures of association are available for such a contingency table.

4.2.1 Chi-squared and Fisher's exact tests

The chi-squared statistic[112,113] is used to test for independence between row and column in contingency table or between the LHS and RHS of the association rule. The significant difference is determined based on the expected frequencies and the observed frequencies in row and column categories or the LHS and RHS of the rule. The chi-squared distribution with 1 degree of freedom (in a 2×2 contingency table) has a critical value of 3.84 at 0.05 significant level. The chi-squared values of range lie in $[0, \infty]$. A higher value (e.g., >3.84 in the case of 2×2 contingency table) indicates that the row (LHS) and the column (RHS) are not independent. Although the chi-squared test is an excellent measure of the significance of the association, it is not at all useful as a measure of the degree of association.[30] Actually, the chi-squared test is often used for goodness-of- fit testing instead of being used as a measure of association because it depends on the sample size.[6,13] Although Fisher's exact test is valid for all sample sizes, when sample sizes are small, i.e., cells have low expected values less than 5, in practice Fisher's exact test might be more appropriate. For example, in infant microbiome and human milk study the chi-squared test was used to assess the significance of the association between infant's gender, mother's secretor status (positive, negative), and microbial cluster type.[114] Fisher's exact test was used to model occurrence data.[115]

4.2.2 Phi coefficient

Phi coefficient is also called mean square contingency coefficient. For two binary variables, correlation can be measured using the phi coefficient. This measure was introduced by Karl Pearson.[116] Phi is a symmetric measure. This measure is analogous to Pearson's product-moment correlation coefficient for continuous variables. In fact, it is its application of a Pearson's correlation coefficient for estimating two binary variables.[117] The phi coefficient can be calculated by two ways.

First, in 2×2 contingency table, phi coefficient is calculated by comparing the product of the diagonal cells to the product of the off-diagonal cells. If most of the data fall along the diagonal cells, then the two binary variables are considered positively associated; in other words, if most of the data fall off the diagonal, then the two binary variables are considered negatively associated. The range of the phi coefficient is $[-1, +1]$. The general rule of thumb for Pearson's correlation coefficient also applies for interpretation of the phi coefficient: -1.0 to -0.7 suggesting strong negative correlation, -0.7 to -0.3 weak negative correlation, -0.3 to $+0.3$ little or no correlation, $+0.3$ to $+0.7$ weak positive correlation, and $+0.7$ to $+1.0$ strong positive correlation.

Second, for tables larger than 2×2, the phi coefficient is equal to the square root of value of the chi-squared statistic divided by sample size. Thus, the phi coefficient is closely related to the chi-squared statistic. The values of phi-coefficient lie in $0 \leq \phi \leq \min\left(\sqrt{R-1}, \sqrt{C-1}\right)$[19] (p. 13), i.e., when the marginal distributions are not equal, the maximum <1[30] (p. 99). The value close to 0 indicates very little association, and close to 1 indicates nearly perfect predictability. As a rule of thumb, any value less than 0.30 or 0.35 may be taken to indicate no more than trivial association.[30]

The phi coefficient as a measure of association has been widely discussed in previous publications[30] (pp. 98–100). (1) It is interpretable as a correlation coefficient, which makes it especially suitable to measure association in behavioral sciences and psychometrics. (2) It is free of the influence of the total sample size, although phi coefficient is derived from chi-squared statistic. (3) It is able to establish equations to explore association testability, and hence provides an analytical solution.[118] However, phi coefficient has serious deficiencies. For example, the value of phi coefficient depends strongly on the cut-off and lack of invariance if the binary variables are dichotomized from a continuous distribution. Also, the range of phi coefficient depends on the dimensions of the table because phi-squared measure is derived from the chi-squared statistic. Thus, the phi-squared statistic is thought not suitable to measure association without modification.[19,119] The phi coefficient was used in microbiome studies to analyze occurrence data.[115]

4.2.3 Cramér's V

Cramér's V (or Cramér's phi) was proposed by Harald Cramér [116] to measure association between two nominal variables. It is the intercorrelation of two discrete variables[120] and may be used with variables having two or more levels. It is a symmetrical measure and does not matter which variable in the columns/rows or the order of rows/columns. Thus, it may be used with nominal and higher (notably ordered or numerical) data types.

Cramér's V is based on Pearson's chi-squared statistic and can be applied to goodness-of-fit chi-squared models when there is a $1 \times k$ table (in this case $r = 1$). For a 2×2 contingency table, phi-squared statistic is identical to the Pearson's product-moment correlation coefficient squared statistic, and Cramér's V is equal to the phi coefficient. The Cramér's V has a value between $[0, +1]$ (inclusive). The value of $V = 0$ suggests that these two variables are independent (no association), while $V = 1$ suggesting perfect or complete association.[19]

Cramér's V has been used in microbiome studies. For example, a modified version of Cramér's V was used as the effect size in power and sample size calculations,[121] in which the value of 0 and 1 denoting the taxa frequencies are the same and are maximally different in both groups, respectively. However, as an estimate of association, Cramér's V has some deficiencies: likely to be inflated where evenness is low, and underestimates association for high levels of informedness.[108]

4.2.4 Goodman and Kruskal's lambda

It is also known as the index of predictive association. If at least one of the two nominal variables has more than two levels, Goodman and Kruskal's lambda[119,122] is a useful measure of association. While the chi-squared statistic and related measures are all motivated to measure "lack of independence" between two categorical variables, Goodman and Kruskal's lambda measures the relative usefulness of one variable in improving the ability of prediction of a second variable.[19] The measure is defined as the proportional amount of reduction in the prediction error. The intuition behind this measure is that if two variables are highly dependent on each other, then whenever knowing the value of one variable, the error in predicting the other variable would be small.[6] The value of this measure lies between 0 and 1; values close to 1 suggest a strong association. Goodman and Kruskal tau has been wrapped as associate () function (Cross Correlation Wrapper) in Package "microbiome" (see manual on January 28, 2020: Version 1.9.19, Date 2019-11-01).

4.2.5 Odds ratio (OR)

OR (also known as the cross-product ratio) was originally proposed by Cornfield[123] as a measure of the degree of association between an antecedent factor and an outcome event. Now, it becomes a well-known measure for studying associations in biomedical sciences and epidemiology. OR has several advantages, compared to other measures by Mosteller.[124] The advantages have been considered by Edwards[125] to be so great that he recommended that only the odds ratio or functions of it be used to measure association in 2×2 tables.[30] The confidence intervals for odds ratios can be calculated[126] and are very often reported in literature. More specifically, the odds ratio lower bound (ORLB) of the 95% confidence interval around sample OR is used as an important measure for assessing statistical significance or lack thereof. ORLB >1 indicates a statistically significant association with higher values indicating stronger associations.[15]

However, using OR and the rate ratio as a measure of association was also strongly criticized as losing the level of the rates.[127,128] The values of odds ratio range lie in $[0, \infty]$, with 0 indicating for perfect negative association between the two variables or characteristics, ∞ for perfect positive association, and 1 indicating the independence or lack of association between them. In practice, the log odds ratio is often used instead of odds ratio due to its nice mathematical properties such as a symmetric measure: the values of log odds ratio range lie in $[-\infty, \infty]$, with $\log 1 = 0$ indicating independence. For example, ORs were used to calculate taxa,[129] to rank the relative enrichment or underrepresentation of COG (Clusters of Orthologous Groups) and

KEGG (Kyoto Encyclopedia of Genes and Genomes) categories,[130] and to estimate the association between microbial community and clinical risk factors.[131]

4.2.6 Yule's Q and Y-coefficients

Yule proposed two measures of association: Yule's Q[132] and Y-coefficients.[118] They are both functions of the odds ratio. The values of odds ratio range lie in [0, ∞], while both Yule's Q and Y coefficients are normalized variants of the odds ratio. They are both symmetric measures, and thus both are invariant under the row and column scaling operations to the contingency table. Both have values between [−1, +1]. When the odds ratio = 0, ∞, then both measures equal −1 and 1, respectively; when the odds ratio = 1, they are 0, suggesting independence.[19] The Yule association coefficients were reviewed in microbiome literature.[115,133]

4.2.7 Kappa (κ)

Cohen's κ-coefficient[134] measures the degree of agreement between a pair of variables, frequently used as a metric of interrater agreement, i.e., kappa most often deals with data that are the result of a judgment, not a measurement. Kappa compares the probability of agreement to that expected if the ratings are independent. The values of range lie in [−1, 1] with 1 presenting complete agreement and 0 meaning no agreement or independence. A negative statistic implies that the agreement is worse than random. The standard for a "good" or "acceptable" kappa value is arbitrary. Fleiss' arbitrary guidelines (0.75 is excellent)[25] seem to be cited most often. Kappa is intrinsically nonlinear, does not account for error well, and retains an influence of bias, so kappa has been reviewed that would not be preferable to correlation as a standard independent measure of agreement.[108] A comprehensive discussion of the kappa statistic from the methodological perspectives can be found in the paper.[135] Examples of the kappa statistic use in microbiome studies include: to assess the level of agreement between metagenomics and 16S rRNA for the bacterial signals,[136] and to account for spurious high classifiability due to class imbalances.[76] Other examples of reporting the kappa statistic can be found in these studies.[137,138]

4.2.8 Cosine similarity

Cosine similarity is a measure of similarity between two nonzero vectors of an inner product space that measures the cosine of the angle between the two space vectors. Cosine is a symmetric measure. It is used to judge the orientation and not magnitude: a cosine similarity of 1 (the cosine of 0°) indicating the two vectors with the same orientation, a similarity of 0 indicating two vectors oriented at 90°, while a similarity of −1 indicating the two vectors diametrically opposed for the similarities range between [0, −1] having angle in the interval (0, 180°). However, the cosine similarity is particularly used in positive space, where the outcome is neatly bounded in [0, 1]. Cosine similarity is a measure that is less influenced by the sparsity resulting from the zeros of nonsignificant associations. Thus, it was used to measure the distance matrices between diseases and between microbiota traits.[139,140] Other examples of using cosine similarity in microbiome studies are available from these reports.[51,141–143]

4.2.9 Jaccard similarity

It is also known as Jaccard similarity coefficient, Jaccard index, or Intersection over Union. Originally, Jaccard similarity was given the French name *coefficient de communauté* (community coefficient) by Paul Jaccard.[144] It is used for measuring the similarity and diversity between finite sample sets and is used extensively in information retrieval to measure the similarity between documents.[145] The Jaccard similarity is defined as the size of the intersection divided by the size of the union of the sample sets. The Jaccard similarity has values between [−1, 1], with 0 suggesting for independence. The measure that is complementary to the Jaccard similarity is called *Jaccard distance*, which measures dissimilarity between sample sets. It is obtained by subtracting the Jaccard coefficient from 1, or, equivalently, by dividing the difference of the sizes of the union and the intersection of two sets by the size of the union. Both cosine and Jaccard measures have the property of null invariance. This property is useful for sparse datasets, where co-presence of items is more important than co-absence.[6] This may be one of reasons that cosine and particularly Jaccard distance measure have been often used in ecology and microbiome datasets, where the data are often sparse. Although some measures we reviewed above are initially defined for two variables, most, such as cosine similarity and Jaccard coefficient, can be extended to more than two variables using the frequency tables tabulated in a multidimensional contingency table. We introduced Jaccard similarity in the Chapters "Community diversity measures and calculations" and "Multivariate community analysis" of Xia et al.'s book "Statistical Analysis of Microbiome Data with R."[146,147] Other examples of Jaccard similarity applications in microbiome studies can be found in these works.[148–150]

4.3 Commonly used measures based on the probability distribution of variables

4.3.1 Entropy

Ludwig Boltzmann developed the definition of entropy in statistical mechanics in the 1870s by analyzing the statistical behavior of the microscopic components of the system. The concept of entropy in information theory is related to the concept of entropy in statistical mechanics. Entropy [151] is a measure of the uncertainty of a random variable, i.e., the variance of a probability distribution. A uniform distribution has large entropy, whereas a skewed distribution has small entropy. The entropy of a random variable X with a probability mass function p(x) is defined by $H(X) = -\sum p(x)\log_2 p(x)$. By using logarithms to base 2, the entropy will then be measured in bits: the number of bits on average required to describe the random variable. Examples of reporting entropy can be found in these microbiome studies.[152–154]

4.3.2 Maximal information coefficient (MIC)

MIC is also called mutual information. MIC is a maximal information-based nonparametric correlation detection method based on Shannon entropy for evaluating the dependencies among variables.[155] It is a measure of the amount of information one random variable contains about another.[151] It presents the reduction in uncertainty of one variable when the other is observed. Entropy represents the self-information of a random variable, while mutual information is more general, called relative entropy, which is a measure of the distance between two probability distributions. It represents how the amount of the entropy reduces in a variable given we know the value of a second variable. Thus, the stronger the association of the two variables, the more the amount in entropy reduces, i.e., its mutual information is higher. Mutual information is an asymmetric measure. The values of mutual information range in [0, 1] with 0 for independence. The attractive characteristics of MIC lie on its capacity of generally measuring either linear or nonlinear dependencies between variables,[44] and providing an R^2 (coefficient of determination in regression) equivalent measure score for functional relationships; thus, it is often used for large-scale datasets such as gene expression and human microbiome to identify known and novel relationships.[90,156] MIC was used in cross-sectional microbiome studies to capture the correlations between taxa.[157–159] It was also utilized to construct a temporal microbial correlation network.[103]

4.3.3 Gini index

Gini index,[160] also called Gini coefficient or Gini ratio, is another measure defined based on the probability distribution of variables. It was proposed by Corrado Gini in his 1912 paper *Variability and Mutability*[161] and named after him. The Gini index was originally proposed as a measure of statistical dispersion intended to represent the income or wealth distribution and is the most commonly used measurement of inequality among values of a frequency distribution.[162] When all values are the same, then the Gini index is 0 expressing perfect equality; a Gini index of 1 expresses maximal inequality among values. A value greater than 1 may occur when negative contribution happens. However, values close to or above 1 are very unlikely in practice. There are some issues in interpreting a Gini index. For example, the same value may result from many different distribution curves. Examples of using Gini index in microbiome studies are available from these works.[163–165]

As we know, some traditional statistical methods for analyzing association are suitable for symmetric binary data, whereas others are suitable for asymmetric binary data. Some are appropriate for pairwise associations, while others can be moved beyond pairs of binary variables to more than two variables. For more complicated datasets, such as multidimensional contingency tables, more sophisticated statistical models or other statistical techniques are required for analyzing association among variables. At least, there are two reasons: (1) partial associations often present in multidimensional data because conditioned on other confounding factors (i.e., hidden variables) the observed association between a pair of variables may disappear or reverse its direction of association. This phenomenon is known as Simpson's paradox. Therefore, the association between variables should be interpreted with caution.[18] (2) Multidimensional data often accompany other special features, such as sparsity and compositionality (see next section for details).

5. Newly developed univariate correlation and association-based methods—Targeting the specific characteristics of microbiome data

In Section 3, we describe the statistical issues in microbiome data. Many efforts have been made to solve these problems from both data processing procedures and statistical methods to modeling perspectives. One critical data processing procedure is normalization. Various normalization methods have been proposed in microbiome and particularly in RNA-sequencing literature.

5.1 Normalization—A data processing step to target the issues of compositionality, variability, heterogeneity, and outliers

Because the counts of abundance reads only carry the relative information about the taxa rather than the true abundances of molecules in the underlying community attributable to the taxa. In microbiome study, to reduce experimental biases due to sampling depth, typically taxon or OTU count data are normalized: each taxon or OTU count is divided by the total sum of counts generated across all taxa in the sample (total library sizes or sequencing depth). The normalized values of taxon or OTU count present the relative abundances of taxa or OTUs. Due to sum constraint posed on the conversion, each component of taxa or OTUs is not independent of each other. Normalization is one of first data steps to deal with heterogeneity due to the variation of sequencing depths and thus is the first attempt to deal with the statistical issues of compositional data.

5.1.1 Traditional normalization methods

Traditional normalization methods are relatively simple, for example, either to transform taxa abundance counts into relative abundances (i.e., proportions) of the taxa, or to rarefy the counts by subsampling without replacement from each sample until all samples have the same number of total counts across taxa, or using the total read count for each sample as the size factor to estimate the library size (the latter way is called total sum scaling (TSS)).

5.1.2 RNA-Seq data-based normalization methods

The popular normalization methods mainly focus on differential abundance analysis. Most of them were originally proposed for analysis of RNA-seq data, which can be called RNA-Seq data-based normalization methods. We have introduced RNA-Seq data-based normalization methods in Chapter "Modeling over-dispersed microbiome data" by Xia et al.[166] The interested readers can refer this book chapter for details. The normalization methods implemented in edgeR package are: (1) Trimmed Mean of M-values (TMM) normalization[167–169]; (2) Upper quantile normalization[170]; (3) Total read count normalization[168]; and (4) Relative Log Expression (RLE) normalization.[171] DESeq2 uses the "median ratio method" described in Anders and Huber[171] to estimate the size factors, so the DESeq normalization is equivalent to RLE normalization implemented in edgeR. Other RNA-Seq data-based normalization methods include variance stabilizing transformation.[172]

Regarding which approach is more appropriate for microbiome data is arguable in literature. For example, on the one hand, rarefying microbiome data has been criticized as "inadmissible" and the approaches of DESeq2 and edgeR have been recommended to use.[173] Rarefaction has one key issue: it maintains the mean of the taxonomic proportions, while it ignores the variation of the proportions.[51] The normalization methods associated with edgeR and DESeq2 have been shown to outperform other methods, particularly when expressed RNA varies across biological conditions or in the presence of highly expressed genes[174] (p. 406). On the other hand, the traditional proportion- and rarefaction-based normalization approaches have been considered to provide more accurate community-level comparisons.[175]

5.1.3 Microbiome data-based normalization methods

5.1.3.1 Cumulative sum scaling (CSS)

CSS normalization[48] is a normalization method specifically designed for microbiome data. The goal of CSS technique is to correct the bias in the assessment of differential abundance introduced by total-sum normalization. In order to capture the relatively invariant count distribution for a dataset, the proposed method divides raw counts into the cumulative sum of counts, up to a percentile determined using a data-driven approach. CSS was criticized because the percentiles could not be determinated for microbiome datasets with high variable counts.[176]

5.1.3.2 Geometric mean of pairwise ratios (GMPR)

GMPR[176] was developed specifically for zero-inflated sequencing data (e.g., microbiome sequencing data). Microbiome sequencing data are more overdispersed and have many zeros compared to RNA-Seq data. The motivation of developing GMPR is that the previously available normalization methods including traditional and RNA-Seq data-based normalization methods cannot address overdispersed and zero-inflated microbiome data; and hence they are inappropriate to be applied to normalize microbiome sequencing data.[176] For example, TSS is not robust to outliers and cannot remove compositionality, instead has been criticized creating compositional effects: making nondifferential features appear to be differential due to the constant-sum constraint.[49,176–178] The percentiles that CSS proposed to obtain using a data-driven approach could not be determinated when microbiome datasets have high variable counts.[176] It was demonstrated that GMPR has several benefits including: (1) is robust to differential and outlier OTUs, (2) improves the performance of differential abundance analysis,

(3) reduces the intersample variability of normalized abundances, and (4) improves the reproducibility of normalized abundance.[176] In summary, it was shown that GMPR outperforms RNA-Seq normalization methods and yields better performance than CSS, another microbiome data-based normalization method as well as applicable to other sequencing data with excess zeros (e.g., single-cell RNA-Seq data).[179] However, GMPR method also has limitations: it is mainly applied to taxon-level analysis of differential abundance and reproducibility to distinguish the "truly" differential from "falsely" differential taxa due to compositional effects. It is also computationally complicated and inefficient for large samples sizes.[176] Examples of using GMPR in microbiome studies can be found in these works.[115,180–182]

5.2 Mitigate the issues of high dimensionality

To target the specific features of microbiome data, both parametric statistical methods for estimating the true covariance matrix in compositional context and nonparametric statistical methods have currently been developed. The parametric approach of inferring covariance matrix has focused on two different features of microbiome data: dimensionality and compositionality.

Both "covariance" and "correlation" measure the relationship and the dependency between two variables (taxa or OTUs for microbiome data). "Covariance" indicates the direction of the linear relationship between taxa or OTUs, whereas "correlation" measures both the strength and direction of the linear relationship between them. In terms of mathematics, correlation is a function of the covariance. By definition, the correlation coefficient of two taxa or OTUs can be obtained dividing the covariance of the two taxa or OTUs by the product of their standard deviations. The *correlation values* essentially are scaled or standardized to a range of $[-1, +1]$.

Correlation is unit-free, dimensionless, and is not influenced by the change in scale of the values, while covariance assumes the units from the product of the units of the two taxa or OTUs and the value of covariance is affected by the change in scale of the taxa or OTUs. Thus, covariance matrix estimation is affected by the sparsity of the taxa or OTUs. Covariance can take any value between $-\infty$ and $+\infty$. Textbooks generally advise us to use the covariance matrix when the variables are on similar scales, and to use the correlation matrix when the scales of the variables differ because scale matters.

One big challenge for statistical analysis of 16S rRNA gene data is high dimensionality. The large p and small n and sparse problems severely reduce the power for inferencing OTU-OTU association analysis, and require additional assumptions for accurate inference. Compared to correlation and association analyses in other study fields, correlation and association analyses in microbiome have several distinct features: (1) Correlation and association are multiple dimensional; the analyses move beyond bivariate to multivariate, from a pair of variables to dynamic interaction, such as correlation networks are getting more popular recently. (2) The measures not only focus on individual values; the summarized values or various measures for diversity are often the staring points of a microbiome study. Then an association analysis is applied to analyzing the diversities. (3) Unlike other fields, correlation and association analyses in microbiome are not only conducted within microbiome, but also performed beyond microbiome to host, from intraomics to interomics. A popular trend we noticed is to analyze the correlation or association between microbiome and metabolism. (4) Various association methods targeting one or more specific features of microbiome data have been developed in current years.

5.3 Mitigate the issues of compositionality

Microbiome data are compositional. The abundance count of a taxon in a sample only reflects the relative abundance of the taxon compared against all other taxa, rather than the absolute count of molecules in the underlying community attributable to the taxon.

To address the challenges due to compositionality, we observed two trends of moving beyond traditional correlation analysis: one is proportionality analysis, which goes further in the framework of compositional analysis; another remains in the framework of correlation or covariance analyses, but adopts the log-ratio transformation technique to form compositional analysis.

5.3.1 Proportionality analysis

The concept of proportionality is the benchmarker of the first approach. In Chapter "Compositional analysis of microbiome data" by Xia et al.,[39] we introduced the proportionality concept and illustrated proportionality analysis. However, as we summarized and discussed in the cited book, proportionality analysis faces several challenges. Among them, the biggest one probably is lack of interpretation.

5.3.2 Log-ratio transformation

Several methods have been specifically designed and developed for estimating correlations of compositional microbiome data and naturally extend bivariate correlation to correlation network. Among them, SparCC[40] and CCREPE (Compositionally Corrected by REnormalization and Permutation)[49] were designed to use log-ratio transformations for solving compositionality under the assumption of sparsity of taxa.

5.3.2.1 SparCC

SparCC (Sparse Correlations for Compositional data) is one of the earliest inferring correlation network methods. It uses iterative bootstrap for estimating the covariance matrix. SparCC was specifically designed to estimate the linear Pearson's correlations between the log-transformed components from compositional data.[40] To remove compositionality, SparCC uses a log-ratio transformation to transform every pair being correlated of taxa so that the ratio of the abundances of two taxa is independent given other taxa are included in the analysis, which is a property of subcompositional coherence termed in compositional data analysis.

Although the approximation method utilized by SparCC assumes that (1) the number of different components (e.g., OTUs or genes) is large and (2) the true correlation network is "sparse" (i.e., most components are not strongly correlated with each other), SparCC does not depend on any particular distribution or make any parametric assumptions of the testing variables.[40,183] First, SparCC has some pros: it performs well under low diversity and high sparsity; thus, SparCC can infer correlations with high accuracy even in the most challenging datasets. SparCC can also simultaneously remove compositional effects from OTU tables, while calculating correlation matrices. In summary, it is useful for correcting spurious correlations and identifying true associations missed by Pearson's correlation.[49] Due to these pros, SparCC was used for defining community structure of metabolite-associated microbes.[93] The authors of SparCC used the HMP datasets to show its inferring taxon-taxon interaction networks of sparse and compositional ecological data. However, SparCC has been reviewed having two fundamental arguable issues. First, it assumes that there are a sufficiently large number of taxa, and that these taxa on average are uncorrelated with each other leading to a sparse network, which potentially has overestimated the underlying association networks. Second, it eliminates zero fractions by adding small pseudo-counts, which obviously simplifies the complication of zero problem.[39] SparCC also has some other cons or weaknesses including: (1) It has poor performance when sample diversity is high which is contrary to its design assumption for high data sparsity. (2) It is biased toward positive correlation. (3) It only measures linear relationship.[184] (4) It may reduce the estimation accuracy due to without considering the influence of errors in compositional data. (5) The inferred covariance matrix has no guarantee to be positively defined. (6) It may even result in the correlation coefficients out of the range [−1, 1].[185] Examples of SparCC use can be found in these microbiome studies.[139,186,187]

5.3.2.2 CCREPE

The R package CCREPE was designed for detecting statistically significant associations between sparse and high-dimensional compositional data using permutation-based methods.[188] It is a correlation-based method and uses permutation and bootstrap to infer the correlated significance. The methods are implemented through ccrepe() and nc. score() functions. The first function calculates similarity measures, P values, q values, and Z scores for all pairwise correlations within one dataset or between two datasets for relative abundances using bootstrap and permutation matrices of the data, while the second function calculates species-level co-variation and co-exclusion patterns based on an extension of the checkerboard score to ordinal data. The package takes the sum to one constraint into account when assigning P values to similarity measures between the taxa. CCREPE was used to integrate the gut microbiome and the gut metabolome of infants in type 1 diabetes study.[189] Although CCREPE's permutation approach is useful for correcting spurious correlations and improving similarity measure, it has been reviewed as: difficult to explain the difference between the permutation and bootstrap samples[185] and fails to adequately control for compositional effects. In addition, it leads to "false confidence" in the observed correlations.[39,40,49] Examples of CCREPE use can be found in these microbiome studies.[189–191]

5.4 Mitigate the issues of compositionality and sparsity as well as dimensionality

5.4.1 LASSO regression methods

LASSO method is appealing to this kind of data because it deals with sparsity by shrinking the coefficients to solve the large p small n problem. LASSO (least absolute shrinkage and selection operator) regressions (another kind of parametric methods) were designed to use LASSO techniques for penalizing sparsity of taxa for solving dimensionality in compositional context. It penalizes excessively complex microbial interaction networks. LASSO shrinks the coefficients associated with covariates toward zero (or limits/shrinks the absolute value toward the mean) and thus is capable of solving variable selection problem under the context of the high likelihood of multicollinearity.

LASSO was proposed by Tibshirani.[192] However, the original 1996 version of LASSO only handles the linear regression and assumes that the response is normally distributed. In 2010, Friedman et al. generalized the linear regression to the binomial and multinomial regression models to handle categorical responses.[193] The generalized regularization multinomial regression model of LASSO can be implemented via the R package *glmnet*. LASSO regression analysis was used to determine the specific metabolic pathways from microbiome that most likely differentiate between prostate cancer and no cancer disease.[194] LASSO also was used for feature selection in meta-analysis of fecal metagenomes.[195] Examples of using or referencing the generalized LASSO are available from the papers.[196–198]

The group LASSO regularization to the binomial regression model was proposed by Meier et al.[199] and can be implemented via the R package *grplasso*.[200] The benefit of group LASSO regularization is on the fact that we can group predictors in a regression model according to different taxonomic rank, and thus incorporating evolutionary information of taxa in the modeling. However, since the group LASSO is a logistic (binomial) regression model, the response variable is limited to only two values. Examples of using or referencing this method can be found in these studies.[201–204]

A hybrid version LASSO, called sparse group LASSO regularization, was proposed by Simon et al. in 2013.[205] The sparse group LASSO combines both algorithms from linear and logistic regression LASSO in its variable selection; thus, it removes the unimportant predictors from the model both individually and in groups. The sparse group LASSO regularization method can be implemented in the R software package *SGL*.[206] However, as group LASSO regularization and the *grplasso* package, the sparse group LASSO regularization method and its software are limited to the binomial responses. The sparse group LASSO regularization method was used or cited in these research papers.[207–209]

All these current three versions of LASSO regularization methods have fundamental limitations: they focus on the variable selection and prediction problems, and are exploratory. As exploratory tools, they do not conduct hypothesis testing; thus, no *P* values are provided. They even do not have components to measure the strength of the association between the selected predictors and the response (in microbiome case, the selected taxa and the groups).

5.4.2 Incorporating LASSO and other techniques

All the following methods were designed to apply LASSO technique and incorporate other techniques to improve covariance estimation.

GLASSO (graphical LASSO)[210] is one early solution to the large p and small n and sparse problems. The authors of GLASSO method illustrated its application using cell-signaling data from proteomics. Graphical LASSO method is still used as a standard in estimating the sparse precision matrix. Nonetheless, GLASSO is a Gaussian graphical model. The normality assumption inhibits its appropriate use to microbiome data (i.e., 16S rRNA sequencing data) because the count-based data are often with many zeros and compositional. The GLASSO method was used or cited in these publications.[211–213]

CCLasso (Correlation inference for Compositional data through Lasso)[185] was proposed based on least squares with penalty to infer the correlation network for latent variables of log ratio transformation from microbiome compositional data using LASSO technique. The proposed method was also applied to the HMP datasets to show its capability in estimating correlation network of microbe species. CCLasso has some pros[185]: (1) by using LASSO technique, correlation matrix estimation is more accurately compared to SparCC; and (2) outperforms SparCC in edge recovery for compositional data; the additional benefit is that the estimated correlation matrix of the latent variables is guaranteed positive definite. However, similar to SparCC, CCLasso also has some cons including only measures linear relationship. The CCLasso method was used or cited in these publications.[214–217]

REBACCA, BAnOCC, and MPLasso. Other LASSO-based methods include REBACCA (regularized estimation of the basis covariance based on compositional data),[218] BAnOCC (Bayesian Analysis of Compositional Covariance),[219] and MPLasso (Microbial Prior Lasso).[220] All these methods were designed to improve covariance estimation via applying LASSO technique or incorporating other techniques. Among them, REBACCA uses global optimization procedures to estimate the correlation network of all species while imposing an explicitly compositional constraint and a sparsity constraint on the network. Examples of REBACCA citations in microbiome studies can be found in these publications.[70,221,222] Examples of BAnOCC citations in microbiome studies and research can be found in these publications.[223–225]

SPIEC-EASI (Sparse Inverse Covariance Estimation for Ecological Association Inference, pronounced speakeasy) represents another approach that combines log-ratio transformations developed for compositional data analysis and LASSO techniques for shrinking sparsity. The statistical method developed by SPIEC-EASI was used for the inference of microbial ecological networks from amplicon sequencing. The purposes of the development were just to address the aforementioned two issues for statistical analysis of

16S rRNA gene data: (1) the spurious correlation results from using traditional correlation methods to analyze the compositional OTUs; and (2) the severely underpowered inference of OTU-OTU association networks due to the large p and small n and sparse problems. To address these two issues, SPIEC-EASI and package SpiecEasi take advantage of the proportionality invariance of relative abundance data and assume that the number of taxa in the dataset is larger than the number of sampled communities. SPIEC-EASI combines data transformations developed for compositional data analysis with a graphic model inference framework to improve covariance estimation and runs GLASSO on the centered log-ratio transformed counts.[226] SPIEC-EASI method was applied to the data from the American Gut project to predict previously unknown microbial associations. SPIEC-EASI uses covariance matrix (inverse covariance) instead of correlation matrix. Thus, the ranges of coefficients are not limited in $[-1, 1]$.

The advantages of SPIEC-EASI include: (1) addresses underpowering issues for inference of OTU-OTU association networks in compositional data; (2) improves edges recovery compared to other state-of-the-art methods; (3) corrects for indirect taxa-taxa associations; (4) outperforms SparCC and CCREPE in terms of recovery of taxon-taxon interactions or constructs more highly reproducible association networks[39,211]; and (5) it may be more important that SPIEC-EASI framework allows statistical inference of cross-domain or -omics associations (e.g., between bacteria and fungi).[227,228] However, a recent study showed that using SPIEC-EASI to identify microbe-metabolite interactions (i.e., microbe-fungal interactions) does not work due to differences in measurement units between sequencing and mass spectrometry measurements.[229] Examples of SPIEC-EASI citations in microbiome studies and research can be found in these publications.[49,230,231]

Currently, You et al.[90] evaluated the performance of six correlation methods including Pearson's product-moment correlation, Spearman's rank-order correlation, SparCC, CCLasso, MIC, and cosine similarity that are used for detecting metabolite-microbe correlation using simulation and real metabolome and microbiome data. Two main points they made regarding specificity, sensitivity, similarity, accuracy, and stability with different sparsity are (1) the newly designed CCLasso and SparCC methods were not significantly superior to the traditional Pearson's and Spearman's methods, and (2) Spearman's correlation and MIC outperform the other four methods in terms of their overall performances. Another study showed that SparCC and proportionality are scale invariant for analyzing a single dataset, but lose scale invariance for analyzing multiomics datasets and SPIEC-EASI cannot handle different types of multiomics datasets either.[229]

6. Interaction analysis in microbiome and multiomics

6.1 Detect microbiome interactions using network analysis—Concept shift of correlation and association analyses

Identifying correlations among taxa within ecological communities is a common goal of genomic survey data analysis. In this book chapter, we exchangeably use taxon-taxon correlation, OTU-OTU correlation, and microbe-microbe correlation; all these terms represent the intracorrelation within microbes and could be called within microbiota correlation. Interomics correlation signifies the correlation between or among two or more omics. One such example is the correlation between microbiota and metabolites or between OTU-metabolite in microbiome-metabolomic analysis.

With the large-scale datasets generated by high-throughput DNA sequencing technologies and methodologies development, statistical analyses in microbiome studies have moved beyond basic descriptions of the composition structure and microbial community diversities. Correlation and association analyses are not only performed among a pair of microbial taxa to explore intertaxa correlation or association, but also are conducted to investigate potential interactions between microbial taxa to find significant taxon co-occurrence patterns. However, it may be more difficult to detect the patterns in large, complex datasets using the standard alpha/beta diversity metrics widely used in microbial ecology.[232] Detecting and characterizing microbial interactions or accurately inferencing microbial ecological interactions from population-level data is one goal of microbiome studies.[45] The terms such as pairwise correlation, positive or negative correlation, or bidirectional could not accurately describe the relationships of complicated microbial taxa. It is more appropriate to use the more general relationship term called "interactions" to describe intertaxa correlation and association. "Interactions" are inferred by detecting significant (typically nondirectional) associations between samples, e.g., by measuring frequency of co-occurrence probabilities[211,232,233] instead of correlations. The interaction network is a measure of microbial association for exploring co-occurrence patterns between microbial taxa in complex communities. The interactions between taxa could be direct or indirect.

Network analysis is an appropriate tool to be used for detecting the complicated interactions. A network essentially consists of a set of nodes (or vertices, also called actors) that are connected to one another via some types of relationships called edges (also called ties). In microbial interaction network, the nodes represent the biological features, such as microbial taxa (i.e., species or OTUs), genes, metabolites, transcripts, microbes, and proteins, as well as represent environmental or host features, such as pH and markers of immune status, while the edges denote functional interactions or an association between the nodes, such as a correlation between the abundance of two taxa, which may suggest a dependency between the taxa. Typically, five microbial interactions between taxa in terms of classical ecological relationships can be defined by the sign (e.g., positive, negative, or neutral) and the magnitude (e.g., strong, weak) of the interaction: mutualism (+, +), competition (−, −), parasitism (+, −), amensalism (−, 0), and commensalism (+, 0) of the interaction (e.g., strong, weak).[45,51,223] The number of connections between a node and its surrounding nodes is measured by node degree, which is one commonly used measure in terms of network analysis. Highly connected nodes (i.e., hubs) likely have greater influence upon the network.[234]

On the one hand, we move beyond correlation to interaction intending to detect the complicated intertaxa relationship; on the other hand, currently most available or commonly used methods to measure the strength of interactions still use the simplest and most familiar measure of correlation. For example, in order to detect an interaction between microbes, the typical way is to compute Pearson's correlation coefficient among all pairs of OTU samples, and the interaction is determined if the absolute value of the correlation coefficient is sufficiently high.[235,236] To construct an interaction network of the cecum and sigmoid, pair-wise Spearman's correlation between microbiome and metabolome data, i.e., Spearman interomic correlation analysis, was conducted.[93] Using correlation analysis to infer a network is paradox and problematic. First, having been shown a correlation between the abundances of two OTUs (e.g., species) does not imply that those OTUs are interacting.[237] Second, correlations can arise between OTUs that are indirectly connected in an ecological network,[211,226] i.e., a spurious correlation can occur. Currently, no solid evidence has been shown that correlation is the proper measure of association in terms of interactions.[211]

6.2 Identify microbe-metabolite interactions

6.2.1 Functional analysis of microbiome

Integrative analysis of multiomics data remains a significant challenge in current microbiome studies. Although how to simultaneously measure the gut microbiome and the gut metabolic state is still not obvious, integrative analysis of metabolome and microbiome data has been increasing in recent years.

This is most due to that the microbiome has been considered playing a major role in metabolite production and metabolism, and also partially due to high-throughput sequencing technologies. For example, shotgun metagenomic sequencing can indirectly assess the metabolic activity of microbiota by identifying marker genes associated with metabolic functions; metabolomics technology plays a critical role in connecting host phenotype and microbiome function.[238,239] As a research field, metabolomics is the systematic study of all small molecules within a biological system. Unlike other metaomics, metabolites and metabolic pathways are relatively conserved across species,[240] thus enabling invaluable insights into the structure and functional potential of the microbiome.[130,241] Therefore, recently various correlation or association studies have been moved beyond the microbiome domain to detect the interactions between microbe-metabolite including in HIV infection,[242] immunity and cancer treatment,[243] and inflammation.[181,244]

The microbe-metabolite correlation analysis is getting popular because (1) emerging data in the field of microbiome research indicate that various forms of microbiome including bacterial and fungal do not exist isolatedly but in complex communities that engage in microbe-microbe-host interactions.[181,244] Specifically, species composition is associated with healthy individuals with tremendous variation and host phenotypes (i.e., obesity, IBD, diabetes), as well as with external factors (i.e., diet).[245] Thus, compositional studies alone cannot provide the information about mechanisms of individuals' healthy and disease. (2) Metagenomic-based studies exhibit that the complex co-occurrence patterns among taxa in the human gut microbiome are driven by metabolic interactions and external factors.[246–248] For example, with advances in acquiring "omic" data, researches show that microbiota, genetics, environmental factors, diet, and drugs interplay each other. Among these interactions, the common denominator is metabolites[243]; thus, metabolomics captures net interactions between genome, microbiome, and the environment. (3) Directly assessing the roles of microbiome in health and disease is difficult because many roles are undefined and cannot be studied in culture systems. Such cases highlight the importance for analyzing microbe-metabolites interactions to understand the host and bacterial processes in therapeutic design for diseases, such as cancers. And (4) the metabolite abundance profiles can be predicted using microbe abundance profiles as metabolites and microbe are interacted.[249,250]

Analysis of the functional capacity of the microbiome could be done directly through shotgun metagenomics or through collecting 16S rRNA gene profiles then indirectly inferring the abundance of functional genes.

There are alternative methods to shotgun metagenome sequencing. These methods either characterize microbial metabolic pathways and functional modules directly from short sequence reads, or predict the relative change in the production of a metabolite by a microbial community rather than assemble short reads from hundreds of different organisms to single-organism genomes. Here, we will introduce alternative functional analysis methods for metagenomic data and methods for 16S rRNA data. Various software tools have been developed toward this end. We briefly introduce some of methods for identifying the microbe-metabolite interactions below.

6.2.2 Software tools for metagenomic data

PRMT (predicted relative metabolomic turnover),[251] a network-based tool, was developed to predict community metabolic functions from metagenomic data. It can be used to conduct the hypothesis that bacterial population diversity is associated to the metabolic capacity of the community. Metagenomic analyses perform functional profiling to explore the functional potential of an ecosystem by describing the changes in the abundance of genes annotated with unique enzyme functions. Different from this paradigm of an environmental metagenomic analysis, PRMT predicts the relative change in the production or consumption of a metabolite by a microbial community. The PRMT method was illustrated in analysis to the metagenome. It is equally applicable to metatranscriptomic data.

However, in PRMT method biological testing and evaluation has remained a bottleneck, while predictive modeling remains speculative if it has not been validated by extensive functional data (e.g., metatranscriptomic, metabolic, or proteomic measurements).[252] Examples of PRMT use can be found in these publications.[250,253–256]

HUMAnN[257] was developed to infer the functional and metabolic potential of microbial metagenomes as alternative approach to shotgun metagenome sequencing, which is challenging to assemble comparably short reads from hundreds of different organisms to single-organism genomes. HUMAnN method characterizes microbial metabolic pathways and functional modules directly from high-throughput short sequence reads. Examples of HUMAnN method use are available from these papers.[258–260]

MicrobiomeAnalyst,[73] a web-based tool for comprehensive statistical, visual, and meta-analysis of microbiome data, contains the Shotgun Data Profiling module allowing functional profiling and metabolic network visualization of shotgun metagenomics or metatranscriptomics data.

The enrichment analysis can be performed and the results can be visually explored within a metabolic network. The framework was developed based on the KEGG global metabolic network using the KEGGscape.[261] The network visualization is displayed with nodes and edges representing metabolites and KEGG Orthology (KO) members, respectively. As an integrated platform available in the website, MicrobiomeAnalyst has the advantages: allowing analysis and storage of microbiome data and reducing infrastructure problems (e.g., storage, data deposition, and fault tolerance).[262] Examples of using MicrobiomeAnalyst can be found in these studies.[181,263–265]

6.2.3 Software tools for 16S rRNA data

Several software tools have been developed for predicting functional profiles of microbial communities from 16S rRNA marker gene sequencing data. Profiling 16S rRNA marker gene does not directly identify metabolic or other functional capabilities of the microorganisms under study. Thus, to infer the metabolic state of a community, these approaches are used in combination with KEGG metabolic gene annotations.

PICRUSt (phylogenetic investigation of communities by reconstruction of unobserved states),[266] a bioinformatics algorithm and software package, was designed to predict the KEGG Ortholog (KO) functional potential of the microbiome using marker gene (e.g., 16S rRNA) data and a database of reference genomes.[266] This predictive metagenomic approach assumes that there exists the strong correlation between phylogeny and biomolecular function,[266] and infers which functions are likely associated with a marker gene based on its sequence similarity with a reference genome.[267] PICRUSt infers unknown gene content by an extended ancestral state reconstruction algorithm. The algorithm links a phylogenetic tree of 16S rRNA gene sequences to operational taxonomic units (OTUs) with gene content.[266,267] Thus, PICRUSt predictions depend on the topology of the tree and the distance to the next sequenced organism. Although it was demonstrated that typical 16S survey data are more than sufficient to generate high-quality predictions from PICRUSt and the generated annotated table of predicted gene family counts for each sample from metagenome prediction is directly comparable to those generated by metagenome annotation pipelines such as HUMAnN[257] or MG-RAST[268], however, PICRUSt has several limitations. First, the topology dependence of prediction algorithm may be the main limitation for PICRUSt. Because a nearest neighbor within the tree topology always exists, PICRUSt links all OTUs, even if distances are large. When microbial communities with a large proportion of phyla have not been well-characterized, the problem will occur.[269] Second, PICRUSt depends on phylogenetic tree reconstruction, which is uncertain and time consuming.[270] PICRUSt uses IMG as its reference genome dataset, with each genome corresponding 16S rRNA in the reference tree. As the number of sequenced genomes continuously increases, phylogenetic tree reconstruction is not only computationally time consuming but also is open to argument. Third, PICRUSt predicts the functional attributes of a microbiome rather than identifying them directly from DNA sequences.[267] Additionally, currently PICRUSt only limits the predictions of 16S marker gene sequences corresponding to bacterial and archaeal genomes (it does not infer viral or eukaryotic components of a metagenome) and its ability to detect patterns also depends on the input data used: it cannot distinguish variation at the strain level.[266] PICRUSt is often applied to infer functional categories associated with taxonomic composition. Typically, once the microbiome metabolic pathways are identified, the microbial metabolic functions can be treated as continuous variables and the PICRUSt values can be analyzed via either parametric test or nonparametric permutation difference test. For example, PICRUSt was used to produce predicted metagenomes from the 16S rRNA gene sequence data.[271] PICRUSt was also used to identify Kyoto Encyclopedia of Genes and Genomes (KEGG) metabolic pathways potentially affected by groups of bacteria in a microbiome and prostate cancer association study.[194] Other examples of PICRUSt use can be found in these studies.[272–274]

Since PICRUSt's publication, two additional software tools for functional inference from 16S rRNA data have been released: Tax4Fun[269] and PiPhillin.[270]

Tax4Fun, an R package, was developed to predict the functional capabilities of microbial communities based on 16S rRNA amplicon data. It uses a linear combination of precomputed genomic reference profiles to predict the metabolic profiles of a metagenome from its taxonomic abundance.[269,275] Tax4Fun uses a nearest neighbor identification algorithm based on a minimum 16S rRNA sequence similarity to link 16S rRNA gene sequences with the functional annotation of sequenced prokaryotic genomes.[269] Unlike PICRUSt, which uses the database from the Joint Genome Institute Integrated Microbial Genomes (JGI IMG)[276,277] as the default database, Tax4Fun uses SILVA ontology[278] as a reference database for 16S rRNA sequences. Tax4Fun not only requires an association matrix between the prokaryotic organisms in the KEGG database and the SILVA reference database, but also needs precomputation of functional profiles for all prokaryotic genomes in KEGG as well as to convert the input sequence data to a SILVA-based profile.[269] As PICRUSt, Tax4Fun showed that the functional predictions and the metagenome profile are highly correlated. Although Tax4Fun cannot replace whole metagenome profiling, it can provide a good approximation to functional profiles obtained from whole metagenomic shotgun sequencing approaches. It was demonstrated that Tax4Fun outperforms PICRUSt on the tested datasets.[269] Tax4Fun may also have the advantage: it is potentially more accurate for communities with a large proportion of poorly characterized phyla.[267] As PICRUSt, the implementation of Tax4Fun relies on outdated functional databases and uncertain phylogenetic trees and requires very specific data preprocessing protocols.[270] Examples of Tax4Fun use are available from these publications in microbiome studies.[279–282]

Piphillin, another newly developed metagenomics inference tool, differs from both PICRUSt and Tax4Fun in that it does not dependent on a phylogenetic tree or a functional database of 16S rRNA sequences and also does not require any singular data preprocessing protocol.[270] Instead, Piphillin uses a nearest-neighbor algorithm to quickly map 16S rRNA sequences to reference genomes to predict the represented genomes. Piphillin uses three different datasets that were sequenced by both shotgun metagenomics and 16S rRNA to examine identity cutoffs for determining the nearest-neighbor genome; thus, it can easily work with any current genome database.[270] It has been showed that Piphillin outperforms both PICRUSt and Tax4Fun in predicting gene composition in human clinical samples regarding correlation to corresponding actual shotgun metagenomics and Piphillin improves the prediction accuracy compared to PICRUSt.[270] In summary, Piphillin has two major advantages[270]: (1) it has fewer parameter requirements and more flexibility compared to PICRUSt and Tax4Fun. Piphillin requires minimal reference genome databases because it does not need a phylogenetic tree or reference OTUs. (2) It is capable to receive data inputs from any upstream 16S rRNA amplicon sequence preprocessing pipeline. Piphillin requires only two input files: an OTU abundance table and a fasta file with OTU representative sequences. Thus, unlike PICRUSt or Tax4Fun, which is restricted to QIIME's or Silva's assignments of counts, Piphillin can also be used in conjunction with Mothur,[283] RDP,[284] DADA[2,285] or UPARSE.[286] However, Piphillin also has limitations, such as it lacks nearest-neighbor reference genomes in understudied environments.[270] Examples of Piphillin use can be found in these studies.[287–289]

The above indirect methods take different approaches to analyze the functional capacity of the microbiome. They may be suitable for different contexts. They all assume that the marker genes and biomolecular functions are strongly associated. However, metaomics analyses (i.e., metatranscriptomic, metaproteomic, and metagenome) have shown that marker genes for metabolism identified in metagenomic data may be consistently underexpressed, even may not be expressed at all; in other words, only a small number of functional modules were consistently transcriptionally activated well beyond their metagenomic abundance.[290–292] Thus, genes inferred from 16S rRNA sequencing may not be present at all, and metagenomics-only approaches will tend to underestimate the functional potential and activities of microbiome. Particularly when the communities have a large proportion of 16S sequences not closely related to existing reference genomes, predictions from all of these methods should be interpreted extremely cautiously.[267] The rigorous experiment studies are needed to better understand the functional roles of microbiome. In literature, integrative multiomics approaches are available for analyzing the microbe-metabolite interactions. Typically, the way these approaches have taken is first to measure the metabolic state of the gut directly using an appropriate method, such as nuclear magnetic resonance spectroscopy or mass spectrometry, and then integrate metabolic measures with microbiome data.[293]

Several statistical methods have been developed but most are targeting other omics, such as gene expression, with few direct applications of metagenomic and metabolomic data.[240,248,250]

6.2.4 Neural networks approach

In order to integrate multiomics datasets for inferring interactions across omics datasets, Morton et al.[229] use neural networks to estimate the conditional probability that each molecule is present given the presence of a specific microorganism. They showed that the learning representations of microbe-metabolite interaction method are able to recover microbe-metabolite relationships between microbially produced metabolites in IBD. Microbiome sequencing data are the count-based data, while untargeted mass spectrometry data are based on "spectral" information; thus, the metabolite signals are continuous; i.e., the data distributions from microbiome and metabolome are different. However, both microbiome and metabolite datasets are proportional (i.e., microbiome data are proportional to total reads in samples, while metabolite data are proportional to their concentrations in samples[90]), they are both disparate and compositional. Thus, to appropriately analyze these data, the method should have the capability to integrate and handle such disparate omics data while treating them as compositional.[215]

Instead of using correlations, the proposed method called "mmvec" (representing microbe-metabolite vectors) inferences the conditional probability of observing a metabolite given that a microbe was observed; specifically, it uses co-occurrence probabilities to identify the most likely microbe-metabolite interactions. The advantages of mmvec method[229] lie on: (1) enabling interpretable findings by ranking the microbe-metabolite interactions and through visualizing reduced standard dimensionality interfaces; (2) enabling scalable inference on large multiomics datasets by using modern graphics processing unit architectures for computations; and (3) both simulations and experimental data showed that mmvec outperforms existing statistical methods including Pearson's, Spearman's, proportionality, SparCC, and SPIEC-EASI correlations in specificity and sensitivity and for inferring microbe-metabolite interactions from multiomics datasets.

The methodology of inferring microbe-metabolite interactions combined with networking strategies to identify microbial metabolites has been reviewed in this study.[294] However, recently Tolosana-Delgado et al.[295] showed that neural networks do not solve the problems of compositional data analysis and practitioners should use them carefully. Furthermore, Quinn and Erb did not agree the claim by the authors of mmvec method that "mmvec is the only method robust to scale deviations" and showed that correlation and proportionality can actually outperform the mmvec neural network for identifying microbe-metabolite interactions when using a correct log-ratio transformation.[296]

The authors of mmvec method responded to above comments and showed that their proposed mmvec method outperforms the correlation and proportionality on real microbiome and metabolome data due to the latter method's inability to deal with sparsity.[229,297] The disagreement and discussion between mmvec method and correlation and proportionality methods not only highlight the importance of choosing an appropriate log-ratio transformation method when analyzing compositional data, but also raise the question again that whether or not the approach of compositional data analysis is more suitable than count-based approach to analyze microbiome and other omics data (e.g., metabolomics) because real-world microbiome and omics data are high dimensional, sparse, and often have many zeros.[4,16] The mmvec method has been used to identify microbe-metabolite interactions in these studies.[298,299]

6.2.5 Nonparametric methods

The parametric methods are only reliable for detecting linear dependencies between microbes. Thus, to detect nonlinear interactions between microbes various nonparametric methods have been proposed. In general, nonparametric methods are able to capture noisy and highly nonlinear microbial interactions; however, they do not necessarily suggest the direction of an interaction: the detected microbes or microbes with environmental factors are positively or negatively associated. Thus, the interactions detected by nonparametric methods may be ambiguous or difficult to model.[223] Here, we will briefly introduce two more nonparametric methods.

6.2.5.1 Molecular ecological network analyses (MENA)

MENA[235] is a nonparametric tool for molecular ecological network analyses. The motivation of MENs development was to examine network interactions in a microbial community based on high-throughput metagenomics data. It was based on methods through random matrix theory, which is a set of statistical tools borrowed from statistical physics and materials science to construct ecological association networks. The remarkable features of MENs approach is that the network generated by MENs analysis is automatically defined and robust to noise, thus very well solving several common issues associated with high-throughput metagenomics data.[235] The MENA method was reviewed in these papers[46,300] and has been cited in more than 400 publications, such as in these studies.[301–303]

6.2.5.2 Microbial co-occurrence relationships in the human microbiome (CoNet)

To determine co-occurrence and co-exclusion relationships among the relative abundances of microbial taxa across all individuals, inferring a microbiome-wide microbial interaction network, CoNet combined multiple parametric and nonparametric similarity measures (ensemble approach) with generalized boosted linear models (GBLM approach) to predict microbial network interactions.[45] The produced microbial interaction networks from the HMP data with combination of these two approaches were shown preventing spurious correlations due to compositional structure of relative abundance data, thus resulting in a significantly lower false-positive rate than other single methods. CoNet method has been cited in nearly 800 publications, such as in these studies.[189,304–306]

A recent study compared MENA, CoNet, MIC, SparCC, Bray-Curtis, LSA and eLSA (local similarity analysis and extended LSA, which will be introduced in Section 11.4), Pearson's and Spearman's correlations in terms of sensitivity, specificity, precision, and ability to provide interpretable results. The results showed that LSA, MIC, Spearman, and SparCC are robust to distributions, and all other methods are sensitive to several distributions. LSA, Spearman, and SparCC also performed well in correctly identifying competitive relationships as mutual exclusions and in their designs.[46] For example, LSA was able to detect both linear and nonlinear ecological associations, even under sparse conditions, while SparCC had a best performance in the high compositionality.

7. Multivariate correlation and association-based methods—Exploratory, interpretive, and discriminatory analyses and classification

Multivariate approaches are important for multiomic microbiome studies not only because these approaches take into account the correlation and dependency of features within omics (i.e., taxa, metabolites) and are more robust for large and noisy datasets, but also because the correlation and association among different factors tend to be multivariate by nature. Multivariate correlation and association can be conducted among factors to identify the associations of microbiome, environment, and host. From single measure of correlation and association to multivariate correlation and association methods, the correlation and association are performed between a pair of variables to a pair of tables.

Multivariate approaches are much more complex than univariate methods, but can simultaneously consider interactions between and within data matrices. Multivariate approaches are available for analysis of multiomics data; however, clearly separating them into different categories is very difficult due to various reasons.

Multivariate analysis methods are practically organized into three categories: exploratory, interpretive, and discriminatory analyses based on the primary research objective.[31] Here, we organize the presentation of multivariate correlation and association-based methods into four subsections: (1) exploratory correlation and association methods, (2) interpretive correlation and association methods, (3) discriminatory correlation and association methods, and (4) classification methods.

7.1 Exploratory correlation and association methods

Exploratory methods are used to discern patterns among samples based on the values of variables (features) measured in those samples. These methods provide a useful visualization of sample similarities. Examples of exploratory analysis include different unconstrained ordination techniques and cluster analyses.[307] Most exploratory methods are performed to identify correlation and association within one omics dataset.

7.1.1 Most widely used methods in microbial ecology

Commonly used unconstrained ordinations in microbial ecology are principal component analysis (PCA), correspondence analysis (CA), principal coordinate analysis (PCoA), nonmetric multidimensional scaling (NMDS), and constrained analysis of principal coordinates (CAP). These multivariate statistical techniques are considered as multivariate exploratory association analyses. We introduced these methods in our book.[307] Here, we briefly introduce them and focus on their applications in microbiome and other omics.

7.1.1.1 Principal component analysis (PCA)

PCA was invented in 1901 by Karl Pearson,[308] and was later independently developed and named by Harold Hotelling in the 1930s.[309,310] PCA uses an orthogonal transformation to convert a set of observations of possibly correlated variables into new synthetic variables called principal components (PCs), which are linearly uncorrelated and orthogonal. Through sample and variable representations, PCA reduces the data dimension, visualizes the similarities between the biological samples, and filters noise.[311] Thus, PCA is considered as one of well-established techniques for dimensionality reduction and visualization. However, PCA approach suffers from several limitations: (1) The components identified by PCA are Gaussian components, and PCA is inconsistent when the number of variables is much higher than the number of samples,[312,313] which may limit its application in the high-throughput data. (2) PCA uses Euclidean distance to measure dissimilarity among samples, and when it is used in a dataset with many zeros, which is often the case in microbiome and other omics data, severe artifacts such as horseshoe visualization effect could be generated.[314–316] To extenuate the horseshoe effect, a chord or Hellinger transformation is often performed before running PCA,[315] and the use of correspondence analysis (CA) is usually advocated for such datasets.[31,316] (3) PCA is inappropriate for data integration for two reasons[317]: first, PCA works with one set of variables at a time. Its main objective is to construct principal components (PCs) to maximize the variance within the sets of variables rather than consider the correlation between different sets of variables; second, the latent variables represented by the PCs are linear combinations of the entire set of variables (all the features) under consideration. This makes the interpretation and visualization of the composite measures in large-scale studies difficult.[318] (4) Additionally, the axes generated by PCA are difficult to interpret because the relationship between the phylogenetic structure and the loadings of the variables on the principal axes is not detected by PCA.[319] In general, traditional PCA has poor performance with both high-dimensional data and functional data.[312,313,320,321] Examples of PCA use in microbiome studies are available from these works.[322–325]

7.1.1.2 Correspondence analysis (CA)

CA is a multivariate exploratory method connecting correlation and contingency proposed by Herman Otto Hartley[326] and later developed by Jean-Paul Benzécri.[327] It was designed to find correspondence between rows and columns of a contingency table and to represent it in an ordination space. Conceptually, CA is similar to PCA, while it is used to analyze nominal or categorical data instead of continuous data.

Three important differences between CA and PCA methods were described by Paliy and Shankar[31]: (1) PCA maximizes the amount of explained variance among measured variables, while CA maximizes the correspondence (measure of similarity of frequencies) between rows (represent measured variables) and columns (represent samples) of a table.[328] (2) PCA assumes a linear relationship among variables, while CA expects a unimodal model. (3) CA uses weighted Euclidean distance (a variant of Euclidean distance) or chi-square distance to estimate the distances among samples in full CA ordination space. One drawback of CA is that its ordination output can often produce a noticeable mathematical artifact called "arch" effect.[31,316] Examples of CA use in microbiome studies are available from these works.[329–332]

7.1.1.3 Principal coordinate analysis (PCoA)

PCoA, a conceptual extension of the PCA technique described above, is a metric multidimensional scaling method. It was designed to explore and to visualize similarities or dissimilarities of data based on projection. It uses spectral decomposition to approximate a matrix of distances/dissimilarities by the distances to reduce dimensions of data points.[333] Different from PCA, which is based on eigen analysis of a correlation or covariance matrix, PCoA can use any distance (dissimilarity) matrix to organize samples.[334] Recently, it is due to its ability to use phylogenetic distance (e.g., UniFrac distance) and community composition (e.g., Bray-Curtis distance) measures to calculate (dis)similarity among microbial populations, PCoA has gained popularity in microbiome and other omics studies. Examples of PCoA use in microbiome studies are available from these works.[335–339]

7.1.1.4 Nonmetric multidimensional scaling (NMDS)

MDS (multidimensional scaling) explicitly chooses a (small) number of ordination axes and then fits the data to those dimensions prior to the analysis. Similar to PCoA, MDS first needs to calculate a matrix of sample (objects) dissimilarities using a chosen distance metric, while NMDS calculates the ranks of these distances among all samples (objects). The intuitive ideas and general procedure of NMDS were provided by Shepard[340]; however, it was Kruskal who developed formal "goodness-of-fit" hypothesis testing procedure of NMDS.[341–343] NMDS represents n samples (objects) geometrically by n points to obtain the interpoint distances corresponding to the dissimilarities between samples (objects). Its aim is to find n points whose interpoint distances closely agree with the given dissimilarities between n samples (objects). NMDS has been developed based on the fundamental hypothesis that dissimilarities and distances are monotonically related. However, in contrast to metric MDS, NMDS instead finds both a nonparametric monotonic relationship between the dissimilarities and the Euclidean distances. Thus, NMDS focuses mainly on the ranking of dissimilarities rather than their numerical values. Examples of using NMDS in microbiome studies can be found in these publications.[337,344–346]

7.1.1.5 Constrained analysis of principal coordinates (CAP)

CAP (also called constrained analysis of proximities in vegan package)[347] is an ordination method similar to redundancy analysis (RDA).[348] It is simply a redundancy analysis of results of principal coordinates analysis (or metric multidimensional scaling).[347] CAP allows non-Euclidean dissimilarity indices, such as Manhattan or Bray-Curtis distance. If Euclidean distance is specified as the ordination method, the results will be identical to RDA. The statistical significance can be obtained in CAP using a permutation test. Examples of using CAP in microbiome studies can be found in these reports.[349–351]

These classical multivariate methods are often used for analyzing microbiome and other omics data including microarray (PCA, CA)[352,353] and genomics (PCA).[354] However, these classical multivariate methods sometimes fail to identify significant pattern in groups of interest because their assumptions and underlying algorithms are not appropriate. For example, PCA assumes the multivariate normal distribution of measurement variables, which is rare true in omics studies.[355,356] Also the underlying algorithms of PCA to decompose the data by maximizing its variance were reviewed as not always appropriate in microbiome and other omics studies.[357]

7.1.2 Sparse principal component analysis (sparse PCA)

The sparse PCA methods[312,313,358] were proposed to address the problem of biological interpretability that PCA suffers from the fact that each principal component is a linear combination of all the original variables while reducing dimensionality. Zou et al.'s sparse PCA[358] uses the lasso (elastic net) by incorporating variable selection to produce modified principal components with sparse loadings and the adaptive lasso for penalizing different coefficients in the L_1 penalty.[359] Johnstone and Lu's sparse PCA[312,313] and Journée's sparse PCA[360] use two single-unit and two-block optimization formulations of the sparse PCA problem to extract a single sparse dominant principal component of a data matrix, or more components at once, respectively. Although sparse PCA methods perform dimensionality reduction by incorporating variable selection which results in sparse solutions; however, the sparse components are determined only within a single data source and hence the sparse PCA methods are not intended for analyzing the relationships between two or more sources of data.[317] Currently to account for the sparse compositional nature of microbiome datasets, robust Aitchison PCA was proposed.[361]

7.1.3 Independent extension of component analysis and principal component analysis

7.1.3.1 Independent component analysis (ICA)

The concept and algorithm of ICA was first proposed by Comon as an extension of PCA,[362] with subsequent developments by Bell and Sejnowski.[363] One of the initial motivations for ICA was sound signal separation. Detailed descriptions of algorithms and applications about ICA were given in the paper[363] and books.[308,309] ICA assumes that the observed data are determined by some unknown fundamental but independent factors or independent components (ICs), which means that different ICs represent different nonoverlapping information. Thus, ICA minimizes both the second-order and higher-order dependencies of data, and strives to generate components as independent as possible. Different from PCA, ICA identifies non-Gaussian components, which makes ICA more flexible in applying to the high-throughput data, for example, in microarray data,[364] metabolomics data.[365]

The successful application of ICA essentially depends on first using PCA to reduce the high dimension of the dataset. It has been reviewed that the capability of generating independent components makes ICA outperform PCA at separating different biological groups[366] and more suitable than PCA for more important applications in biomedical studies including gene expression.[310]

However, ICA has some fundamental limitations including instability, hard for choosing the number of components to extract, and difficult for dealing with high dimensionality.[366] Among them, the primary limitation is not easy using ICA to reduce high dimensionality. ICA usually is effective to datasets with a large number of samples and only a small number of variables,[367] which contrast to the data feature of microbiome and other omics including metabolism: a large number of variables but a relatively small number of samples. In other words, ICA cannot effectively solve larger p and smaller n problem. Thus, directly applying ICA to the high-dimensional dataset is questionable and often results in no practical relevance.[367] In practice, ICA is often used following with PCA.

7.1.3.2 Independent principal component analysis (IPCA)

IPCA[311] was proposed to solve the problems of both the high dimensionality of high-throughput data and noisy characteristics of biological data in omics studies. Omics data have the problems: the data are extremely noisy, and large p and small n, i.e., the number of variables (the biological entities that are measured) is much larger than the number of samples (or observations). As we reviewed above, the particular power of PCA lies in its dimensions—reduction by loading the highest variance that is related to the biological question. As an alternative to PCA, the power of ICA lies in optimizing an independence condition to give more meaningful components. IPCA combines the advantages of both PCA and ICA, and uses ICA to denoise the loading vectors produced by PCA to better cluster and visualize the biological samples with a smaller number of components than PCA. Additionally, the sparse IPCA (sIPCA) can internally perform variable selection to identify biologically relevant features (e.g., taxa, genes) with respect to the biological experiment. The better performances of IPCA over PCA and ICA in the super-Gaussian and in high-dimensional cases were demonstrated by simulation studies, real datasets in microarray, and metabolomics.

7.1.4 Kernel principal component analysis (KPCA)

As reviewed above, PCA is a popular tool for linear dimensionality reduction and feature extraction; however, PCA uses the linear combinations of original features to replace the original features, and hence only allows linear dimensionality reduction. Thus, PCA is not capable to do dimensionality reduction for the data with more complicated structures because the data cannot be well represented in a linear subspace. To overcome this limitation, various nonlinear dimension reduction methods have been proposed. Among them, Kernel PCA[368] is one of the most used methods. As the nonlinear form of PCA, Kernel PCA is able to better exploit the complicated spatial structure of high-dimensional features. It generalizes PCA to nonlinear dimensionality reduction.[369,370] Examples of using Kernel PCA in microbiome studies can be found in these publications.[371,372]

7.1.5 Generalized, sparse generalized, and adaptive generalized extensions of principal component analysis

7.1.5.1 Generalized principal component analysis (gPCA)

One limitation of PCA is relied on unimodal linear transformation. However, high-dimensional space data typically are multimodal, heterogeneous representing with significantly different geometric structures or statistical characteristics, which pose challenges for subspace clustering. To remedy the limitation of PCA and capture the features of high-dimensional data, generalized principal component analysis (gPCA)[373–376] has been proposed. The gPCA method uses polynomial differentiation rather than polynomial factorization to solve the subspace clustering problem. Its algorithm is based on estimating a collection of polynomials from data and then evaluating their derivatives at points on the subspaces. By using the algorithm, gPCA can create structured low-dimensional data representations. The distinctive feature of gPCA is that it does not explicitly assume any statistical distribution for the data. It originally was proposed for solving segmentation problems in computer vision. Now it has been used in high-dimensional omics data.

7.1.5.2 Sparse generalized principal component analysis (sparse gPCA)

The sparse gPCA method was developed for unsupervised dimension reduction for data from an exponential family distribution.[377] A generalized version of PCA (gPCA) to address the application to the exponential family of distributions was developed[373] (which was first available in 2015 as a technical report). Various penalties on regression coefficients to induce sparsity are available in statistical literature including the L_1 (LASSO-type penalty),[192] L_2 (ridge-type penalty)[378] penalties and the combination of these two (elastic net),[379] as well as the Smoothly Clipped Absolute Deviation (SCAD) penalty.[380] The sparse gPCA method was an extension of this exponential family version of gPCA by adding L_1 and SCAD penalties to introduce sparsity. Although the sparse gPCA method was shown achieving sparse dimension reduction for non-Gaussian data, its application was mainly illustrated to Poisson-based techniques in the analysis of text data[377]; more studies are needed to verify its application to other high-dimensional and non-Gaussian data such as microbiome data.

7.1.5.3 Sparse nonnegative generalized PCA (sparse nonnegative gPCA)

Sparse nonnegative gPCA (a modified version of PCA) was developed to solve the two problems of gPCA: (1) high dimensionality and (2) nonnegativity of the underlying spectra measured by nuclear magnetic resonance (NMR) spectroscopy in metabolomics.[381] This method incorporates feature selection through sparsity and constrains the loadings to be nonnegative to ensure PCA direction vectors interpretable. As a modification of PCA, sparse nonnegative gPCA proposes a framework in high-dimensional settings to: (1) incorporate sparsity, structural dependencies, and nonnegativity into the principal component (PC) loadings, and (2) develop a fast, computationally efficient algorithm to compute these parameters.[381] It was shown that sparse nonnegative gPCA has the capabilities for dimension reduction, pattern recognition, sample exploration, and feature selection in spectroscopy data. Specifically, this method has more power to reduce dimension than sparse nonnegative PCA and can be used to understand important biological patterns in the multidimensional data such as NMR data.

7.1.5.4 Generalized least squares matrix decomposition (GMD)

GMD was proposed to address the problems of variables in big high-dimensional datasets which are both sparse and structured.[382] Although sparse and functional PCA methods can consistently recover the matrix factors in the settings of high-dimensional and functional data,[312,313,321] however, when extending these techniques to regularize both the row and column factors of a matrix, they all can fail to capture relevant aspects of structured high-dimensional data, or data in which variables are associated with a location.[382] To directly account for structural dependencies and incorporate two-way regularization for exploratory analysis of big structured datasets, GMD not only generalizes PCA to reduce dimensionality but also regularizes matrix decomposition by adding two-way penalties for sparsity or smoothness to get off the irrelevant variables or noise in high-dimensional settings. The computational algorithms were developed to perform gPCA, sparse gPCA, and functional gPCA on big datasets. In summary, GMD is a general and flexible framework for PCA conjuncting with formulations for sparse gPCA to obtain low-dimensional representations of the variables which are both sparse and structured.

7.1.5.5 Adaptive generalized principal components analysis (adaptive gPCA)

Adaptive gPCA is another structured dimensionality reduction method. As described above, microbiome data often have large p and small n problem, and the results produced by PCA are unstable or inconsistent and difficult to interpret. The goal of adaptive gPCA was to mitigate the large p and small n problem.[319] The adaptive gPCA is obtained by extending gPCA through the following procedures: first including a prior in the model to encode the intuition so that species close together on the tree will behave similarly, and then performing gPCA on the posterior estimate and taking into account the variance structure of the posterior. The proposed method includes phylogenetic information in exploratory data analysis as other phylogenetic methods such as double principal coordinates analysis (DPCoA)[383] and edge PCA.[384] The distinctive feature of adaptive gPCA is that it could provide more easily and biologically relevant interpretations because it uses the phylogenetic relationships between the bacterial species (the low-dimensional representation of the samples) instead of grouping together species at a very high taxonomic level, and using distance-based methods.

7.1.6 Multilevel, double, and edge extensions of principal component analysis

7.1.6.1 Multilevel principal component analysis (multilevel PCA)

The classical PCA has been applied to taxa relative abundances at any taxonomic level (i.e., Phylum, Class, Order, Family, Genus, and Species) in microbiome study to identify possible similarities of the samples within the groups. However, PCA is sensitive to the relative scaling of the variables and may result in misleading treatment effect when interindividual variability (within-subject variation) is larger than intraindividual variability. To focus on within-subject variation, the approach of multilevel PCA is to split within-subject variation from between-subject variation before using PCA, generating different multivariate submodels for the between- and the within-subject variation in the data. Multilevel PCA has a major advantage to avoid different levels of variation sources being confounded with each other because through variation splitting each submodel can be analyzed separately.[363] Thus, multilevel PCA can detect large variations between the subjects due to their biological differences, e.g., age, gender.[385] This method was originally proposed for analyzing the omics data derived from crossover designed nutritional intervention studies using NMR-based metabolomics. The method can be considered as a multivariate extension of a paired *t*-test.

7.1.6.2 Double principal coordinates analysis (DPCoA)

DPCoA originally described in Ref. 383 is another method for giving a low-dimensional representation of ecological count data (such as at the species level) by incorporating phylogenetic information on the structure of the variables (i.e., the similarities between species). It has been shown that DPCoA is equivalent to gPCA in the cases of using tree-structured variables[386] and using any Euclidean distance measures on the variables.[319] Here are two examples of using DPCoA in human microbiome and metagenomic data.[386,387] The drawback of DPCoA is its very limited applications. For example, it can fail to capture much of the true latent structure in the data compared to adaptive gPCA or standard PCA.[319]

7.1.6.3 Edge principal components analysis (edge PCA)

Edge PCA was developed to leverage how the microbial community data of nucleic acid sequence samples sit on a phylogenetic tree.[384] Although phylogenetics-based methods, such as weighted UniFrac, compute sample distances accounting for the natural hierarchical structure of the data, however, the classical ordination methods (e.g., PCA, PCoA) ignore the fact that the underlying distances were calculated based on a phylogenetic tree. As a result, using PCA to explore the distances is difficult to explain what the axes represent. This motivates the authors of edge PCA to propose this new ordination method for comparing microbial sequence samples by accounting for the underlying phylogenetic structure of the data. In edge PCA, each PC axis is a collection of signed weights on the edges of the phylogenetic tree, and these weights can be readily visualized and understood by a suitable thickening and coloring of the edges.[384] However, as an exploratory technique, edge PCA cannot be used for hypothesis testing.

7.1.7 Clustering methods

Clustering is a common exploratory technique to describe the proximity between objects (depending on the fields, also called subjects or samples). Clustering was originated in psychology by Zubin[388] and Tryon[389] and in anthropology by Driver and Kroeber.[390] Because in cluster analysis any distance metric can be used to generate (dis)similarity measures, the flexibility makes this method equally well suited to microbiome and other omics data. Cluster analysis seeks to divide variables into groups based on the similarity of variables across all objects so that variables within each group (cluster) are more homogeneous to one another than to variables in other groups.[390,391]

7.1.7.1 Standard clustering methods

Various standard clustering methods are available, including hierarchical cluster analysis (HCA),[392] k-means clustering,[393] Calinski-Harabasz's pseudo-F-statistic,[394] k-medoids,[395] partitioning around medoids (PAM),[396] Rousseeuw's Silhouette index,[397] and model-based clustering methods.[398]

HCA was reviewed in 2009 as the most widely used form of clustering in practice[399,400] by far. All these clustering methods share the same algorithm that minimizes the within-group distances/similarities and maximizes between-group distances similarities. HCA organizes variables using a joint tree with length of branches indicating the relative similarity of different variables. HCA is usually accomplished with an agglomerative method to join and order nodes into a "hierarchy" tree (called dendrogram). This is called agglomerative hierarchical clustering, in which various linkage methods[307,399] can be used, including single linkage,[401] complete linkage,[402,403] and average linkage.[404]

These different linkage methods use different algorithms to compute the distances between nodes and merge groups: single linkage and complete linkage merge groups based on the minimum distance and maximum distance between two objects in two groups, respectively, while average linkage merges groups based on the average distance of all the objects in one group to all the objects in the other. Ward hierarchical grouping[405] is another important hierarchical clustering method. Ward's method begins with N clusters, each containing one object, which is similar to the linkage methods. However, this method does not use cluster distances to group objects. Instead, it computes the total within-cluster sum of squares to determine the next two groups merged at each step of the algorithm. These different distances-computing and groups-merging methods were assessed influencing relative node positioning and tree branching.[399]

The clustering algorithms have been compared on various types of data.[399,406–409] These studies found that Ward's method almost always performs better overall than other hierarchical methods, complete linkage was most similar to Ward's method and performs well in general, and single linkage does the worst overall. However, Milligan[410] demonstrated that complete linkage and Ward's method have bad performance when outliers present.

HCA is attractive in exploratory high-throughput data because HCA provides a convenient tool to visualize the similarities of variables, and inference on the grouping of variables based on the dendrogram structure; hence, HCA facilitates the interpretation of microbiome and other omics data. More important, the biclustering (two-way clustering), a special case of HCA, can incorporate a correlation method (e.g., Spearman's rank correlation) to cluster rows and columns of the data matrix simultaneously. Thus, biclustering can be used to find features (microbial taxa, genes, metabolites, etc.) that correlate only in a subset of objects but not in the rest of the dataset.[31,197,411,412] Examples of applying HCA in microbiome studies can be found in the research papers[413–417] and book chapters.[39,307]

K-means clustering was used in microbiome example to identify responsive and nonresponsive groups based on butyrate concentrations before and during the fiber intervention.[418] Other microbiome examples using k-means clustering are provided in studies.[416,419,420] Examples of using Rousseeuw's Silhouette index in microbiome studies were provided in the research papers.[421–423] Calinski-Harabasz's pseudo-F-statistic was used in this microbiome example.[421] In another microbiome studies, PAM were used to identify enterotype clusters.[423,424]

7.1.7.2 Newly developed clustering methods for microbiome data and specifically for OTU inferences

In recent years, several clustering methods have been proposed specifically to target microbiome data and specifically for OTU inferences. Here, we briefly introduce some of these new methods.

7.1.7.2.1 Neighborhood co-regularized multiview spectral clustering (NCMSC) NCMSC (neighborhood co-regularization of the clustering) algorithm[425] was proposed to extend the spectral clustering (SC) method[426,427] to a multiview setting. The goal of NCMSC is to adapt cluster assignments for different views (datasets) using neighborhood co-regularization technique. It is different from other spectral clustering methods in that NCMSC method promotes consistent cluster assignments across multiple views and penalizes solutions that differ significantly rather than aggregate clusters based on the individual data representations. NCMSC method was developed by adopting the co-regularized multiview spectral clustering (CMSC) algorithm which applies co-regularization framework to clustering task.[428] However, co-regularization approach of NCMSC is geared toward solutions that capture local/neighborhood-based relations in the dataset, which makes these two approaches fundamentally different from each other. The proposed NCMSC algorithm was compared with the standard clustering methods of k-means, HCA, SC, and CMSC in two performance measures: clustering accuracy (ACC) and stability (normalized mutual information (NMI))[429] and found that NCMSC algorithm notably outperforms other clustering methods.

The better performance of NCMSC algorithm is mainly due to three reasons[425]: (1) the employed neighborhood-based cluster assignment models having the capability to capture additional "local" relations in the data as compared to CMSC approach; (2) multiview algorithms in general tend to perform better than their single-view counterparts; and (3) k-means or hierarchical clustering tends to perform poorer than SC-based approaches. It was demonstrated that NCMSC method can identify distinct clusters within the group of women with a similar microbial compositions in a microbiome dataset.[425] Personalized microbial network inference via co-regularized spectral clustering was also developed by the same research team.[430] Examples of the use of NCMSC for analysis of microbiome datasets are available in the works by Biesbroek et al.,[431] Borgdorff et al.,[432] Gautam et al.,[433] Borgdorff et al.,[434] Kootte et al.,[435] and Botschuijver et al.[436]

7.1.7.2.2 Multiseeds based clustering (MSClust) MSClust algorithm[437] was inspired by the HCA method and a modified greedy network clustering algorithm called SPICI[438] for OTU inferences. The goal is to find the balance between inference accuracy and computational efficiency, while do not sacrifice inference accuracy due to analysis of large numbers of sequences. The development of this method was motivated to overcome the sensitivity of seed selection in traditional heuristic clustering methods, as well as the large memory usage of hierarchical clustering methods. MSClust was developed by adopting the SPICI framework which generates clusters from local density community based on a greedy clustering method and incorporating an adaptive seed-selection procedure and a greedy heuristic clustering procedure. Because the clustering results that produced by most heuristic clustering methods are sensitive to the selected seeds that represent the clusters, to void this limitation, MSClust takes two steps[437]: first adaptively selects multiple seeds for each cluster rather than one single seed, and uses a greedy clustering strategy to process the reads.

The performance of MSClust was compared with seven commonly used OTU inference methods in terms of two clustering quality performance scores NID and NMI[439,440] based on a number of benchmark datasets: Mothur,[283] CD-HIT,[441] Uclust,[442] GramCluster,[443] DNAClust,[444] ESPRIT,[445] and ESPRIT-Tree.[446] It was demonstrated that MSClust can achieve a balance between cluster quality and time complexity; as a similar or equivalent method to hierarchical clustering-based methods, MSClust requires much less memory usage.[437] Thus, MSClust is a nice alternative to existing OTU-based clustering methods. Examples of MSClust application in microbiome studies can be found in the reports.[447–451]

Other clustering methods for OTU inferences include MtHc,[452] DBH,[453] DMclust,[454] ESPRIT-Forest,[455] and DMSC.[456]

7.1.7.2.3 Motif-based hierarchical clustering method (MtHc) MtHc was proposed for clustering massive 16S rRNA sequences into OTUs. The goal of MtHc is to address the challenges of the balance between inference accuracy and computational efficiency. MtHc was developed under the framework of a complete weighted network where all the 16S rRNA sequences are viewed as nodes, each pair of sequences is connected by an imaginary edge, and the distance of a pair of sequences represents the weight of the edge.[452] MtHc consists of three main phrases: first heuristically search the motif, then use the motif as a seed to form candidate clusters, and finally hierarchically merge the candidate clusters to generate the OTUs. Examples of MtHc use are available in the studies.[457,458]

7.1.7.2.4 de Bruijn graph-based heuristic clustering method (DBH) DBH clustering method was proposed for clustering large-scale 16S rRNA sequences into OTUs, which is similar to MSClust. The goal of DBH is to address the sensitivity issue that traditional heuristic clustering methods just select one sequence as the seed for each cluster and the results are sensitive to the selected sequences that represent the clusters. DBH uses a novel seed selection strategy and greedy clustering approach for clustering massive 16S rRNA sequences into OTUs.[453] Examples of DBH citation were provided in the studies.[456,459,460]

7.1.7.2.5 Density-based modularity clustering method (DMclust) DMclust method was proposed for accurate OTU picking of 16S rRNA sequences. The goal of DMclust is to obtain an appropriate balance between clustering accuracy and computational efficiency. DMclust uses a novel density-based modularity clustering method to bin 16S rRNA sequences into OTUs with high clustering accuracy.[454] Examples of DMclust application can be found in the studies.[459,461]

7.1.7.2.6 ESPRIT-Forest ESPRIT-Forest, a parallel hierarchical clustering algorithm, was developed for parallel hierarchical clustering of massive amplicon sequence data in subquadratic time.[455] The standard hierarchical clustering methods suffer from computationally expensive to perform extremely large sequence datasets due to its quadratic time and space complexities. The goal of ESPRIT-Forest is to achieve subquadratic time and space complexity, while maintaining a high clustering accuracy comparable to the standard method. To achieve the goal, ESPRIT-Forest first organizes sequences into a pseudo-metric via partitioning tree for sublinear time searching of nearest neighbors, and then uses a new multiple-pair merging criterion to construct clusters in parallel using multiple threads.[455] Examples of ESPRIT-Forest use or cited for comparison of methods are available in the studies.[451,462–464]

7.1.7.2.7 Dynamic multiseeds clustering method (DMSC) DMSC is a dynamic multiseeds method for clustering 16S rRNA sequences into OTUs.[456] Similar to DBH, the goal of DMSC is to address the issues of overestimation of OTUs number or sensitivity to sequencing errors that traditional heuristic clustering methods just select one single seed sequence to represent each cluster. The strategies of DMSC used to pick OTUs consist of three steps: first heuristically generates clusters based on the distance threshold; then selects the multicore sequences (MCS) as the seeds based on predefined minimum distance threshold; and finally assign new sequence to the respective cluster based on the distance to MCS and its standard deviation within the MCS. The performance of DMSC was compared with seven state-of-the-art OTUs clustering algorithms: CD-HIT,[441] Uclust,[442] DBH,[453] DySC,[465] ESPRIT-Forest,[455] AL clustering algorithm implemented in mothur,[283] and CROP,[466] and especially with the traditional heuristic methods such as CD-HIT, Uclust, DySC, and DBH in terms of the inferred OTUs number, normalized mutual information (NMI), and Matthew correlation coefficient (MCC) metrics. It was demonstrated that DMSC can produce OTUs with higher quality and reduce OTUs overestimation with low memory usage. Additionally, DMSC is also robust to the sequencing errors.[456] DMSC paper was cited in one study.[467]

7.2 Interpretive correlation and association methods

7.2.1 Identifying correlation and association between two omics datasets

Interpretive methods are "constrained" techniques which are used to explain variation in a set of dependent variables (measured variables, called response variables) by another set of independent variables (explanatory variables) aiming to find axes in the multidimensional dataset space that maximize the association between the explanatory variable(s) and the measured variables.[31] Constrained ordination analysis provides the tools not only to allow revealization of sample similarities but also interpretation of the relationships between the datasets of explanatory and response variables. Thus, constrained ordination techniques actually move beyond the exploratory analysis and can be considered as specific hypothesis testing of how environmental variables determine values of response variables.[307]

Interpretive methods are sometimes differentiated each other based on whether they distinguish variables in two datasets as explanatory and response variables, respectively. If both sets of variables are treated as equal, then the methods are called symmetric approaches; otherwise, they belong to asymmetric approaches. In general, correlation-based methods are symmetric, while association-based methods are asymmetric.

Ecological studies often require studying the common structure of a pair of data tables. Previously, the widely used methods for studying the common structure of a pair of data tables in ecology are redundancy analysis (RDA) and canonical correspondence analysis (CCA). In microbiome literature, these multivariate methods are often used for studying interactions between environmental variables and microbiome communities. In Chapter "Exploratory analysis of microbiome data and beyond" by Xia et al.,[307] we introduce and illustrate them with microbiome data. Here, we will just briefly describe RDA and CCA, while provide more details on CANCOR (canonical correlation analysis) and CIA (Co-inertia analysis).

7.2.1.1 Canonical correlation analysis (CANCOR)

CANCOR (also often abbreviated as CCA) is another multivariate correlation method often employed for omics integration. Suppose that there exist correlations among the variables from two vectors of random variables, then CANCOR will find the linear combinations of these two vectors in terms of canonical (latent) variables to maximize the correlation with each other.[310,468] By using Pearson's correlation to measure the association between two linear combinations, CANCOR achieves the goal of summarizing the variables in these two vectors and reduces the dimensions of them. Although the goal of CANCOR is to find correlations and does not assume which variables are predictive and which are responsive, however, to perform CANCOR, variables within the dataset should be linearly independent, and the number of samples should be more than the number of variables. Both assumptions are usually not met in omics data.

Canonical correlation analysis is also often abbreviated as CCA; thus, it is often confused with canonical correspondence analysis. Both CCAs resembles each other in that they both can be used to detect the multivariate relationships between two datasets (e.g., an environmental dataset and a taxa abundance dataset). However, these two CCAs have distinctive assumptions: canonical correspondence analysis assumes a reasonable unimodal response curve of taxa to environmental variables,[469] while canonical correlation analysis assumes linear responses. CANCOR has been used to detect the relationships between bacterial community composition and chemical parameters,[470,471] and between microbial communities and epidemiological background variables (metadata).[472] Although CANCOR was reviewed that can be used to detect the relationships between microbiome composition and environmental factors,[351,473,474] however, CANCOR has several fundamental limitations: (1) when applying CANCOR method to high-dimensional data, sample covariance matrices are singular; thus, it is inappropriate method for high-dimensional data (it is usually applied to stationary scenarios)[475]; (2) CANCOR (also PLS and CIA) tends to include linear combinations of every variable under consideration or all the features that are generated from the data; hence when applying this method to high-dimensional data, it will be linear combinations of thousands of variables, making difficult to biological interpretability since determining whether or not a variable contributes to the canonical correlations is difficult[317,318] and also difficult to interpret their effect sizes[472]; (3) is its linear assumption between taxa to environmental variables, which is likely to be violated in nature; (4) is its symmetrically treating both sets of variables (does not consider one set of variables as "independent variables," and the other set as "dependent variables"); consequently, canonical correlation analysis is essentially a correlation method, not a flexible association method; (5) it is difficult to identify weak correlations; (6) it is difficult to visualize the inherently complicated results of CANCOR.[476] Among other limitations, canonical correlation analysis has limited applications in general[477] and especially in microbiome studies.[478]

7.2.1.2 Sparse extensions of canonical correlation analysis

To address the limitation of linear combination and especially to deal with CANCOR's applications to high-dimensional data, various versions of penalized CCA or sparse CCA have been developed to extend canonical correlation analysis via introducing sparsity to the linear combinations. Although the approaches could be different, all sparse CCAs are based on the principle of "bet on the sparsity" aiming to identify sparse linear combinations of the two sets of variables with high correlations and shrinking the coefficients for variables with less contribution to zero.[317,318,475,479–483] We briefly introduce some of sparse CCA below.

7.2.1.2.1 Penalized matrix decomposition (PMD) Witten et al.[482] showed that applying their developed penalized matrix decomposition (PMD) using an L_1 penalty on the rows and columns of the decomposition to a cross-products matrix yields a new method for penalized canonical correlation analysis (CCA). They showed that this penalized CCA method works well for genomic dataset consisting of gene expression and DNA copy number measurements on the same set of samples.

7.2.1.2.2 Sparse canonical correlation analysis (sparse CCA) Parkhomenko et al.[317,479] proposed a sparse canonical correlation analysis (sparse CCA) for high-dimensional data and applied to genomic integration. Instead of using L_1 penalty, sparse CCA solves the high dimensionality by selecting sparse subsets of variables while maximizing the correlation between the subsets of variables of different types. An extension of sparse CCA called adaptive sparse CCA was also presented. The algorithms behind sparse CCA[317] are: (1) analyzing and targeting sparsity in both sets of variables simultaneously; (2) reducing dimensionality to improve biological interpretability by providing sparse sets of associated variables. Hardoon and Shawe-Taylor[318] proposed a sparse CCA using a convex least squares approach which seeks a semantic projection that uses as few relevant features as possible to explain as much correlation as possible.

Most widely used sparse CCA methods use the model fit criteria to determine the level of sparsity of the canonical vectors, although these sparse CCAs extend the classical method to high-dimensional settings; however, theoretically they lack guarantees for recovering an appropriate sparsity level of the sparse CCA solution. Thus, Gossmann et al.[484] proposed a FDR-corrected sparse CCA for canonical vectors to determine an appropriate sparsity level for a sparse CCA solution. The proposed FDR criterion generalizes the conventional FDR[485] to canonical correlation analysis.

To discover interpretable associations in very high-dimensional multi-dataset, i.e., observations of multiple sets of variables on the same subjects, Solari et al.[486] proposed a two-stage approach to sparse CCA problem via concave minimization: first, the sparsity pattern of the canonical directions is computed via a fast, convergent concave minimization program; then shrink observations on the sets of covariates to two drastically smaller matrices enabling regular CCA methods can be used. In addition, directed sparse CCA and multiview sparse CCA were also introduced: the former is used to find associations for a specified experiment design, and latter is used to discover associations between multiple sets of covariates. The proposed methods can be used to identify the associations between metabolomics, transcriptomics, and microbiomics in high-dimensional, omic datasets.

To implement sparse CCA, several software packages were also developed. Of them, the R package PMA (Penalized Multivariate Analysis) is a widely used implementation[487] and FlashPCA: an R package flashpcaR and within the standalone commandline tool flashpca was another substantial improved and fast implementation of sparse CCA, enabling rapid analysis of hundreds of thousands of metabolomic, transcriptomic, or any other quantitative set of measurements together with thousands of phenotypes.[488,489]

7.2.1.3 Kernel extension of canonical correlation analysis (kernel CCA)

Another way to deal with high-dimensional data is the nonlinear approach of CCA extensions. Kernel CCA and neural networks-based CCA are reviewed as the two main kinds of methods for identifying nonlinear canonical correlations.[486] We briefly introduce some of them below.

Kernel methods are more popular for analyzing nonlinear associations. The popularity of using kernel methods in analysis of nonlinear associations is due to two reasons[486]: mainly because the vast theoretical literature on kernel methods and especially Support Vector Machines (SVMs) literature are available[490–494] and also because Kernel methods need estimating the significantly fewer number of parameters compared to neural networks-based methods.[495,496] Fukumizu et al.[497] mathematically prove the statistical convergence and consistency of kernel CCA, and hence providing a theoretical justification for the method.

The attractive feature of Support Vector regression is its ability to perform a nonlinear mapping of the dataset into some high-dimensional feature space, and then linear operations may be performed.[498] In Lai and Fyfe,[498] the linear CCA was extended to a kernel CCA by nonlinearly transforming the data to a feature space and then performing linear CCA in this feature space. Melzer et al.[499] use a nonlinear feature extraction technique to regression and object recognition, in which the nonlinear transformation of the input data was performed using kernel methods. Bach and Jordan[500] describe the algorithms for independent component analysis (ICA) using CCA-based methods in a reproducing kernel Hilbert spaces. Hardoon et al.[501] present a general method using kernel CCA to learn a semantic representation to web images and their associated text and review some of the methods that have been developed for learning the feature space. Larson et al.[502] propose a kernelized version of CCA (KCCA) to detect gene-gene interactions and apply it to identify complex multiloci disease-associated single-nucleotide polymorphisms (SNPs) related to ovarian cancer. It was shown that KCCA has the advantages in gene-gene interaction analysis and genetic association studies. A regularized version of CCA (RCCA) has been also developed to extend CCA.[503]

7.2.1.4 Neural and deep networks-based canonical correlation analysis

7.2.1.4.1 Neural networks-based CCA Neural networks are well known for their capability of performing powerful transformations. Lai and Fyfe[504] implemented neural networks of canonical correlation analysis to find nonlinear canonical correlation. In Lai and Fyfe,[498] a nonlinear extension to the neural method was developed and then a nonlinear CCA was performed.

7.2.1.4.2 Deep CCA Another nonlinear extension of the linear CCA method is deep CCA. It was developed by Andrew et al.[505] as an alternative to the nonparametric method kernel CCA for learning correlated nonlinear transformations. However, unlike kernel CCA, deep CCA does not require an inner product, and as a parametric method deep CCA has the advantages: training timescales well with data size and computing the representations of unseen instances without referencing the training data.

7.2.1.5 Extending CCA to functional data analysis

Another nonparametric extension of CCA is functional data analysis. It is useful when the data are themselves curves or functions.[506,507]

Sparse additive models, a new class of methods for high-dimensional nonparametric regression and classification,[508] were developed. These methods are essentially variants of a functional version of the grouped lasso, and component selection and smoothing in multivariate nonparametric regression. The methods combine ideas from sparse linear modeling and additive nonparametric regression. The algorithm can be used for effectively fitting the models even when the number of covariates is larger than the sample size. Balakrishnan et al.[509] extend linear-CCA to high-dimensional nonparametric CCA from the two approaches: sparse additive functional CCA that uses Sobolev spaces without a reproducing kernel and sparse additive kernel CCA that uses reproducing kernel Hilbert spaces. Both these methods can be used to identify nonlinear relationships of two sets of high-dimensional data on the same set of samples, such as DNA copy number and gene expression.

7.2.1.6 Co-inertia analysis (CIA)

CIA is also often abbreviated as COIA. Similar to CANCOR, CIA is a multivariate method initially developed for the analysis of ecological data for simultaneously analyzing two sets of variables[510] and then has often been used for omics integration. Different from CANCOR, CIA uses the covariance instead of a correlation to analyze the relationships of two sets of variables. The inertia is a term of ecology, simply means a sum of variances. The total inertia is a global measure of the variability of the data. CIA works on a covariance matrix (taxa by environment) instead of a correlation matrix. CIA measures the co-structure (i.e., concordance) between two datasets by maximizing the covariance between components, i.e., measuring the co-structure of samples in the environmental and taxa hyperspaces. By using covariance to represent the similarity with no variance constraint, CIA achieves the goal of summarizing the variables in these two vectors and reduces the dimensions of them.

The underlying algorithm behind it is when the two structures vary simultaneously and also vary inversely, CIA is high and when they vary independently or do not vary, then it is low. CIA first performs data reduction using PCA or correspondence analysis on the two datasets separately, and then constrains the resulting components so that the squared covariances between the two datasets are maximized. In a simpler sense, CIA is linked to partial least-squares regression and can be considered as a PCA of the joint covariances of two datasets.[511]

Previously, it was not commonly used compared to redundancy analysis (RDA) and canonical correspondence analysis (CCA). In 2003 Dray et al.[512]

presented the co-inertia criterion for measuring the adequacy between two datasets and compared the co-inertia approach with CCA and RDA methods using simulated ecological data. The results showed that CCA and RDA are very efficient only in the case of few uncorrelated (i.e., orthogonal) variables, while if the variables are correlated CIA is very stable and thus a good alternative. Since then, CIA was getting popular in ecology and showed its great potential in the era of omics. Nowadays CIA has been adapted in microarray, microbiome, and other omics studies. CIA bases on the same measuring criterion as CCA or CANCOR, but is very flexible and suitable for quantitative and/or qualitative or fuzzy environmental variables. Comparing to CCA and CANCOR, CIA is more appropriate for multidimensional data (e.g., microbiome and other omics data) because in CIA the connecting tables (taxa and/or samples) could have either similar (even low) or different numbers of variables, avoiding the multicollinearity problem; whereas CANCOR requires that the number of taxa and the number of environmental variables must be much lower than the number of samples, and to predict the taxa structure, CCA requires that the number of environmental variables must be smaller than the number of sampling sites otherwise the multicollinearity problem may occur. Based on the authors of CIA and others,[510,512,513] CIA has no built-in procedure for variable selection; consequently, it may be redundant if we use CIA for taxa selection.[514]

However, CIA has several noteworthy advantages. For example, (1) compared to CANCOR, CIA has fewer assumptions and does not constraint the number of variables in the datasets compared to the number of samples.[314] (2) CIA is the simplest and most robust approach for matching two tables and can be extended to analyze more than two datasets simultaneously,[515,516] such as extended to three-way tables (environment by sample by time) to study the dynamics of the taxa-environment relationship. Hence, CIA is more suitable for detecting the association analysis between environmental variables and microbiome. (3) It is able to maximize the covariance of projected scores. (4) It enables coupling with several dimension reduction approaches, including PCA, correspondence analysis (CA), and multiple correspondence analysis, such that it can accommodate both discrete count data and continuous data. (5) CIA provides a more readily interpretable quantitative assessment of the strength of the global association (RV coefficient).[31] The coefficient of CIA lies between 0 and 1 with higher numbers indicating more global similarity between two datasets.

CIA has been used in microbiome and other omics, such as in genomics for comparing data from two microarray datasets,[513,517] proteomics,[517] in integrative analysis of the metabolome and microbiome,[93,518–520] in evaluating the covariance between the metabolites and genes of obese and lean humans,[521] in detecting mRNA and microRNA from gene expression data,[522,523] for integration of proteomic and gene expression data,[524] in diet study to identify potential shared biological trends between groups,[525] and in integrating microbiome and metabolomic datasets.[526]

Recently, CIA has been extended to sparse co-inertia analysis (SCIA)[527] by introducing sparsity to the linear combinations. SCIA weights coefficients based on contribution of variables, in which the coefficients for variables with less contribution are shrunk to zero.

7.2.1.7 Procrustes analysis (PA), redundancy analysis (RDA), and canonical correspondence analysis (CCA)

7.2.1.7.1 Procrustes analysis (PA) PA is a statistical method that utilizes data-reduction methods such as PCA and CANCOR for comparing the distributions of multiple sets of corresponding samples and visual integration of omics data.[528,529] PA is a fast and simple visualization technique for comparison of shapes in a multidimensional space, aiming to superimpose structures and then move, rotate, and scale them until it achieves the best match (the smallest difference in shapes). For example, it superimposes the principal components of two datasets at the low-dimensional space, allowing researchers to quickly examine the congruency of their multiomics datasets.

The advantages of PA[31] include (1) like CIA, compared to CANCOR, PA has fewer assumptions and does not constraint the number of variables in the datasets compared to the number of samples.[314] (2) PA can be run on a wider range of datasets. (3) Like CIA, PA can be used in more than two datasets simultaneously.[515,516] One drawback of PA is that it should be complimented with other multivariate methods such as CANCOR to draw strong conclusions.[530] Examples of PA in metabolome and microbiome data can be found in studies.[93,411,530–533] Other examples of using PA are from these studies.[518,520]

7.2.1.7.2 Redundancy analysis (RDA) RDA was also named as principal component analysis with instrumental variables.[534] As a constrained ordination, RDA was developed to assess how much of the variation in one set of variables can be explained by the variation in another set of variables. However, as a multivariate extension of simple linear regression into sets of variables,[534] RDA summarizes the linear relations between multiple dependent variables and multiple independent variables in a matrix, which is then incorporated into PCA. RDA assumes that variables from two datasets (e.g., an environmental dataset and a taxa abundance dataset) play different roles: one set of variables can be considered the "independent variables," and the other set is considered the "dependent variables." In other words, the variables in these two sets are asymmetrical.

RDA is different from canonical correlation analysis (CANCOR, also often abbreviated as CCA) in that CCA puts both sets of variables equally or treat them symmetrically. RDA has limitations such as its assumption of linear relationships among variables. RDA uses the similar principles as PCA, which is actually a canonical version of PCA where the principal components are constrained to be linear combinations of the explanatory variables. Thus, RDA is inappropriate when relationship between response and environmental variables is unimodal rather than linear. RDA was used to investigate the association between log relative abundance and different human milk consumption patterns while controlling various explanatory variables.[114] Other examples of using RDA are from studies.[531,535–537]

7.2.1.7.3 Canonical correspondence analysis (CCA) Since its introduction in 1986,[540] CCA has become one of the popular multivariate methods in community ecology and adopted by microbiome researchers. Similar to RDA, CCA aims to find the relationship between two sets of variables. However, different from RDA which constructs a linear relationship among variables, CCA assumes a unimodal relationship and measures the separation based on the eigen values produced by CCA.[539] Thus, CCA is a canonical form of CA of the response variable set that is constrained by the set of explanatory variables.[540] CCA can be used both for detecting taxa-environment relations and for investigating specific questions about the response of taxa to environmental variables. Examples of CCA use in microbiome studies can be found in Refs. 541–543.

7.2.2 Identifying correlation and association among more than two omics datasets

Although a few of above multivariate correlation and association methods can be extended to analyze more than two datasets, however, most of them are limited to the analysis of two datasets. Several multivariate methods have been proposed for conducting correlation and association analyses for more than two datasets. The common goal of these methods is to find a linear combination of variables within each dataset so as to maximize the sum of squared pairwise correlations or the sum of squared covariances between each linear combination. We briefly introduce some of them below.

7.2.2.1 Generalized canonical correlation analysis (GCCA)

GCCA is an extension of CANCOR from only two sets (blocks) of variables to several sets (blocks) of variables. Several GCCAs have been proposed[544,545] since Hotelling developed CANCOR in 1936.[310] Of them, the version proposed by Carroll[546] has been considered as more appropriate for the analysis of multiple-set data. Basically, GCCA first extends CANCOR to several sets of random variables after removing dependencies within each set, then derives the canonical variables which are a new linear combination of the variables, and finally reconstitutes each set and estimates the correlation between canonical variables.[547] Either correlation or covariance matrices can be used to conduct GCCA. GCCA can be used to analyze the associations of multiple omics data. For example, currently GCCA was adopted to analyze the associations between SNP (single-nucleotide polymorphisms) block, phenotype block, and disease block.[548]

7.2.2.2 Multiple co-inertia analysis (MCIA)

Chessel and Hanafi extended CIA enabling the analysis of two tables to the co-inertia analysis of K tables (multiple Co-inertia analysis).[549] The extension has been applied in the field of environmental science and phylogenetics.[515] Meng et al. further introduced MCIA to multiomics datasets.[550] As we stated above, a typical omics dataset is a contingency table or matrix often with the number of taxa exceeding the number of measurements or samples, which results in multidimensional problem. As a CIA, MCIA can address the multidimensional problem using a two-step process. First step is to transform data into comparable lower dimensional spaces by conducting an ordination such as PCA and correspondence analysis (CA) on each dataset separately.[513] The second step is to generalize CIA.[513,549] After completing these two steps, MCIA simultaneously ordinates columns (samples) and rows (taxa) of multiple tables on the same dimensional space, closely projects the taxa or samples that share similar trends, and provides a simple graphical representation of concordance between datasets. Such that MCIA can identify co-relationships between multiple high-dimensional datasets, such as transcriptomic, proteomic, and metabolomic data. MCIA was reviewed as providing comparable analysis results of RCGGA.[550] However, MCIA does not impose any sparsity in selecting taxa, resulting that more taxa could be selected. It is difficult to interpret without considering other methods.

7.2.2.3 Consensus principal components analysis (CPCA)

CPCA method is one of ad hoc extensions of NIPALS (nonlinear iterative partial least squares). It was introduced by Wold et al.[551] and widely investigated in the chemometrics.[552–554] Similar to MCIA, CPCA also takes two steps: the first step computes the parameters: block scores, block loadings, global scores, and global loadings through the iterative procedure; the second step aims at computing subsequent components that have scores and loadings of higher order than 1. Compared to generalized canonical correlation analysis, CPCA is less vulnerable to multicollinearity within each dataset.[517]

7.2.2.4 Penalized canonical correlation analysis (PCCA)

Witten et al.[555,556] proposed a new framework for computing a rank-K approximation for a matrix called penalized matrix decomposition and applied to sparse principal components and canonical correlation, which results in a method for penalized canonical correlation analysis (PCCA). The penalized CCA method was demonstrated by simulated data and a genomic dataset. PCCA originally focuses on the variable selection from multiple datasets, but also can be used for correlation analysis of multiple tables.

7.2.2.5 Regularized generalized canonical correlation analysis (RGCCA)

RGCCA is a general framework combining many multiblock data analysis methods.[547,557–559] RGCCA generalizes the regularized canonical correlation analysis to three or more sets of variables. It models the linear relationships between several blocks of variables for the same set of individuals to find linear combinations of block variables or components.[560,561] RGCCA has two objectives: one is to block components in order to explain their own block well; another is to block components that are assumed to be connected and highly correlated. RGCCA has several advantages: first, it is very flexible. For example, the flexibility of the Partial Least Squares (PLS) path modeling algorithms is very similar to the PLS algorithm proposed by Herman Wold.[562] Second, it does not assume that blocks are necessarily fully connected so that it allows a large variety of hierarchical models such as Carroll's GCCA and Chessel and Hanafi's MCIA to be included in the RGCCA framework. For example, recently Garali et al.[563] introduced a strategy for integrative exploratory analysis of multiomics data under the framework of RGCCA and its sparse counterpart (SRGCCA). RGCCA or SRGCCA encompasses other omics integration techniques (e.g., PCA, PLS, CCA) and their extensions and allows for the incorporation of prior knowledge through the use of a design matrix. Third, to avoid spurious relationships between blocks in a high-dimensional block setting or in the presence of multicollinearity within blocks, RGCCA regularized various correlation-based methods to stably analyze ill-conditioned data blocks as possible. Moreover, to identify significant variables in each block which are active in the relationships between blocks, a variable selection method was proposed by sparse generalized canonical correlation analysis (SGCCA) combining RGCCA with an L_1 penalty.[556]

7.2.3 The family of partial least squares

The family of partial least squares (PLS) is another type of multivariate interpretive correlation and association methods. As an approach of dimension reduction, the PLS family uses a small number of linear combinations to summarize the variables in the two feature types so as to maximize the association between these two feature types as demonstrated by these linear combinations. The common method of measuring association in this family is to use covariance to quantify the association with the constraint that the linear combination from one feature type has a unit variance.[51] This family is appealing because the family methods have been developed to solve multidimensional and sparse problems.

7.2.3.1 Partial least squares (PLS)

PLS technique was developed by Herman Wold in the 1970s by extending the multiple linear regression model.[564–567] It takes a latent variable approach to model the covariance structures in two spaces (i.e., the X and Y spaces) so that both variables X and Y are projected to a new space, which is called projection to latent (hidden) structures.[568] Thus, PLS is alternatively called projection to latent structures. PLS regression aims to predict response variable(s) Y from a (large) set of predictor variables X through reducing the set of predictor variables to a smaller set of uncorrelated components and then performs least-squares regression on these components.

A PLS model is either to find the fundamental relations between two sets of variables observed on the same set of individuals or to find a linear relationship by projecting the predicted variables and the observable variables to a new space or to find the multidimensional direction in the X space that explains the maximum multidimensional variance direction in the Y space.

PLS was originally applied in the social sciences. Today, it is most widely used in chemometrics and related areas and also used in bioinformatics, microbiome, and other omics. Compared to multivariate regression, one advantage of PLS is that it has fewer assumptions and hence it can use predictor variables that are collinear and not independent.[569] PLS is particularly suitable for: (1) modeling the data matrix of predictors having more variables (taxa) than observations (samples): large p and small n problem; and (2) when there is multicollinearity among predictors. However, PLS has several disadvantage: first, by using PLS, the users are required to define X and Y as response or predictive variables, but without considering inherent systematic variation that may exist within each dataset that does not correspond with the outcome.[570]

Second, similar to PCA and CANCOR, the potential disadvantage of PLS is that the learned projections are a linear combination of all variables or features in the two datasets, making it difficult to interpret the solutions.[318] Additionally, the validation of PLS and also OPLS models still needs to be confirmed in more real works.[571] Examples of using PLS to analyze the associations in multiomics studies include gut microbiota, gene expression, and metabolomic data.[572]

7.2.3.2 Sparse partial least squares (sPLS)

Sparse PLS is a regression or a canonical correlation framework, taking the sparse approach to integrate several datasets.[573–575] It includes a built-in procedure to select variables while integrating data. Thus, it is suitable in genomic studies to simultaneously analyze the mutual interactions between the different datasets (i.e., transcriptomics, proteomics, and metabolomics data). sPLS was evaluated as outperformed CIA in selecting highly relevant features (i.e., genes), while CIA tended to select redundant information.[576]

7.2.4 The family of orthogonal projections to latent structures

Orthogonal projections to latent structures are another class of interpretive correlation and association analyses.

7.2.4.1 Orthogonal projections to latent structures (O-PLS)

O-PLS[570,577] is a linear regression method. In two blocks of X (predictor or descriptor variables) and Y (response or property variables), structured noise, which is defined as the system variation of X (or Y) not linearly corrected with Y (or X), often exists. O-PLS removes systematic variation in X that is orthogonal (i.e., not correlated) to Y. The presence of structured noise will make latent variable methods (e.g., PLS) have weakened score-loading corresponding beyond the first component, which makes it difficult to correctly interpret scores and loadings and other model parameters.

7.2.4.2 Two-way orthogonal partial least squares (O2-PLS)

O2-PLS, based on the method of orthogonal projections to latent structures (O-PLS),[577,578] is another integrative data analysis method. O2-PLS partitions the systematic variability in each dataset of X and Y into three blocks, joint, unique, and residual, i.e., separates the variation in each dataset into three parts: the X/Y joint predictive variation, the Y-orthogonal variation in X, and the X unrelated variation in Y, and is then able to identify significantly predictive features of the joint variation.[570,579–581] It was derived from the prediction approach using basic partial least squares projections to latent structures (PLS).[570] O2-PLS was designed to model and predict both X (descriptor variables) and Y (property variables) with an integral orthogonal signal correction (OSC) filter to remove the structured noise in both X and Y from their joint X-Y covariation for the prediction.[570,578] O2-PLS thereby efficiently leads to a minimal number of predictive components with full score-loading correspondence. And it also provides an opportunity to interpret the structured noise. O2-PLS is a generalization of PLS and O-PLS, which moves PLS and O-PLS from one directional method (X ➔ Y) to be bidirectional, i.e., X $\leftrightarrow$ Y; O2-PLS disregards matrix assignment and treats X and Y as equals, i.e., both X and Y can be the predictor. O2-PLS can handle moderate amounts of missing data and especially suited for situations where both X and Y blocks have many noisy and correlated (multicollinear) variables.[582,583]

A simulation study showed that the O2-PLS estimates were close to the true parameters in both low and higher dimensions and more noise (50% of the data) only affected the systematic part estimates but not systematically affected the joint estimates.[579] O2-PLS method was evaluated as highly useful in omics data integration analysis, e.g., for comparing and contrasting blocks of data compiled using different spectral methods or different "omics" platforms (microarray data, electrophoresis data, etc.). O2-PLS can be used to identify interactions between and within omics datasets, such as metabolome and transcriptome data and microbiome-metabolite interactions. O-PLS and O2-PLS have been applied in various biological and biochemical data for modeling and interpretation of linear relationships between a descriptor matrix and response matrix[580–584] and for pairwise integration of metabolomic, transcriptomic, and metagenomic data.[585]

7.2.4.3 Kernel-based orthogonal projections to latent structures (K-OPLS)

K-OPLS[586,587] is a Kernel-based classification and regression method, which is a reformulation of the original OPLS method to its kernel equivalent. Developed based on successful applications of O-PLS and O2-PLS methods in various chemical and biological systems for modeling and interpretation of linear relationships between a descriptor matrix and response matrix, K-OPLS aims to combine the strengths of kernel-based methods to model nonlinear structures in the data while maintaining the ability of the OPLS method to model structured noise.[587] K-OPLS is reformulated by replacing the descriptor matrix with the kernel Gram matrix. Kernel-based pattern recognition methods including Support Vector Machines (SVMs), Kernel-PCA, and Kernel-PLS have the computational benefit: allowing the kernel matrix to be treated as dot products in a high-dimensional feature space,[586] thus enabling the usage of the "kernel trick"[588] to efficiently transform the data into a higher-dimensional feature space where predictive and response-orthogonal components are calculated. In other words, in Kernel-based setting, common kernel functions either polynomial or Gaussian functions can be used to implicitly perform the transformation to higher-dimensional spaces. The strategy of leveraging both strengths of kernel-based methods and O-PLS-based methods is expected to improve predictive performance considerably especially when strong nonlinear relationships exist between descriptor and response variables while retaining the O-PLS model framework.[587]

7.3 Discriminatory correlation and association methods

Discriminatory methods, usually called discriminant analyses (DA), are an extension of the interpretive multivariate techniques. DA is used in situations where the classes (clusters) are known a priori. The aim of DA is to classify an observation, or several observations, into these known groups[589] or to separate samples between different classes based on the values of measured variables aiming to define discriminant functions (synthetic variables) or hyperspace planes that will maximize the separation among different classes[31] or to construct a classifier (or classification rule) that will separate the predefined classes as much as possible.[590]

7.3.1 Linear discriminant and its effect size analyses

7.3.1.1 Linear discriminant analysis (LDA)

LDA, also called canonical discriminant analysis (CDA), presents a group of ordination techniques that find linear combinations of observed variables that maximize the grouping of samples into separate classes.[591,592] It was designed to use the measured variables (serve as the predictor variables) to predict sample classes (also called grouping variable, the response variable). A discriminant analysis companies with a model of class prediction.

Thus, LDA is often utilized to classify and predict response (the predictive approach is an extension of DA algorithms). LDA technically is similar to PCA/PCoA. However, LDA differentiates from PCA and PCoA due to its prediction capability: LDA is able to predict a new (unknown) sample based on the values of measured variables in that sample.[593] Two differences between PCA/PCoA and LDA are: (1) PCA and PCoA aim to find the linear combinations of the measurements that maximize variance, while LDA aims to find the linear combinations of the measurements that best describe the separation between the groups. (2) After performing PCA/PCoA, the separation of groups may occur, but they do not require to know the groups of interest a priori; in contrast, in LDA, the groups of interest are known a priori. Examples of the use of LDA to separate dietary groups based on metabolic or microbiome data are available in studies.[533,594] LDA was also used to separate locations on the basis of environmental data[595] and used in anti-PD-1 and microbiome study.[336]

7.3.1.2 Linear discriminant analysis effect size (LefSe)

LEfSe[596] is an algorithm for high-dimensional biomarker discovery and explanation that identifies genomic features (genes, pathways, or taxa) characterizing the differences between two or more biological conditions (or classes) with its emphasizing on statistical significance, biological consistency, and effect relevance. The algorithm is specifically developed using: (1) the nonparametric factorial Kruskal-Wallis (KW) sum-rank test[597] to detect features with significant differential abundance with respect to the class of interest; (2) the Wilcoxon rank-sum test[598,599] to investigate subsequent biological consistency by a set of pairwise tests among subclasses of different classes; (3) the LDA[591] to estimate the effect size of each differentially abundant feature and to perform dimension reduction. Although LefSe algorithm uses LDA, it is used for performing class comparison rather than class prediction. Thus, the estimated LefSe analysis is not directly related to its predictive ability.

LefSe analysis can be used to support high-dimensional class comparisons of metagenomic data to determine the features (organisms, taxa, OTUs, genes, or functions) most likely to explain differences between classes based on a null hypothesis of no difference between classes. The appealing feature is that it is coupling standard tests for statistical significance with additional tests encoding biological consistency and effect sizes. The visualization tools provide an effective way to statistically summarize and visualize the hierarchical relationships inherent in 16S-based taxonomies/phylogenies or in ontologies of pathways and biomolecular functions.[596]

One drawback of LEfSe analysis is that it is performed via the univariate-based analyses on a per-dataset basis due to its development via Kruskal-Wallis test and Wilcoxon rank-sum test. Thus, the adjustments for multiple comparisons should be conducted, which may be costed on the power of testing. LEfSe can be implemented via a convenient graphical interface incorporated in the Galaxy framework[600,601] or online (http://huttenhower.sph.harvard.edu/lefse/). LEfSe analysis can be used to perform on a multilevel basis (phylum, class, order, family, genus, and species level)[602] or differentiate enriched bacterial functions among groups. For example, LEfSe analysis has been performed on OTU-table using the online Galaxy interface to identify bacterial taxa that were differentially abundant in groups,[603] to identify features that were statistically different among groups, and to estimate their effect size.[604] Others have reported examples of utilizing LEFSe analysis.[272,351,418,605,606]

7.3.2 *Extensions of LDA to partial least squares and orthogonal projections to latent structures*

PLS-DA (partial least squares discriminant analysis) and OPLS-DA (orthogonal projections to latent structures discriminant analysis) are extensions of LDA. The common feature of PLS-DA and OPLS-DA is that they both may utilize weaker sources of variation to separate groups.

7.3.2.1 Partial least square-discriminant analysis (PLS-DA)

PLS-DA is a popular multivariate tool performing three kinds of functions: (1) used as a multivariate dimensionality-reduction tool[607,608]; (2) also used for feature selection method[609]; and (3) used for classification method.[6,181,610–614] The beneficial properties of PLS-DA lie on its capabilities of dealing with large p and small n problem. Thus, it can be used in big datasets allowing to treat multicollinearity, noise and missing values, and few samples (observations) and many features (variables). PLS-DA adapts PLS regression methods to the problem of supervised clustering and is considered as a "supervised" version of PCA (principal component analysis) in the sense that it plays dimensionality reduction and classification double roles. PLS-DA previously has been widely used in the field of chemometrics,[607,615] and now has been getting more popular in analyzing metabolomics and other integrative omics data.[616–619] For microbiome study, both PCA and PLS-DA can be used for analyzing bacterial relative abundances at each taxonomic level (Phylum, Class, Order, Family, Genus, and Species) to identify possible similarities of the samples within the groups or integratedly analyzing the association of multiomics, such as metagenomics, metabolomics, and environmental data.[620]

Due to the unsupervised nature of the data and sample variance, PCA was reviewed as a less satisfactory method to discriminate between the omics data distributions. PLS-DA can be applied to predict measurements taken from two or more experimental groups (e.g., healthy vs. disease).[607,621] PLS-DA is able to measure linear/polynomial correlation between variable matrices via reducing the model dimensions, and allowing easy distribution of the samples and the omics features. Comparet to PCA, PLS-DA is more capable. Thus, it was used following PCA in large datasets to discriminate between the omics (e.g., metabolite) distributions. However, PLS-DA tends to construct overly complex models when the measurement variation exists but does not correlate to an experimental group.[577,584] Thus, although PLS-DA is a powerful method, relying on weaker sources of variability in the dataset may force group separation at the expense of model validity.[622] Examples of PLS-DA utilization can be found in microbiome study.[351]

7.3.2.2 Orthogonal projections to latent structures-discriminant analysis (OPLS-DA)

OPLS-DA was developed based on the OPLS method for the purpose of discriminant analysis and the principle of partial least-squares (PLS) regression. Within the framework of the OPLS-DA method, the strengths of PLS-DA and soft independent modeling of class analogy (SIMCA) classification are combined.[584] OPLS-DA uses class-orthogonal variation to augment classification performance in cases where the divergent individual classes exhibit in within-class variation. If no such variation is present in the classes, OPLS-DA will have largely equivalent prediction results to traditional supervised classification using PLS-DA. As other orthogonal projections to latent structures (OPLS) methods including OPLS, OPLS-DA, K-OPLS, and O2-PLS, OPLS-DA facilitates the separation between experimental groups based on high-dimensional measurements and interpretation of the different types of variation in the data.[623]

Compared to PLS-DA, the main benefit of OPLS-DA lies in its ability to separate predictive from nonpredictive (Y-orthogonal) variation. OPLS-DA may be applied to analyze discrete variables, as in classification and biomarker studies. For example, this method was used to assess the in situ chemical composition of two different cell types of mouse liver samples in chemometrics. It was showed that OPLS-DA was able to separate predictive variation (between cell type) from variation that is uncorrelated to cell type so that it facilitated understanding of different sources of variation.[623] OPLS-DA also has been used in analysis of metabolomics data.[624] However, OPLS-DA could not identify small and subtle treatment effects when they appear differently among human subjects.[625] A simulation study showed that PCA, PLS, and OPLS may dangerously lead to statistically unreliable conclusions when used without validation because these tools aggressively force separations between experimental groups at the expense of model validity. Especially, it is even dangerous using OPLS-DA as an alternative method when PCA fails to expose group separation because if a PCA model fails to achieve group separation, a subsequent OPLS-DA model, despite any appearance of group separation, is often unreliable or invalid.[622] OPLS-DA also eliminates variation that is unrelated to the separation of groups, creating a less complex model.[622] Examples of OPLS-DA use in microbiome studies are available in the works.[532,626,627]

7.4 Classification methods

Machine learning is mainly about learning from data. A typical supervised learning problem is: for an outcome (output variable (Y)), usually quantitative or categorical, the purpose is to predict the outcome based on a set of features (such as diet and clinical measurements). More generally, the main goal of supervised learning is to build a model from a set of categorized data points to predict the correct category membership of unlabeled future data. The prediction model that is built based on this set of data (input variables (X)) is called learner which can be used to predict the future outcome variable. A good prediction model (learner) is one that accurately predicts such an outcome. This learning problem is called "supervised" because the outcome variable is present to guide the learning process[628] (p. 2). In other words, the process of an algorithm learning from the training dataset can be considered as a teacher supervising the learning process.

If only the features are observed and have no measurements of the outcome, then it is an unsupervised learning problem. The unsupervised problem is less developed in the literature. For full cover of learning problems, the interested readers can reference these two books.[628,629]

Both classification and regression problems belong to supervised machine learning. The goal of classification problem is to predict discrete (qualitative) values. In other words, the output variable is a category (e.g., "disease" and "no disease," "IBD" and "Non-IBD"). The goal of regression is to predict continuous (quantitative) values. In other words, the output variable is a real value, such as "age" or "weight."

The goal of building predictive models is very different from the traditional goal of fitting an explanatory model to one particular dataset. The traditional model fitting focuses on how well the model fits the particular training dataset, whereas predictive model building focuses on how well the model will generalize to future novel input dataset. Here, we introduce two classification methods: random forest and support vector machines.

7.4.1 Random forest (RF)

RF is an ensemble machine learning method based on the use of classification and decision regression trees. Decision tree learning seeks to construct a statistical prediction model to predict the values of response variable(s) based on the given values of predictor variables. The model is obtained by recursively partitioning the space of predictor variables and establishing a value of the response variable within each partition, which results in a decision tree.[630] The decision tree contains a set of if-then logical conditions.

Breiman[631] proposed random forests based on the two well-known ensemble learning methods of classification trees: boosting[632] and bagging.[633] He then added a layer of randomness to bagging. The two methods differentiate each other in this way: in boosting, successive trees depend on earlier trees: giving extra weight to points incorrectly predicted by earlier predictors. In the end, prediction takes from a weighted vote. In contrast, in bagging, each of successive trees is independently constructed using a bootstrap sample of the dataset. In the end, prediction takes from a simple majority vote.[634]

RF has implemented two new strategies: (1) constructing each tree using a different bootstrap sample of the data; (2) in a random forest, splitting each node using the best among a subset of predictors randomly chosen at that node rather than splitting each node using the best split among all variables in standard trees. Thus, RF actually substantially modifies and then averages the bagging that builds a large collection of decorrelated trees[628] (p. 587).

Compared to other common classification methods, such as linear discriminant analysis, support vector machines, neural networks, classification trees, and logistic regression, the advantages of RF include:

(1) Robust against overfitting and usually less sensitive to the input values[631,634]; (2) very user-friendly because only two parameters are required (the number of variables in the random subset at each node and the number of trees in the forest)[634]; (3) flexibility to perform several types of statistical data analysis, including regression, classification, survival

analysis, and unsupervised learning[635]; (4) very high discriminating power and classification accuracy[635]; (5) no distributional assumptions about the predictor or response variables[635]; and (6) can handle situations in which the number of predictor variables greatly exceeds the number of observations.[635] RF was reviewed as having the best performance among the compared several different discriminatory and classification techniques when applied to the set of taxa data, with SVM closely followed.[636] With this range of capabilities, RF offers powerful alternatives to traditional parametric and semiparametric statistical methods for the analysis of microbiome and other omics data. However, RF does not explicitly perform feature (e.g., taxa, genes) selection and may perform poorly when large numbers of irrelevant features are present.[637] Many examples of using RF in microbiome studies are available.[532,638–641] Examples of RF use in microbiome and metabolism research are provided by these reports.[260,642–645]

7.4.2 Support vector machines (SVMs)

SVM[646] is a supervised machine learning algorithm that can be used for both classification and regression. The basic model of SVMs was described in 1995 by Cortes and Vapnik. The goal of the SVM algorithm is to use a training set of objects (samples) separated into classes to find a hyperplane in the data space that produces the largest minimum distance (called margin) between the objects (samples) that belong to different classes.[646] So the hyperplane is known as the maximum margin hyperplane. SVM only uses the objects (samples) on the edges of the margin (called support vectors) to separate objects (samples) rather than using the differences in class means. Since the separating hyperplane is supported (defined) by the vectors (data points) nearest the margin, so the algorithm is called SVM.

SVM has been shown to perform well in many real learning problems with a variety of settings and is often considered one of the best "out-of-the-box" classifiers.[628] SVM has the advantages of increasing class separation and reducing expected prediction error. Additionally, SVM is flexible for both linear and nonlinear-based discriminatory analyses,[647] and it is suitable for analysis of high-dimensionality datasets with small sample size when combining with feature selection approaches.[636,648] Like the RF methods and sparse regression models, SVM is efficient in addressing the high-dimensionality problem; however, due to its limited ability to exploit the phylogenetic structure of the microbiome data, SVM may not be optimal to detect clustered microbiome signals, similar to the RF methods and sparse regression models.[198,649] Examples of using SVMs in microbiome studies can be found in the cited works.[636,644,650,651]

8. Hypothesis testing of univariate and multivariate regression-based association methods

Association analyses can be formulated by a regression with a statistical significance testing. Regression-based methods are widely used to construct association analysis mainly because: (1) regressions are relatively straightforward to implement hypothesis testing association; (2) regression models have the advantage of being able to adjust for relevant covariates compared to exploratory, interpretive, and discriminatory analyses and classification methods; (3) regression framework can be flexibly equipped with many well-studied statistical tools to handle specific analytical needs,[51] such as incorporating random effects within the framework of linear mixed model to account for intersubject (sample) correlation between subjects (samples) due to study design[652–656] or correcting for data heterogeneity due to unobserved confounders[657] or incorporating a penalized or sparse approach as well as a variable selection method to handle high-dimensional data and hence facilitating biological interpretability.[94,358,359,380,658–660]

However, regression-based methods typically need to choose the response variable(s) and the predictor variable(s) before running the regression analysis, which is not only a nontrivial effort when the underlying biology is poorly understood for the system being studied,[51] but also a real challenge in association analysis of microbiome and other high-dimensional omics data. Because in microbiome and other omics variables (i.e., taxa, genes) could be thousand and the number samples are often much less than the number of variables, resulting in large p and small n problem, either choosing one feature set of variables as the response variables or another feature set of variables as the predictor variables could be a challenge to fit the regression model. Thus, a summary statistics or a distance/dissimilarity measure is often performed with a multivariate regression method.

We can roughly divide regression-based hypothesis testing of association between the microbiome and its variables of interest into three approaches: (1) alpha and beta diversities-based association analysis; (2) count (or sequencing read counts/absolute abundance)-based association analysis; and (3) relative (compositional) abundance-based association analysis.

8.1 Alpha and beta diversities-based association analysis

The alpha and beta diversities belong to summary statistics. They are not metrics for measuring association directly. However, microbial diversities have been successfully applied to profile overall microbiome composition. The strategy of this approach is: first to calculate diversities or distance metrics to measure the phylogenetic or taxonomic dissimilarity between each pair of samples, and then to test the association between microbiome diversity or taxonomic distance/dissimilarity and an outcome variable of interest by a statistical method. This strategy is an aggregate-based or distance- or dissimilarity-based analysis. We here just refer to as alpha and beta diversities-based analysis for simplicity. The basic idea of using alpha and beta diversities prior to association analysis is to summarize data and to reduce the dimensions of taxa. We can consider the alpha and beta diversities-based analyses as indirect association analysis.

8.1.1 Classic statistical tests

Conducting statistical hypothesis testing of the association between microbiome and environmental covariates (e.g., treatment) as well as host factors is one important basic step in microbiome study.[194] Various alpha and beta metrics have been developed. The most often used alpha diversities are Shannon diversity, Chao 1 richness, and Simpson diversity. Depending on the number of groups for the testing and the distributions of the alpha diversities, typically Welch's *t*-test, Wilcoxon rank-sum test, or Kruskal-Wallis test will be used for the association test and followed by a multiple-comparison correction.[661]

8.1.1.1 Student's *t*-test

The "*t*-statistic" (abbreviated from "hypothesis test statistic") and *t*-test were introduced in 1908 by William Sealy Gosset who used "Student" as his pen name,[662] thus the name "Student's *t*-test" . A two-sample *t*-test is used to test that the means of two populations are equal. It is most commonly applied when the test statistic would follow a normal distribution. Student's *t*-test was used to test Shannon diversity between two groups.[663]

8.1.1.2 Welch's *t*-test

Welch's *t*-test,[664] an unequal variances *t*-test, is a generalized version of Student's *t*-test when several different population variances are involved. When the two samples have unequal variances and unequal sample sizes, Welch's *t*-test is considered as more reliable.[665] Welch's *t*-test was utilized to compare differences of OTUs abundance, taxa richness, and proteins.[666–669]

8.1.1.3 Wilcoxon rank-sum test and Wilcoxon signed-rank test

Wilcoxon rank-sum test and Wilcoxon signed-rank test were proposed by Frank Wilcoxon in a single paper.[599] Wilcoxon rank-sum test is used to compare two independent samples, while Wilcoxon signed-rank test is used to compare two related samples, matched samples, or to conduct a paired difference test of repeated measurements on a single sample to assess whether their population mean ranks differ. They are nonparametric alternatives to the unpaired and paired Student's *t*-tests (also known as "*t*-test for matched pairs" or "*t*-test for dependent samples"), respectively. The two nonparametric tests do not assume that the samples are normally distributed. The Wilcoxon unpaired two-sample test statistic is a technique equivalent to the statistic proposed by the German Gustav Deuchler in 1914. However, Deuchler incorrectly calculated the variance.[670] Wilcoxon formulated a test of significance with a point null hypothesis against its complementary alternative in his 1945 paper. However, in this paper the null hypothesis was only given for the equal sample size case and only a few points were tabulated (though Wilcoxon gave larger tables in a later paper). A thorough analysis of the statistic was provided by Henry Mann and Donald Ransom Whitney in their 1947 paper.[598] This is the reason that Wilcoxon rank-sum test is also called Wilcoxon-Mann-Whitney test and Mann-Whitney *U* test is equivalent to Wilcoxon rank-sum test. Wilcoxon rank-sum test and Wilcoxon signed-rank test were used to compare the median differences in alpha-diversity measures, proportion of core genera, and abundance of specific genera for categorical variables and variables in the case of matched samples, respectively, in the microbiome study by Falony et al.[671] Other examples of using Mann-Whitney U test or Wilcoxon rank-sum test in microbiome studies are provided in the reports.[272,533,606,672–675] For within-group comparison of alpha diversity generally Wilcoxon signed-rank test can be applied to analyze each pairwise within-group comparison of gut microbiota diversity (gene richness)[676] and relative abundance of microbial phyla.[672] Other examples of Wilcoxon signed-rank test use in microbiome studies can be found in these papers.[415,606,677,678]

Mann-Whitney *U* test and Wilcoxon rank-sum test are also often used to identify association between taxa or OTUs and covariates. However, these approaches conduct the association analysis based on the ranks of observed relative abundances, resulting in information loss and high false-negative rates.

8.1.1.4 One-way ANOVA

One-way ANOVA (analysis of variance) was proposed by Ronald Fisher.[679] ANOVA proposes the null hypothesis that all the means of compared groups are equal. ANOVA assumes that the underlying analysis data are normally distributed. However, most of microbiome community composition data, especially multivariate data, are not normally distributed; thus, the application of ANOVA needs to be careful in microbiome studies. In the case that the microbiome data are not normally distributed, either the nonparametric alternative Wilcoxon rank-sum test or other suitable statistical methods are applied. One example of ANOVA was used to identify significant differences in phylogenetic diversity and species richness indexes.[680]

8.1.1.5 Kruskal-Wallis test

Kruskal-Wallis test, proposed by Kruskal and Wallis in 1952, is a nonparametric method for testing whether samples are originated from the same distribution.[597,681] It extends the Mann-Whitney U test to more than two groups. The null hypothesis of the Kruskal-Wallis test is that the mean ranks of the groups are the same. As the nonparametric equivalent one-way ANOVA, Kruskal-Wallis test is called one-way ANOVA on ranks. Unlike the analogous one-way ANOVA, the nonparametric Kruskal-Wallis test does not assume a normal distribution of the underlying data. Thus, Kruskal-Wallis test is more suitable for analysis of microbiome data. Because the postsequencing microbiome data are often not normally distributed and contain some strong outliers, it is more appropriate to use ranks rather than actual values to avoid the testing being affected by the presence of outliers or by the nonnormal distribution of data.

Examples of Kruskal-Wallis test utilization in microbiome studies are available from the works.[415,682–685]

Pearson's correlation coefficient method can also be employed to evaluate the differences of the alpha diversity within and between groups of interest[194,686] by the Mann-Whitney U test or PERMANOVA.[687] For example, in skin microbiome studies, the Pearson's correlation coefficient was used to compare skin microbial diversity (Chao 1 richness) with host characteristics and environmental factors to detect the strength of a linear association between them.[688]

8.1.2 Adaptive microbiome α-diversity-based association analysis (aMiAD)

aMiAD[689] is an adaptive microbiome α-diversity-based association analysis method. The development of aMiAD was motivated by the performance of current α-diversity-based and microbial community-level association tests: on the one hand, current α-diversity-based association tests are item by item based. Because the nature of true association is unknown, it is sensitive and challenge to choose α-diversity metric and the test results are unpredictable.

Thus, the validity of smallest *P* value or the largest effect size picked from multiple item-by-item analyses is arguable due to the inherent multiplicity issue. On the other hand, most microbial community-level association tests (e.g., MiRKAT,[659] MiSPU,[690] and OMiAT[691]) only provide statistical significance, but do not estimate effect size and effect direction; hence, their practical use is limited. The goal of aMiAD is to simultaneously test the significance and estimate the effect size of the microbial diversity, as well as ensure the estimation to be accurate and valid while having high statistical power.

Two major steps have been implemented in the framework of aMiAD. First, aMiAD uses the score test[692] to formulate both linear and logistic regression models to test the association between each of the α-diversity metrics and a host trait while adjusting for covariates, respectively. Then, aMiAD adaptively takes the minimum *P* value from multiple candidate item-by-item α-diversity-based association analyses as its test statistic and estimates its own *P* value and microbial diversity effect size using a residual-based permutation method. In aMiAD, three nonphylogenetic α-diversity metrics and three phylogenetic α-diversity metrics are selected as the candidate α-diversity metrics because of their distinguished features which properly modulate abundance and phylogenetic information. The three nonphylogenetic metrics are: (1) Richness (also known as Observed), (2) Shannon index,[693] and Simpson index.[694] These nonphylogenetic metrics are based solely on abundance information, weight relatively rare, mid-abundant, and abundant species, respectively. The three phylogenetic metrics are: (1) phylogenetic diversity (PD),[695] phylogenetic entropy (PE),[696] and phylogenetic quadratic entropy (PQE).[697,698] These phylogenetic metrics are based on both abundance and phylogenetic information, weight relatively rare, mid-abundant, and abundant species, respectively. It was demonstrated by the author of aMiAD that aMiAD has better performance in terms of type I error rates and power in most conditions compared to multiple item-by-item α-diversity-based association analyses, and has better type I error control and better or comparable high power compared to three adaptive community-level association tests (OMiRKAT, aMiSPU, and OMiAT).

However, aMiAD also suffers some limitations: (1) the approach that aMiAD used is based on the minimum *P* value statistic and a residual-based permutation method. Although this approach has also been widely used in other studies,[659,690,691,699,700] its validity is still arguable and more researches are needed to evaluate the validity and applications in microbiome studies. (2) Currently aMiAD is limited to model linear and logistic regression because it assumes independent samples and hence not suitable for analysis of correlated microbiome data.[701] To model correlated or dependent (e.g., family-based or longitudinal) data, it needs to be extended to the linear mixed effects model[702] or generalized linear mixed effects model.[703] The paper was cited in the studies.[704,705]

8.1.3 Most widely used multivariate statistical tests in microbial ecology and microbiome studies

Mantel test, ANOSIM (analysis of similarities), PERMANOVA (permutational multivariate analysis of variance), and MRPP (multiresponse permutation procedures) are most widely used multivariate statistical tests of significance in microbial ecology and have been adopted in microbiome studies.[147]

8.1.3.1 Mantel test

Mantel test, a permutational testing procedure, was initiated by Mantel[706] and further developed by Mantel and Valand.[707] Similar to that of CANCOR, CIA, and PA, the goal of Mantel test is to test the correlation between two data matrices.[708] Mantel test typically compares two distance (dissimilarity) matrices and performs statistical testing significance of the linear relationship between matrices through a permutation testing of objects (samples). Mantel test has the advantage to be able to use different types of variables such as categorical, rank, or interval-scale data in the analysis because it uses a distance (dissimilarity) matrix as its input data. Mantel's Pearson test was used to identify the correlation between the metabolomics data and microbiome compositional data.[533] Other microbiome studies that used Mantel test can be found in these works.[709–711]

8.1.3.2 ANOSIM

ANOSIM test is simply a modified version of the Mantel test based on a standardized rank correlation between two distance matrices. It was developed by Clarke.[712] ANOSIM test is a nonparametric procedure for testing the hypothesis of no difference between two or more classes of objects (groups of samples) based on permutation test of among- and within-class (group) similarities.[712] ANOSIM ranks values based on their similarity/dissimilarity, which minimizes the effects of outliers in the data, making it useful for analysis of highly skewed data. Examples of applying ANOSIM in microbiome studies can be found in these reports.[713–715]

8.1.3.3 PERMANOVA

PERMANOVA is a nonparametric multivariate ANOVA method proposed by Anderson[687,716] to conduct hypothesis testing of no difference between two or more classes of objects (groups of samples) based on the analysis and partitioning sums of square distances. PERMANOVA obtains its significance levels (*P* values) through a permutational procedure. As ANOSIM, PERMANOVA may be implemented with any distance/dissimilarity metric, whereas classical ANOVA and multivariate ANOVA methods use Euclidean distance. PERMANOVA differs from ANOSIM in that instead of using the ranks values based on their similarity/dissimilarity, PERMANOVA uses the raw data. Examples of PERMANOVA use in microbiome studies are available from these works.[272,606,673]

8.1.3.4 MRPP

MRPP is a nonparametric procedure proposed by Mielke[717] to conduct hypothesis testing of no difference between two or more classes of objects (groups of samples) based on permutation test of among- and within-class (group) dissimilarities.[717,718] MRPP may test the differences in mean (location) or differences in within-class (group) distance (spread).[719] In both concept and method, MRPP is similar to PERMANOVA in that it is allied with ANOVA: comparing dissimilarities within and among classes (groups). MRPP shares the same underlying idea with PERMANOVA too. The test statistic of MRPP is based on the difference of weighted mean between- and within-class (group) dissimilarities.

8.1.3.5 MRBP

MRBP (blocked multiresponse permutation procedure) is a multivariate permutation method based on distance functions proposed by Mielke.[718] MRBP is a blocked version of MRPP[717] that focuses on within-group differences after accounting for block differences and was developed to conduct hypothesis testing of no difference between two or more groups.[718] The test statistic of MRBP was the average pairwise distance between blocks within each group. Similar to other permutational procedures such as PERMANOVA, MRBP generates its significance levels (*P* values) through permutation: the *P* value was determined by the proportion of all possible values of the test statistics that were less than or equal to the observed test statistic based on a Pearson type III distribution.[720] An effect size known as the chance-corrected within-group agreement (A) provided by MRBP is useful in microbial ecology and microbiome studies. A-value (range of [0, 1]) is calculated by comparing within-group homogeneity to random expectation with an A-value of 1 indicating that all components within a group are identical, while a zero occurs when within-group heterogeneity is greater than would be expected by chance.[721] MRPP was used to test the differences between groups for fecal bacterial communities[414] and fungal communities.[713] MRPP was even implemented in a pairwise manner to analyze the differences for within- and between-treatment groups.[722]

Various standard statistical methods, including nonparametric permutation tests of Mantel test, ANOSIM, PERMANOV, and MRPP, have been utilized to conduct association of beta diversity with environmental factors or groups of interest in microbiome studies.[147] Most popular beta diversity in microbiome studies are Bray-Curtis distance and unweighted/weighted UniFrac distances metrics.[194] The differences of the beta-diversity distances within and between groups of interest are often evaluated employing the parametric tests, such as the Pearson's correlation coefficient method.[194,686] Examples of MRBP use in microbiome studies to test for significance of group differences are available from these works.[414,723,724]

Although processing a correlation and/or association based on various diversity measures is still a way to identify the relationship between microbiome and host, this approach suffers from several challenges or limitations. For example, (1) overall reduction of multivariate taxa counts into a single number diversity measure is difficult to interpret.[725] (2) What alpha and beta measures are appropriate? Many different alpha and beta distance metrics have been developed, which are designed to measure different things. The challenge is: how to choose a particular metric to fit a particular dataset based on the research hypothesis. (3) The association analysis conducted via alpha and beta diversities does not directly regress the outcome of interest on the microbiome profiles. It is a secondary association analysis, which could lose information of microbiome taxa and hence the association test is less powerful.

8.2 Count-based association analysis

Microbiome sequencing read data are "really discrete count data."[726] Thus, we can conduct regression-based association analysis based on sequencing read counts or absolute abundance (AA), in which the counts could be either independent or outcome variables.

8.2.1 Differential abundance analysis

Simulation studies showed that count models are statistically more powerful to detect differential expression than approximate normal models.[169] Thus, the high-throughput sequencing data were advised to be treated as count data.[727,728] Differential abundance analysis of microbiome sequencing read counts adopts from differential expression analysis for RNA-seq data. A number of statistical tools were originally developed for differential analysis of RNA-seq data. Among these tools, these two packages edgeR[168] and DESeq (DESeq2)[171,172] have been most widely used for analysis of RNA-seq data and were suggested for use to identify differentially abundant OTUs directly.[173] Xia et al. introduced Poisson and negative binomial models as well as the packages in modeling high-throughput sequencing data and implemented microbiome data analysis using the edgeR and DESeq2 packages.[166]

Several challenges need to be addressed when modeling count, for example: (1) How to handle overdispersed and zero-inflated microbiome abundance data. The negative binomial (NB) methods implemented in the packages edgeR and DESeq2 may not perform well in identifying the differential abundant OTUs, as the counts at the OTU level are very sparse and the models cannot account for many zeros observed. Although the NB formulations in edgeR and DESeq2 are adjusted to ensure the over dispersion parameter locally to fit the unobserved heterogeneity of RNA sequence data, these approaches may not be appropriate if the overdispersion is due to excess zeros because they underestimate the probability of zeros and consequently underestimate the variability present in the outcome.[166] (2) How to model the high-dimensional structures of microbial taxa. (3) Differential abundance analysis cannot adjust for covariates to rule out the effects of potential confounders.

8.2.2 Classic overdispersed and zero-inflated models

NB models and the modified versions of NB models implemented in edgeR and DESeq2 may not be suitable for modeling the overdispersion due to excess zeros in microbiome count data. When such situations happened, the good alternatives are the classic zero-inflated/hurdle models. Based on my knowledge, it was Xu et al. in 2015 who first comprehensively assessed the performance of various zero-inflated and zero-hurdle models for analyzing overdispersed and zero-inflated microbiome data through an extensive simulation study and application to a gut microbiome data[729]; it was Wang et al. in 2016 who first used the hurdle model with NB distribution for analyzing species (97% similarity threshold OTUs) based on Xu et al.'s evaluation.[50] The model comparisons conducted by Xu et al. are wide including standard parametric and nonparametric models, Poisson hurdle (PH) and negative binomial hurdle (NBH) models, and zero-inflated Poisson (ZIP) and Zero-inflated negative binomial (ZINB) models. The assessments are very comprehensive: covering different perspectives such as type I error, power of the test, the precision and efficiency of parameter estimations for the covariate effect on both the counts and the (structural) zeros, the goodness of fit, and the relative bias of prediction for zeros. Thus, it deserves to summarize the main findings regarding the overdispersed, zero-hurdle, and zero-inflated models here.[729]

(1) NB model could deflate the overall type I error rates and may be prone to reduced power. NB also underestimates the probability of the zeros. For the covariate effect, NB model has the pattern of estimation bias varying across different scenarios. This model also is unstable for a high degree of zero inflation. (2) The zero-hurdle and zero-inflated models have well-controlled type I error rates and consistently better power to detect the significance of the overall covariate effect across different proportion of zero inflation. They also show unbiased estimation of the probability of the zeros. (3) For the covariate effect on the log scale of count data, the performance of the zero-hurdle and zero-inflated models is consistent across different covariate effect scenarios and different degrees of zero inflation; for the covariate effect on the probability of (structural) zeros, zero-hurdle models give unbiased and stable estimates in all simulation scenarios. ZINB has unbiased estimation for both ZIP and ZINB distributed data, while ZIP is only unbiased for ZIP distributed data. ZINB is unstable when the proportion of zero inflation is low, and it gets more stable when either zero inflation proportion increases or sample size increases. (4) For model selection, zero-hurdle and zero-inflated models in general have smaller AICs than one part NB model. The AIC values produced by NBH and ZINB model are very close (if not exactly identical) and are the smallest among all fitted models. The Vuong test favors ZINB over NB.

The main findings summarized above are consistent with previous studies. For example, results regarding to zero-inflated models confirmed those from Xia et al. in 2012 who compared four competing statistical models for count responses (i.e., Poisson, NB, ZIP, and ZINB),[730] and Feng et al. in 2015 who specifically conducted theoretical comparisons of NB and ZIP distributions.[731] Xia et al. in 2018[174] comprehensively introduced and discussed the zero and sparsity issues in microbiome and other omics data: zero sources, overdispersion due to zero-inflated, concept adjustments for use of zero-inflated and zero-hurdle models in microbiome studies and then illustrated how to model zero-inflated and overdispersed microbiome data including zero-inflated Poisson (ZIP), zero-inflated negative binomial model (ZINB), zero-hurdle Poisson (ZHP), and zero-hurdle negative binomial (ZHNB) using the edgeR and DESeq2 packages.

One big problem of classic overdispersed and zero-inflated models is that they treat the taxa as independent, which ignores the compositionality of microbiome data.

8.2.3 Dirichlet-multinomial models and their extensions

As we presented above, typically because the majority of taxa can be observed in only a very small subset of samples which causes the data table to be highly sparse and the within-group heterogeneity among samples leading to pronounced overdispersion in taxa proportions. Dirichlet-multinomial (DM) models are alternative approaches to analyze highly sparse and often overdispersed microbiome data. DM was originally proposed by Mosimann[732,733] and was introduced into the microbiome context by Holmes et al.[734] and La Rosa et al.[121] The framework of DM models is multivariate setting. DM in nature is suitable for analyzing multiple response variables simultaneously because we can assume that a vector of taxon counts follows the multinomial distribution with underlying proportion parameters sampled from a Dirichlet distribution. Many of the existing methods are nonparametric and hence are widely applicable, but they are limited in interpretability and can be inefficient for analyzing microbiome data. Compared to the multinomial distribution, the DM has been shown more appropriate to fit microbiome data, and hence, the DM framework has been adopted for development of new statistical methods.[121,735–739]

However, the DM and specially the original setting of DM may not be adequate for microbiome data because the DM model intrinsically has the limitations to address the issues arised from targeting characteristics of microbiome data. For example, (1) DM in its originality lacks a component or function to link to covariates or predictors to form a hypothesis testing of the association between environmental covariates (e.g., treatment effects) and microbiome: either considering microbiome as response variables (e.g., diet alters the human gut microbiome) or treating microbiome as predictor variables (e.g., dysfunctional microbiome leads to disease). (2) DM model is unable to address the compositionality instead intrinsically imposes a negative correlation among taxon counts[740–743]; thus when the taxa present both negative and positive correlations, the DM model is not adequate for characterizing microbiome data. (3) In addition to not accommodating complex correlation among taxa, DM is also not flexible enough to accommodate the high level of zero inflation, thus limiting the power of the DM-based test.[740,743,744] (4) DM model has only one dispersion parameter and thus is not flexible to handle various dispersion patterns and zero-inflation levels among multiple taxa.[745] (5) DM model has a limited number of parameters to adequately model the variances and covariances of the composition.[746]

To address the modeling issues along with other drawbacks of the original version of DM as well as to target specific characteristics of microbiome data, various DM extensions including those following to be introduced have been currently proposed. Even a Dirichlet-multinomial framework for multivariate count outcomes in genomics was proposed.[747] Currently Harrison et al.[748] conducted a comprehensive simulation to compare several alternative models including ALDEx2,[37] ANCOM,[177] DESeq2,[172] edgeR,[168] mvabund,[749] and repeated Wilcoxon rank-sum tests with a Benjamini-Hochberg false discovery rate (FDR) correction.[750] It demonstrated that DM modeling is likely to be broadly useful and sensitive for analyses of microbiomes, other DNA barcoding, gene expression, metabolomics, and other applications in molecular ecology. Thus, we expect that DM model will be getting more applications in microbiome and other omics data and more extensions of DM model will be developed. In the following section, we will introduce some extensions of DM model in various directions.

8.2.3.1 Extension #1: Reformulate and reparameterize DM model to make it suitable to perform hypothesis testing the association of microbiome data

DMM (Dirichlet multinomial mixtures), a parametric probability model, was introduced by Holmes et al.[734] to analyze microbial metagenomics data. DMM model is formulated in multivariate multinomial framework that can account for the discrete nature, sparsity, and variable size of the sample datasets. Under multivariate multinomial framework, the unobserved community parameter vectors are presented by the probability generated by the model; DMM model takes the advantage of the Dirichlet prior and extends the Dirichlet prior to a mixture of Dirichlets[751–753] so that making DMM model can fit the dataset with a mixture of multiple metacommunities rather than a single metacommunity. In other words, DMM model has the flexible feature as Dirichlet model and also has a means to cluster communities.[734] Based on Holmes et al.'s DMM models, La Rosa et al. further proposed a multivariate statistic method[121] for hypothesis testing and power calculations of taxonomic-based human microbiome data. Within the DMM framework, both location (mean) and scales (variance/dispersion) are reparameterized to make the model suitable to perform hypothesis testing for comparing microbiome populations between two or more groups of subjects using the Wald-type test. This parametric test actually is an independence test between a categorical variable and the microbiome composition. However, except for the general limitations suffered from DM model, this test has some specific drawbacks including: first, the homogeneous dispersion assumption for differential mean levels between groups can be deviated in many scenarios.[745] Second, this test can be underpowered when analyzing the categorical variable with too many levels and it is not applicable to a continuous variable.[754] Examples of using DMM for hypothesis testing and power calculations in microbiome data are available in these studies.[671,755,756]

8.2.3.2 Extension #2: Impose a sparse group L_1 penalty to DM model in variable selection to account for overdispersion of taxa abundance

Variable selection for sparse DM regression was proposed to address overdispersion of microbiome data.[735,736] The proposed method takes the capability of DM regression model using the likelihood ratio test to identify the association between taxa composition and covariates, and imposes a L_1 penalty in variable selection to make DM model to be able to account for overdispersion. The proposed method and DM model in general were criticized to be inappropriate for microbiome data[177,746,750] because DM models intrinsically impose a negative correlation among every pair of taxa,[732] which is in contrast to the fact that microbiome data display both positive and negative correlations. This paper has been cited in more than 100 publications. Examples of using this method can be found in these works.[52,53,176,757]

8.2.3.3 Extension #3: Re-equip DM model with the phylogenetic tree to capture the local signal and evolutionary information of taxa to make better inferences about associations between multivariate taxa (OTUs) and covariates (outcomes) of interest

Although DM and its extensions of DMM models are flexible to perform hypothesis testing across groups, however, the testing suffers from many drawbacks including inability to localize any signal to a subgroup of taxa and reduced test power when a large number of taxa are present.[758] Wang and Zhao[738,739] extend DMM to the Dirichlet-tree multinomial (DTM), which was first proposed by Dennis[759] with the name "hyper-Dirichlet type 1 distribution."[759] DTM does not place a single global DM on all taxa instead treats the whole taxa as a collection of independent local DMs, each corresponding to a particular internal node on the phylogenetic tree.[738,739] Tang et al.[758] further introduce the phylogenetic scan test (PhyloScan) for investigating cross-group differences in microbiome compositions using the DTM model. The PhyloScan captures the evolutionary information of taxa over the phylogenetic tree to allow nodes to borrow signal strength from their parents and children. The benefits of extending the traditional DM onto phylogenetic trees are: (1) it provides greater flexibility to naturally incorporate sequencing depth, overdispersion, and is easily adapted to deal with localized signals[745]; (2) it more accurately links clades in compositional taxonomic data to covariates[760]; and (3) it is more effective to detect phenotype-microbiome associations[758] and in prediction accuracy.[738,739] However, as DTM takes absolute abundance modeling approach, overdispersion and zero-inflated and high-dimensional structures need to be appropriately and fully addressed in this case.[761]

8.2.3.4 Extension #4: Re-equip DM model to model zero-inflated microbiome compositional data

ZIGDM (zero-inflated generalized Dirichlet multinomial) was developed by Tang and Chen.[745] The goal of ZIGDM is to handle excessive zero observations in taxon counts and flexibly accommodate complex correlation structures and dispersion patterns among taxa in modeling multivariate taxon counts.

Two strategies were used in the development of ZIGDM regression model: (1) It added additional parameters to the DM model for flexibly accommodating the over-dispersion and zero-inflation of the data. Thus, ZIGDM was developed so that both mean and dispersion levels of the taxa abundance could be linked to the covariates of interest as well as a hypothesis association testing could be performed to detect both differential mean and dispersion. (2) It generalized the generalized DM (GDM) distribution[762] by not only allowing more general covariance structure but also using a fast expectation-maximization (EM) algorithm to estimate the parameters so that make ZIGDM suitable for modeling microbiome data and handling excessive zeros in taxon counts. It was demonstrated that both ZIGDM and GDM models outperform the DM model in terms of detecting differential mean/dispersion and robustness to the underlying distributions.[745] Example of ZIGDM use can be found in the paper.[763]

8.2.3.5 Extension #5: Re-equip DM model with mixed effects to fit longitudinal microbiome data

Mixed effect DTM (mixed effect Dirichlet-tree multinomial) model[764] was proposed to extend DM model's functionalities in naturally incorporating sequencing depth, overdispersion, and dealing with localized signals. The main works done in this development include: (1) extending the DTM framework to longitudinal data setting with mixed effect for modeling covariates; (2) using empirical Bayes shrinkage on each node to enhance inference of taxon proportions; and (3) applying random forest along with covariate information for prediction.[765]

8.2.3.6 Extension #6: Re-equip DM model with Bayesian technique

Bayesian DM (Bayesian Dirichlet-multinomial regression model)[737] was proposed to identify significant associations between potential covariates (environmental predictors) and taxa from a microbiome abundance table using spike-and-slab priors. Bayesian DM uses a log-linear regression parameterization of the Dirichlet-multinomial likelihood parameters to directly incorporate the covariates and implement a Markov Chain Monte Carlo (MCMC) algorithm for posterior inference. Bayesian DM model was used to evaluate interactions and shift in fecal bacterial communities over time.[766]

mLDM (metagenomic Lognormal-Dirichlet-Multinomial)[767] was proposed to use a hierarchical Bayesian model (lognormal-Dirichlet-multinomial) on the compositional counts for inferencing associations between environmental factors and taxa and among taxa. This model imposes sparsity constraints to estimate a sparse inverse covariance matrix between taxa through maximizing the L_1 penalized posterior distribution. Thus, this model can estimate absolute taxa abundance and simultaneously infer both conditionally dependent associations among taxa and direct associations between taxa and environmental factors. Considering the conditional dependency structure between taxa to distinguish between direct and indirect associations is one main advantage of mLDM method. This method also allows for measured covariates to be included as network nodes.[768] However, the mLDM method has been criticized as: (1) cannot address unmeasured sources of heterogeneity (i.e., latent factors)[769]; (2) introduces too many parameters, which results in limiting its scalability and efficiency to dimensionality cases.[770] The relative abundance approach was discussed in this study.[771]

BDMMA (Bayesian Dirichlet-multinomial regression meta-analysis)[53,54] is a meta-analysis method that simultaneously models the batch effects and detects the association between microbial taxa and covariates (e.g., phenotypes). The goal of BDMMA is to correct batch effects while considering the microbial taxa interactions and the overdispersion of the microbiome data. BDMMA was developed by adopting the Bayesian approach: impose a spike-and-slab prior in its posterior estimation to select the taxa significantly associated with the covariates of interest. BDMMA automatically models the dependence among microbial taxa and is robust to the high dimensionality of the microbiome and their association sparsity.[53,54] BDMMA was reviewed in these papers.[57,772]

8.2.4 Kernel-based methods for association tests

In recent years in microbiome and omics studies, the developments and applications of distance/kernel-based methods for association tests have been increasing due to their capabilities of handling multidimensional data and covariates. For example, kernel-based methods are robust for testing mixture effects (positive and negative) of genetic variants and the rare variants, and flexible for adjusting covariates when fit within the framework of regression models (e.g., mixed models).[773] Due to their flexibility and computational feasibility, kernel methods are particularly appealing for microbiome and large-scale genetic studies. The recent advances of kernel methods, particularly hypothesis testing, various traits of interest, and underlying kernel design, were reviewed in the article.[773]

The listed software tools in this review article are mainly based on kernel statistics for genetic analyses including: sequence kernel association test (SKAT),[423,774,775] gene-centric gene-gene interaction with smoothing-spline ANOVA (3G-SPA),[776] implementation of gene-environment set association test (GESAT) for GxE interaction kernel testing (iSKAT),[777] computing kernel and burden statistics for pedigree data based on retrospective likelihood (pedgene),[778] family-based rare variant association test (FARVAT),[779] family-based association kernel test for both rare and common variants (famSKAT-RC),[780] boosting the power of SKAT by properly estimating the null distribution of SKAT (iGasso),[781] implementing various statistical methods for testing multitrait variant set association (MSKAT),[782] recalibrated lightweight SKAT (RL-SKAT) with small-sample adjustment,[783] implementing small-sample adjusted kernel machine association tests for univariate, multivariate, and correlated outcomes (SSKAT),[784–786] computationally efficient calculation of SKAT P values for large data (fastSKAT),[787] gene-based association test combining SKAT-type methods for complex traits and their corresponding optimal association tests between a set of genetic variants and familial, multivariate, longitudinal, or survival traits (KMgene),[788] and association test between microbiome composition and a continuous, dichotomous, multivariate, survival, and structured outcome (MiRKAT).[659,785,789]

Recently, several multivariate kernel machine regression-based association tests have been proposed using relative abundance. The kernel machine regression framework has been extended in several paths to test the association between the outcomes and the microbiome community. We introduce them below.

8.2.4.1 Extension #1: Extend a single kernel to multiple kernels to optimally select (dis)similarity measures into association test

MiRKAT (microbiome regression-based kernel association test) is a semiparametric kernel machine-based method to directly test the association between the outcome variable and the microbiome community or community diversity while adjusting for covariates.[659] MiRKAT was previously developed based on the kernel machine regression framework for genotyping data.[735,736] MiRKAT can work on both continuous and dichotomous outcome variables through using the linear and logistic kernel machine models, respectively. The association is suggested by the high correspondence when the pairwise similarity in the outcome variable is compared to the pairwise similarity in the microbiome profiles. MiRKAT on a single Kernel base can construct the kernel matrix via transformation of the phylogenetic or taxonomic distance metrics such as weighted or unweighted UniFrac, and Bray-Curtis metrics. While optimal MiRKAT, which is based on multiple kernels, extends MiRKAT to simultaneously consider multiple possible kernels, such as unweighted UniFrac, weighted UniFrac, generalized UniFrac, and Bray-Curtis kernels. The advantages of MiRKAT method[659] lie on: (1) MiRKAT is closely related to existing association approaches such as the PERMANOVA method, while it is more flexible to adjust confounding variables; (2) the optimal MiRKAT is able to incorporate diverse distance-based measures or multiple candidate kernels simultaneously, which is robust to poor kernel choice and has reasonable power to facilitate the results interpretation when individual distance metrics yielded disparate results. However, same as other microbial community-level association tests such as PERMANOVA,[716,790] MiSPU,[690] OMiAT,[691] and aMiAD,[689] MiRKAT assumes that samples are independent. This was criticized to be not suitable for correlated microbiome studies.[701]

8.2.4.2 Extension #2: Optimally combine and adapt both sum of powered score tests (SPU) and MiRKAT to select (dis)similarity measures into association test

OMiAT (optimal microbiome-based association test) is a data-driven microbiome-based group association testing method.[691] OMiAT is optimal because this test takes through diverse tests from both the sum of powered score tests (SPU) and microbiome regression-based kernel association test (MiRKAT). As we stated above, attractiveness of MiRKAT is its capability to incorporate diverse distance-based measures which results in robust test results. The SPU framework[700] was developed based on two different versions of the generalized taxon proportion, unweighted and weighted generalized taxon proportion.[690] Thus, the SPU framework has the advantages[690]: (1) using these two newly defined suitable measures for discovering rare and common/abundant taxa, respectively; (2) the data-driven approach of MiSPU, adaptive MiSPU (aMiSPU) comprehensively takes highly adaptive test and uses the variable selection/weighting of the SPU framework based on the two generalized taxon proportions. Thus, this method is robust and powerful. Due to concerning the accuracy of the generalized taxon proportions, the SPU used for OMiAT is based on standardized compositional data with no phylogenetic information incorporation.[691] To incorporate OMiAT, MiCAM (microbiome comprehensive association mapping) was proposed to discover microbial taxa through hierarchically testing all taxa and applying multiple testing correction per taxonomic rank. OMiAT was developed within a generalized linear model framework so that it is flexible to handle different types of outcome variables (e.g., continuous and binary outcome variables) and adjust for potential covariates. Under a generalized linear model framework, OMiAT was formulated as this way: first, fitting a multiple linear and logistic regressions for continuous and binary outcome variables, respectively, and to relate taxa (OTUs) with an outcome variable while adjusting for covariates. Second, performing a score test. Third, implementing the SPU method on the standardized compositional data to sum individual score components to be powered with diverse choices of value. Fourth, incorporating MiRKAT and optimal MiRKAT methods as well as driven by the data to choose diverse distance-based measures, including Bray-Curtis dissimilarity, unweighted UniFrac, weighted UniFrac, and generalized UniFrac measures with different parameter values. Then, taking the minimum P value from all the score tests for SPU and MiRKAT as its test statistic and reporting the P values that estimated from the test statistic and calculated for the test statistics using a permutation-based method. Finally, building the microbial taxa discovery framework MiCAM to finc-map diverse microbial taxa from the highest to the lowest taxonomic rank. Importantly, the false discovery rate (FDR) is controlled by the Benjamini-Hochberg method.[485,791,792]

OMiAT has several advantages including: (1) due to taking an optimal test from all different tests leveraging the combined strength of SPU for varying microbial abundances and MiRKAT for different relative contributions from microbial abundance and phylogenetic information; thus, OMiAT is highly robust and powerful; (2) OMiAT estimates *P* values for the aggregated data using the score test statistic and permutation-based method. This semiparametric approach avoids the limitations of other aggregate-based methods.[691] For example, the nonparametric Kruskal-Wallis test[597] used in LEfSe[596] and STAMP[793] was designed for one-way layout data structure, which is suitable for univariate analysis, and thus, it is difficult to handle covariate adjustments (e.g., environmental factors). Moreover, this method limits its use for categorical variables, not suitable for analyzing continuous outcome variables. The assumption that the testing taxa are independently used in the parametric methods of DESeq2[172] and ZIG[48] may not be validated due to the issue of relative abundance, which can result in inflated type I error rates. (3) The FDR controlled by the Benjamini-Hochberg method is considered as valid robustly whenever the multiple tests are independent or correlated in various scenarios.[794] (4) The simulation study showed that the type I error rates of OMiAT were mostly well controlled ($\leq \sim 5\%$) as optimal MiRKAT, aMiSPU, and the aggregate-based methods in both linear and logistic models. (5) OMiAT is clearly more powerful than the other methods under most of the scenarios and is highly comparable to optimal MiRKAT in situations where abundant OTUs are associated, and to the aggregate-based method in situations where rare OTUs are associated. (6) Real studies showed that OMiAT discovered the greatest number of taxa. However, as MiRKAT, OMiAT assumes that samples are independent, which was criticized to be not suitable for microbiome association studies with related outcomes.[701,786]

8.2.5 Adaptive methods for association tests

aMiSPU (its R package called *MiSPU*)[690] is a multivariate association testing method based on multiple phylogenetic distance measures to conduct an overall association test between the composition of a microbial community and an outcome of interest. The existing multivariate distance- or dissimilarity-based tests (e.g., PERMANOVA) and the microbiome regression-based kernel association test (MiRKAT), which is closely related to distance-based methods (e.g., PERMANOVA), either need a prior to choose a specific distance measure to measure the dissimilarity between each pair of samples, or need to choose kernels by the end user. And more importantly, both distance- or dissimilarity-based and kernel-based tests do not implement automatic taxon selection or weighting.[690] The goals of MiSPU test and its adaptive version (aMiSPU) are to jointly and highly adapt over all observed taxa, and thus to obtain more power across various scenarios, alleviating the issue with the choice of a phylogenetic distance.

The input data of MiSPU and aMiSPU tests are a rooted phylogenetic tree, a sample of OTU counts, an outcome of interest, and possibly some covariates. The framework of MiSPU and aMiSPU tests consists of three key steps: Step #1: calculating a generalized taxon proportion for each taxon; Step #2: calculating the test statistics; and Step #3: applying a residual permutation scheme to obtain the *P* values. The benefits of MiSPU tests and aMiSPU[690] are: (1) introducing variable selection or variable weighting into association tests to leverage the effects of relative abundances of microbial taxa and that of branch lengths in a phylogenetic tree through using the two versions of the generalized taxon proportion. (2) The proposed aMiSPU test can be used to select and rank the important taxa, whereas MiRKAT and other distance-based methods have no such ability. However, the approach of combining OTU-specific tests with a global test is often criticized to have poor performance because many of the OTU-specific tests only contribute noise.[691,795] The adaptive approach uses the minimum *P* value of multiple tests as test statistics to maintain robust performance. Although it has also been widely used in other fields,[796] such as gene-based tests for rare variants[700] or Single Nucleotide Polymorphisms (SNPs),[797] pathway-based tests,[798] the assumption that the minimum *P* value is robust or better is arguable. A common misleading thought behind adaptive approach is that adapting means to take the smallest *P* value rather than to take the best fitting measure. Additionally, the proposed methods treat OTUs or taxa as independent instead of compositional, which is inconsistent with the nature of microbiome data. Example of MiSPU use can be found in this study.[799]

Adaptive independence test[754] is an adaptive parametric method based on likelihood ratio test by learning a good partition or slicing scheme from the data. It was motivated by the inability of testing the association between the community composition and a continuous variable using DM parametric test proposed by La Rosa et al.[121] and the lack of interpretability and inefficiency of permutations-based multivariate nonparametric methods for analyzing microbiome data. Thus, the goal of proposed method is to test for independence between the microbial community composition and both categorical and continuous variables while providing a meaningful biological interpretability. The adaptive independence test uses the Dirichlet-multinomial (DM) distribution to account for multivariate and overdispersed count data. It is distinct from DM model in that: the proposed method first partitions the range of the variable into a few slices, and hence formulates the testing problem as a problem of comparing multiple groups of microbiome samples, with each group indexed by a slice. Then it regularizes the log-likelihood ratio by penalizing slices over all possible slicing schemes. Finally, it applies an adaptive test to adapt slices.[754] It was demonstrated that the proposed method outperforms La Rosa et al.'s DM method and other approaches such as PERMANOVA, the distance-covariance test, and the microbiome regression-based kernel association test. The authors of this method[754] also discussed three limitations of this proposed approach: (1) failing to consider phylogenetic information. (2) The DM model that this method based on only allows model negatively correlated components due to its dependence structure, so when presenting both negative and positive correlations, the model is not adequate to characterize microbiome data. (3) The proposed method is not able to handle excess zeros, which is an intrinsic feature of sequencing microbiome data.

8.2.6 Other absolute abundance-based association analyses

8.2.6.1 Microbial association mapping (massMap)

massMap[800] is a two-stage microbial association mapping framework. It was developed to strengthen statistical power in association tests to detect taxa at the target rank through grouping information from the taxonomic tree. The target rank usually refers mapping outcome variables to taxa at the lowest definable taxonomic rank, such as genus or species.[800] The massMap framework is built by three components. The massMap is processed in two stages.

In the first stage, massMap screens a preselected higher taxonomic rank (component #1) and uses the microbial group test OMiAT to identify the taxonomic groups that contain the associated taxa (component #2). At the screening rank, the group association test is conducted to examine the association between each group of taxa and the outcome variable adjusting for covariates. A linear or logistic regression model is used for a continuous or binary outcome, respectively. *In the second stage, massMap tests the association for each candidate taxon at the target rank within the significant taxonomic groups.* At the target rank, similarly a linear or logistic regression model is used for a continuous or binary outcome, respectively. However, instead a group association test is conducted between each group of taxa and the outcome variable as at the screening rank; this time the association test is conducted between each taxon and the outcome variable. As in OMiAT, the association between each taxon and the outcome variable is tested by the nonparametric score test statistic and P values of statistics is calculated using the residual-permutation method.

In the two-stage structured tests, both Hierarchical BH (HBH)[801,802] and selected subset testing (SST) procedures[803,804] are conducted to resolve the dependency among taxa (component #3).

The massMap framework has several advantages[800] including: (1) massMap uses two advanced FDR-controlling procedures: the hierarchical BH (HBH) and the selected subset testing with BH (SST) to accommodate the hierarchically structured hypotheses. The merit of HBH is its capability of discoveries,[801,802] but sometimes has higher FDR than the nominal level, whereas SST is more conservative than HBH and can control the FDR at the desired level if tests between two stages are independent. Additionally, the application of SST procedure in microarray data analysis has showed it permits greater discovery than the traditional BH procedure.[805] Thus, combining HBH and SST procedures could enhance the association mapping power. (2) Because of using two-stage strategy which includes tree information, and incorporating OMiAT and the advanced FDR controlling methodologies, massMap largely alleviates the multiplicity issue and hence has more power than the traditional one-stage association method. Because both HBH and SST reach the highest power at screening family rank with OMiAT and the results for HBH and SST are similar, conducting the screening at the family rank using massMap was recommended. (3) Real data studies showed that massMap more efficiently detects more biologically meaningful taxa than the traditional BH method with much smaller FDR-adjusted *P* values. The massMap paper was cited in the research papers.[749,806]

8.2.6.2 Logistic normal multinomial (LNM)

LNM (logistic normal multinomial) regression model[746] was proposed to identify the association between environmental/biological covariates and bacterial taxa. The goal of LNM is to account for dispersion and excess zeros of taxa count data in microbiome studies. Both multinomial logistic (ML)[807,808] and DM regression models can handle multivariate count or compositional data. However, DL model does not allow for overdispersion of count data, while DM model can allow for overdispersion, but suffers several drawbacks as we cited above such as: lacking sufficient parameters for analysis of variances and covariances of the composition; mean values of Dirichlet variates cannot be independently determined due to dependence structure between Dirichlet variates; and particularly Dirichlet variates are always negatively correlated, which is not consistent to microbiome data in nature. The strategies of LNM are to leverage the strength of the ML model to link covariates with taxonomic counts, and take the advantage of a group L_1 penalty function in variable selection to develop a group-penalized likelihood estimation procedure for the LNM model. However, LNM method for estimating the dependence structure of taxa was criticized to be infeasible for high numbers of unique taxa, and only suitable for small collections of taxa[809] as well as cannot exploit the tree structure information.[739]

8.2.6.3 Inference for absolute abundance (IFAA)

The development of *IFAA*[810] was motivated by avoiding two problems which use relative abundance (RA) for inference in microbiome analyses: (1) the RA is calculated by dividing absolute abundances (AA) over the common denominator (CD), the summation of all AA (i.e., library size), which results in compositional structure of RA; (2) the confounding effect of library size and zero-inflated data structures. The authors of IFAA showed that IFAA can handle high-dimensional microbiome data and high-dimensional covariates data due to its incorporating regularization methods. Thus, IFAA can be used to obtain the robust association identification of human microbiome with exposure variables and clinical outcomes. The distinctive features of IFAA lie on: drawing inference directly on the AA of microbial taxa in an ecosystem instead of the RA, and without requiring imputing zero. The algorithm behind IFAA is to use two-phases modeling strategy based on the ratios of nonzero AA observed in the samples: first, it uses Phase 1 to identify the taxa whose AA are associated with the covariates of interest and then uses Phase 2 to estimate the association parameters. The simulation study carried out by the authors of IFAA showed that IFAA has a best performance to identify association in terms of precision rate, recall rate, and type I error rate compared to the established existing approaches: ANCOM,[177] DESeq2,[172] edgeR,[168] Wilcoxon rank-sum test, and ZIG.[48] Real data studies showed that IFAA could detect the associated taxa or OTUs, while ANCOM could not detect any taxa or OTUs at the same FDR rate in the datasets. It also showed that the nonparametric Spearman's correlation test is more likely to overidentify the taxa or OTUs. However, given IFAA focuses on studying the association of nonzero taxa with exposures, it ignores analysis of the presence/absence of the microbial taxa; not to say differentiate structural zeros and sampling zeros.

8.3 Relative (or compositional) abundance-based association analysis

In relative abundance approach, regression is conducted based on relative abundance (RA): the proportion of microbial taxa observed in the sample.

RA could be used as either independent or outcome variables. Typical compositional data analysis approach belongs to relative abundance-based approach.

Compositional-based association analysis not only conducts regression based on RA, but also further tries to solve the compositionality by log-ratio transformation of compositional variables to ensure standard statistical methods can be used. To avoid a zero relative abundance or a log-zero, in practice, the compositional-based approach usually replaces a zero count by a small pseudo count[807] (e.g., an arbitrary value of 0.5 or 1) or a small random count generated from an appropriate probability distribution prior to the relative abundance normalization and log-ratio transformations[39] (p. 389).

The fundamental theoretical works started with Aitchison's compositional research[36,807]; the present methods in applications of microbiome and omics data are ALDEx2[41] and ANCOM.[177] We comprehensively reviewed and introduced the "Compositional Analysis of Microbiome Data" in Chapter 10 of the book[811] (pp. 331–396). Here, we introduce four newly developed compositional-based association analysis models based on relative abundance microbiome data. The first is zero-inflated beta regression, which is a univariate association method; the second is multivariate two-part zero-inflated logistic-normal model; the third is adaptive multivariate two-sample test for microbiome differential analysis; and the fourth is the robust regression with compositional covariates. We also briefly discuss the challenges of inferencing microbiome data based on relative abundance.

8.3.1 Zero-inflated beta regression (ZIBSeq)

ZIBSeq,[812] a zero-inflated beta regression approach for differential abundance analysis of metagenomics data, was proposed to identify differentially abundant features between multiple clinical conditions while considering the features of metagenomics data with small sample size, high dimensionality, sparsity, often with a large number of zeros and skewed distribution under the compositions (proportions) setting.

ZIBSeq consists of four steps:

Step 1: Feature screening. Remove any features with total counts less than 2 times larger than the sample size.

Step 2: Data normalization. ZIBSeq uses a simple normalization procedure to convert the raw abundance measure to a proportion by dividing each feature read count by the total feature read counts in the sample, which results in relative abundance measure ranging [0, 1]. After normalization, a square root or cube root transformations are performed to ensure that the proportion data are better fitting a beta distribution if the distributions of the proportion are extremely left skewed.

Step 3: Zero-inflated beta regression. Perform zero-inflated beta regression[813] to predict each normalized feature (response variable) with outcome (explanatory variable); the P value of the regression coefficient in each regression is obtained.

Step 4: Multiple hypothesis testing correction. Use the FDR algorithm proposed by Storey and Tibshirani[814] to estimate a conservative q value based on P values obtained in Step 3 under the assumption that P values are uniformly distributed.

It has been shown[812] that ZIBSeq has better performance in terms of large AUC values and outperformed zero-inflated Gaussian (ZIG) model[48] as well as can identify biologically important taxa in a real microbiome data application. However, although ZIBSeq method can handle zero-inflated proportion data, it cannot deal with repeatedly measured proportion data or longitudinal data.[815] Also these kinds of analyzing relative abundance approach of individual taxa one by one at a time with a multiple testing correction procedure to control for type I error rate were criticized as being not able to incorporate the intertaxa correlation[761,816] and cannot provide P values and correct statistical inferences for the selected taxa.[816] This kind of two-part beta regression model was also criticized that cannot provide a straightforward interpretation of covariate effects on the overall marginal (unconditional) mean.[817] Examples of ZIBSeq use can be found in these publications.[818,819]

8.3.2 Multivariate two-part zero-inflated logistic-normal model (MZILN)

MZILN regression model[761] was proposed to analyze the association between covariates (e.g., disease risk factors) and individual microbial taxa and overall microbial community composition. MZILN distribution was developed based on the multivariate logistic-normal distribution[36,807] to account for the zero-inflated structure for the relative abundance of microbiome data. The goal of this method is to appropriately address the issues arised from the characteristics of microbiome data: excess zeros, high dimensionality, the hierarchical phylogenetic tree, and compositional structure.

The proposed method has three important components: (1) uses a zero-inflated logistic-normal model to handle the zero-inflated data structure and the compositional structure; (2) borrows the estimating equations approach from GEE[820] to address the intertaxa correlation structure induced by the hierarchical phylogenetic tree structure and the compositional data structure; and (3) incorporates regularization approaches such as LASSO,[192] SCAD,[380] and MCP[821] to address high dimensionality of the data. The simulation study demonstrated that MZILN model outperformed the established existing approaches[761]: the sparse Dirichlet-multinomial (DM) regression,[736] kernel-penalized regression (KPR),[822] zero-inflated beta (ZIB) regression,[812] and Spearman's (SP) correlation test in terms of recall rate, and F1 score as well as precision rate except when data sparsity level is high where ZIB has higher precision rate. However, ZIB does not provide effect size estimates, and consequently cannot identify the direction of association. In this sense, MZILN is also superior to ZIB model. A real study found that MZILN identified more genera than ZIB, SP, and Wilcoxon rank-sum test, and less genera than DM as expected. MZILN had good overlap with DM. However, MZILN treats the zero-part parameters as nuisance parameters and does not consider the zero part in the estimation.[761] This is the main limitation of this method and somehow is not consistent with name of "zero-inflated logistic-normal model." Also although MZILN explicitly incorporates zero-inflated logistic normal distribution to simulate individual bacterial taxon and hence ensure the model validation and comparison based on both the desired sample-level and taxa-level properties (e.g., sparsity and overdispersion), however, it ignores the taxa-taxa relationships.[823] Examples of using this method can be found in these studies.[824,825]

8.3.3 Multivariate two-sample test for adaptive microbiome differential analysis (AMDA)

AMDA[826] is a taxa set-based (group-based) multivariate method for differential abundance analysis. Different from other differential abundance analyses, AMDA method was developed based on the centered log-ratio transformed relative abundances. Thus, it is a compositional-based association analysis.

The goal of AMDA is to examine whether the composition of a taxa-set is different between two conditions, without adjusting for multiple testing correction using individual taxon-based univariate differential abundance analysis. The taxon-based individual analyses have been reviewed suffering from three inherent limitations.[826] First, the type I error of a taxon-based individual analysis either may not be correct or may fail to control FDR in the presence of negative correlations[827] because FDR control procedures assume individual tests are either independent or under positive dependence,[485,794] while negative correlation among taxa abundance is common in microbiome data, especially for compositional data. Second, it is much more challenging to perform the multiple testing correction on high dimensionality nature of microbiome data and hence reducing the power of detecting differentially abundant taxa. Third, most taxon-based individual differential analysis methods heavily rely on the normalization and/or transformation, resulting in even more challenges in independent replication studies.[173,750,828] The most existing multivariate microbiome differential analysis methods are the global test; they are unable to identify specific taxon in the set of taxa that are differentially abundant. Thus, these existing methods make difficult to interpret the results[826] and may also jeopardize the power to test those not differentially abundant taxa in the set of taxa.[829] Motivated to overcome limitations from existing approaches of both individual and multivariate microbiome differential analysis, the goal of the proposed method is to enhance both interpretation and power under the framework of multivariate microbiome differential analysis. The proposed AMD takes a two-stage adaptive steps: Step #1 is to select some putative taxa that are more likely to be differentially abundant between two conditions. Step #2 is to examine the differential abundances of the selected set of taxa using a multivariate two-sample kernel-based maximum mean discrepancy (MMD) test.[830,831] Then, a permutation test is applied to the more likely differentially abundant subset of taxa to obtain statistical significance to avoid inflated type I error. However, MMD test equally utilizes information in all dimensions and typically is underpowered for analysis of high-dimensional sparse data. Because the true underlying biological scenario is never known, AMDA takes adaptive two-sample test of high-dimensional means. The AMDA adaptive method is different from other common adaptive approaches in the sense that AMDA tests the hypothesis in a selected subset of microbiome features rather than to assign different weights to variables or do another loop of permutations to combine multiple sets of weights.[826] It was demonstrated that AMDA method outperforms several competing methods (e.g., MiRKAT, OMiAT) in terms of statistical power and correct type I error rate.

8.3.4 Robust regression with compositional covariates (RobRegCC)

RobRegCC[832] is a sparse robust log-contrast regression framework and package. The goal of RobRegCC is to consider compositional and noncompositional measurements as predictors and identify outliers in continuous response variables. RobRegCC model was built on the framework of Aitchison and Bacon-Shone's seminal log-contrast model,[833] in which the outcome variable is modeled as linear combination of log-ratios derived from the compositional covariate data. RobRegCC model is a robust extended log-contrast regression. The extensions include: (1) Formulating a mean shift vector into log-contrast regression to enable modeling outliers in the response variable in high-dimensional setting. (2) Using sparsity-promoting convex and nonconvex regularizers (e.g., adaptive L_1 penalty) to select parsimonious model resulting in a family of robust estimators. (3) Using data-driven robust initialization procedure and cross-validation scheme specifically tailored to robust model selection and adapted to the compositional setting. With the development of RobRegCC, the toolbox of statistical regression analysis of compositional microbiome data was added another family member especially for modeling compositional covariates. However, current RobRegCC framework only models continuous outcome variables; a robust logistic regression needs to be extended. This method was cited in this study.[834]

As count-based approach, inferencing relative abundance (RA) also has some challenges or special issues. First, microbiome data are high dimensional. Converting the sequencing count reads to RA cannot remove the dimensionality. Second, converting the sequencing count reads to RA rather than make the microbial taxa even suffer the issue of dependency (as we reviewed above, the structure of microbial taxa is initially constrained due to technique and sampling process. The RA processing adds one more constraint: the common factor). Third, inferencing on RA cannot remove the issue of the zero-inflated structure of the sequencing rather than make the RA data range [0, 1]. How to deal with the boundary [0, 1] is also a real challenge. Although the newly developed methods have tried to address microbiome data properties (e.g., sparsity and overdispersion) within compositional data setting, it still lacks statistical methods to fully address the issues of high dimensionality and zero-inflated and boundary microbiome data.

9. Phylogenetic tree-based association analysis

Microbiome data are structured as a phylogenetic tree. Phylogenetic trees describe the evolutionary relationships between species. Thus, a phylogenetic tree relates all the microbial species, containing the evolution information of the species, which is useful for incorporating biological structure. Given the unique data characteristic, the phylogenetic tree-based association analysis methods are deserved as an independent category for introduction.

9.1 Taxonomic tree-based general framework for association analysis of taxa

Motivated by challenge of appropriately analyzing sparse compositional microbiome data and assumptions of parametric methods on the data structure, Tang et al.[743] developed general framework for association analysis of taxa based on taxonomic tree. The generality of this framework lies: (1) no assumptions of distribution, so the proposed method can perform robust association tests for the microbiome data with arbitrary intertaxa dependencies; (2) overall association of the whole microbiome community it allows for adjusting the confounding covariates and accommodating excess zeros; and (3) incorporation of taxonomic tree, which localizes lineages with taxonomic information in association tests of covariates. However, as in other multivariate analyses of microbiome absolute abundance data (e.g., differential analysis), in this case, the issues of overdispersion, zero-inflated, and high dimensional structures of microbiome data should be appropriately and fully addressed.[761] Besides these limitations, the drawbacks of this kind of global test are: (1) unable to identify specific taxon from the set of differentially abundant taxa[826]; (2) may also jeopardize the power of the test when many taxa in the set of taxa are not differentially abundant.[829]

9.2 Generalized mixed model framework (glmmTree)

glmmTree[649] is a predictive method based on a generalized mixed model framework. It was developed to capture clustered and dense microbiome signals. As most microbiome models, the proposed model uses "OTU" as a basic analysis unit. However, in glmmTree model, OTUs are used in the context on a phylogenetic tree and a phylogeny-induced correlation structure among OTUs is introduced to capture the evolutionary information. First, the patristic distance between OTU (i.e., the length of the shortest path linking two OUTs on the tree) and the correlation of the traits between these two OTUs are calculated. Then the distance and correlation are modeled using trait evolutionary model.[835] Next, a generalized linear mixed model is built based on the trait evolutionary model, in which the outcome variable of interest can be binary (e.g., disease status) or continuous (e.g., BMI), and the normalized abundance vector of OTUs (the OTU effects are assumed as random) incorporating with the trait evolutionary model as predictor and the demographic and other environmental or clinical variables serve as covariates.

The association between outcome and OTUs is constructed or estimated in two ways to capture both clustered and dense microbiome signals. One is

OTU clusters of different sizes: the outcome-associated OTUs are clustered based on different phylogenetic depths. Another is the signal density (number of associated clusters), which can also vary depending on the outcome. It was demonstrated that the glmmTree model outperforms existing methods[649] such as Sparse Inverse Correlation Shrinkage method (SICS), Lasso,[192] MCP,[821] Elastic Net,[379] and Random Forest in the dense and clustered signal scenarios. Phylogenetic trees have been reviewed as a useful way for incorporating biological structure. The glmmTree model using a phylogenetic tree describing the relationship between microorganisms is able to improve predictions of covariates on microbiome data.[836] However, the glmmTree model also has some limitations including: (1) the multivariate normal distribution assumption of OTU random effects needs to be validated by more studies; (2) current version of glmmTree model is mainly suitable for independent microbiome data; (3) the method that incorporates tree structure into the LASSO model is an ad-hoc method[837,838]; (4) without performing variable selection in model building, its prediction performance is subpar for sparse signal scenarios (i.e., only a subset of OTUs are associated with outcome).[198]

9.3 Phylogenetic tree-based microbiome association test (TMAT)

TMAT[839] is a phylogenetic tree-based microbiome association test. Human microbiome data are sparse due to high intersubject variation and have few OTUs shared across individuals. Thus, the association between OTUs and host traits is typically analyzed by a nonparametric test such as the Mann-Whitney *U* test or Wilcoxon rank-sum test to overcome these issues. However, the nonparametric approaches are prone to be information loss and increase the false-negative rates of nonparametric statistics. Motivated by this, the goal of TMAT is to use patterns of similarity among different OTUs to develop a quasi-scores-based test statistics for each internal node of a phylogenetic tree, and then those statistics are combined into a single statistic with a minimum *P* value to identify mutations associated with host traits. TMAT and glmmTree methods use the relative abundances of OTUs as analysis unit. However, TMAT is different from the glmmTree method in the sense that TMAT uses the log-transformed read count per million (CPM) as the response variable, while the glmmTree method considers the OTUs as predictors (random effects).

It was demonstrated that TMAT has several advantages,[839] including: (1) is generally the most efficient method compared to the available methods (e.g., optimal MiRKAT, MiSPU, OMiAT, ANCOM, edgeR, and the

Wilcoxon rank-sum test) in terms of type I error rate control, power, sample size (the library sizes), and the mean sparsity (OMiAT is the second powerful method overall, and Wilcoxon, edgeR, and ANCOM usually had the worst performances). (2) TMAT is computationally efficient due to the utilization of a distribution-based *P* value, while the permutation-based approaches (e.g., OMiAT, optimal MiRKAT, and aMiSPU) are computationally very intensive in small significance level. (3) TMAT is also superior to Wilcoxon rank-sum test and ANCOM as well as edgeR in the sense that both Wilcoxon rank-sum test and ANCOM are not able to adjust for the effect of covariates, edgeR does not correctly control the nominal significance level, and ANCOM has deflated type I error rates in some cases.

However, TMAT method also has several limitations[839]: first, its statistical power is dependent on the tree data quality. Second, it is arguable that whether TMAT (same in other models) adopts the minimum *P* value method for significant testing is appropriate or not. Third, statistical power and type I error are affected by the number of leaf nodes in TMAT. Fourth, TMAT assumes the absolute read count for each leaf node distributed with Poisson and assumes the internal nodes are independent, which may be violated in real datasets. Fifth, the current version of TMAT remains controversial to accurately identify association between taxa and host traits at the species level of OTUs and cannot be applied to the detection of functional genes in metagenomics data. Additionally, log-CPM transformation is widely used in RNA sequencing data analyses (e.g., in the edgeR package).[840] TMAT considers phylogenetic tree structures and uses log CPM transformation, which may be its superiority.[839] However, compared to RNA-Seq data, microbiome sequencing data are more overdispersed and contain a vast number of zeros[52,176]; hence, the usage of log-CPM transformation in microbiome study is needed to further discussion. Furthermore, the assumption of normal distribution for log CPM is arguable even if it has been shown in this case it is almost true. Also adding 1 to make log-CPM transformation positive is a limitation too.[39]

Other classification and regression models that incorporate the tree information into prediction include these studies.[738,739,822,841–843]

10. Microbiome-based association test for survival outcomes

So far the statistical methods for microbiome-based association test of survival outcomes are rare. The two methods are deserved for brief introduction here: MiRKAT-S and OMiSA. These two methods treat OTUs as counts; however, it was stated that OMiSA is usable for either count or composition data.[699]

10.1 Microbiome regression-based kernel association test for censored survival outcomes (MiRKAT-S)

MiRKAT-S is a community-level analysis of microbiota with survival outcomes.[789] MiRKAT-S was developed under framework of MiRKAT: it first transforms distance metrics into its kernel (similarity) matrices. Then MiRKAT-S extends MiRKAT framework to accommodate censored survival outcomes to compare similarity in the microbiota to similarity in survival times between individuals.

It was demonstrated that MiRKAT-S has three main advantages.[789] (1) MiRKAT-S has substantially better power than other distance-based approaches, including Cox proportional hazards regression,[844] PCoA,[845] and Ward's agglomerative hierarchical clustering method.[405] (2) MiRKAT-S allows survival outcomes in the kernel machine regression framework with kernels that appropriately encode microbiome data. (3) MiRKAT-S is able to perform the test using various kernels or similarity matrices, such as the UniFrac and generalized UniFrac distances, and the Bray-Curtis dissimilarity, which provides robustness to the nature of the true association between the microbiota and survival.

However, MiRKAT-S has been reviewed having three critical issues.[699] First, MiRKAT-S performs poorly when associated OTUs are rare in abundance,[789] which suggests that MiRKAT-S may simply ignore the information of numerous rare or mid-abundant taxa. Second, MiRKAT-S handles distance metrics individually and not adaptively tests several methods. This approach may result in poor type I error rate control or could lead to a substantial loss of power due to multiple testing correction in MiRKAT-S. Third, as with MiRKAT, MiRKAT-S can assess only the entire community and is not currently applicable to higher-level taxa. Probably, the major limitation of MiRKAT-S is that it is limited to analysis of association between the entire microbial community and survival outcome. Thus, we cannot use MiRKAT-S to identify individual taxa that are associated with the outcome. MiRKAT-S also cannot provide information about relationships among taxa within a microbial community.[789] The distance-based survival association OMiSA method, which we will introduce below, has the same limitation. MiRKAT-S was used to test the association of overall bacterial composition with progression-free survival, adjusting for covariates in the study.[846,847] Other example of MiRKAT-S use can be found in the study.[846]

10.2 Optimal microbiome-based survival analysis (OMiSA)

OMiSA is adaptive microbiome-based association test for survival (i.e., time-to-event) outcomes.[699] As we reviewed above, as the adaptive test of MiRKAT, optimal MiRKAT optimally adapts the test results from multiple MiRKAT tests using different distance metrics including unweighted/ weighted UniFrac, generalized UniFrac, and Bray-Curtis. OMiAT further optimally adapts the tests from the sum of powered score tests (SPU) and MiRKAT tests, enabling OMiAT robustly discovers rare, mid-abundant, and abundant associated lineages along with the functionality of optimal MiRKAT. Two association tests were developed in OMiSA: one is MiSALN presenting microbiome-based survival analysis using linear and nonlinear bases of OTUs; another is optimally adapted MiRKAT-S. By using MiSALN, OMiSA aims to be able to weigh rare, mid-abundant, and abundant OTUs, and through optimally adapting MiRKAT-S, it incorporates different distance metrics (e.g., UniFrac distance and Bray-Curtis dissimilarity).

Compared to MiRKAT-S, OMiSA has made progresses in following three directions[699]: (1) OMiSA can powerfully discover microbial taxa regardless of whether the microbial taxa in association analysis are rare or abundant and phylogenetically related or not. (2) OMiRKAT-S (the optimal MiRKAT-S), which is an adaptive version of MiRKAT-S can incorporate different distance or dissimilarity metrics. (3) The usability of MiRKAT-S has been extended from assessing only the entire community to be a general microbial group analytic method, which enables the analysis of higher-level taxa. Although adaptive tests could avoid the limitation of one-by-one approach due to the unknown true association pattern, they are not the perfect methods: (1) actually any approaches for effective association testing either one-by-one tests or adaptive tests need to know the measures they used. (2) Users of adaptive tests need to know the algorithm behind the tests. Examples of OMiSA use can be found in these studies.[846,847]

11. Longitudinal analysis of microbiome and omics data

11.1 Targeting the dependence of microbiome and omics data in longitudinal setting

There are two approaches for detecting microbial interactions or inferring dependence between taxa (OTUs) in microbiome and omics: cross-sectional analysis or longitudinal analysis. The microbiome is inherently dynamic, driven by interactions with the host and the environment, and varies over time. Thus, longitudinal microbiome data analysis provides rich information on the profile of microbiome with host and environment interactions. The distinguishing feature of longitudinal studies is that the subjects are measured repeatedly during the study, allowing the direct assessment of changes in response variable over time.[848,849] Longitudinal study also captures between-individual differences (heterogeneity among individuals) and within-subject dynamics. The longitudinal analysis of microbiome and omics data will enhance our understanding of short- and long-term trends of microbiome and other omics by intervention, such as diet, and the development and persistence of chronic diseases caused by microbiome and other omics. Generally speaking, developing longitudinal methods are more challenge due to the dependence of microbiome and omics data.

The topic of longitudinal data analysis in microbiome studies has been comprehensively reviewed and introduced by Xia et al.[72] Three categories of models were covered including: (1) standard longitudinal models, such as the generalized estimating equations (GEEs) and generalized linear mixed-effects model (GLMM), ZINB mixed-effects model; (2) newly developed overdispersed and zero-inflated longitudinal models, such as zero-inflated Gaussian (ZIG) mixture model, extensions of negative binomial mixed-effects and zero-inflated negative binomial models, Bayesian semiparametric generalized linear regression model, zero-inflated beta regression model with random effects, differential distribution analysis-based, zero-inflated negative binomial model, mixed-effects Dirichlet-tree multinomial (DTM) model; and (3) regression-based time series models, such as time series clustering method (e.g., microbiome counts trajectories infinite mixture engine (MC-TIMME)), dynamical systems theory model (e.g., lotka-volterra (LV) and generalized lotka-volterra (gLV) models), time-dependent generalized additive models, nonautoregressive microbial time series model. The interested readers can refer these models in Xia et al.[72] for details. Here, we introduce some other longitudinal models in microbiome and omics data.

11.2 Standard longitudinal models—Linear mixed effects models (LMMs)

The classic LMMs are equipped with fixed- and random-effect components, which provide a standardized and flexible approach to model both fixed and random effects. In the area of microarray study, LMMs were an established methodology. Currently, LMMs have been widely used in metabolomics research because LMMs can flexibly remove the effects of fixed- and random-effect confounding variables from metabolomic data.[850] LMMs have been increasingly applied in analysis of genome-wide association studies (GWAS) and metabolomics data.[851,852] Recently, LMMs were used for analysis of differential microbiome feature abundance and multiomics (i.e., various microbial measurement types: metagenomes, metatranscriptomes, proteomes, metabolomes, 16S rRNA).[71] Multivariable LMMs were conducted for per-feature with specifying recruitment sites and subjects as random effects to account for the correlations in the repeated measures.

When LMMs are used to fit multiomics abundance data, raw abundances should be transformed or normalized by an appropriate method. For instance, (1) using arcsine square root to transform microbial taxonomic and functional relative abundances, (2) using log-transform (with pseudo count 1 for zero values) for metabolite profiles and protein abundances, and (3) using log-transform with no pseudo-count for expression ratios (nonfinite values removed). It is arguable whether or not these transformation/normalization methods are appropriate to each outcomes of interest. Especially adding a pseudo count 1 for zero values may not even make sense because it forces the metabolites from "nothing" (absence) to "being" (presence). LMMs also cannot address the sparsity issue.[853]

11.3 Static analysis of longitudinal microbiome data

Although some microbiome data have been designed in longitudinal setting, the analytical techniques that they used are still static rather than in dynamic ways. Typically, the static approach analyzes dynamic microbiome data or changes microbial structure at each time point in several steps: first using community diversity measures including alpha and beta diversities,[146] then using graphic summary plots (e.g., heatmap, network, and phylogenetic tree), clustering, ordination techniques (e.g., PCA, PCoA, NMDS),[307] finally performing either univariate community analysis via Wilcoxon rank-sum test, Kruskal-Wallis test, or other tests[661] and/or multivariate community analysis via distance-based methods (e.g., weighted UniFrac) and a permutation test (e.g., PERMANOVA, Mantel test, ANOSIM) to assess the association between the overall microbiome composition and the outcome of interest at each time point.[147] For example, the study on multiomics of the gut microbial ecosystem was a longitudinal design, in which 132 subjects were followed for 1 year each during inflammatory bowel diseases (up to 24 time points each).[71] In this study, to understand the etiology of the IBD-associated gut microbiome in a systems-level, statistical association analysis was assessed at several time points in terms of both alpha and beta diversities: (1) Gini-Simpson alpha diversity was compared using a Wald test between IBD and non-IBD. (2) PCoA was performed based on Bray-Curtis dissimilarity matrices of normalized absolute abundances from 16S rRNA gene amplicon sequencing, human RNA sequencing, metabolomes, and functional profiles (metagenomes, metatranscriptomes, and proteomes). The species-level abundances were used for sequencing data; functional profiles were first summarized to the KO level (Kyoto Encyclopedia of Genes and Genomes (KEGG) Orthologues (KOs)) using HUMAnN2. Then the association tests between measurement type pairs were conducted by Mantel test, and the association tests between measurement type pairs or disease groups adjusted/not adjusted by covariates were conducted with PERMANOVA. Other examples of using static analytical techniques to analyze dynamic microbiome data include assessing the associations of antibiotics use in early life and the composition of intestinal,[854] microbiota disruption and metabolic consequences,[855] smoking, and the oral microbiome.[104] However, those analyses are not adequate to capture the dynamic nature of longitudinal microbiome data due to ignoring inherent ordering and other temporal dependencies of the microbiome data.[103]

11.4 Regression-based time series models

In recent years, we observed increasing applications of time series approaches in microbiome data.

We will briefly introduce some models for time series microbiome data below.

11.4.1 Local similarity analysis (LSA) and extended LSA (eLSA)

LSA was developed for discovering local and potentially time-delayed co-occurrence and association patterns in time series data.[856,857] An association is considered locally if it only occurs in a subinterval of the time of interest, and an association is time-delayed if there is a time lag for the response of one organism to the change in another organism. LSA is a technique that does not require significant data reduction, while it can be used to identify more complex dependence associations among taxa as well as associations between taxa and environmental factors.[856] This method can be adjusted to work with a static data as well such as a presence-absence matrix. However, LSA was originally designed for time series data without replicates, which cannot measure the variability of local similarity score and thus cannot obtain its confidence interval. With LSA extended to time series data with replicates, the extended LSA (eLSA) technique is more efficient to capture subinterval and time-delayed associations, providing insights to the real dynamics of biological systems.[857] eLSA is more approximate to test the statistical significance of its inferred pairwise local similarity analysis.[858]

11.4.2 Learning interactions from microbial time series (LIMITS)

Fisher and Mehta[237] think there are three major obstacles that hinder detecting ecological interactions between taxa from metagenomic studies: (1) correlation between the abundances of taxa does not imply interaction of taxa; (2) the sum constraint problem results in difficult inference in time series models; and (3) various errors due to experiment, misassignment of sequencing, and bias inferences of taxa interactions. To overcome these obstacles and model the microbial dynamics for identifying keystone taxa, the proposed LIMITS uses sparse linear regression with bootstrap aggregation to infer a discrete-time Lotka-Volterra model.

11.4.3 Metagenomic microbial interaction simulator (MetaMIS)

MetaMIS[859] is the Lotka-Volterra model-based software platform. MetaMIS was designed to analyze the time series metagenomic data of microbial community profiles. It uses a partial least square regression to identify the interaction terms, and implements network visualization with a graphical user interface. It first infers underlying microbial interactions based on OTUs—abundance tables and then interprets interaction networks using the Lotka-Volterra model. MetaMIS evaluates the similarity to biological reality via an embed Bray-Curtis dissimilarity method.

11.4.4 Microbial dynamical systems inference engine (MDSINE)

MDSINE[860] is a Bayesian-based probabilistic method for microbiome time series analyses, which uses Bayesian algorithm for inferring dynamical systems models from time series data and predicting temporal behaviors. It constitutes a comprehensive toolbox for dynamical systems inference. MDSINE is capable of accurately forecasting microbial dynamics, predicting stable subcommunities, and identifying most crucial bacteria in the dynamical community.

11.4.5 Temporal insights into microbial ecology (TIME)

TIME[861] is a webtool that was developed specifically to provide a suite of analysis and visualization tools for time series analysis of microbial ecology, and relies on a Granger-LASSO model to identify causal relationships.

11.4.6 Dynamic interaction network inference

Dynamic interaction network inference[862] is a computational pipeline. The goal of the pipeline is to enable the integration of data across individuals to reconstruct dynamic models from time series microbiome data. The pipeline first aligns the data collected for all individuals and then uses the aligned profiles to learn a dynamic Bayesian network which represents causal relationships between taxa and clinical variables.

11.4.7 SplinectomeR

SplinectomeR[863] is an R package that uses smoothing splines to summarize data for straightforward hypothesis testing in longitudinal studies. The goal of splinectomeR is to enable group comparisons in longitudinal microbiome studies while avoiding loss of longitudinal power due to suffering from subject dropout, irregular sampling, and biological variation, which may cause the data not being normally distributed.

In summary, time series models are widely used techniques in other fields, but in recent years we observed increasing applications of time series approaches for modeling microbial dynamics. When using time series models, we need to carefully design and choose the appropriate analytical tools. Otherwise, the results can be extremely misleading.[864] We need particularly pay attention to time series data: (1) We need emphasize that the microbiome data are temporal. Thus, we cannot treat the time series data as a static time point and test them by a simple statistical procedure (e.g., *t*-test). We cannot treat the time points as independent samples, which could overestimate differences between groups. (2) We cannot average the abundances of mixed populations, especially average those abundances in sequence-based microbiome data analyses. For example, we cannot aggregate two OTUs or species with opposite population dynamics. If we aggregate them, the temporal information could be lost and thus obtaining wrong microbiome profile. (3) Time series models typically work on the high frequent time series of microbial data. Thus, they are subject to the irregular sampling times and modeling and implementation are complicated.[103]

11.5 Principal trend analysis

11.5.1 Principal trend analysis (PTA)

PTA[865] was proposed to extract principal trends of time course or underlying dominant time course patterns of gene expression data from a group of patients, and to identify important genes that make dominant contributions to the principal trends or significant patterns from a genome-wide longitudinal dataset in genomic and translational medicine. The authors of PTA demonstrated that it can be used for dimension reduction, time course signal recovery, and feature selection in high-dimensional longitudinal data, such as high-throughput longitudinal genomic and proteomic data (i.e., gene expression) and microarray. Different from the classical dimension-reduction techniques (e.g., PCA), which usually ignores the temporal structures and simply treats data from individual time points as independent observations,[866] the proposed method leverages the benefits of the spline-based methods on time course data analysis[867] and principal component analysis for dimension reduction.

11.5.2 Joint principal trend analysis (JPTA)

JPTA[475] was proposed to simultaneously analyze two longitudinal high-dimensional datasets to extract shared latent trends (latent time course patterns) and identify relevant important features. As we reviewed in Section 7.2, the correlation methods for determining the relationship between two sets of variables include classical canonical correlation analysis[310] and its extensions in two directions: penalized/sparse CCA and from identifying linearity correlations to nonlinear relationships (i.e., nonparametric approaches). The classical CCA method is inappropriate for high-dimensional data due to singular covariance matrices, while CCA extensions were reviewed as simply applying CCA or sparse CCA to sample covariance matrices or applying sparse additive kernel/functional CCA to the corresponding inner product matrices. All these established methods cannot capture the underlying longitudinal trajectories shared by the two datasets.[475] To address this issue, the proposed JPTA employs latent factor models with shared principal trends (PTs) to preserve the time order of longitudinal datasets and characterize those embedded temporal trajectories. In addition, JPTA uses L_1-norm regularization incorporating feature selection scheme to identify features contributing to the underlying PTs. It was showed that JPTA outperforms CANCOR and PTA in providing more biologically meaningful results and better interpretations of the identified features. JPTA is also robust in term of noisy features, sample sizes, and signal-to-noise ratios.[475] However, there are certain issues that JPTA has not considered, such as missing values, dropouts, and measurement errors.

11.6 Newly developed univariate overdispersed and zero-inflated longitudinal models

11.6.1 Two-part zero-inflated beta regression model with random effects (ZIBR)

ZIBR[815] was proposed to test the association between microbial abundance and clinical covariates in longitudinal microbiome data setting.

Both zero-or-one-inflated beta and zero-inflated beta regression models have been developed for proportion data.[812,813]

However, they cannot handle longitudinal or repeatedly measured proportion data. The goal of ZIBR is to extend zero-inflated beta regression model to longitudinal data setting for analyzing microbiome data. Through this extension, ZIBR is able to jointly model the covariates that affect the taxon in terms of both the presence/absence of the taxon in the samples (via a logistic regression component) and nonzero abundance of the taxon (i.e., its abundance) (via a beta regression component). The correlations among the repeated measurements on the same subject are also accounted by a random effect of each component.[174] Thus, the ZIBR model enables studying the taxa based on longitudinal or recurrent measures.[868] The two-part model and zero-inflated models including ZIB are the useful tools for dealing "zeros" separately (i.e., presence/absence and abundance levels separately), and particularly for analyzing the complex multiomics data.[52,869]

However, ZIBR also has several limitations. For example, (1) These kinds of method[815,870,871] focus on the temporal pattern of single microbial taxa and never intend to study the interdependence of the microbial taxa[103] or are not able to incorporate the intertaxa correlation[761,816] or do not address time trends and within-subject correlations.[872] (2) Because these methods analyze individual taxa one by one at a time with a multiple testing correction procedure to control for type I error rate, they cannot provide *P* values and correct statistical inferences for the selected taxa,[816,873] also ignore the multivariate nature of the data[874] or the microbial community as a whole and hence loss statistical power after multiple testing correction.[701] (3) Such models often implicitly assume that all zeros can be explained by a common probability model, which is not always true because these models do not differentiate potentially three different sources of zeros (i.e., outlier zeros, structural zeros, and sampling zeros) in microbiome data,[875] instead focus on the inference for the nonzero taxa. Actually to formally describe the distribution of microbiome data, the methods that can handle high-dimensional parameters from both the nonzero part and zero part of microbial data are expected.[761]

As we comprehensively reviewed and discussed in our book[811] (see details on pp. 36–37, 339–341, and 453–454), we can differentiate rounded, structural, and sampling zeros in microbiome and omics data. The zero issues are very complicated and dealing with zeros is one of the biggest challenges in microbiome research. (4) Although ZIBR and other zero-inflated models make inference on relative abundance of microbiome data, they directly model the probability of producing excess zeros and make an implicit assumption that microbial composition is identical among individuals. Thus, such models cannot capture the effects of individual differences in microbial composition.[876] Actually, they consider a single taxon at a time and are not designed for differential composition analysis.[744] (5) Although these methods have the capabilities to capture the microbial dynamics and identify the time dependent taxa, it is difficult to always justify the assumption that the abundance of an individual taxon changes either at a fixed rate or in an autoregressive pattern.[877] (6) ZIBR also assumes that all subjects can provide samples at the same time points with no missing measurements or requires that at the same time points the samples in every condition are provided. Such a stringent assumption is often incompatible with real study cohort even after the missing samples are imputated.[227,878] (7) ZIBR and other zero-inflated models for differential abundance analysis[351,879,880] treat the dispersion as a nuisance parameter, their focus is to test mean abundance change of the nonzero component and/or probability (prevalence) of the nonzero component. These models also assume a common fixed dispersion parameter, which has been reviewed as very restrictive and is inconsistent with observations in real dynamic and heterogeneous microbiome studies.[761,881] Additionally, these models cannot provide a straightforward interpretation of covariate effects on the overall marginal (unconditional) mean,[817] and are difficult to justify parametric assumptions on the density of the response and also the probability of boundary values (because the relative abundance of microbiome data is bounded in [0, 1]) in practice when simulating data to evaluate the performances of these methods.[882] Examples of ZIBR use are available from these studies.[209,837,883–887]

11.6.2 Fast zero-inflated negative binomial mixed modeling (FZINBMM)

FZINBMM[853] was developed to analyze high-dimensional longitudinal metagenomic count data, including both 16S rRNA and whole-metagenome shotgun sequencing data. The goal of FZINBMM is to simultaneously address the main challenges of longitudinal metagenomics data (i.e., high dimensionality, dependence among samples, and zero inflation of observed counts).

FZINBMM takes two advantages: (1) FZINBMM is built on zero-inflated negative binomial mixed models (ZINBMMs). Thus, it has the capabilities to analyze overdispersed and zero-inflated longitudinal metagenomic count data. (2) FZINBMM uses a fast and stable EM-iterative weighted least squares (IWLS) model-fitting algorithm to fit the ZINBMMs, which takes advantage of fitting linear mixed models (LMMs).

Thus, FZINBMM can handle various types of fixed and random effects and within-subject correlation structures and analyze many taxa fast.

It was demonstrated[853] that FZINBMM outperformed in computational efficiency and statistically comparable with other two ZINBMMs: glmmTMB[888] and GLMMadaptive[889] as well as that FZINBMM outperformed linear mixed models (LMMs), negative binomial mixed models (NBMMs),[872,890] and zero-inflated Gaussian mixed models (ZIGMMs). However, FZINBMM also has several limitations, such as (1) analyzing one taxon at a time, and (2) assuming subject-specific effects (random effects) are followed as a multivariate normal distribution. Additionally, FZINBMM also shares most other limitations of ZIBR we described above.

11.7 Multivariate distance/kernel-based longitudinal models

Recently, the kernel machine regression framework has been extended to test the association between the outcomes and the microbiome community in longitudinal setting. Below, we introduce some longitudinal multivariate distance/kernel-based association tests of microbiome data.

11.7.1 Nonparametric microbial interdependence test (NMIT)

NMIT[103] is a multivariate distance-based nonparametric test framework. The goal of NMIT is to test the overall microbial interdependence group differences over time. NMIT is used to compare temporal microbial interdependence structures between groups and test its association with covariates which are either binary outcome (e.g., disease status, gender, or case-control group indicator), or disease-associated quantitative variables (e.g., age, BMI, blood pressure, or biomarker measurement). The NMIT framework consists of three components with two major steps of statistical analysis or testing: Step #1(core part of the proposed test): Perform pairwise correlation analysis of taxa for each subject in which correlation matrix is constructed for each subject to summarize their microbial interdependent correlation structures. Step #2: Perform permutation MANOVA[716,790] to test whether the correlation structure is different between groups or associated with an interested outcome or not. Then the microbial dependency in different groups and the differences of the temporal correlation structure between groups can be visualized using network analysis and heatmap, respectively.

Currently, longitudinal models for discovering dynamic nature of the microbiome data are still rare. Thus, this proposed test and other longitudinal models could help microbiome research community to obtain some first insights into the dynamic microbiome data. However, the proposed methods also have several drawbacks: (1) NMIT is a distance-based testing method, providing an overall assessment of the group difference in terms of the interdependent relationship (microbial interdependence similarity) among taxa.[103] However, researchers are often more interested in identifying specific key taxa that are associated with outcomes or biological covariates. (2) Currently, the proposed test includes three correlation methods (Pearson's correlation, Kendall's rank correlation, and maximal information coefficient (MIC)). However, using correlation coefficients to detect dependencies of microbial taxa suffers from limitations of detecting spurious correlations due to compositionality[815] and being severely underpowered owing to the relatively low number of samples.[891] (3) The study demonstrated Kendall method is always comparable or has a slight power edge over Pearson's method, while MIC method has less power than Pearson's and Kendall's methods.[103] This is controversy to microbiome literature. More studies are needed to confirm this claim. Since the interspecies correlations were calculated based on the original abundance data rather than log-ratio transformated abundance data in the comparisons of three methods, we do not know whether or not the better performance of Pearson's method than MIC method is due to spurious correlation. (4) Current version of NMIT cannot handle time-varying covariate.[103] Recently, NMIT was built in QIIME 2 as one of the two q2-longitudinal plugins to facilitate streamlined analysis and visualization of longitudinal and paired sample datasets.[892] Another is linear mixed-effects models (LMMs).[893] When readers perform statistical analysis through QIIME 2, it is better to know the advantages and disadvantages of the underlying methods they are using. Other examples of NMIT use can be found in these studies.[894,895]

11.7.2 Correlated sequence kernel association test (cSKAT)

cSKAT[786] was proposed to directly test microbiome association with related outcomes (i.e., those outcomes from longitudinal and family studies) using the linear mixed effects models (LMMs). The cSKAT methods have several characteristics: (1) using random effects in LMMs to account for the outcome correlations and the effect of variables of interest (e.g., a set of OTUs or a single-nucleotide polymorphism (SNP)-set); (2) using correlated sequence kernel association test (cSKAT) (a small-sample adjusted kernel variance component score test) to account for high dimensionality and addressing the problem of far fewer samples than the number of association tests. Although it was demonstrated that cSKAT is flexible to be fitted by using different softwares and allowing for different types of correlated data such as longitudinal data and family data, cSKAT has two major limitations[701]: first, cSKAT is based on the linear mixed effects model, hence limited to handling a continuous outcome; second, cSKAT is limited to the item-by-item use of the ecological distances. Additionally, it is not a perfect exact test because a conservative test may occur, and hence correcting estimation error remains in future development.[786] Examples of cSKAT review are available in these studies.[773,826,868]

11.7.3 Generalized linear mixed model and its data-driven adaptive test (GLMM-MiRKAT and aGLMM-MiRKAT)

GLMM-MiRKAT[701]is a distance-based kernel association test based on the generalized linear mixed model (GLMM)[703] and aGLMM-MiRKAT[701] is its data-driven adaptive test. The design of GLMM-MiRKAT and aGLMM-MiRKAT was based on two frameworks of statistical methods: GLMM and ecological distance/dissimilarity for microbial community analysis to model data dependency due to clusters (repeated measures), i.e., to account for the within cluster correlation in responses. Under the framework of GLMM, the diverse host traits of interest (e.g., Gaussian, Binomial, and Poisson) can be handled. Thus, GLMM-MiRKAT can be considered as an extension of cSKAT[786] to handle non-Gaussian host traits. By using framework of ecological distances (e.g., Jaccard/Bray-Curtis dissimilarity, unique fraction distance), the multiple features of taxa can be handled in a multivariate manner. The proposed methods have several characteristics: (1) by using the framework of GLMM, the association between microbial composition and various host trait adjusting for covariates can be tested; (2) the large p and small n problem due to high-dimensional nature of the data was addressed by applying the kernel trick[896] and performing a variance component test[897]; (3) as other data-driven adaptive tests, aGLMM-MiRKAT, which is based on the test statistic of the minimum *P* value from multiple item-by-item GLMM-MiRKAT analyses, was proposed to avoid to choose the optimal distance measures from both nonphylogeny-based distances (e.g., Jaccard and Bray and Curtis dissimilarities) and phylogeny-based distances (e.g., unweighted/weighted UniFrac distances, and generalized UniFrac distance). Although the study demonstrated that aGLMM-MiRKAT is a useful analytical tool to detect diverse types of host traits of interest with robust power and valid statistical inference,[701] as other count-based models, the proposed methods treat taxa or OTUs as independent rather than compositional, which ignores one important feature of microbiome data structure. Additionally, in the illustrated examples, the criteria of excluding measurements with low sequencing depth (i.e., <10,000 total reads) and removing OTUs with average relative abundance $<10^{-5}$ are also arbitrary.

11.7.4 Paired and longitudinal UniFrac ecological dissimilarity (Pldist)

Pldist[898] is a newly developed dissimilarities for longitudinal microbiome association analysis. It was developed to summarize within-individual (or within-pair) shifts in microbiome composition and then compare these compositional shifts across individuals (or pairs). The pldist method takes the approach of distance-based analysis and modifies the distance metric to accommodate related samples in order to reduce intersubject variation. The input data of this method are a taxon counts table and possibly a phylogenetic tree, regardless of what sequencing and quantification methods are used to generate the data. The pldist consists of two paired and two longitudinal UniFrac dissimilarities: unweighted PUniFrac distance/dissimilarity, generalized PUniFrac distance dissimilarity, unweighted LUniFrac distance/dissimilarity, and generalized LUniFrac distance dissimilarity, with LUniFrac dissimilarities extending from PUniFrac dissimilarities, respectively. Some characteristics of pldist method include: (1) it is flexible to incorporate phylogenetic and nonphylogenetic dissimilarities, such as Gower's distance,[899] Bray-Curtis dissimilarity,[900] and Jaccard distance[901]; (2) it uses the centered log-ratio transformation (CLR) to account for data compositionality; (3) similarly as standard UniFrac dissimilarities, the PUniFrac and LUniFrac dissimilarities may be utilized in any analysis where a beta-diversity matrix is required including ordination (e.g., PCA, PCoA), classification and clustering, and global hypothesis testing, such as permutation-based methods (e.g., PERMANOVA) and kernel machine regression-based association tests such as MiRKAT[735,736] and MiRKAT-S.[659,690,785,789] With the development of pldist, the UniFrac family was added another tool to measure dissimilarities with paired and longitudinal data allowing longitudinal analysis with a wide variety of outcome types. The pldist method is differentiated from the approach of linear mixed models. The difference lies on its explicitly considering changes in the microbiome over time,[898] which may be more suitable for fitting longitudinal microbiome data. However, as other compositional methods, pldist replaces zeros with a small pseudocount; this is still arguable in the field of microbiome research.

11.7.5 Longitudinal microbiome data simulation

microbiomeDASim[902] is an R package for microbiome differential abundance simulation. It was proposed to simulate longitudinal differential abundance for microbiome data. The development of microbiomeDASim was motivated by the importance of longitudinal analysis of microbiome data and statistical challenges for simulating longitudinal microbiome data to evaluate the increasing number of longitudinal statistical models. The simulation framework implemented in the microbiomeDASim package allows for appropriate comparison between methods while taking into account of two challenges on data generating process: (1) nonnegative restriction and (2) the presence of missing data/high percentage of zero reads; and three logistical challenges on data collection: (1) low number of repeated measurements, (2) asynchronous repeated measures, and (3) small number of subjects. The microbiomeDASim package can facilitate comparing and validating statistical methods for analysis of longitudinal microbiome data[902]; however, the assumptions such as multivariate (truncated) normal distribution and taxa independence limit its use in simulation study of microbiome models.

12. Features and trends of correlation and association analyses in microbiome and omics

What are statistical correlation and association in microbiome and other omics? In general, even in traditional statistics, accurately defining correlation and association is difficult because too many methods have been designed to measure the magnitude of correlation or association between two variables. It is much more challenge to define correlation and/or association and measure them in the fields of microbiome and omics. In general, a paradigm shift is occurring from static, single dimension to dynamic and multiple dimensions in correlation and association analyses of microbiome and omics. We summarize some features or trends of the paradigm shift as below.

(1) Organizational frameworks of association have been shifted from the levels of measurement to the levels of domain.

Although whether or not measurement of association should be categorized and based on three or four levels (or scales) of measurement: nominal, ordinal, interval/ratio is arguable, the fact is that measures of association have historically been constructed for nominal-level (categorical), ordinal-level (ranked), and interval-level variables,[19] and nowadays using the nominal, ordinal, interval/ratio typology as a pragmatic organizational framework is still considered as a simple and convenient way.[29] In addition, the data for measures of association have been constructed or cross-classified into contingency tables or simple bivariate lists of response measurements.

However, as we depict in Figs. 1 and 2, in microbiome study, the statistical framework of association analysis involves three domains: environment, microbiome, and host, as well as other omics. The relationships of data that are determined are far beyond between two variables and involve three different domains. Not only the data types in microbiome study are totally different from those of traditional fields, but also within each domain of microbiome data, the data are often multiple dimensions. Thus, in microbiome studies, the structure of association is very complicated. Here, we want to emphasize that association structure including domain and data features in microbiome study is much different from those in traditional association analysis.

(2) Dimensions of association have been shifted from the association of variables to the association of domains.

Another important concept is association dimension. Historically when measuring association researchers typically have considered four dimensions of association[29]: (1) Whether the association is symmetric (without a specified independent or dependent variable) or asymmetric (with well-defined independent and dependent variables)? (2) Whether it is one-way association (one variable implies the other, but not vice versa) or two-way association (two variables imply each other)? (3) What kinds of models of association (e.g., maximum-corrected, chance-corrected, or proportional reduction-in-error measures)? (4) Do the measures of association variously measure correlation, association, or agreement?

The association analysis in microbiome studies could be among variables within each domain: environment, microbiome, or host. However, association analyses are also conducted among environment, microbiome, and host. The coverage and scope of association analyses among domains have increased rapidly in microbiome studies so that metagenome-wide association studies (MWASs) have been used more recently.[903–905] We can anticipate that MWASs will be a main focus of association analyses in microbiome studies.

In systems biology, genomics, a structural framework of four-dimensional data has been proposed. Commenting on annotation of genomes, Palsson[353] adopted a two-dimensional matrix (systemic annotation) in analyzing, interpreting, and predicting the genotype-phenotype relationships: a list of the biochemical components (1 dimension (D)) and biochemical reaction networks (2D). Bork and Serrano[352] used "cellular parts lists" (1D) describing the data generated from various omics communities, and stated that the parts lists need to be organized in interaction networks (2D) and a structural framework of interaction networks generating a spatial framework (3D) integrated with temporal data (4D). Reed et al.[354] reviewed and described four-dimensional genome annotation with network components (1D), component interactions (2D), genome spatial orientation (3D), and evolutionary changes (4D) in the study of changes in genome sequences. In a review of molecular ecosystems biology, Raes and Bork[906] described the current status of parts list (1D), connectivity between the parts (2D), and their variations in the spatial (3D) and temporal (4D) contexts from protein to environments. Actually, if we want to functionally understand the gut microbiome, we need to connect parts lists (1D) to networks (2D) in a spatial (3D) and temporal (4D) context.[906] We described a statistical framework of associations and hypotheses in microbiome study.[4,16,39,72,146,147,166,174,307,661,811] The associations of microbiome data involve multiple dimensions: it could have occurred between a pair of taxa (1D), among network interactions (2D) in the context of space (3D) and time (4D). In this chapter, we have extended the framework in terms of the four-dimensions when we introduced association analysis above.

(3) Types of association have been shifted from an overarching concept to dependence and particularly co-occurrence.

Traditionally, association is used as an overarching concept including all types of measures of correlation, association, and agreement between two variables at all levels of measurement (wider domain), and more specifically used as a measure of relationship between two nominal-level variables, two ordinal-level variables, or some combination of the two (narrower domain).[29] Correlation, association, and agreement correlate each other. Measures of agreement (or concordance or reproducibility) is defined to ascertain the identity of two (sometimes more than two) variables at any level of measurement, i.e., $X_i = Y_i$ for all i.[29] The test statistics of agreement are used to assess inter-rater variability or whether different techniques produce similar results.[88]

However, in microbiome and omics, correlation and association in terms of dependence and especially co-occurrence are two more important concepts. We have presented more details about the relationship between association and correlation when we discussed the mining association rules.

(4) More dynamic association analyses are needed in microbiome and omics studies.

First, microbiome is dynamic because a pair of taxa within microbiome could be associated (associations of individual taxa). Second, microbiome is dynamic not only because a pair of individual taxa is associated, but also because microbiome is associated with various external factors (host and environmental factors): (1) microbiome compositions could be associated with host factors (e.g., biologic and genetic); (2) microbiome compositions could be associated with environmental factors or covariates including clinical or experimental conditions. Third, microbiome is also intrinsically dynamic because microbiome have two intrinsic factors: microbiome is evolutionary and temporal (e.g., microbiome's state follows maturation during life span and varies based on host health and disease) (see Fig. 2).

In summary, to better understand the dynamic and complicated system of microbiome and its functions, we should review and treat associations of microbiome data as different traditional approaches in both concepts and methodologies.

For measures of association and association analyses in microbiome studies, we need consider at least four conditions: first, microbiome data structure and features, such as tree-structured, multidimensional, compositional, sparse, and often have many zeros. Second, if the researcher wants to measure the association between different domains, such as microbiome and host, then both data features in microbiome and host are needed to be considered. For example, when we measure the correlation between microbiome and metabolome, then we not only need to account for the data features of microbiome, but also account for the data features of metabolome. Third, another consideration is whether or not a model to estimate differential conditions such as a treatment, a disease of interest, or different genotypes. For example, the model is capable of estimating a single co-occurrence networks or capable of estimating separate taxa co-occurrence networks for groups defined by a binary variable. Fourth, we need consider whether the measure or model of association can be easily generalized to multivariate data structures. Microbiome data are multivariate; multivariate data analysis and modeling are more important and extremely useful in microbiome studies.

(5) Trends and limitations of application and development of correlation and association methods in microbiome and omics studies.

We have observed several general trends regarding application and development of correlation and association methods: (1) researchers start either borrowing or developing univariate methods, move to most likely use multivariate techniques; (2) in early microbiome studies, alpha and beta diversities were most widely used, and currently various multivariate distance/kernel/adaptive-based association approaches are getting popular; (3) in early days, the statistical methods and models were developed based on taxa abundance data and nowadays, several statistical methods and models have been proposed incorporating phylogenetic tree information; (4) at the beginning, count-based approach and relative abundance/compositional data analysis approach were parallelly developed in literature, now statisticians and researchers begin to discuss their appropriateness and respective statistical issues; and (5) at the beginning, new methods were developed by reparameterizing classic methods or validating their applications in microbiome data, such as negative binomial, zero-inflated and zero-hurdle models, more currently various specifically designed methods have been developed to target characteristics of microbiome data including compositionality, high-dimensionality, sparsity using Lasso, regularization, and sparse techniques.

The existing correlation and association methods have limitations. For example, the methods for power analysis, mediation analysis, and longitudinal data analysis are still in infancy stage. The methods for integrating multiple omics data are still rare.

13. Further discussion regarding association analysis in microbiome and integrating multiomics studies

We believe that several topics need to be further discussed and to be validated.

(1) *Which association analysis approach is more appropriate: count-based vs. relative (or compositional)-based association analyses?*

In the statistical method development and data analysis, generally microbiome taxa are either inferenced based on absolute abundance (count reads) or relative abundance. When relative values were used in the analysis, some researchers treat them as compositional and others treat them just proportional. Microbiome data are structured as tree, high dimensional, compositional, sparse, and often have many zeros. Either approach chosen to inference microbial taxa will face the statistical issues of dependency, compositionality, sparsity, overdispersion, and zero-inflated. However, the strategies of addressing these challenges are different when count-based or relative (or compositional)-based approaches are used. When the count-based approach is taken, the methods more focus on taking account of overdispersion and zero-inflated issues as well as dependency of taxa, while the relative (or compositional)-based approach is chosen, the direct challenges that the methods faced are how to replace the zero values prior to normalization and how to deal with the boundary [0, 1]. Actually, the count-based approach also normalizes the taxa abundance, such as DESeq2 and edgeR have various normalization methods, and standard count models such as zero-inflated and zero-hurdle models use off-set to normalize the taxa abundance.

Researchers choose the count-based approach or use the relative abundance of taxa to perform the association analysis instead take compositional-based association analysis are mainly based on following three reasons: first, the spurious correlation concern is originated in ecology studies (i.e., relatively low-dimensional data); however, microbiome studies are high dimensional, usually having the large number of taxa. Thus, when the samples have large diversity, the compositional effect (e.g., the spurious correlation) is mild[40,103] or as the number of taxa increases, the compositional effect is attenuating.[800] Second, the biological interpretation. That is, association analysis is to detect which taxa (instead of ratio of taxa) are associated with the outcome[800] because an association between ratio of taxa and the outcome is difficult to interpret biologically. Third, technological developments including the estimation of absolute cellular abundances from microbiome sequence data in microbiome data science[69] may help to correct for data compositionality. Therefore, these researchers prefer to use the original absolute abundance data to calculate the interspecies correlations[103] or take the relative abundance for better interpretation when develop their statistical methods.[800]

Count-based methods have been widely used in rRNA expression studies and adopted to analyze microbiome data.[166,171–174,727] In recent years, another link of count-based approach, Dirichlet models have been developed and increasingly applied in microbiome data analysis. In the meantime, several compositional-based methods were also proposed in the literature. Thus, more researches and discussions on the appropriateness of count-based and compositional-based methods will be helpful. If compositional data structure is not big problem in microbiome data analysis, then it is very meaningful because: (1) Many classical statistical methods can be used. (2) Log-ratio transformation used in compositional studies, which is often biologically not interpretable, could be avoided. (3) An arbitrary small constant count does not need to be added to a zero value when using log-ratio transformation. This approach ignores the zero values in the analysis and does not differentiate different kinds of zero sources, such as structure zeros and sampling zeros. (4) Researchers can develop new methods or use the established count-based models, such as NB, ZINB, Hurdle models to equip with multivariate structure and dependency and multiple-comparisons correction. (5) Either count-based approach or relative abundance approach may be appropriate. (6) It is also not necessary to use only positive values for estimation as to avoid compositional issues in the development of methods (e.g., as did in the multivariate zero-inflated logistic-normal and inference for absolute abundance models).

(2) *Whether traditional correlation methods are appropriate?*

The development of correlation and association methods has been moved from mainly focused on linear methods to nonlinear and generalized models. We observed that traditional correlation methods (e.g., Pearson's and Spearman's correlations) are still used in microbiome field. Controversy is that more researches have suggested that Pearson's correlation analysis is not suitable for microbiome data due to its assumptions of independence and linearity, which could result in spurious correlation when applied to analyze compositional microbiome data. The paradox is that some researchers and statisticians still used Pearson's correlation as the building block of the methods in developing their new methods. For example, to inference interaction network, first Pearson's correlation coefficient among all pairs of OTU samples is commonly computed, and then an interaction between microbes is determined if the absolute correlation coefficient is sufficiently high.[235,236]

Application of classic correlation analysis to microbiome and omics data will result in spurious correlation; thus, classic correlation analysis does not fit the data features and properties of microbiome and omics. To take account of compositionality, sparsity, and heterogeneity, we need to redefine the concepts of correlation and association in microbiome and omics studies. Actually, the challenge of the traditional concepts also provides an opportunity to rethink correlation and association to meet new changes of research fields.

(3) *Which approach is more powerful: univariate analysis vs. multivariate analysis?*

Univariate and multivariate correlation and association take different approaches. Univariate analysis directly detect the correlation or association among individual taxa or between taxa and environment or host factors. Then an adjusting method (e.g., FDR) is applied to adjust multiple comparisons or testing. Multivariate analysis first uses a summary statistic or a distance/dissimilarity to summarize measure of the taxa between samples and then employs a statistical method or model to perform correlation and association analysis. In microbiome literature, in general multivariate approach is thought as more powerful than univariate approach. However, it still lacks of real convincing comparisons between these two approaches.

(4) *Which methods are more appropriate: parametric vs. nonparametric correlation and association methods?*

Both parametric and nonparametric methods have been used for estimating correlation and association. In general, both methods have their own pros and cons. For example, parametric methods are often more efficient than the corresponding nonparametric methods, and have the analysis results easily interpretable and useful for downstream data analysis. The main reason that nonparametric methods are growing in popularity is that we do not need to assume the distributions about the population as what we have to make with a parametric method. When the data do not meet the distributional requirements of parametric methods, such as sparse or skewed data, it is more beneficial to use nonparametric methods. For example, in microbiome study, there is no prior information for true correlation network of taxon-taxon interaction in real data; in such case, the nonparametric methods can be used to detect nonlinear relationships among taxa, while parametric methods such as CCLasso and SparCC are only reliable for exploring linear dependencies among them.[185]

(5) *Whether the methods that are able to detect more taxa (or OTUs) association with outcomes are better than those detect less taxa or OTUs association?*

Almost all newly developed methods considered the capability of detecting more taxa (or OTUs) association with outcomes as a criterion of method validation. This may be not true. The validation of methods should be based on whether the proposed models fit the data. Specially whether the models can fit the microbiome data features, such as multivariate, sparse with many zeros. For example, it has been shown that under the overdispersed and zero-inflated situations, the statistical significance testing results based on the low level of Poisson model are more significant than using NB and ZINB, but with worse model fitting. NB and ZINB are more fitted to the data, but have inflated P values.[730] Thus, the models that capture more taxa do not necessarily indicate that the models are better than those captured less taxa.

(6) *Which kind of methods and models have the more priority to be developed?*

Microbiome data have several features such as structured as tree, high-dimensional, compositional, sparse, and often have many zeros. These features challenge almost all standard statistical methods and impact the development of new statistical methods, which even make the traditional correlation and association methods invalid or redefine the concepts of correlation and association. Thus, no model can fit all features of microbiome data. In microbiome literature, there exist the phenomena: different approaches against each other (e.g., count-based vs. compositional-based), later proposed methods against previously proposed methods, even same authors against their previous developed methods when they proposed new methods. All these phenomena highlight the challenges of analyzing microbiome data. The question is which kind of methods and models are more fitted to microbiome data and need to be developed first?

(7) *How to simulate microbiome data to assess and compare statistical methods and which real microbiome data are better to be used for evaluation of new proposed methods?*

We still lack research on how to simulate microbiome data to assess and compare different statistical methods and evaluate their flexibilities in real data. To show the performances of method comparisons, the simulated data are critical. One model in one simulation that has better performance does not exactly means that this model will have better performance in other simulated data. For example, it is important: what models are used to simulate the data, how to set the parameters of sample size, dimensions (the number of taxa), the ratio of sample size and the number of taxa, percentage of zeros, and sparsity when the data are simulated. Also different real data also impact the analysis results. Otherwise, the arguments of one method that is better than others are not fully convincing. Microbiome data have unique characteristics; it is a real challenge to simulate microbiome abundances with high fidelity to the real data. Currently, few simulation methods can simulate microbiome data that capture all the desired sample-level and taxa-level properties (e.g., sparsity and overdispersion) as well as the taxa-taxa relationships.[823]

14. Closing comments

In this book chapter, we centralize on correlation and association methods for analyzing microbiome and omics data. We comprehensively reviewed and introduced various statistical methods for detecting correlation and association among environment, microbiome, and host factors, as well as among other omics. We started with traditional definitions of correlation and association and depicted the new implication of the association concept in microbiome data analysis.

In summary, the development of correlation and association analysis methods in microbiome and omics has focused on the unique data features that reflect the research advances in the fields: (1) from normal (Pearson's product-moment correlation) and monotonic (Spearman's rank-order) relationships to sparse and compositional relationships (SparCC, CCLasso, SPIEC-EASI); (2) from the correlation focusing on two variables to correlation network and co-occurrence networks; (3) from single co-occurrence network to differential co-occurrence networks; and (4) the developing correlation and methods have moved beyond one-omics (i.e., taxa abundances and diversity of microbiome) to interomic relationships.

References

1. Beale DJ, Karpe AV, Ahmed W, Beale DJ, Kouremenos KA, Palombo EA. Beyond metabolomics: a review of multi-omics-based approaches. In: *Microbial Metabolomics: Applications in Clinical, Environmental, and Industrial Microbiology*. Cham: Springer International Publishing; 2016:289–312.
2. Zhang X, Figeys D. Perspective and guidelines for metaproteomics in microbiome studies. *J Proteome Res*. 2019;18(6):2370–2380.
3. Spor A, Koren O, Ley R. Unravelling the effects of the environment and host genotype on the gut microbiome. *Nat Rev Microbiol*. 2011;9(4):279–290.
4. Xia Y, Sun J. Hypothesis testing and statistical analysis of microbiome. *Genes Dis*. 2017;4(3):138–148.
5. Rodgers JL, Nicewander WA. Thirteen ways to look at the correlation coefficient. *Am Stat*. 1988;42(1):59–66.
6. Tan P-N, Kumar V, Srivastava J. Selecting the right objective measure for association analysis. *Inf Syst*. 2004;29(4):293–313.
7. Bonett DG, Price RM. Inferential methods for the tetrachoric correlation coefficient. *J Educ Behav Stat*. 2005;30(2):213–225.
8. Agrawal R, Imieliński T, Swami A. Mining association rules between sets of items in large databases. In: *Proceedings of the 1993 ACM SIGMOD International Conference on Management of Data—SIGMOD '93*; 1993:207.
9. Brossette SE, Sprague AP, Hardin JM, Waites KB, Jones WT, Moser SA. Association rules and data mining in hospital infection control and public health surveillance. *J Am Med Inform Assoc*. 1998;5(4):373–381.
10. Greenblum S, Turnbaugh PJ, Borenstein E. Metagenomic systems biology of the human gut microbiome reveals topological shifts associated with obesity and inflammatory bowel disease. *Proc Natl Acad Sci USA*. 2012;109(2):594–599.
11. Qin J, Li Y, Cai Z, et al. A metagenome-wide association study of gut microbiota in type 2 diabetes. *Nature*. 2012;490:55–60.
12. Khamis H. Measures of association: how to choose? *J Diagn Med Sonogr*. 2008;24:155–162.
13. Ordonez C, Ezquerra N, Santana CA. Constraining and summarizing association rules in medical data. *Knowl Inf Syst*. 2006;9(3):1–2.
14. Wright A, Chen ES, Maloney FL. An automated technique for identifying associations between medications, laboratory results and problems. *J Biomed Inform*. 2010;43(6):891–901.
15. Abar O, Charnigo RJ, Rayapati A, Kavuluru R. On interestingness measures for mining statistically significant and novel clinical associations from EMRs. *ACM BCB*. 2016;2016:587–594. ACM Conference on Bioinformatics, Computational Biology and Biomedicine.
16. Xia Y, Sun J, et al. *What Are Microbiome Data? Statistical Analysis of Microbiome Data with R*. Singapore: Springer; 201829–41.
17. Hahsler M. A Probabilistic Comparison of Commonly Used Interest Measures for Association Rules. http://michael.hahsler.net/research/association_rules/measures.html; 2015.
18. Tan P-N, Michael S, Kumar V. Chapter 6. Association analysis: basic concepts and algorithms. In: *Introduction to Data Mining*. Addison-Wesley; 2005.
19. Liebetrau AM. *Measures of Association (Sage University Papers Series on Quantitative Applications in the Social Sciences, 07-004)*. CA, Sage: Newbury Park; 1983.
20. Pearson K. Mathematical contributions to the theory of evolution—on a form of spurious correlation which may arise when indices are used in the measurement of organs. *Proc R Soc Lond*. 1896–1897;60(359–367):489–498.
21. Baldi P, Brunak S, Chauvin Y, Andersen CA, Nielsen H. Assessing the accuracy of prediction algorithms for classification: an overview. *Bioinformatics*. 2000;16(5): 412–424.

22. Shadish W, Cook T, Campbell D. *Experimental and Quasi-Experimental Designs for Generalized Causal Inference*. Boston: Houghton Mifflin Compary; 2002.
23. Al-Katib WA, Dennis SM. Epididymal and testicular lesions in rams following experimental infection with Actinobacillus seminis. *N Z Vet J*. 2007;55(3):125–129.
24. Sheldon IM, Cronin J, Goetze L, Donofrio G, Schuberth H-J. Defining postpartum uterine disease and the mechanisms of infection and immunity in the female reproductive tract in cattle. *Biol Reprod*. 2009;81(6):1025–1032.
25. Fleiss JL. Measuring nominal scale agreement among many raters. *Psychol Bull*. 1971;76(5):378–382.
26. Cook TC, Campbell DT. *Quasi-Experimentation*. Boston: Houghton Mifflin; 1979.
27. Locke J. *An Essay Concerning Human Understanding Oxford*. England: Clarendon Press; 1975.
28. Moe KK, Yano T, Misumi K, Kubota C, Nibe K, Yamazaki W. Detection of antibodies against fusobacterium necrophorum and Porphyromonas levii-like species in dairy cattle with papillomatous digital dermatitis. *Microbiol Immunol*. 2010;54(6):338–346.
29. Berry KJ, Johnston JE, Paul J, Mielke W. Chapter 1. Introduction. In: *The Measurement of Association: A Permutation Statistical Approach*. Springer Nature: Switzerland; 2018.
29a. Reynolds HT. The Analysis of Cross-Classifications. New York: The Free Press; 1977.
30. Fleiss JL, Levin B, Paik MC. *Statistical Methods for Rates and Proportions*. Hoboken, New Jersey: John Wiley & Sons, Inc; 2003.
31. Paliy O, Shankar V. Application of multivariate statistical techniques in microbial ecology. *Mol Ecol*. 2016;25(5):1032–1057.
32. Joyce AR, Palsson BØ. The model organism as a system: integrating 'omics' data sets. *Nat Rev Mol Cell Biol*. 2006;7(3):198–210.
33. Martín L, Anguita A, Maojo V, Crespo J. Integration of omics data for cancer research. In: Cho WCS, ed. *An Omics Perspective on Cancer Research*. Dordrecht, Netherlands: Springer; 2010:249–266.
34. Clarke R, Ressom HW, Wang A, et al. The properties of high-dimensional data spaces: implications for exploring gene and protein expression data. *Nat Rev Cancer*. 2008;8(1):37–49.
35. Smyth GK. Linear models and empirical bayes methods for assessing differential expression in microarray experiments. *Stat Appl Genet Mol Biol*. 2004;3:3.
36. Aitchison J. *The Statistical Analysis of Compositional Data*. Chapman & Hall; 1986. reprinted in 2003, with additional material, by The Blackburn Press.
37. Fernandes AD, Reid JNS, Macklaim JM, McMurrough TA, Edgell DR, Gloor GB. Unifying the analysis of high-throughput sequencing datasets: characterizing RNA-seq, 16S rRNA gene sequencing and selective growth experiments by compositional data analysis. *Microbiome*. 2014;2(1):15.
38. Lovell D, Müller W, Taylor J, Zwart A, Helliwell C. Proportions, percentages, PPM: do the molecular biosciences treat compositional data right? In: Pawlowsky-Glahn V, Buccianti A, eds. *Compositional Data Analysis: Theory and Applications*. Chichester, UK: John Wiley & Sons, Ltd; 2011.
39. Xia Y, Sun J, Chen D-G. Compositional analysis of microbiome data. In: *Statistical Analysis of Microbiome Data with R*. Singapore: Springer; 2018:331–393.
40. Friedman J, Alm EJ. Inferring correlation networks from genomic survey data. *PLoS Comput Biol*. 2012;8(9):e1002687.
41. Fernandes AD, Macklaim JM, Linn TG, Reid G, Gloor GB. ANOVA-like differential expression (ALDEx) analysis for mixed population RNA-seq. *PLoS One*. 2013;8(7): e67019.
42. Lovell D, Pawlowsky-Glahn V, Egozcue JJ, Marguerat S, Bähler J. Proportionality: a valid alternative to correlation for relative data. *PLoS Comput Biol*. 2015;11(3): e1004075.

43. Eaton ML. *Multivariate Statistics: A Vector Space Approach*. 605 Third Ave., New York, NY 10158, USA: John Wiley & Sons, Inc.; 1983:512
44. Steuer R, Kurths J, Daub CO, Weise J, Selbig J. The mutual information: detecting and evaluating dependencies between variables. *Bioinformatics*. 2002;18(Suppl. 2): S231–S240.
45. Faust K, Sathirapongsasuti JF, et al. Microbial co-occurrence relationships in the human microbiome. *PLoS Comput Biol*. 2012;8(7):e1002606.
46. Weiss S, Van Treuren W, Lozupone C, et al. Correlation detection strategies in microbial data sets vary widely in sensitivity and precision. *ISME J*. 2016;10:1669.
47. Sohn MB, Li H. A GLM-based latent variable ordination method for microbiome samples. *Biometrics*. 2018;74(2):448–457.
48. Paulson JN, Stine OC, Bravo HC, Pop M. Differential abundance analysis for microbial marker-gene surveys. *Nat Methods*. 2013;10(12):1200–1202.
49. Tsilimigras MC, Fodor AA. Compositional data analysis of the microbiome: fundamentals, tools, and challenges. *Ann Epidemiol*. 2016;26(5):330–335.
50. Wang J, Thingholm LB, Skieceviciene J, et al. Genome-wide association analysis identifies variation in vitamin D receptor and other host factors influencing the gut microbiota. *Nat Genet*. 2016;48(11):1396–1406.
51. Jiang D, Armour CR, Hu C, et al. Microbiome multi-omics network analysis: statistical considerations, limitations, and opportunities. *Front Genet*. 2019;10:995.
52. Chen L, Garmaeva S, Zhernakova A, Fu J, Wijmenga C. A system biology perspective on environment–host–microbe interactions. *Hum Mol Genet*. 2018;27(R2): R187–R194.
53. Dai Z, Coker OO, Nakatsu G, et al. Multi-cohort analysis of colorectal cancer metagenome identified altered bacteria across populations and universal bacterial markers. *Microbiome*. 2018;6(1):70.
54. Dai Z, Wong SH, Yu J, Wei Y. Batch effects correction for microbiome data with Dirichlet-multinomial regression. *Bioinformatics*. 2018;35(5):807–814.
55. Gibbons SM, Duvallet C, Alm EJ. Correcting for batch effects in case-control microbiome studies. *PLoS Comput Biol*. 2018;14(4):e1006102.
56. Randall DW, Kieswich J, Swann J, et al. Batch effect exerts a bigger influence on the rat urinary metabolome and gut microbiota than uraemia: a cautionary tale. *Microbiome*. 2019;7(1):127.
57. Wang Y, LêCao K-A. Managing batch effects in microbiome data. *Brief Bioinform*. 2019;bbz105. https://doi.org/10.1093/bib/bbz105.
58. Costea PI, Zeller G, Sunagawa S, et al. Towards standards for human fecal sample processing in metagenomic studies. *Nat Biotechnol*. 2017;35(11):1069–1076.
59. Kennedy NA, Walker AW, Berry SH, et al. The impact of different DNA extraction kits and laboratories upon the assessment of human gut microbiota composition by 16S rRNA gene sequencing. *PLoS One*. 2014;9(2):e88982.
60. Maukonen J, Simões C, Saarela M. The currently used commercial DNA-extraction methods give different results of clostridial and actinobacterial populations derived from human fecal samples. *FEMS Microbiol Ecol*. 2012;79(3):697–708.
61. McOrist AL, Jackson M, Bird AR. A comparison of five methods for extraction of bacterial DNA from human faecal samples. *J Microbiol Methods*. 2002;50(2): 131–139.
62. Salonen A, Nikkila J, Jalanka-Tuovinen J, et al. Comparative analysis of fecal DNA extraction methods with phylogenetic microarray: effective recovery of bacterial and archaeal DNA using mechanical cell lysis. *J Microbiol Methods*. 2010;81(2): 127–134.
63. Smith B, Li N, Andersen AS, Slotved HC, Krogfelt KA. Optimising bacterial DNA extraction from faecal samples: comparison of three methods. *Open Microbiol J*. 2011;5:14–17.

64. Wesolowska-Andersen A, Bahl MI, Carvalho V, et al. Choice of bacterial DNA extraction method from fecal material influences community structure as evaluated by metagenomic analysis. *Microbiome*. 2014;2:19.
65. Sinha R, Abu-Ali G, Vogtmann E, et al. Assessment of variation in microbial community amplicon sequencing by the Microbiome Quality Control (MBQC) project consortium. *Nat Biotechnol*. 2017;35(11):1077–1086.
66. Schmidt TSB, Raes J, Bork P. The human gut microbiome: from association to modulation. *Cell*. 2018;172(6):1198–1215.
67. Hang J, Desai V, Zavaljevski N, et al. 16S rRNA gene pyrosequencing of reference and clinical samples and investigation of the temperature stability of microbiome profiles. *Microbiome*. 2014;2(1):31.
68. Song SJ, Amir A, Metcalf JL, et al. Preservation methods differ in fecal microbiome stability, affecting suitability for field studies. *mSystems*. 2016;1(3):e00021–16.
69. Vandeputte D, Tito RY, Vanleeuwen R, Falony G, Raes J. Practical considerations for large-scale gut microbiome studies. *FEMS Microbiol Rev*. 2017;41(1):S154–S167.
70. Mallick H, Ma S, Franzosa EA, Vatanen T, Morgan XC, Huttenhower C. Experimental design and quantitative analysis of microbial community multiomics. *Genome Biol*. 2017;18(1):228.
71. Lloyd-Price J, Arze C, Ananthakrishnan AN, et al. Multi-omics of the gut microbial ecosystem in inflammatory bowel diseases. *Nature*. 2019;569(7758):655–662.
72. Xia Y, Sun J, Chen D-G. Introductory overview of statistical analysis of microbiome data. In: *Statistical Analysis of Microbiome Data with R*. Singapore: Springer; 2018:43–75.
73. Dhariwal A, Chong J, Habib S, King IL, Agellon LB, Xia J. MicrobiomeAnalyst: a web-based tool for comprehensive statistical, visual and meta-analysis of microbiome data. *Nucleic Acids Res*. 2017;45(W1):W180–W188.
74. Ho NT, Li F, Wang S, Kuhn L. metamicrobiomeR: an R package for analysis of microbiome relative abundance data using zero-inflated beta GAMLSS and meta-analysis across studies using random effects models. *BMC Bioinf*. 2019;20(1):188.
75. Duvallet C. Meta-analysis generates and prioritizes hypotheses for translational microbiome research. *J Microbial Biotechnol*. 2018;11(2):273–276.
76. Duvallet C, Gibbons SM, Gurry T, Irizarry RA, Alm EJ. Meta-analysis of gut microbiome studies identifies disease-specific and shared responses. *Nat Commun*. 2017;8:1784. https://doi.org/10.1038/s41467-017-01973-8.
77. Pasolli E, Truong DT, Malik F, Waldron L, Segata N. Machine learning meta-analysis of large metagenomic datasets: tools and biological insights. *PLoS Comput Biol*. 2016;12(7):e1004977.
78. Galton F. Regression towards mediocrity in hereditary stature. *J Anthropol Inst G B Irel*. 1886;15:246–263.
79. Pearson K. Notes on the history of correlation. *Biometrika*. 1920;13(1):25–45.
80. Pearson K. Mathematical contributions to the theory of evolution. III. Regression, heredity, and panmixia. *Philos Trans R Soc Lond Ser A*. 1896;187:253–318.
81. Blum S, Mazuz M, Brenner J, Friedgut O, Stram Y, Koren O. Sample-based assessment of the microbial etiology of bovine necrotic vulvovaginitis. *Theriogenology*. 2007;68(2):290–293.
82. Lobb DA, Loeman HJ, Sparrow DG, Morck DW. Bovine polymorphonuclear neutrophil-mediated phagocytosis and an immunoglobulin G2 protease produced by Porphyromonas levii. *Can J Vet Res*. 1999;63(2):113–118.
83. Theriot CM, Koenigsknecht MJ, Carlson Jr PE, et al. Antibiotic-induced shifts in the mouse gut microbiome and metabolome increase susceptibility to Clostridium difficile infection. *Nat Commun*. 2014;5:3114.
84. Weir TL, Manter DK, Sheflin AM, Barnett BA, Heuberger AL, Ryan EP. Stool microbiome and metabolome differences between colorectal cancer patients and healthy adults. *PLoS One*. 2013;8(8):e70803.

85. Turnbaugh PJ, Ley RE, Hamady M, Fraser-Liggett CM, Knight R, Gordon JI. The human microbiome project. *Nature*. 2007;449:804–810.
86. Spearman C. The proof and measurement of association between two things. *Am J Psychol*. 1904;15(1):72–101.
87. Borkowf CB. Computing the nonnull asymptotic variance and the asymptotic relative efficiency of Spearman's rank correlation. *Comput Stat Data Anal*. 2002;39(3):271–286.
88. Kendall MG. *Rank Correlation Methods*. London: Charles Griffin & Co; 1955.
89. Yule GU, Kendall MG. *An Introduction to the Theory of Statistics*. London: Charles Griffin & Co; 1950.
90. You Y, Liang D, Wei R, et al. Evaluation of metabolite-microbe correlation detection methods. *Anal Biochem*. 2019;567:106–111.
91. Ammons MCB, Morrissey K, et al. Biochemical association of metabolic profile and microbiome in chronic pressure ulcer wounds. *PLoS One*. 2015;10(5):e0126735.
92. Gilbert JA, Quinn RA, Debelius J, et al. Microbiome-wide association studies link dynamic microbial consortia to disease. *Nature*. 2016;535(7610):94.
93. McHardy IH, Goudarzi M, Tong M, et al. Integrative analysis of the microbiome and metabolome of the human intestinal mucosal surface reveals exquisite inter-relationships. *Microbiome*. 2013;1(1):17.
94. Wu C, Zhou F, Ren J, Li X, Jiang Y, Ma S. A selective review of multi-level omics data integration using variable selection. *High Throughput*. 2019;8(1):4.
95. Kendall MG. A new measure of rank correlation. *Biometrika*. 1938;30(1/2):81–93.
96. Kendall MG. *Rank Correlation Methods*. Oxford: England, Griffin; 1948.
97. Kendall MG. *Rank Correlation Methods*. London: Griffin; 1970.
98. Zar JH. *Biostatistical Analysis*. Englewood Cliffs, NJ: Prentice-Hall; 2010.
99. Stuart A. The estimation and comparison of strengths of association in contingency tables. *Biometrika*. 1953;40(1/2):105–110.
100. Berry KJ, Johnston JE, Zahran S, Mielke Jr PW. Stuart's tau measure of effect size for ordinal variables: some methodological considerations. *Behav Res Methods*. 2009; 41(4):1144–1148.
101. Somers RH. A similarity between Goodman and Kruskal's Tau and Kendall's Tau, with a partial interpretation of the latter. *J Am Stat Assoc*. 1962;57(300):804–812.
102. Goodman LA, Kruskal WH. Measures of association for cross classifications. II: further discussion and references. *J Am Stat Assoc*. 1959;54(285):123–163.
103. Zhang Y, Han SW, Cox LM, Li H. A multivariate distance-based analytic framework for microbial interdependence association test in longitudinal study. *Genet Epidemiol*. 2017;41(8):769–778.
104. Wu J, Peters BA, Dominianni C, et al. Cigarette smoking and the oral microbiome in a large study of American adults. *ISME J*. 2016;10(10):2435–2446.
105. Matthews BW. Comparison of the predicted and observed secondary structure of T4 phage lysozyme. *Biochim Biophys Acta Protein Struct Mol Enzymol*. 1975;405(2):442–451.
106. Fisher RA. *Statistical Methods for Research Workers*. New York: Hafner; 1958.
107. Hutchinson TP. Kappa muddles together two sources of disagreement: tetrachoric correlation is preferable. *Res Nurs Health*. 1993;16(4):313–316.
108. Powers D, Ailab. Evaluation: from precision, recall and F-measure to ROC, informedness, markedness & correlation. *J Mach Learn Technol*. 2011;2:37–63.
109. Boughorbel S, Jarray F, El-Anbari M. Optimal classifier for imbalanced data using Matthews Correlation Coefficient metric. *PLoS One*. 2017;12(6):e0177678.
110. Westcott SL, Schloss PD. De novo clustering methods outperform reference-based methods for assigning 16S rRNA gene sequences to operational taxonomic units. *PeerJ*. 2015;3:e1487.
111. Schloss PD, Westcott SL. Assessing and improving methods used in operational taxonomic unit-based approaches for 16S rRNA gene sequence analysis. *Appl Environ Microbiol*. 2011;77(10):3219–3226.

112. Pearson K. X. On the criterion that a given system of deviations from the probable in the case of a correlated system of variables is such that it can be reasonably supposed to have arisen from random sampling. *Lond Edinb Dubl Phil Mag J Sci.* 1900;50(302):157–175.
113. Plackett RL. Karl pearson and the chi-squared test. *Int Stat Rev.* 1983;51(1):59–72.
114. Borewicz K, Gu F, Saccenti E, et al. Correlating infant faecal microbiota composition and human milk oligosaccharide consumption by microbiota of one-month old breastfed infants. *Mol Nutr Food Res.* 2019;24(10):201801214.
115. Cougoul A, Bailly X, Vourc'h G, Gasqui P. Rarity of microbial species: in search of reliable associations. *PLoS One.* 2019;14(3):e0200458.
116. Cramér H. Chapter 21. The two-dimensional case. In: *Mathematical Methods of Statistics.* Princeton: Princeton University Press; 1946:282.
117. Guilford J. *Psychometric Methods.* New York: McGraw-Hill Book Company, Inc.; 1936.
118. Yule GU. On the methods of measuring association between two attributes. *J R Stat Soc.* 1912;75(6):579–652.
119. Goodman LA, Kruskal WH. Measures of association for cross classifications. *J Am Stat Assoc.* 1954;49(268):732–764.
120. Sheskin DJ. *Handbook of Parametric and Nonparametric Statistical Procedures.* Boca Raton: Chapman and Hall/CRC; 2011.
121. La Rosa PS, Brooks JP, Deych E, et al. Hypothesis testing and power calculations for taxonomic-based human microbiome data. *PLoS One.* 2012;7(12):e52078.
122. Goodman LA, Kruskal WH. *Measures of Association for Cross Classifications.* New York, NY: Springer; 19792–34.
123. Cornfield J. A method of estimating comparative rates from clinical data; applications to cancer of the lung, breast, and cervix. *J Natl Cancer Inst.* 1951;11(6):1269–1275.
124. Mosteller F. Association and estimation in contingency tables. *J Am Stat Assoc.* 1968;63(321):1–28.
125. Edwards AWF. The measure of association in a 2×2 table. *J R Stat Soc Ser A.* 1963;126:109–114.
126. Morris JA, Gardner MJ. Calculating confidence intervals for relative risks (odds ratios) and standardised ratios and rates. *Br Med J (Clin Res Ed).* 1988;296(6632):1313–1316.
127. Berkson J. Smoking and lung cancer. *Am Stat.* 1963;17(4):15–22.
128. Feinstein AR. Clinical biostatistics; xx. The epidemiologic trohoc, the ablative risk ratio, and 'retrospective' research. *Clin Pharmacol Ther.* 1973;14(2):291–307.
129. Ahn J, Sinha R, Pei Z, et al. Human gut microbiome and risk for colorectal cancer. *J Natl Cancer Inst.* 2013;105(24):1907–1911.
130. Gill SR, Pop M, Deboy RT, et al. Metagenomic analysis of the human distal gut microbiome. *Science (New York, NY).* 2006;312(5778):1355–1359.
131. Schmitt FCF, Brenner T, Uhle F, et al. Gut microbiome patterns correlate with higher postoperative complication rates after pancreatic surgery. *BMC Microbiol.* 2019;19(1):42.
132. Yule GU. On the association of attributes in statistics: with illustrations from the material of the childhood society, &c. *Philos Trans R Soc Lond Ser A.* 1900;194:257–319.
133. Egozcue JJ, Pawlowsky-Glahn V, Gloor GB. Linear association in compositional data analysis. *Aust J Stat.* 2018;47(1):3–31.
134. Cohen J. A coefficient of agreement for nominal scales. *Educ Psychol Meas.* 1960;20: 37–46.
135. McHugh ML. Interrater reliability: the kappa statistic. *Biochem Med.* 2012;22(3): 276–282.
136. de Goffau MC, Lager S, Sovio U, et al. Human placenta has no microbiome but can contain potential pathogens. *Nature.* 2019;572(7769):329–334.
137. Kim H-N, Joo E-J, Cheong HS, et al. Gut microbiota and risk of persistent non-alcoholic fatty liver diseases. *J Clin Med.* 2019;8(8):1089.

138. Meier R, Thompson Jeffrey A, Chung M, et al. A Bayesian framework for identifying consistent patterns of microbial abundance between body sites. *Stat Appl Genet Mol Biol.* 2019;18(6). https://doi.org/10.1515/sagmb-2019-0027.
139. Jackson MA, Bonder MJ, Kuncheva Z, et al. Detection of stable community structures within gut microbiota co-occurrence networks from different human populations. *PeerJ.* 2018;6:e4303.
140. Jackson MA, Verdi S, Maxan M-E, et al. Gut microbiota associations with common diseases and prescription medications in a population-based cohort. *Nat Commun.* 2018;9(1):2655.
141. de Meij TG, Budding AE, de Groot EF, et al. Composition and stability of intestinal microbiota of healthy children within a Dutch population. *FASEB J.* 2016;30(4):1512–1522.
142. Drell T, Štšepetova J, Simm J, et al. The influence of different maternal microbial communities on the development of infant gut and oral microbiota. *Sci Rep.* 2017;7(1): 9940.
143. Jiang P, Green SJ, Chlipala GE, Turek FW, Vitaterna MH. Reproducible changes in the gut microbiome suggest a shift in microbial and host metabolism during spaceflight. *Microbiome.* 2019;7(1):113.
144. Jaccard P. Nouvelles recherches sur la distribution orale. *Bull Soc Vaud Sci Nat.* 1908;44:223–270.
145. van Rijsbergen CJ. *Information Retrieval.* London: Butterworths.; 1979
146. Xia Y, Sun J, Chen DG. Community diversity measures and calculations. In: *Statistical Analysis of Microbiome Data with R.* Singapore: Springer; 2018:167–190.
147. Xia Y, Sun J, Chen DG. Multivariate community analysis. In: *Statistical Analysis of Microbiome Data with R.* Singapore: Springer; 2018:285–330.
148. Boutin S, Graeber SY, Weitnauer M, et al. Comparison of microbiomes from different niches of upper and lower airways in children and adolescents with cystic fibrosis. *PLoS One.* 2015;10(1):e0116029.
149. Mainali KP, Bewick S, Thielen P, et al. Statistical analysis of co-occurrence patterns in microbial presence-absence datasets. *PLoS One.* 2017;12(11):e0187132.
150. Wang Z, Lou H, Wang Y, Shamir R, Jiang R, Chen T. GePMI: a statistical model for personal intestinal microbiome identification. *NPJ Biofilms Microbiomes.* 2018;4:20.
151. Cover TM, Thomas JA. *Elements of Information Theory.* Hoboken, New Jersey: John Wiley & Sons, Inc.; 2006.
152. Li J, Convertino M. Optimal microbiome networks: macroecology and criticality. *Entropy.* 2019;21(5):506.
153. Martín MÁ. Enterotype-like microbiome stratification as emergent structure in complex adaptive systems: a mathematical model. *bioRxiv.* 2018. https://doi.org/10.1101/402701.
154. Menon R, Ramanan V, Korolev KS. Interactions between species introduce spurious associations in microbiome studies. *PLoS Comput Biol.* 2018;14(1):e1005939.
155. Reshef DN, Reshef YA, Finucane HK, et al. Detecting novel associations in large data sets. *Science (New York, NY).* 2011;334(6062):1518–1524.
156. Daub CO, Steuer R, Selbig J, Kloska S. Estimating mutual information using B-spline functions—an improved similarity measure for analysing gene expression data. *BMC Bioinf.* 2004;5:118.
157. Cho I, Blaser MJ. The human microbiome: at the interface of health and disease. *Nat Rev Genet.* 2012;13(4):260–270.
158. Maurice CF, Haiser HJ, Turnbaugh PJ. Xenobiotics shape the physiology and gene expression of the active human gut microbiome. *Cell.* 2013;152(1–2):39–50.
159. Pinto AJ, Schroeder J, Lunn M, Sloan W, Raskin L. Spatial-temporal survey and occupancy-abundance modeling to predict bacterial community dynamics in the drinking water microbiome. *mBio.* 2014;5(3):e01135–01114.

160. Breiman L, Friedman JH, Olshen RA, Stone CJ. *Classification and Regression Trees*. New York: Chapman & Hall; 1984.
161. Ceriani L, Verme P. The origins of the Gini index: extracts from Variabilità e Mutabilità (1912) by Corrado Gini. *J Econ Inequal*. 2012;10(3):421–443.
162. Malmuthuge N, Guan LL. Gut microbiome and omics: a new definition to ruminant production and health. *Anim Front*. 2016;6(2):8–12.
163. Janzon A, Goodrich JK, Koren O, Waters JL, Ley RE. Interactions between the gut microbiome and mucosal immunoglobulins A, M, and G in the developing infant gut. *mSystems*. 2019;4(6):e00612–e00619.
164. Kobayashi T, Andoh A. Numerical analyses of intestinal microbiota by data mining. *J Clin Biochem Nutr*. 2018;62(2):124–131.
165. Piñero F, Vazquez M, Baré P, et al. A different gut microbiome linked to inflammation found in cirrhotic patients with and without hepatocellular carcinoma. *Ann Hepatol*. 2019;18(3):480–487.
166. Xia Y, Sun J, Chen DG. Modeling over-dispersed microbiome data. In: *Statistical Analysis of Microbiome Data with R*. Singapore: Springer; 2018:395–451.
167. McCarthy DJ, Chen Y, Smyth GK. Differential expression analysis of multifactor RNA-Seq experiments with respect to biological variation. *Nucleic Acids Res*. 2012;40(10):4288–4297.
168. Robinson MD, McCarthy DJ, Smyth GK. edgeR: a bioconductor package for differential expression analysis of digital gene expression data. *Bioinformatics*. 2010;26(1): 139–140.
169. Robinson MD, Oshlack A. A scaling normalization method for differential expression analysis of RNA-seq data. *Genome Biol*. 2010;11(3):R25.
170. Bullard JH, Purdom E, Hansen KD, Dudoit S. Evaluation of statistical methods for normalization and differential expression in mRNA-Seq experiments. *BMC Bioinf*. 2010;11(1):94.
171. Anders S, Huber W. Differential expression analysis for sequence count data. *Genome Biol*. 2010;11(10):R106.
172. Love MI, Huber W, Anders S. Moderated estimation of fold change and dispersion for RNA-seq data with DESeq2. *Genome Biol*. 2014;15:550.
173. McMurdie PJ, Holmes S. Waste not, want not: why rarefying microbiome data is inadmissible. *PLoS Comput Biol*. 2014;10(4):e1003531.
174. Xia Y, Sun J, Chen DG. Modeling zero-inflated microbiome data. In: *Statistical Analysis of Microbiome Data with R*. Singapore: Springer; 2018:453–496.
175. McKnight DT, Huerlimann R, Bower DS, Schwarzkopf L, Alford RA, Zenger KR. Methods for normalizing microbiome data: an ecological perspective. *Methods Ecol Evol*. 2019;10(3):389–400.
176. Chen L, Reeve J, Zhang L, Huang S, Wang X, Chen J. GMPR: a robust normalization method for zero-inflated count data with application to microbiome sequencing data. *PeerJ*. 2018;6:e4600.
177. Mandal S, Van Treuren W, White RA, Eggesbo M, Knight R, Peddada SD. Analysis of composition of microbiomes: a novel method for studying microbial composition. *Microb Ecol Health Dis*. 2015;26:27663.
178. Morton JT, Sanders J, Quinn RA, et al. Balance trees reveal microbial niche differentiation. *mSystems*. 2017;2(1):e00162–00116.
179. Vallejos CA, Risso D, Scialdone A, Dudoit S, Marioni JC. Normalizing single-cell RNA sequencing data: challenges and opportunities. *Nat Methods*. 2017;14(6): 565–571.
180. Giraldez MD, Spengler RM, Etheridge A, et al. Phospho-RNA-seq: a modified small RNA-seq method that reveals circulating mRNA and lncRNA fragments as potential biomarkers in human plasma. *EMBO J*. 2019;38(11):e101695.

181. Lee DM, Battson ML, Jarrell DK, et al. SGLT2 inhibition via dapagliflozin improves generalized vascular dysfunction and alters the gut microbiota in type 2 diabetic mice. *Cardiovasc Diabetol.* 2018;17(1):62.
182. Lee SC, Chua LL, Yap SH, et al. Enrichment of gut-derived Fusobacterium is associated with suboptimal immune recovery in HIV-infected individuals. *Sci Rep.* 2018;8(1):14277.
183. Biswas S, McDonald M, et al. Learning microbial interaction networks from metagenomic count data. *J Comput Biol.* 2016;23(6):526–535.
184. Linden SK, Sutton P, Karlsson NG, Korolik V, McGuckin MA. Mucins in the mucosal barrier to infection. *Mucosal Immunol.* 2008;1:183–197.
185. Fang H, Huang C, et al. CCLasso: correlation inference for compositional data through Lasso. *Bioinformatics.* 2015;31(19):3172–3180.
186. Deshpande NP, Riordan SM, Castaño-Rodríguez N, Wilkins MR, Kaakoush NO. Signatures within the esophageal microbiome are associated with host genetics, age, and disease. *Microbiome.* 2018;6(1):227.
187. Yoon G, Gaynanova I, Müller CL. Microbial networks in SPRING—semi-parametric rank-based correlation and partial correlation estimation for quantitative microbiome data. *Front Genet.* 2019;10:516.
188. Schwager E, Bielski C, Weingart G. ccrepe: ccrepe_and_nc.score. R package version 1.22.0. https://bioconductor.org/packages/ccrepe/; 2019.
189. Kostic AD, Gevers D, Siljander H, et al. The dynamics of the human infant gut microbiome in development and in progression toward type 1 diabetes. *Cell Host Microbe.* 2015;17(2):260–273.
190. Daquigan N, Seekatz AM, Greathouse KL, Young VB, White JR. High-resolution profiling of the gut microbiome reveals the extent of Clostridium difficile burden. *NPJ Biofilms Microbiomes.* 2017;3:35.
191. Esan EO, Abbey L, Yurgel S. Exploring the long-term effect of plastic on compost microbiome. *PLoS One.* 2019;14(3):e0214376.
192. Tibshirani R. Regression shrinkage and selection via the Lasso. *J R Stat Soc Ser B Methodol.* 1996;58(1):267–288.
193. Friedman J, Hastie T, et al. Regularization paths for generalized linear models via coordinate descent. *J Stat Softw.* 2010;33(1):1–22.
194. Liss MA, White JR, Goros M, et al. Metabolic biosynthesis pathways identified from fecal microbiome associated with prostate cancer. *Eur Urol.* 2018;74(5):575–582.
195. Wirbel J, Pyl PT, Kartal E, et al. Meta-analysis of fecal metagenomes reveals global microbial signatures that are specific for colorectal cancer. *Nat Med.* 2019;25(4):679–689.
196. Forslund, K., F. Hildebrand, T. Nielsen, G. Falony, E. Le Chatelier, S. Sunagawa, E. Prifti, S. Vieira-Silva, V. Gudmundsdottir, H. Krogh Pedersen, M. Arumugam, K. Kristiansen, A. Yvonne Voigt, H. Vestergaard, R. Hercog, P. Igor Costea, J. Roat Kultima, J. Li, T. Jørgensen, F. Levenez, J. Dore, H. Bjørn Nielsen, S. Brunak, J. Raes, T. Hansen, J. Wang, S. Dusko Ehrlich, P. Bork, O. Pedersen and H. I. T. c. Meta (2015). "Disentangling type 2 diabetes and metformin treatment signatures in the human gut microbiota." Nature 528(7581): 262-266.
197. Shankar J, Szpakowski S, Solis NV, et al. A systematic evaluation of high-dimensional, ensemble-based regression for exploring large model spaces in microbiome analyses. *BMC Bioinf.* 2015;16(1):31.
198. Xiao J, Chen L, Yu Y, Zhang X, Chen J. A phylogeny-regularized sparse regression model for predictive modeling of microbial community data. *Front Microbiol.* 2018;9:3112.
199. Meier L, van de Geer S, Bühlmann P. The group LASSO for logistic regression. *J R Stat Soc B.* 2008;70(1):53–71.

200. Meier L. *Grplasso: Fitting User-Specified Models with Group Lasso Penalty. R Package Version 0.4-6.* 2018.
201. Bickel PJ, Ritov YA, Tsybakov AB. Simultaneous analysis of Lasso and Dantzig selector. *Ann Stat.* 2009;37(4):1705–1732.
202. Muenchhoff M, Adland E, Karimanzira O, et al. Nonprogressing HIV-infected children share fundamental immunological features of nonpathogenic SIV infection. *Sci Transl Med.* 2016;8(358):358ra125.
203. Ravikumar P, Wainwright MJ, Lafferty JD. High-dimensional Ising model selection using 1-regularized logistic regression. *Ann Stat.* 2010;38(3):1287–1319.
204. van de Geer S, Buhlmann P, Ritov Y, Dezeure R. On asymptotically optimal confidence regions and tests for high-dimensional models. *Ann Stat.* 2014;42(3):1166–1202.
205. Simon N, Friedman J, Hastie T, Tibshirani R. A sparse-group Lasso. *J Comput Graph Stat.* 2013;22(2):231–245.
206. Simon N, Friedman J, Hastie T, Tibshirani R. *Fit a GLM (or Cox Model) with a Combination of Lasso and Group Lasso Regularization. R Package Version 1.2.* 2018.
207. Garcia TP, Müller S, Carroll RJ, Walzem RL. Identification of important regressor groups, subgroups and individuals via regularization methods: application to gut microbiome data. *Bioinformatics.* 2013;30(6):831–837.
208. Liquet B, de Micheaux PL, Hejblum BP, Thiébaut R. Group and sparse group partial least square approaches applied in genomics context. *Bioinformatics.* 2015;32(1):35–42.
209. Zhai J, Kim J, Knox KS, Twigg HL, Zhou H, Zhou JJ. Variance component selection with applications to microbiome taxonomic data. *Front Microbiol.* 2018;9:509.
210. Friedman J, Hastie T, et al. Sparse inverse covariance estimation with the graphical lasso. *Biostatistics.* 2007;9(3):432–441.
211. Kurtz ZD, Müller CL, Miraldi ER, Littman DR, Blaser MJ, Bonneau RA. Sparse and compositionally robust inference of microbial ecological networks. *PLoS Comput Biol.* 2015;11(5):e1004226.
212. Lo C, Marculescu R. MPLasso: Inferring microbial association networks using prior microbial knowledge. *PLoS Comput Biol.* 2017;13(12):e1005915.
213. McGregor K, Labbe A, Greenwood CMT. MDiNE: a model to estimate differential co-occurrence networks in microbiome studies. *Bioinformatics.* 2020;36(6):1840–1847.
214. Bálint M, Bahram M, Eren AM, et al. Millions of reads, thousands of taxa: microbial community structure and associations analyzed via marker genes. *FEMS Microbiol Rev.* 2016;40(5):686–700.
215. Knight R, Vrbanac A, Taylor BC, et al. Best practices for analysing microbiomes. *Nat Rev Microbiol.* 2018;16(7):410–422.
216. Layeghifard M, Hwang DM, Guttman DS. Disentangling interactions in the microbiome: a network perspective. *Trends Microbiol.* 2017;25(3):217–228.
217. Silverman JD, Washburne AD, Mukherjee S, David LA. A phylogenetic transform enhances analysis of compositional microbiota data. *Elife.* 2017;6:e21887.
218. Ban Y, An L, Jiang H. Investigating microbial co-occurrence patterns based on metagenomic compositional data. *Bioinformatics (Oxford, England).* 2015;31(20):3322–3329.
219. Schwager E, Mallick H, Ventz S, Huttenhower C. A Bayesian method for detecting pairwise associations in compositional data. *PLoS Comput Biol.* 2017;13(11):e1005852.
220. Dethlefsen L, McFall-Ngai M, et al. An ecological and evolutionary perspective on human-microbe mutualism and disease. *Nature.* 2007;449:811–818.
221. Cardona C, Weisenhorn P, Henry C, Gilbert JA. Network-based metabolic analysis and microbial community modeling. *Curr Opin Microbiol.* 2016;31:124–131.
222. Faust K, Lima-Mendez G, Lerat J-S, et al. Cross-biome comparison of microbial association networks. *Front Microbiol.* 2015;6:1200.

223. Dohlman AB, Shen X. Mapping the microbial interactome: statistical and experimental approaches for microbiome network inference. *Exp Biol Med (Maywood).* 2019;244(6):445–458.
224. Abu-Ali GS, Mehta RS, Lloyd-Price J, et al. Metatranscriptome of human faecal microbial communities in a cohort of adult men. *Nat Microbiol.* 2018;3(3):356–366.
225. Chiquet J, Mariadassou M, Robin S. *Variational Inference for Sparse Network Reconstruction From Count Data.* arXiv Preprint; 2018, arXiv:1806.03120.
226. Gevers D, Kugathasan S, et al. The treatment-naive microbiome in new-onset Crohn's disease. *Cell Host Microbe.* 2014;15(3):382–392.
227. Tipton L, Cuenco KT, Huang L, et al. Measuring associations between the microbiota and repeated measures of continuous clinical variables using a lasso-penalized generalized linear mixed model. *BioData Min.* 2018;11(1):12.
228. Tipton L, Müller CL, Kurtz ZD, et al. Fungi stabilize connectivity in the lung and skin microbial ecosystems. *Microbiome.* 2018;6(1):12.
229. Morton JT, Aksenov AA, Nothias LF, et al. Learning representations of microbe–metabolite interactions. *Nat Methods.* 2019;16:1306–1314.
230. Gloor GB, Macklaim JM, Pawlowsky-Glahn V, Egozcue JJ. Microbiome datasets are compositional: and this is not optional. *Front Microbiol.* 2017;8:2224.
231. Mahana D, Trent CM, Kurtz ZD, et al. Antibiotic perturbation of the murine gut microbiome enhances the adiposity, insulin resistance, and liver disease associated with high-fat diet. *Genome Med.* 2016;8(1):48.
232. Barberán A, Bates ST, Casamayor EO, Fierer N. Using network analysis to explore co-occurrence patterns in soil microbial communities. *ISME J.* 2012;6(2):343–351.
233. Fuhrman J, Steele J. Community structure of marine bacterioplankton: patterns, networks, and relationships to function. *Aquat Microb Ecol.* 2008;53:69–81.
234. Agler MT, Ruhe J, Kroll S, et al. Microbial hub taxa link host and abiotic factors to plant microbiome variation. *PLoS Biol.* 2016;14(1):e1002352.
235. Deng Y, Jiang Y-H, Yang Y, He Z, Luo F, Zhou J. Molecular ecological network analyses. *BMC Bioinf.* 2012;13:113.
236. Steele JA, Countway PD, Xia L, et al. Marine bacterial, archaeal and protistan association networks reveal ecological linkages. *ISME J.* 2011;5(9):1414–1425.
237. Fisher CK, Mehta P. Identifying keystone species in the human gut microbiome from metagenomic timeseries using sparse linear regression. *PLoS One.* 2014;9(7):e102451.
238. Fiehn O. Metabolomics—the link between genotypes and phenotypes. *Plant Mol Biol.* 2002;48(1–2):155–171.
239. Patti GJ, Yanes O, Siuzdak G. Metabolomics: the apogee of the omics trilogy. *Nat Rev Mol Cell Biol.* 2012;13(4):263–269.
240. Chong J, Xia J. Computational approaches for integrative analysis of the metabolome and microbiome. *Metabolites.* 2017;7(4):62.
241. Human Microbiome Project, C. Structure, function and diversity of the healthy human microbiome. *Nature.* 2012;486(7402):207–214.
242. Cribbs SK, Uppal K, et al. Correlation of the lung microbiota with metabolic profiles in bronchoalveolar lavage fluid in HIV infection. *Microbiome.* 2016;4:3.
243. Johnson CH, Spilker ME, Goetz L, Peterson SN, Siuzdak G. Metabolite and microbiome interplay in cancer immunotherapy. *Cancer Res.* 2016;76(21):6146–6152.
244. Lee K, Pletcher SD, Lynch SV, Goldberg AN, Cope EK. Heterogeneity of microbiota dysbiosis in chronic rhinosinusitis: potential clinical implications and microbial community mechanisms contributing to sinonasal inflammation. *Front Cell Infect Microbiol.* 2018;8:168.
245. Levy R, Borenstein E. Metabolic modeling of species interaction in the human microbiome elucidates community-level assembly rules. *Proc Natl Acad Sci USA.* 2013;110(31):12804–12809.

246. Kundu P, Manna B, Majumder S, Ghosh A. Species-wide metabolic interaction network for understanding natural lignocellulose digestion in termite gut microbiota. *Sci Rep*. 2019;9(1):16329.
247. Levy R, Borenstein E. Metagenomic systems biology and metabolic modeling of the human microbiome: from species composition to community assembly rules. *Gut microbes*. 2014;5(2):265–270.
248. Sung J, Kim S, Cabatbat JJT, et al. Global metabolic interaction network of the human gut microbiota for context-specific community-scale analysis. *Nat Commun*. 2017;8(1):15393.
249. Mallick H, Franzosa EA, McLver LJ, et al. Predictive metabolomic profiling of microbial communities using amplicon or metagenomic sequences. *Nat Commun*. 2019;10(1):3136.
250. Noecker C, Eng A, Srinivasan S, et al. Metabolic model-based integration of microbiome taxonomic and metabolomic profiles elucidates mechanistic links between ecological and metabolic variation. *mSystems*. 2016;1(1):e00013–e00015.
251. Larsen PE, Collart FR, Field D, et al. Predicted Relative Metabolomic Turnover (PRMT): determining metabolic turnover from a coastal marine metagenomic dataset. *Microb Inf Exp*. 2011;1(1):4.
252. Segata N, Boernigen D, Tickle TL, Morgan XC, Garrett WS, Huttenhower C. Computational meta'omics for microbial community studies. *Mol Syst Biol*. 2013;9(1):666.
253. Casero D, Gill K, Sridharan V, et al. Space-type radiation induces multimodal responses in the mouse gut microbiome and metabolome. *Microbiome*. 2017;5(1):105.
254. Garza DR, van Verk MC, Huynen MA, Dutilh BE. Towards predicting the environmental metabolome from metagenomics with a mechanistic model. *Nat Microbiol*. 2018;3(4):456–460.
255. Larsen PE, Dai Y. Metabolome of human gut microbiome is predictive of host dysbiosis. *GigaScience*. 2015;4:42.
256. Mason OU, Scott NM, Gonzalez A, et al. Metagenomics reveals sediment microbial community response to Deepwater Horizon oil spill. *ISME J*. 2014;8(7):1464–1475.
257. Abubucker S, Segata N, Goll J. Metabolic reconstruction for metagenomic data and its application to the human microbiome. *PLoS Comput Biol*. 2012;8(6):e1002358.
258. Aagaard K, Ma J, Antony KM, Ganu R, Petrosino J, Versalovic J. The placenta harbors a unique microbiome. *Sci Transl Med*. 2014;6(237):237ra265.
259. Caspi R, Altman T, Billington R, et al. The MetaCyc database of metabolic pathways and enzymes and the BioCyc collection of pathway/genome databases. *Nucleic Acids Res*. 2013;42(D1):D459–D471.
260. Zeller G, Tap J, Voigt AY, et al. Potential of fecal microbiota for early-stage detection of colorectal cancer. *Mol Syst Biol*. 2014;10(11):766.
261. Nishida K, Ono K, Kanaya S, Takahashi K. KEGGscape: a Cytoscape app for pathway data integration. *F1000Res*. 2014;3:144.
262. Vázquez-Baeza Y, Callewaert C, Debelius J, et al. Impacts of the human gut microbiome on therapeutics. *Annu Rev Pharmacol Toxicol*. 2018;58(1):253–270.
263. Starr AE, Deeke SA, Li L, et al. Proteomic and metaproteomic approaches to understand host–microbe interactions. *Anal Chem*. 2018;90(1):86–109.
264. Stinson LF, Boyce MC, Payne MS, Keelan JA. The not-so-sterile womb: evidence that the human fetus is exposed to bacteria prior to birth. *Front Microbiol*. 2019;10:1124.
265. Stull VJ, Finer E, Bergmans RS, et al. Impact of edible cricket consumption on gut microbiota in healthy adults, a double-blind, randomized crossover trial. *Sci Rep*. 2018;8(1):10762.

266. Langille MGI, Zaneveld J, Caporaso JG, et al. Predictive functional profiling of microbial communities using 16S rRNA marker gene sequences. *Nat Biotechnol.* 2013;31(9):814–821.
267. Douglas GM, Beiko RG, Langille MGI. Predicting the functional potential of the microbiome from marker genes using PICRUSt. In: Beiko RG, Hsiao W, Parkinson J, eds. *Microbiome Analysis: Methods and Protocols.* New York: Springer Nature; 2018:169–177.
268. Meyer F, Paarmann D, D'Souza M, et al. The metagenomics RAST server—a public resource for the automatic phylogenetic and functional analysis of metagenomes. *BMC Bioinf.* 2008;9:386.
269. Aßhauer KP, Wemheuer B, Daniel R, Meinicke P. Tax4Fun: predicting functional profiles from metagenomic 16S rRNA data. *Bioinformatics (Oxford, England).* 2015;31(17):2882–2884.
270. Iwai S, Weinmaier T, Schmidt BL, et al. Piphillin: improved prediction of metagenomic content by direct inference from human microbiomes. *PLoS One.* 2016;11(11):e0166104.
271. Goodrich JK, Waters JL, Poole AC, et al. Human genetics shape the gut microbiome. *Cell.* 2014;159(4):789–799.
272. Carmody RN, Gerber GK, Luevano JM, et al. Diet dominates host genotype in shaping the murine gut microbiota. *Cell Host Microbe.* 2015;17(1):72–84.
273. Sampson TR, Debelius JW, Thron T, et al. Gut microbiota regulate motor deficits and neuroinflammation in a model of Parkinson's disease. *Cell.* 2016;167(6): 1469–1480.e1412.
274. Thompson LR, Sanders JG, Mcdonald D, Fogliano V, Jurburg S, Larsen P. A communal catalogue reveals Earth's multiscale microbial diversity. *Nature.* 2017;551(7681):457–463.
275. Aßhauer K, Meinicke P. On the estimation of metabolic profiles in metagenomics. In: Beißbarth T, Kollmar M, Leha A, Morgenstern B, Schultz A-K, Waack S, Wingender E, eds. vol. 34. *German Conference on Bioinformatics 2013 (GCB'13).* Germany: Schloss Dagstuhl — Leibniz-Zentrum für Informatik, Dagstuhl Publishing; 2013:1–13. OpenAccess Series in Informatics.
276. Markowitz VM, Chen IMA, Palaniappan K, et al. IMG: the Integrated Microbial Genomes database and comparative analysis system. *Nucleic Acids Res.* 2012;40(Database issue):D115–D122.
277. Markowitz VM, Chen IMA, Palaniappan K, et al. IMG 4 version of the integrated microbial genomes comparative analysis system. *Nucleic Acids Res.* 2014;42(Database issue):D560–D567.
278. Quast C, Pruesse E, Yilmaz P, et al. The SILVA ribosomal RNA gene database project: improved data processing and web-based tools. *Nucleic Acids Res.* 2013;41(Database issue):D590–D596.
279. Bautista QM, Schroeder J, Sevillano-Rivera M, et al. Emerging investigators series: microbial communities in full-scale drinking water distribution systems—a meta-analysis. *Environ Sci Water Res Technol.* 2016;2(4):631–644.
280. Bian X, Chi L, Gao B, Tu P, Ru H, Lu K. Gut microbiome response to sucralose and its potential role in inducing liver inflammation in mice. *Front Physiol.* 2017;8:487.
281. Camarinha-Silva A, Maushammer M, Wellmann R, Vital M, Preuss S, Bennewitz J. Host genome influence on gut microbial composition and microbial prediction of complex traits in pigs. *Genetics.* 2017;206(3):1637–1644.
282. Mukherjee A, Chettri B, Langpoklakpam JS, et al. Bioinformatic approaches including predictive metagenomic profiling reveal characteristics of bacterial response to petroleum hydrocarbon contamination in diverse environments. *Sci Rep.* 2017; 7(1):1108.

283. Schloss PD, Westcott SL, Ryabin T, Hall JR, Hartmann M, Hollister EB. Introducing mothur: open-source, platform-independent, community-supported software for describing and comparing microbial communities. *Appl Environ Microbiol.* 2009;75(23):7537–7541.
284. Cole JR, Wang Q, Fish JA, et al. Ribosomal Database Project: data and tools for high throughput rRNA analysis. *Nucleic Acids Res.* 2014;42(Database issue):D633–D642.
285. Callahan BJ, McMurdie PJ, Rosen MJ, Han AW, Johnson AJA, Holmes SP. DADA2: high-resolution sample inference from Illumina amplicon data. *Nat Methods.* 2016;13(7):581–583.
286. Edgar RC. UPARSE: highly accurate OTU sequences from microbial amplicon reads. *Nat Methods.* 2013;10(10):996.
287. Abia ALK, Alisoltani A, Keshri J, Ubomba-Jaswa E. Metagenomic analysis of the bacterial communities and their functional profiles in water and sediments of the Apies River, South Africa, as a function of land use. *Sci Total Environ.* 2018;616–617:326–334.
288. Bates KA, Clare FC, O'Hanlon S, et al. Amphibian chytridiomycosis outbreak dynamics are linked with host skin bacterial community structure. *Nat Commun.* 2018;9(1):693.
289. Mullish BH, Pechlivanis A, Barker GF, Thursz MR, Marchesi JR, McDonald JAK. Functional microbiomics: evaluation of gut microbiota-bile acid metabolism interactions in health and disease. *Methods.* 2018;149:49–58.
290. Franzosa EA, Morgan XC, Segata N, et al. Relating the metatranscriptome and metagenome of the human gut. *Proc Natl Acad Sci USA.* 2014;111(22):E2329–E2338.
291. Gosalbes MJ, Durbán A, Pignatelli M, et al. Metatranscriptomic approach to analyze the functional human gut microbiota. *PLoS One.* 2011;6(3):e17447.
292. Verberkmoes NC, Russell AL, Shah M, et al. Shotgun metaproteomics of the human distal gut microbiota. *ISME J.* 2009;3(2):179–189.
293. Perez-Cobas AE, Gosalbes MJ, Friedrichs A, et al. Gut microbiota disturbance during antibiotic therapy: a multi-omic approach. *Gut.* 2013;62(11):1591–1601.
294. Chang PV. Chemical mechanisms of colonization resistance by the gut microbial metabolome. *ACS Chem Biol.* 2020. https://doi.org/10.1021/acschembio.9b00813.
295. Tolosana Delgado R, Talebi H, Khodadadzadeh M, Boogaart K. *On Machine Learning Algorithms and Compositional Data.* 2019.
296. Quinn TP, Erb I. Another look at microbe–metabolite interactions: how scale invariant correlations can outperform a neural network. *bioRxiv.* 2019. https://doi.org/10.1101/847475.
297. Morton JT, McDonald D, Aksenov AA, et al. Revisiting microbe-metabolite interactions: doing better than random. *bioRxiv.* 2019. https://doi.org/10.1101/2019.12.10.871905.
298. Baker JL, Morton JT, Dinis M, et al. Deep metagenomics examines the oral microbiome during dental caries, revealing novel taxa and co-occurrences with host molecules. *bioRxiv.* 2019. https://doi.org/10.1101/804443.
299. Mu A, Carter GP, Li L, et al. Microbe-metabolite associations linked to the rebounding murine gut microbiome post-colonization with vancomycin resistant Enterococcus faecium. *bioRxiv.* 2019. https://doi.org/10.1101/849539.
300. Banerjee S, Schlaeppi K, van der Heijden MGA. Keystone taxa as drivers of microbiome structure and functioning. *Nat Rev Microbiol.* 2018;16(9):567–576.
301. Ligi T, Oopkaup K, Truu M, et al. Characterization of bacterial communities in soil and sediment of a created riverine wetland complex using high-throughput 16S rRNA amplicon sequencing. *Ecol Eng.* 2014;72:56–66.
302. Mann E, Schmitz-Esser S, Zebeli Q, Wagner M, Ritzmann M, Metzler-Zebeli BU. Mucosa-associated bacterial microbiome of the gastrointestinal tract of weaned pigs and dynamics linked to dietary calcium-phosphorus. *PLoS One.* 2014;9(1):e86950.

303. Wang J-T, Zheng Y-M, Hu H-W, Zhang L-M, Li J, He J-Z. Soil pH determines the alpha diversity but not beta diversity of soil fungal community along altitude in a typical Tibetan forest ecosystem. *J Soil Sediment.* 2015;15(5):1224–1232.
304. Dutilh BE, Cassman N, McNair K, et al. A highly abundant bacteriophage discovered in the unknown sequences of human faecal metagenomes. *Nat Commun.* 2014;5(1):4498.
305. Ridaura VK, Faith JJ, Rey FE, et al. Gut microbiota from twins discordant for obesity modulate metabolism in mice. *Science.* 2013;341(6150):1241214.
306. Stein RR, Bucci V, Toussaint NC, Buffie CG, Ratsch G, Pamer EG. Ecological modeling from time-series inference: insight into dynamics and stability of intestinal microbiota. *PLoS Comput Biol.* 2013;9(12):e1003388.
307. Xia Y, Sun J, Chen D-G. Exploratory analysis of microbiome data and beyond. In: *Statistical Analysis of Microbiome Data with R*. Singapore: Springer; 2018:191–249.
308. Pearson K. LIII. On lines and planes of closest fit to systems of points in space. *Lond Edinb Dubl Phil Mag J Sci.* 1901;2(11):559–572.
309. Hotelling H. Analysis of a complex of statistical variables into principal components. *J Educ Psychol.* 1933;24(6):417–441.
310. Hotelling H. Relations between two sets of variates. *Biometrika.* 1935;28:321–377.
311. Jolliffe I. *Principal Component Analysis.* New York: Springer; 2002.
312. Johnstone IM, Lu AY. On consistency and sparsity for principal components analysis in high dimensions. *J Am Stat Assoc.* 2009;104(486):682–693.
313. Johnstone IM, Lu AY. *Sparse Principal Components Analysis.* arXiv e-prints; 2009.
314. Legendre P, Legendre L. *Numerical Ecology.* Amsterdam: Elsevier; 2012.
315. Pierre L, Legendre G. Ecologically meaningful transformations for ordination of species data. *Oecologia.* 2001;129:271–280.
316. ter Braak C, Šmilauer P. Topics in constrained and unconstrained ordination. *Plant Ecol.* 2015;216:683–696.
317. Parkhomenko E, Tritchler D, Beyene J. Sparse canonical correlation analysis with application to genomic data integration. *Stat Appl Genet Mol Biol.* 2009;8:1.
318. Hardoon DR, Shawe-Taylor J. Sparse canonical correlation analysis. *Mach Learn.* 2011;83(3):331–353.
319. Fukuyama J. Adaptive gPCA: a method for structured dimensionality reduction with applications to microbiome data. *Ann Appl Stat.* 2019;13(2):1043–1067.
320. Jolliffe IT, Trendafilov NT, Uddin M. A modified principal component technique based on the LASSO. *J Comput Graph Stat.* 2003;12(3):531–547.
321. Silverman BW. Smoothed functional principal components analysis by choice of norm. *Ann Stat.* 1996;24(1):1–24.
322. Clos-Garcia M, Andrés-Marin N, Fernández-Eulate G, et al. Gut microbiome and serum metabolome analyses identify molecular biomarkers and altered glutamate metabolism in fibromyalgia. *EBioMedicine.* 2019;46:499–511.
323. Matson V, Fessler J, Bao R, et al. The commensal microbiome is associated with anti–PD-1 efficacy in metastatic melanoma patients. *Science.* 2018;359(6371):104–108.
324. Sui Y, Lewis GK, Wang Y, et al. Mucosal vaccine efficacy against intrarectal SHIV is independent of anti-Env antibody response. *J Clin Invest.* 2019;129(3):1314–1328.
325. Wakita Y, Shimomura Y, Kitada Y, Yamamoto H, Ohashi Y, Matsumoto M. Taxonomic classification for microbiome analysis, which correlates well with the metabolite milieu of the gut. *BMC Microbiol.* 2018;18(1):188.
326. Hirschfeld HO. A connection between correlation and contingency. *Math Proc Camb Philos Soc.* 1935;31(4):520–524.
327. Benzécri J-P. L'Analyse des Données. In: *L'Analyse des Correspondances*. Paris, France: Dunod; 1973:vol. II.
328. Yelland PM. An introduction to correspondence analysis. *Math J* 2010;12:1–23.

329. Alcaraz LD, Belda-Ferre P, Cabrera-Rubio R, et al. Identifying a healthy oral microbiome through metagenomics. *Clin Microbiol Infect*. 2012;18(s4):54–57.
330. Gomez A, Petrzelkova K, Yeoman CJ, et al. Gut microbiome composition and metabolomic profiles of wild western lowland gorillas (Gorilla gorilla gorilla) reflect host ecology. *Mol Ecol*. 2015;24(10):2551–2565.
331. Jakobsson HE, Jernberg C, Andersson AF, Sjolund-Karlsson M, Jansson JK, Engstrand L. Short-term antibiotic treatment has differing long-term impacts on the human throat and gut microbiome. *PLoS One*. 2010;5(3):e9836.
332. Nogueira VLR, Rocha LL, Colares GB, et al. Microbiomes and potential metabolic pathways of pristine and anthropized Brazilian mangroves. *Reg Stud Mar Sci*. 2015;2:56–64.
333. Gower JC. Principal coordinates analysis. In: *Encyclopedia of Biostatistics*. John Wiley & Sons, Inc.; 2005
334. Gower JC. Some distance properties of latent root and vector methods used in multivariate analysis. *Biometrika*. 1966;53(3-4):325–338.
335. Zhang SL, Bai L, Goel N, et al. Human and rat gut microbiome composition is maintained following sleep restriction. *Proc Natl Acad Sci USA*. 2017;114(8): E1564–e1571.
336. Gopalakrishnan V, Spencer CN, Nezi L, et al. Gut microbiome modulates response to anti–PD-1 immunotherapy in melanoma patients. *Science*. 2018;359(6371):97–103.
337. Jovel J, Patterson J, Wang W, et al. Characterization of the gut microbiome using 16S or shotgun metagenomics. *Front Microbiol*. 2016;7:459.
338. Khine WWT, Zhang Y, Goie GJY, et al. Gut microbiome of pre-adolescent children of two ethnicities residing in three distant cities. *Sci Rep*. 2019;9(1):7831.
339. Ross AA, Müller KM, Weese JS, Neufeld JD. Comprehensive skin microbiome analysis reveals the uniqueness of human skin and evidence for phylosymbiosis within the class Mammalia. *Proc Natl Acad Sci USA*. 2018;115(25):E5786–E5795.
340. Shepard RN. The analysis of proximities: multidimensional scaling with an unknown distance function. I. *Psychometrika*. 1962;27(2):125–140.
341. Kruskal JB. Multidimensional scaling by optimizing goodness of fit to a nonmetric hypothesis. *Psychometrika*. 1964;29(1):1–27.
342. Kruskal JB. Nonmetric multidimensional scaling: a numerical method. *Psychometrika*. 1964;29(2):115–129.
343. Mead A. Review of the development of multidimensional scaling methods. *J R Stat Soc Ser D Stat*. 1992;41(1):27–39.
344. Antharam VC, McEwen DC, Garrett TJ, et al. An integrated metabolomic and microbiome analysis identified specific gut microbiota associated with fecal cholesterol and coprostanol in clostridium difficile infection. *PLoS One*. 2016;11(2):e0148824.
345. Lewis Z, Sidamonidze K, Tsereteli TVD, et al. The fecal microbial community of breast-fed infants from Armenia and Georgia. *Sci Rep*. 2017;7:40932.
346. Ramette A. Multivariate analyses in microbial ecology. *FEMS Microbiol Ecol*. 2007;62(2):142–160.
347. Anderson MJ, Willis TJ. Canonical analysis of principal coordinates: a useful method of constrained ordination for ecology. *Ecology*. 2003;84:511–525.
348. Ter Braak CJF, Prentice IC. A theory of gradient analysis. In: Begon M, Fitter AH, Ford ED, Macfadyen A, eds. vol. 18. *Advances in Ecological Research*. Academic Press; 1988:271–317.
349. Park R, Dzialo MC, Spaepen S, et al. Microbial communities of the house fly Musca domestica vary with geographical location and habitat. *Microbiome*. 2019;7(1):147.
350. Pérez-Jaramillo JE, Carrión VJ, Bosse M, et al. Linking rhizosphere microbiome composition of wild and domesticated Phaseolus vulgaris to genotypic and root phenotypic traits. *ISME J*. 2017;11(10):2244–2257.

351. Zhang C, Derrien M, Levenez F, et al. Ecological robustness of the gut microbiota in response to ingestion of transient food-borne microbes. *ISME J.* 2016;10(9): 2235–2245.
352. Bork P, Serrano L. Towards cellular systems in 4D. *Cell.* 2005;121(4):507–509.
353. Palsson B. Two-dimensional annotation of genomes. *Nat Biotechnol.* 2004;22(10): 1218–1219.
354. Reed JL, Famili I, Thiele I, Palsson BO. Towards multidimensional genome annotation. *Nat Rev Genet.* 2006;7(2):130–141.
355. Lee S, Batzoglou S. Application of independent component analysis to microarrays. *Genome Biol.* 2003;4(11):R76.
356. Purdom E, Holmes S. Error distribution for gene expression data. *Stat Appl Genet Mol Biol.* 2005;4(1). Article 16.
357. van den Berg RA, Hoefsloot HCJ, Westerhuis JA, Smilde AK, van der Werf MJ. Centering, scaling, and transformations: improving the biological information content of metabolomics data. *BMC Genomics.* 2006;7:142.
358. Zou H, Hastie T, Tibshirani R. Sparse principal component analysis. *J Comput Graph Stat.* 2006;15(2):265–286.
359. Zou H. The adaptive Lasso and its oracle properties. *J Am Stat Assoc.* 2006;101(476): 1418–1429.
360. Journée M, Nesterov Y, Richtarik P, Sepulchre R. Generalized power method for sparse principal component analysis. *J Mach Learn Res.* 2008;11:517–553.
361. Martino C, Morton JT, Marotz CA, et al. A novel sparse compositional technique reveals microbial perturbations. *mSystems.* 2019;4(1):e00016–e00019.
362. Hyvärinen A, Oja E. Indepedent component analysis: algorithms and applications. *Neural Netw.* 2000;13(4–5):411–430.
363. van Velzen EJJ, Westerhuis JA, van Duynhoven JPM, et al. Multilevel data analysis of a crossover designed human nutritional intervention study. *J Proteome Res.* 2008;7(10):4483–4491.
364. Aziz R, Verma CK, Srivastava N. A fuzzy based feature selection from independent component subspace for machine learning classification of microarray data. *Genomics Data.* 2016;8:4–15.
365. Steinfath M, Groth D, Lisec J, Selbig J. Metabolite profile analysis: from raw data to regression and classification. *Physiol Plant.* 2008;132(2):150–161.
366. Yao F, Coquery J, Lê Cao K-A. Independent Principal Component Analysis for biologically meaningful dimension reduction of large biological data sets. *BMC Bioinf.* 2012;13(1):24.
367. Frigyesi A, Veerla S, Lindgren D, Höglund M. Independent component analysis reveals new and biologically significant structures in micro array data. *BMC Bioinf.* 2006;7(1):290.
368. Schölkopf B, Smola A, Müller K-R. Kernel principal component analysis. In: *International conference on artificial neural networks*, Springer; 1997.
369. Schölkopf B, Smola A, Müller K-R. Nonlinear component analysis as a kernel eigenvalue problem. *Neural Comput.* 1998;10(5):1299–1319.
370. Schölkopf B, Smola A, Müller K-R. Kernel principal component analysis. In: *Advances in Kernel Methods—Support Vector Learning*. MIT Press; 1999:327–352.
371. Loncar-Turukalo T, Lazic I, Maljkovic N, Brdar S. *Clustering of Microbiome Data: Evaluation of Ensemble Design Approaches*. 2019.
372. Shiokawa Y, Date Y, Kikuchi J. Application of kernel principal component analysis and computational machine learning to exploration of metabolites strongly associated with diet. *Sci Rep.* 2018;8(1):3426.
373. Landgraf AJ, Lee Y. Generalized principal component analysis: projection of saturated model parameters. *Technometrics.* 2019;1–14. https://doi.org/10.1080/00401706.2019.1668854.

374. Vidal R, Ma Y, Piazzi J. A new GPCA algorithm for clustering subspaces by fitting, differentiating and dividing polynomials. In: *IEEE International Conference on Computer Vision and Pattern Recognition*. 2004:510–517. vol. I.
375. Vidal R, Ma Y, Sastry S. Generalized principal component analysis (GPCA). In: *CVPR*. 2003:621–628. vol. 1.
376. Vidal R, Ma Y, Sastry S. Generalized principal component analysis. *IEEE Trans Pattern Anal Mach Intell*. 2005;27(12):1945–1959.
377. Smallman L, Artemiou A, Morgan J. Sparse generalised principal component analysis. *Pattern Recogn*. 2018;83:443–455.
378. Hoerl AE, Kennard RW. Ridge regression: biased estimation for nonorthogonal problems. *Technometrics*. 2000;42(1):80–86.
379. Zou H, Hastie T. Regularization and variable selection via the elastic net. *J R Stat Soc Ser B Stat Methodol*. 2005;67(2):301–320.
380. Fan J, Li R. Variable selection via nonconcave penalized likelihood and its oracle properties. *J Am Stat Assoc*. 2001;96(456):1348–1360.
381. Allen GI, Maletić-Savatić M. Sparse non-negative generalized PCA with applications to metabolomics. *Bioinformatics*. 2011;27(21):3029–3035.
382. Allen GI, Grosenick L, Taylor J. A generalized least-square matrix decomposition. *J Am Stat Assoc*. 2014;109(505):145–159.
383. Pavoine S, Dufour A-B, Chessel D. From dissimilarities among species to dissimilarities among communities: a double principal coordinate analysis. *J Theor Biol*. 2004;228(4): 523–537.
384. Matsen FA, Evans SN. Edge principal components and squash clustering: using the special structure of phylogenetic placement data for sample comparison. *PLoS One*. 2013;8(3):e56859.
385. Savorani F, Rasmussen M, Mikkelsen M, Engelsen S. A primer to nutritional metabolomics by NMR spectroscopy and chemometrics. *Food Res Int*. 2013;54: 1131–1145.
386. Purdom E. Analysis of a data matrix and a graph: metagenomic data and the phylogenetic tree. *Ann Appl Stat*. 2011;5(4):2326–2358.
387. Bik EM, Eckburg PB, Gill SR, et al. Molecular analysis of the bacterial microbiota in the human stomach. *Proc Natl Acad Sci USA*. 2006;103(3):732–737.
388. Zubin J. A technique for measuring like-mindedness. *J Abnorm Soc Psychol*. 1938;33(4):508–516.
389. Tryon RC. *Cluster Analysis; Correlation Profile and Orthometric (Factor) Analysis for the Isolation of Unities in Mind and Personality*. Ann Arbor, Michigan: Edwards Brother, Inc., Lithoprinters and Publishers; 1939.
390. Driver H, Kroeber A. *Quantitative Expression of Cultural Relationships*. Berkeley: University of California Press; 1932.
391. Bailey KD. Cluster analysis. *Sociol Methodol*. 1975;6:59–128.
392. Bridges Jr C. Hierarchical cluster analysis. *Psychol Rep*. 1966;8(3):851–854.
393. MacQueen J. Some methods for classification and analysis of multivariate observations. In: *Proceedings of the Fifth Berkeley Symposium on Mathematical Statistics and Probability, Volume 1: Statistics*, Berkeley, CA: University of California Press; 1967.
394. Caliński T, Harabasz J. A dendrite method for cluster analysis. *Commun Stat*. 1974;3(1):1–27.
395. Kaufman L, Rousseeuw PJ. Clustering by means of medoids. In: *Statistical Data Analysis Based on the L1-Norm and Related Methods*. New York, North-Holland: Y. Dodge, Elsevier; 1987:405–416.
396. Kaufman L, Rousseeuw P. Partitioning around medoids (Program PAM). In: *Finding Groups in Data: An Introduction to Cluster Analysis*. New York: John Wiley; 1990:68–125.

397. Rousseeuw PJ. Silhouettes: a graphical aid to the interpretation and validation of cluster analysis. *J Comput Appl Math*. 1987;20:53–65.
398. Banfield JD, Raftery AE. Model-based Gaussian and non-Gaussian clustering. *Biometrics*. 1993;49(3):803–821.
399. Ferreira L, Hitchcock DB. A comparison of hierarchical methods for clustering functional data. *Commun Stat Simul Comput*. 2009;38(9):1925–1949.
400. Kettenring J. The practice of cluster analysis. *J Classif*. 2006;23:3–30.
401. Sneath PH. The application of computers to taxonomy. *J Gen Microbiol*. 1957;17(1): 201–226.
402. McQuitty LL. Hierarchical linkage analysis for the isolation of types. *Educ Psychol Meas*. 1960;20:55–67.
403. Sokal RR, Sneath PHA. *Principles of Numerical Taxonomy*. W.H. Freeman; 1963.
404. Sokal RR, Michener CD. A statistical method for evaluating systematic relationships. *Univ Kans Sci Bull*. 1958;38:1409–1438.
405. Ward JH. Hierarchical grouping to optimize an objective function. *J Am Stat Assoc*. 1963;58(301):236–244.
406. Blashfield RK. Mixture model tests of cluster analysis: Accuracy of four agglomerative hierarchical methods. *Psychol Bull*. 1976;83(3):377–388.
407. Hands S, Everitt B. A Monte Carlo study of the recovery of cluster structure in binary data by hierarchical clustering techniques. *Multivar Behav Res*. 1987;22(2):235–243.
408. Johnson RA, Wichern DW. *Applied Multivariate Statistical Analysis*. Upper Saddle River, NJ: Prentice Hall; 2007.
409. Kuiper FK, Fisher L. 391: a Monte Carlo comparison of six clustering procedures. *Biometrics*. 1975;31(3):777–783.
410. Milligan GW. An examination of the effect of six types of error perturbation on fifteen clustering algorithms. *Psychometrika*. 1980;45(3):325–342.
411. Shankar V, Homer D, Rigsbee L, et al. The networks of human gut microbe-metabolite associations are different between health and irritable bowel syndrome. *ISME J*. 2015;9(8):1899–1903.
412. Sridharan GV, Choi K, Klemashevich C, et al. Prediction and quantification of bioactive microbiota metabolites in the mouse gut. *Nat Commun*. 2014;5(1):5492.
413. Gajer P, Brotman RM, Bai G, et al. Temporal dynamics of the human vaginal microbiota. *Sci Transl Med*. 2012;4(132):132ra152.
414. Li F, Hullar MA, Beresford SA, Lampe JW. Variation of glucoraphanin metabolism in vivo and ex vivo by human gut bacteria. *Br J Nutr*. 2011;106(3):408–416.
415. Romo-Vaquero M, Cortés-Martín A, Loria-Kohen V, et al. Deciphering the human gut microbiome of urolithin metabotypes: association with enterotypes and potential cardiometabolic health implications. *Mol Nutr Food Res*. 2019;63(4):1800958.
416. Shankar V, Hamilton MJ, Khoruts A, et al. Species and genus level resolution analysis of gut microbiota in Clostridium difficile patients following fecal microbiota transplantation. *Microbiome*. 2014;2:13.
417. Veiga P, Gallini CA, Beal C, et al. Bifidobacterium animalis subsp. lactis fermented milk product reduces inflammation by altering a niche for colitogenic microbes. *Proc Natl Acad Sci USA*. 2010;107(42):18132–18137.
418. Venkataraman A, Sieber JR, Schmidt AW, Waldron C, Theis KR, Schmidt TM. Variable responses of human microbiomes to dietary supplementation with resistant starch. *Microbiome*. 2016;4(1):33.
419. Rahbar S. *K-Means Clustering Method on Microbiome Data Unsupervised Machine-Learning Method to Group Microbime Data of the Same Charactristics*. 2017.
420. Taie WS, Omar Y, Badr A. Clustering of human intestine microbiomes with K-means. In: *2018 21st Saudi Computer Society National Computer Conference (NCC)*; 2018.

421. Kang C, Zhang Y, Zhu X, et al. Healthy subjects differentially respond to dietary capsaicin correlating with specific gut enterotypes. *J Clin Endocrinol Metabol.* 2016;101(12):4681–4689.
422. Volokh O, Klimenko N, Berezhnaya Y, et al. Human gut microbiome response induced by fermented dairy product intake in healthy volunteers. *Nutrients.* 2019;11(3).
423. Wu GD, Chen J, Hoffmann C, et al. Linking long-term dietary patterns with gut microbial enterotypes. *Science.* 2011;334(6052):105–108.
424. Hullar MAJ, Lancaster SM, Li F, et al. Enterolignan-producing phenotypes are associated with increased gut microbial diversity and altered composition in premenopausal women in the United States. *Cancer Epidemiol Biomark Prev.* 2015;24(3):546–554.
425. Tsivtsivadze E, Borgdorff H, Wijgert J, Schuren F, Verhelst R, Heskes T. *Neighborhood Co-regularized Multi-view Spectral Clustering of Microbiome Data.* 2013.
426. Luxburg U. A tutorial on spectral clustering. *Stat Comput.* 2007;17:395–416.
427. Ng A, Jordan M, Weiss Y. On spectral clustering: analysis and an algorithm. *Adv Neural Inf Proces Syst.* 2002;2(14):849–856.
428. Kumar A, Rai P, Daume III H. Co-regularized multi-view spectral clustering. In: *Advances in Neural Information Processing Systems.* 2011.
429. Strehl A, Ghosh J. *Cluster Ensembles—a Knowledge Reuse Framework for Combining Multiple Partitions.* JMLR.org; 2003.
430. Imangaliyev S, Keijser B, Crielaard W, Tsivtsivadze E. Personalized microbial network inference via co-regularized spectral clustering. *Methods.* 2015;83:28–35.
431. Biesbroek G, Tsivtsivadze E, Sanders EAM, et al. Early respiratory microbiota composition determines bacterial succession patterns and respiratory health in children. *Am J Respir Crit Care Med.* 2014;190(11):1283–1292.
432. Borgdorff H, Tsivtsivadze E, Verhelst R, et al. Lactobacillus-dominated cervicovaginal microbiota associated with reduced HIV/STI prevalence and genital HIV viral load in African women. *ISME J.* 2014;8(9):1781–1793.
433. Gautam R, Borgdorff H, Jespers V, et al. Correlates of the molecular vaginal microbiota composition of African women. *BMC Infect Dis.* 2015;15(1):86.
434. Borgdorff H, Armstrong SD, Tytgat HLP, et al. Unique insights in the cervicovaginal Lactobacillus iners and L. crispatus proteomes and their associations with microbiota dysbiosis. *PLoS One.* 2016;11(3):e0150767.
435. Kootte RS, Levin E, Salojärvi J, et al. Improvement of insulin sensitivity after lean donor feces in metabolic syndrome is driven by baseline intestinal microbiota composition. *Cell Metab.* 2017;26(4):611–619.e616.
436. Botschuijver S, Welting O, Levin E, et al. Reversal of visceral hypersensitivity in rat by Menthacarin®, a proprietary combination of essential oils from peppermint and caraway, coincides with mycobiome modulation. *Neurogastroenterol Motil.* 2018;30(6):e13299.
437. Chen W, Cheng Y, Zhang C, Zhang S, Zhao H. MSClust: a Multi-Seeds based Clustering algorithm for microbiome profiling using 16S rRNA sequence. *J Microbiol Methods.* 2013;94(3):347–355.
438. Jiang P, Singh M. SPICi: a fast clustering algorithm for large biological networks. *Bioinformatics.* 2010;26(8):1105–1111.
439. Sun Y, Cai Y, Huse SM, et al. A large-scale benchmark study of existing algorithms for taxonomy-independent microbial community analysis. *Brief Bioinform.* 2012;13(1):107–121.
440. Vinh NX, Epps J, Bailey J. *Information Theoretic Measures for Clusterings Comparison: Variants, Properties, Normalization and Correction for Chance.* JMLR.org; 2010.
441. Li W, Godzik A. Cd-hit: a fast program for clustering and comparing large sets of protein or nucleotide sequences. *Bioinformatics.* 2006;22(13):1658–1659.

442. Edgar RC. Search and clustering orders of magnitude faster than BLAST. *Bioinformatics*. 2010;26(19):2460–2461.
443. Russell DJ, Way SF, Benson AK, Sayood K. A grammar-based distance metric enables fast and accurate clustering of large sets of 16S sequences. *BMC Bioinf.* 2010;11:601.
444. Ghodsi M, Liu B, Pop M. DNACLUST: accurate and efficient clustering of phylogenetic marker genes. *BMC Bioinf.* 2011;12(1):271.
445. Sun Y, Cai Y, Liu L, et al. ESPRIT: estimating species richness using large collections of 16S rRNA pyrosequences. *Nucleic Acids Res.* 2009;37(10):e76.
446. Cai Y, Sun Y. ESPRIT-Tree: hierarchical clustering analysis of millions of 16S rRNA pyrosequences in quasilinear computational time. *Nucleic Acids Res.* 2011;39(14):e95.
447. Flynn JM, Brown EA, Chain FJJ, MacIsaac HJ, Cristescu ME. Toward accurate molecular identification of species in complex environmental samples: testing the performance of sequence filtering and clustering methods. *Ecol Evol.* 2015;5(11): 2252–2266.
448. Franzén O, Hu J, Bao X, Itzkowitz SH, Peter I, Bashir A. Improved OTU-picking using long-read 16S rRNA gene amplicon sequencing and generic hierarchical clustering. *Microbiome*. 2015;3(1):43.
449. Mao Q, Zheng W, Wang L, Cai Y, Mai V, Sun Y. Parallel hierarchical clustering in linearithmic time for large-scale sequence analysis. In: *2015 IEEE International Conference on Data Mining*; 2015.
450. Schmidt TSB, Matias Rodrigues JF, von Mering C. Limits to robustness and reproducibility in the demarcation of operational taxonomic units. *Environ Microbiol.* 2015;17(5): 1689–1706.
451. Zheng W, Mao Q, Genco RJ, et al. A parallel computational framework for ultra-large-scale sequence clustering analysis. *Bioinformatics*. 2018;35(3):380–388.
452. Wei Z-G, Zhang S-W. MtHc: a motif-based hierarchical method for clustering massive 16S rRNA sequences into OTUs. *Mol Biosyst.* 2015;11(7):1907–1913.
453. Wei Z-G, Zhang S-W. DBH: a de Bruijn graph-based heuristic method for clustering large-scale 16S rRNA sequences into OTUs. *J Theor Biol.* 2017;425:80–87.
454. Wei Z-G, Zhang S-W, Zhang Y-Z. DMclust, a density-based modularity method for accurate OTU picking of 16S rRNA sequences. *Mol Inf.* 2017;36(12):1600059.
455. Cai Y, Zheng W, Yao J, et al. ESPRIT-Forest: parallel clustering of massive amplicon sequence data in subquadratic time. *PLoS Comput Biol.* 2017;13(4):e1005518.
456. Wei Z-G, Zhang S-W. DMSC: a dynamic multi-seeds method for clustering 16S rRNA sequences into OTUs. *Front Microbiol.* 2019;10:428.
457. Claesson MJ, Clooney AG, O'Toole PW. A clinician's guide to microbiome analysis. *Nat Rev Gastroenterol Hepatol.* 2017;14(10):585–595.
458. Czaja AJ. Factoring the intestinal microbiome into the pathogenesis of autoimmune hepatitis. *World J Gastroenterol.* 2016;22(42):9257–9278.
459. Igolkina AA, Grekhov GA, Pershina EV, et al. Identifying components of mixed and contaminated soil samples by detecting specific signatures of control 16S rRNA libraries. *Ecol Indic.* 2018;94:446–453.
460. Wei Z-G, Zhang S-W. NPBSS: a new PacBio sequencing simulator for generating the continuous long reads with an empirical model. *BMC Bioinf.* 2018;19(1):177.
461. Humphries A, Daud A. The gut microbiota and immune checkpoint inhibitors. *Hum Vaccin Immunother.* 2018;14(9):2178–2182.
462. Asgari E, Garakani K, McHardy AC, Mofrad MRK. MicroPheno: predicting environments and host phenotypes from 16S rRNA gene sequencing using a k-mer based representation of shallow sub-samples. *Bioinformatics*. 2018;34(13):i32–i42.
463. Zheng W, Yang L, Genco RJ, Wactawski-Wende J, Buck M, Sun Y. SENSE: siamese neural network for sequence embedding and alignment-free comparison. *Bioinformatics*. 2018;35(11):1820–1828.

464. Zou Q, Lin G, Jiang X, Liu X, Zeng X. Sequence clustering in bioinformatics: an empirical study. *Brief Bioinform.* 2020;21(1):1–10.
465. Zheng Z, Kramer S, Schmidt B. DySC: software for greedy clustering of 16S rRNA reads. *Bioinformatics (Oxford, England).* 2012;28:2182–2183.
466. Hao X, Jiang R, Chen T. Clustering 16S rRNA for OTU prediction: a method of unsupervised Bayesian clustering. *Bioinformatics (Oxford, England).* 2011;27:611–618.
467. Feng X, Song L, Wang S, et al. Accurate prediction of neoadjuvant chemotherapy pathological complete remission (pCR) for the four sub-types of breast cancer. *IEEE Access.* 2019;7:134697–134706.
468. Mesuere B, Devreese B, Debyser G, Aerts M, Vandamme P, Dawyndt P. Unipept: tryptic peptide-based biodiversity analysis of metaproteome samples. *J Proteome Res.* 2012;11(12):5773.
469. Muth T, Behne A, Heyer R, et al. The MetaProteomeAnalyzer: a powerful open-source software suite for metaproteomics data analysis and interpretation. *J Proteome Res.* 2015;14(3):1557.
470. Sinkko H, Lukkari K, Jama AS, et al. Phosphorus chemistry and bacterial community composition interact in brackish sediments receiving agricultural discharges. *PLoS One.* 2011;6(6):e21555.
471. Ye R, Wright AL. Multivariate analysis of chemical and microbial properties in histosols as influenced by land-use types. *Soil Tillage Res.* 2010;110(1):94–100.
472. Wang X, Eijkemans MJC, Wallinga J, et al. Multivariate approach for studying interactions between environmental variables and microbial communities. *PLoS One.* 2012;7(11):e50267.
473. Rodriguez-Valera F. Environmental genomics, the big picture? *FEMS Microbiol Lett.* 2004;231(2):153.
474. Zhang X, Ning Z, Mayne J, et al. MetaPro-IQ: a universal metaproteomic approach to studying human and mouse gut microbiota. *Microbiome.* 2016;4(1):31.
475. Zhang Y, Ouyang Z. Joint principal trend analysis for longitudinal high-dimensional data. *Biometrics.* 2018;74(2):430–438.
476. Tofallis C. Model building with multiple dependent variables and constraints. *J R Stat Soc Ser D.* 1999;48:371–378.
477. Cliff N, Krus DJ. Interpretation of canonical analysis: rotated vs. unrotated solutions. *Psychometrika.* 1976;41(1):35–42.
478. Gygi SP, Rochon Y, Franza BR, Aebersold R. Correlation between protein and mRNA abundance in yeast. *Mol Cell Biol.* 1999;19(3):1720.
479. Parkhomenko E, Tritchler D, Beyene J. Genome-wide sparse canonical correlation of gene expression with genotypes. *BMC Proc.* 2007;1(Suppl. 1):S119.
480. Suo X, Minden V, Nelson B, Tibshirani R, Saunders M. *Sparse Canonical Correlation Analysis.* ArXiv Preprint; 2017, ArXiv:1705.10865.
481. Waaijenborg S, de Witt Hamer Philip CV, Aeilko HZ. Quantifying the association between gene expressions and DNA-markers by penalized canonical correlation analysis. *Stat Appl Genet Mol Biol.* 2008;7(1). Article 3.
482. Witten D, Tibshirani R, Hastie T. A penalized matrix decomposition, with applications to sparse principal components and canonical correlation analysis. *Biostatistics.* 2009;10(3):515–534.
483. Witten DM, Tibshirani RJ. Extensions of sparse canonical correlation analysis with applications to genomic data. *Stat Appl Genet Mol Biol.* 2009;8(1):28.
484. Gossmann A, Zille P, Calhoun V, Wang Y-P. FDR-corrected sparse canonical correlation analysis with applications to imaging genomics. *IEEE Trans Medical Imaging.* 2017. https://doi.org/10.1109/TMI.2018.2815583.
485. Benjamini Y, Hochberg Y. Controlling the false discovery rate: a practical and powerful approach to multiple testing. *J R Stat Soc Ser B Methodol.* 1995;57(1):289–300.

486. Solari OS, Brown JB, Bickel PJ. *Sparse Canonical Correlation Analysis via Concave Minimization*. arXiv e-prints; 2019.
487. Witten D, Tibshirani R, Gross S, Narasimhan B. *PMA: Penalized Multivariate Analysis*. 2019.
488. Abraham G, Inouye M. Fast principal component analysis of large-scale genome-wide data. *PLoS One*. 2014;9(4):e93766.
489. Abraham G, Inouye M. FlashPCA: fast sparse canonical correlation analysis of genomic data. *bioRxiv*. 2016. https://doi.org/10.1101/047217.
490. Alam MA, Nasser M, Fukumizu K. Sensitivity analysis in robust and kernel canonical correlation analysis. In: *Proceedings of 11th International Conference on Computer and Information Technology, ICCIT 2008*; 2008.
491. Blaschko M, Lampert C, Gretton A. *Semi-supervised Laplacian Regularization of Kernel Canonical Correlation Analysis*. 2008.
492. Cai J. The distance between feature subspaces of kernel canonical correlation analysis. *Math Comput Model*. 2013;57(3):970–975.
493. Hardoon DR, Shawe-Taylor J. Convergence analysis of kernel Canonical Correlation Analysis: theory and practice. *Mach Learn*. 2009;74(1):23–38.
494. Van Gestel T, Suykens J, De Brabanter J, De Moor B, Vandewalle J. *Kernel Canonical Correlation Analysis and Least Squares Support Vector Machines*. 2001.
495. Akaho S. A kernel method for canonical correlation analysis. In: *Proceedings of the International Meeting of the Psychometric Society (IMPS 2001); Osaka*; 2001:4.
496. Akaho S. *A Kernel Method for Canonical Correlation Analysis*. 2006.
497. Fukumizu K, Bach FR, Gretton A. Statistical consistency of kernel canonical correlation analysis. *J Mach Learn Res*. 2007;8:361–383.
498. Lai PL, Fyfe C. Kernel and nonlinear canonical correlation analysis. *Int J Neural Syst*. 2000;10(5):365–377.
499. Melzer T, Reiter M, Bischof H. *Nonlinear Feature Extraction Using Generalized Canonical Correlation Analysis*. 2001.
500. Bach FR, Jordan MI. Kernel independent component analysis. *J Mach Learn Res*. 2003;3:1–48.
501. Hardoon DR, Szedmak S, Shawe-Taylor J. Canonical correlation analysis: an overview with application to learning methods. *Neural Comput*. 2004;16(12):2639–2664.
502. Larson NB, Jenkins GD, Larson MC, et al. Kernel canonical correlation analysis for assessing gene-gene interactions and application to ovarian cancer. *Eur J Hum Genet*. 2014;22(1):126–131.
503. Bie T, De Moor B. *On the Regularization of Canonical Correlation Analysis*. 2003.
504. Lai PL, Fyfe C. A neural implementation of canonical correlation analysis. *Neural Netw*. 1999;12(10):1391–1397.
505. Andrew G, Arora R, Bilmes J, Livescu K. Deep canonical correlation analysis. In: Sanjoy D, David M, eds. vol. 28. *Proceedings of the 30th International Conference on Machine Learning*. 2013:1247–1255. Proceedings of Machine Learning Research, PMLR.
506. Leurgans SE, Moyeed RA, Silverman BW. Canonical correlation analysis when the data are curves. *J R Stat Soc Ser B Methodol*. 1993;55(3):725–740.
507. Ramsay, J. O. a. S., B. W. *Functional Data Analysis. Encyclopedia of Statistics in Behavioral Science*. New York: Springer-Verlag; 2005.
508. Ravikumar P, Lafferty J, Liu H, Wasserman L. Sparse additive models. *J R Stat Soc Series B Stat Methodology*. 2009;71(5):1009–1030.
509. Balakrishnan S, Puniyani K, Lafferty J. Sparse additive functional and kernel CCA. In: *Proceedings of the 29th International Conference on Machine Learning, ICML 2012*. 2012. vol. 1.

510. Dolédec S, Chessel D. Co-inertia analysis: an alternative method for studying species–environment relationships. *Freshw Biol.* 1994;31(3):277–294.
511. Thioulouse J. Simultaneous analysis of a sequence of paired ecological tables: a comparison of several methods. *Ann Appl Stat.* 2011;5(4):2300–2325.
512. Dray S, Chessel D, Thioulouse J. Co-Inertia analysis and the linking of ecological data tables. *Ecology.* 2003;84(11):3078–3089.
513. Culhane AC, Perrière G, Higgins DG. Cross-platform comparison and visualisation of gene expression data using co-inertia analysis. *BMC Bioinf.* 2003;4(1):59.
514. Zhang X, Nieuwdorp M, Groen AK, Zwinderman AH. Statistical evaluation of diet-microbe associations. *BMC Microbiol.* 2019;19(1):90.
515. Bady P, Dolédec S, Dumont B, Fruget J-F. Multiple co-inertia analysis: a tool for assessing synchrony in the temporal variability of aquatic communities. *C R Biol.* 2004;327(1):29–36.
516. Berge J. Orthogonal procrustes rotation for two or more matrices. *Psychometrika.* 1977;42(2):267–276.
517. Hanafi M, Kohler A, Qannari E-M. Connections between multiple co-inertia analysis and consensus principal component analysis. *Chemom Intel Lab Syst.* 2011;106(1): 37–40.
518. Claesson MJ, Jeffery IB, Conde S, Power SE, O'Connor EM, Cusack S. Gut microbiota composition correlates with diet and health in the elderly. *Nature.* 2012;488:178–184.
519. Hill CJ, Lynch DB, Murphy K, et al. Evolution of gut microbiota composition from birth to 24 weeks in the INFANTMET Cohort. *Microbiome.* 2017;5(1):4.
520. Zhang C, Yin A, Li H, Wang R, Wu G, Shen J. Dietary modulation of gut microbiota contributes to alleviation of both genetic and simple obesity in children. *EBioMedicine.* 2015;2(8):968–984.
521. Liu R, Hong J, Xu X, et al. Gut microbiome and serum metabolome alterations in obesity and after weight-loss intervention. *Nat Med.* 2017;23(7):859–868.
522. Jovanović I, Živković M, Jovanović J, Djurić T, Stanković A. The co-inertia approach in identification of specific microRNA in early and advanced atherosclerosis plaque. *Med Hypotheses.* 2014;83(1):11–15.
523. Raimondi S, Roncaglia L, Lucia M. Bioconversion of soy isoflavones daidzin and daidzein by Bifidobacterium strains. *Appl Microbiol Biotechnol.* 2009;81(5):943–950.
524. Gao J, Lin L, Chen Z, et al. In vitro digestion and fermentation of three polysaccharide fractions from Laminaria japonica and their impact on lipid metabolism-associated human gut microbiota. *J Agric Food Chem.* 2019;67(26):7496–7505.
525. Yuan JP, Wang JH, Liu X. Metabolism of dietary soy isoflavones to equol by human intestinal microflora—implications for health. *Mol Nutr Food Res.* 2007;51(7):765–781.
526. Tap J, Furet JP, Bensaada M, et al. Gut microbiota richness promotes its stability upon increased dietary fibre intake in healthy adults. *Environ Microbiol.* 2015;17(12): 4954–4964.
527. Min EJ, Safo SE, Long Q. Penalized co-inertia analysis with applications to -omics data. *Bioinformatics.* 2018;35(6):1018–1025.
528. Gower JC. Generalized procrustes analysis. *Psychometrika.* 1975;40(1):33–51.
529. Hurley JR, Cattell RB. The Procrustes Program: producing direct rotation to test a hypothesized factor structure. *Behav Sci.* 1962;7(2):258–262.
530. Quinn RA, Navas-Molina JA, Hyde ER, et al. From sample to multi-omics conclusions in under 48 hours. *mSystems.* 2016;1(2), e00038–00016.
531. Chen T, Long W, Zhang C, Liu S, Zhao L, Hamaker BR. Fiber-utilizing capacity varies in Prevotella- versus Bacteroides-dominated gut microbiota. *Sci Rep.* 2017;7(1):2594.

532. Shankar V, Agans R, Holmes B, Raymer M, Paliy O. Do gut microbial communities differ in pediatric IBS and health? *Gut microbes*. 2013;4(4):347–352.
533. Smits SA, Marcobal A, Higginbottom S, Sonnenburg JL, Kashyap PC. Individualized responses of gut microbiota to dietary intervention modeled in humanized mice. *mSystems*. 2016;1(5):e00098–00016.
534. Rao CR. The use and interpretation of principal component analysis in applied research. *Sankhyā: Indian J Stat Ser A (1961-2002)*. 1964;26(4):329–358.
535. Rajilic-Stojanovic M, Maathuis A, Heilig HG, Venema K, de Vos WM, Smidt H. Evaluating the microbial diversity of an in vitro model of the human large intestine by phylogenetic microarray analysis. *Microbiology*. 2010;156(Pt. 11):3270–3281.
536. Ringel-Kulka T, Cheng J, Ringel Y, et al. Intestinal microbiota in healthy U.S. young children and adults—a high throughput microarray analysis. *PLoS One*. 2013;8(5):e64315.
537. Zhang C, Zhang M, Pang X, Zhao Y, Wang L, Zhao L. Structural resilience of the gut microbiota in adult mice under high-fat dietary perturbations. *ISME J*. 2012;6(10):1848–1857.
538. Wilmes P, Bond PL. The application of two-dimensional polyacrylamide gel electrophoresis and downstream analyses to a mixed community of prokaryotic microorganisms. *Environ Microbiol*. 2004;6(9):911.
539. Ram RJ, Verberkmoes NC, Thelen MP, et al. Community proteomics of a natural microbial biofilm. *Science*. 2005;308(5730):1915.
540. ter Braak CJF. Canonical correspondence analysis—a new eigenvector technique for multivariate direct gradient analysis. *Ecology* 1986;67:1167–1179.
541. Akorli J, Gendrin M, Pels NAP, Yeboah-Manu D, Christophides GK, Wilson MD. Seasonality and locality affect the diversity of anopheles gambiae and anopheles coluzzii midgut microbiota from Ghana. *PLoS One*. 2016;11(6):e0157529.
542. Dinleyici EC, Martínez-Martínez D, Kara A, et al. Time series analysis of the microbiota of children suffering from acute infectious diarrhea and their recovery after treatment. *Front Microbiol*. 2018;9:1230.
543. Nie Z, Zheng Y, Xie S, et al. Unraveling the correlation between microbiota succession and metabolite changes in traditional Shanxi aged vinegar. *Sci Rep*. 2017;7(1):9240.
544. Gower JC. Generalized canonical analysis. In: Coppi R, Bolasco S, eds. *Multiway Data Analysis*. Amsterdam: Elsevier (North Holland); 1989:221–232.
545. Kettenring JR. Canonical analysis of several sets of variables. *Biometrika*. 1971;58(3):433–451.
546. Carroll JD. Generalization of canonical correlation analysis to three or more sets of variables. In: *Proceedings of 76th Annual Convention of the American Psychological Association*; 1968.
547. Tenenhaus M, Tenenhaus A, G. PJ. Regularized generalized canonical correlation analysis. *Psychometrika*. 2011;76:257–284.
548. Jun I, Choi W, Park M. Multi-block analysis of genomic data using generalized canonical correlation analysis. *Genome Inform*. 2018;16(4):e33.
549. Chessel D, Hanafi M. Analysis of the co-inertia of K tables Analyses de la co-inertie de K nuages de points. *Rev Stat Appl*. 1996;44(2):35–60.
550. Meng C, Kuster B, Culhane AC, Gholami AM. A multivariate approach to the integration of multi-omics datasets. *BMC Bioinf*. 2014;15:162.
551. Wold S, Hellberg S, Lundstedt T, Sjostrom M, Wold H. *Proceedings of Symposium on PLS: Theory and Application*. Germany: Frankfurt am Main; 1987.
552. Qin SJ, Valle S, Piovoso MJ. On unifying multiblock analysis with application to decentralized process monitoring. *J Chemometr*. 2001;15(9):715–742.
553. Smilde AK, Westerhuis JA, de Jong S. A framework for sequential multiblock component methods. *J Chemometr*. 2003;17(6):323–337.

554. Westerhuis JA, Kourti T, MacGregor JF. Analysis of multiblock and hierarchical PCA and PLS models. *J Chemometr.* 1998;12(5):301–321.
555. Rafii F. The role of colonic bacteria in the metabolism of the natural isoflavone daidzin to equol. *Metabolites.* 2015;5(1):56–73.
556. Tenenhaus A, Philippe C, Guillemot V, Le Cao K-A, Grill J, Frouin V. Variable selection for generalized canonical correlation analysis. *Biostatistics.* 2014;15(3):569–583.
557. Setchell KD, Brown NM, Summer S. Dietary factors influence production of the soy isoflavone metabolite s-(−)equol in healthy adults. *J Nutr.* 2013;143(12):1950–1958.
558. Tenenhaus M, Tenenhaus A, Groenen PJF. Regularized generalized canonical correlation analysis: a framework for sequential multiblock component methods. *Psychometrika.* 2017;82:737–777.
559. Zhan S, Ho SC. Meta-analysis of the effects of soy protein containing isoflavones on the lipid profile. *Am J Clin Nutr.* 2005;81(2):397–408.
560. Liu B, Qin L, Liu A. Prevalence of the equol-producer phenotype and its relationship with dietary isoflavone and serum lipids in healthy Chinese adults. *J Epidemiol.* 2010;20(5):377–384.
561. Xu X, Wang HJ, Murphy PA, Cook L, Hendrich S. Daidzein is a more bioavailable soymilk isoflavone than is genistein in adult women. *J Nutr.* 1994;124(6):825–832.
562. Setchell KD, Brown NM, Lydeking-Olsen E. The clinical importance of the metabolite equol—a clue to the effectiveness of soy and its isoflavones. *J Nutr.* 2002;132(12):3577–3584.
563. Garali I, Adanyeguh IM, Ichou F, et al. A strategy for multimodal data integration: application to biomarkers identification in spinocerebellar ataxia. *Brief Bioinform.* 2017;19(6):1356–1369.
564. Wold H. Estimation of principal components and related models by iterative least squares. In: Krishnaiah PR, ed. *Multivariate Analysis.* New York: Academic Press; 1966:391–420.
565. Wold H. Partial least squares. In: Kotz SJ, Norman L, eds. vol. 6. *Encyclopedia of Statistical Sciences.* New York: Wiley; 1985:581–591.
566. Wold H. Soft modeling: the basic design and some extensions. In: *Systems Under Indirect Observation:Causality, Stracture, Prediction, Part II, Number 139 in Proceedings of the Conference on Systems Under Indirect Observation, Cartigny, Switzerland, North Holland*; October 1982.
567. Wold S, Ruhe A, Wold H, Dunn IWJ. The collinearity problem in linear regression. The partial least squares (PLS) approach to generalized inverses. *SIAM J Sci Stat Comput.* 1984;5(3):735–743.
568. Abdi H. Partial least squares regression and projection on latent structure regression (PLS Regression). *WIREs Comput Stat.* 2010;2(1):97–106.
569. Tobias RD. An introduction to partial least squares regression. In: *Proceedings of the Twentieth Annual SAS Users Group International Conference*, Cary, NC: SAS Institute Inc.; 1995.
570. Trygg J, Wold S. O2-PLS, a two-block (X–Y) latent variable regression (LVR) method with an integral OSC filter. *J Chemometr.* 2003;17(1):53–64.
571. Brereton RG. A short history of chemometrics: a personal view. *J Chemometr.* 2014;28(10):749–760.
572. Dao MC, Sokolovska N, Brazeilles R, et al. A data integration multi-omics approach to study calorie restriction-induced changes in insulin sensitivity. *Front Physiol.* 2018;9:1958.
573. Chun H, Keleş S. Sparse partial least squares regression for simultaneous dimension reduction and variable selection. *J R Stat Soc Ser B Stat Methodol.* 2010;72(1):3–25.
574. Chung D, Keles S. Sparse partial least squares classification for high dimensional data. *Stat Appl Genet Mol Biol.* 2010;9(1):17.

575. Lê Cao K-A, Rossouw D, Robert-Granié C, Besse P. A sparse PLS for variable selection when integrating omics data. *Stat Appl Genet Mol Biol.* 2008;7, Article 35.
576. Lê Cao K-A, Martin PGP, Robert-Granié C, Besse P. Sparse canonical methods for biological data integration: application to a cross-platform study. *BMC Bioinf.* 2009;10, Article 34.
577. Trygg J, Wold S. Orthogonal projections to latent structures (O-PLS). *J Chemometr.* 2002;16(3):119–128.
578. Trygg J. O2-PLS for qualitative and quantitative analysis in multivariate calibration. *J Chemometr.* 2002;16(6):283–293.
579. Bouhaddani SE, Houwing-Duistermaat J, Salo P, Perola M, Jongbloed G, Uh H-W. Evaluation of O2PLS in Omics data integration. *BMC Bioinf.* 2016;17(Suppl. 2):11.
580. Bylesjö M, Eriksson D, Kusano M, Moritz T, Trygg J. Data integration in plant biology: the O2PLS method for combined modeling of transcript and metabolite data. *Plant J.* 2007;52(6):1181–1191.
581. Bylesjö M, Eriksson D, Sjödin A, Jansson S, Moritz T, Trygg J. Orthogonal projections to latent structures as a strategy for microarray data normalization. *BMC Bioinf.* 2007;8(1):207.
582. Cloarec O, Dumas ME, Craig A, et al. Statistical total correlation spectroscopy: an exploratory approach for latent biomarker identification from metabolic 1H NMR data sets. *Anal Chem.* 2005;77(5):1282–1289.
583. Cloarec O, Dumas ME, Trygg J, et al. Evaluation of the orthogonal projection on latent structure model limitations caused by chemical shift variability and improved visualization of biomarker changes in 1H NMR spectroscopic metabonomic studies. *Anal Chem.* 2005;77(2):517–526.
584. Bylesjö M, Rantalainen M, Cloarec O, Nicholson JK, Holmes E, Trygg J. OPLS discriminant analysis: combining the strengths of PLS-DA and SIMCA classification. *J Chemometr.* 2006;20(8–10):341–351.
585. El Aidy S, Derrien M, Merrifield CA, et al. Gut bacteria–host metabolic interplay during conventionalisation of the mouse germfree colon. *ISME J.* 2013;7(4):743–755.
586. Bylesjö M, Rantalainen M, Nicholson JK, Holmes E, Trygg J. K-OPLS package: kernel-based orthogonal projections to latent structures for prediction and interpretation in feature space. *BMC Bioinf.* 2008;9:106.
587. Rantalainen M, Bylesjö M, Cloarec O, Nicholson JK, Holmes E, Trygg J. Kernel-based orthogonal projections to latent structures (K-OPLS). *J Chemometr.* 2007;21(7–9):379–385.
588. Aizerman M, Braverman E, Rozonoer L. Theoretical foundations of the potential function method in pattern recognition learning. *Autom Remote Control.* 1964;25: 821–837.
589. Härdle WK, Simar L. Discriminant analysis. In: *Applied Multivariate Statistical Analysis.* Berlin, Heidelberg: Springer; 2019:395–411.
590. Izenman AJ. Linear discriminant analysis. In: *Modern Multivariate Statistical Techniques: Regression, Classification, and Manifold Learning.* New York, NY: Springer; 2008:237–280.
591. Fisher RA. The use of multiple measurements in taxonomic problems. *Ann Eugen.* 1936;7(2):179–188.
592. Izenman AJ. Linear discriminant analysis. In: *Modern Multivariate Statistical Techniques.* New York, NY: Springer; 2013.
593. Putnam RA, Mohaidat QI, Daabous A, Rehse SJ. A comparison of multivariate analysis techniques and variable selection strategies in a laser-induced breakdown spectroscopy bacterial classification. *Spectrochim Acta, Part B.* 2013;87:161–167.
594. Koenig JE, Spor A, Scalfone N, Fricker AD, Stombaugh J, Knight R. Succession of microbial consortia in the developing infant gut microbiome. *Proc Natl Acad Sci USA.* 2011;108:4578–4585.

595. Werner JJ, Knights D, Garcia ML, et al. Bacterial community structures are unique and resilient in full-scale bioenergy systems. *Proc Natl Acad Sci USA*. 2011;108(10): 4158–4163.
596. Segata N, Izard J, Waldron L. Metagenomic biomarker discovery and explanation. *Genome Biol*. 2011;12, Article R60.
597. Kruskal WH, Wallis WA. Use of ranks in one-criterion variance analysis. *J Am Stat Assoc*. 1952;47(260):583–621.
598. Mann HB, Whitney DR. On a test of whether one of two random variables is stochastically larger than the other. *Ann Math Stat*. 1947;18(1):50–60.
599. Wilcoxon F. Individual comparisons by ranking methods. *Biometrics*. 1945;1(6):80–83.
600. Blankenberg D, Von Kuster G, Coraor N, et al. Galaxy: a web-based genome analysis tool for experimentalists. *Curr Protoc Mol Biol*. 2010; Chapter 19: Unit 19.10.11-21.
601. Goecks J, Nekrutenko A, Taylor J. Galaxy: a comprehensive approach for supporting accessible, reproducible, and transparent computational research in the life sciences. *Genome Biol*. 2010;11, Article R86.
602. Wolf A, Moissl-Eichinger C, Perras A, Koskinen K, Tomazic PV, Thurnher D. The salivary microbiome as an indicator of carcinogenesis in patients with oropharyngeal squamous cell carcinoma: a pilot study. *Sci Rep*. 2017;7(1):5867.
603. Puri P, Liangpunsakul S, Christensen JE, et al. The circulating microbiome signature and inferred functional metagenomics in alcoholic hepatitis. *Hepatology*. 2018;67(4): 1284–1302.
604. Thomas AM, Manghi P, Asnicar F, et al. Metagenomic analysis of colorectal cancer datasets identifies cross-cohort microbial diagnostic signatures and a link with choline degradation. *Nat Med*. 2019;25(4):667–678.
605. Chumpitazi BP, Cope JL, Hollister EB, et al. Randomised clinical trial: gut microbiome biomarkers are associated with clinical response to a low FODMAP diet in children with the irritable bowel syndrome. *Aliment Pharmacol Ther*. 2015;42(4):418–427.
606. Muniz Pedrogo DA, Jensen MD, Van Dyke CT, et al. Gut microbial carbohydrate metabolism hinders weight loss in overweight adults undergoing lifestyle intervention with a volumetric diet. *Mayo Clin Proc*. 2018;93(8):1104–1110.
607. Barker M, Rayens W. Partial least squares for discrimination. *J Chemometr*. 2003;17(3):166–173.
608. Ståhle L, Wold S. Partial least squares analysis with cross-validation for the two-class problem: a Monte Carlo study. *J Chemometr*. 1987;1(3):185–196.
609. Christin C, Hoefsloot HCJ, Smilde AK, et al. A critical assessment of feature selection methods for biomarker discovery in clinical proteomics. *Mol Cell Proteomics*. 2013;12(1):263–276.
610. Botella C, Ferré J, Boqué R. Classification from microarray data using probabilistic discriminant partial least squares with reject option. *Talanta*. 2009;80(1):321–328.
611. Lee LC, Liong CY, Jemain AA. Partial least squares-discriminant analysis (PLS-DA) for classification of high-dimensional (HD) data: a review of contemporary practice strategies and knowledge gaps. *Analyst*. 2018;143(15):3526–3539.
612. Nguyen DV, R. D.M. Classification of acute leukemia based on DNA microarray gene expressions using partial least squares. In: Lin SM, Johnson KF, eds. *Methods of Microarray Data Analysis*. Boston, MA: Springer; 2002.
613. Nguyen DV, Rocke DM. Tumor classification by partial least squares using microarray gene expression data. *Bioinformatics*. 2002;18(1):39–50.
614. Tan Y, Shi L, Tong W, Gene Hwang GT, Wang C. Multi class tumor classification by discriminant partial least squares using microarray gene expression data and assessment of classification models. *Comput Biol Chem*. 2004;28(3):235–243.
615. Gottfries J, Blennow K, Wallin A, Gottfries CG. Diagnosis of dementias using partial least squares discriminant analysis. *Dementia*. 1995;6(2):83–88.

616. Eriksson L, Antti H, Gottfries J, et al. Using chemometrics for navigating in the large data sets of genomics, proteomics, and metabonomics (gpm). *Anal Bioanal Chem.* 2004;380(3):419–429.
617. Rohart F, Gautier B, Singh A, Lê Cao K-A. mixOmics: an R package for 'omics feature selection and multiple data integration. *PLoS Comput Biol.* 2017;13(11):e1005752.
618. Worley B, Halouska S, Powers R. Utilities for quantifying separation in PCA/PLS-DA scores plots. *Anal Biochem.* 2013;433(2):102–104.
619. Worley B, Powers R. Multivariate analysis in metabolomics. *Curr Metabolomics.* 2013;1(1):92–107.
620. Gomez-Alvarez V, Revetta RP, Santo Domingo JW. Metagenome analyses of corroded concrete wastewater pipe biofilms reveal a complex microbial system. *BMC Microbiol.* 2012;12:122.
621. Brereton RG, Lloyd GR. Partial least squares discriminant analysis: taking the magic away. *J Chemometr.* 2014;28(4):213–225.
622. Worley B, Powers R. PCA as a practical indicator of OPLS-DA model reliability. *Curr Metabolomics.* 2016;4(2):97–103.
623. Stenlund H, Gorzsás A, Persson P, Sundberg B, Trygg J. Orthogonal projections to latent structures discriminant analysis modeling on in situ FT-IR spectral imaging of liver tissue for identifying sources of variability. *Anal Chem.* 2008;80(18):6898–6906.
624. Bocca C, Kouassi Nzoughet J, Leruez S, et al. A plasma metabolomic signature involving purine metabolism in human optic atrophy 1 (OPA1)-related disorders. *Invest Ophthalmol Vis Sci.* 2018;59(1):185–195.
625. Westerhuis JA, van Velzen EJJ, Hoefsloot HCJ, Smilde AK. Multivariate paired data analysis: multilevel PLSDA versus OPLSDA. *Metabolomics.* 2010;6(1):119–128.
626. Bennet SMP, Bohn L, Storsrud S, et al. Multivariate modelling of faecal bacterial profiles of patients with IBS predicts responsiveness to a diet low in FODMAPs. *Gut.* 2018;67(5):872–881.
627. Ramadan Z, Xu H, Laflamme D, et al. Fecal microbiota of cats with naturally occurring chronic diarrhea assessed using 16S rRNA gene 454-pyrosequencing before and after dietary treatment. *J Vet Intern Med.* 2014;28(1):59–65.
628. Hastie T, Tibshirani R, Friedman J. *The Elements of Statistical Learning: Data Mining, Inference, and Prediction.* New York, NY: Springer; 2009.
629. James G, Witten D, Hastie T, Tibshirani R. *An Introduction to Statistical Learning with Applications in R.* New York, NY: Springer; 2013.
630. Loh W-Y. Classification and regression trees. *WIREs Data Min Knowl Discovery.* 2011;1(1):14–23.
631. Breiman L. Random forests. *Mach Learn.* 2001;45(1):5–32.
632. Schapire R, Freund Y, Bartlett P, Lee W. Boosting the margin: a new explanation for the effectiveness of voting methods. *Ann Stat.* 2001;26.
633. Breiman L. Bagging predictors "machine learning" *Mach Learn.* 1996;24:123–140.
634. Liaw A, Wiener M. Classification and regression by randomForest. *R News.* 2002;2(3):18–22.
635. Cutler DR, Edwards Jr TC, Beard KH, et al. Random forests for classification in ecology. *Ecology.* 2007;88(11):2783–2792.
636. Knights D, Costello EK, Knight R. Supervised classification of human microbiota. *FEMS Microbiol Rev.* 2011;35(2):343–359.
637. Gashler M, Giraud-Carrier C, Martinez T. Decision tree ensemble: small heterogeneous is better than large homogeneous. In: *2008 Seventh International Conference on Machine Learning and Applications*; 2008.
638. Griffin NW, Ahern PP, Cheng J, et al. Prior dietary practices and connections to a human gut microbial metacommunity alter responses to diet interventions. *Cell Host Microbe.* 2017;21(1):84–96.

639. Lozupone CA, Li M, Campbell TB, et al. Alterations in the gut microbiota associated with HIV-1 infection. *Cell Host Microbe*. 2013;14(3):329–339.
640. Piening BD, Zhou W, Contrepois K, et al. Integrative personal omics profiles during periods of weight gain and loss. *Cell Syst*. 2018;6(2):157–170.e158.
641. Yatsunenko T, Rey FE, Manary MJ, et al. Human gut microbiome viewed across age and geography. *Nature*. 2012;486(7402):222–227.
642. Beck D, Foster JA. Machine learning techniques accurately classify microbial communities by bacterial vaginosis characteristics. *PLoS One*. 2014;9(2):e87830.
643. Chatterjee I, Zhang Y, Zhang J, Xia Y, Sun J. Vitamin D receptor promotes healthy microbial metabolites and microbiome. *Sci Rep*. 2020. https://doi.org/10.1038/s41598-020-64226-7.
644. Papa E, Docktor M, Smillie C, et al. Non-invasive mapping of the gastrointestinal microbiota identifies children with inflammatory bowel disease. *PLoS One*. 2012;7(6):e39242.
645. Roguet A, Eren AM, Newton RJ, McLellan SL. Fecal source identification using random forest. *Microbiome*. 2018;6(1):185.
646. Cortes C, Vapnik V. Support-vector networks. *Mach Learn*. 1995;20(3):273–297.
647. Gu S, Tan Y, He X. Discriminant analysis via support vectors. *Neurocomputing*. 2010;73(10):1669–1675.
648. Gokcen I, Peng J. Comparing linear discriminant analysis and support vector machines. In: *Advances in Information Systems*. Berlin, Heidelberg: Springer Berlin Heidelberg; 2002.
649. Xiao J, Chen L, Johnson S, Yu Y, Zhang X, Chen J. Predictive modeling of microbiome data using a phylogeny-regularized generalized linear mixed model. *Front Microbiol*. 2018;9:1391.
650. Oudah M, Henschel A. Taxonomy-aware feature engineering for microbiome classification. *BMC Bioinf*. 2018;19(1):227.
651. Yang C, Mills D, Mathee K, et al. An ecoinformatics tool for microbial community studies: supervised classification of Amplicon Length Heterogeneity (ALH) profiles of 16S rRNA. *J Microbiol Methods*. 2006;65(1):49–62.
652. Holscher HD, Bauer LL, Gourineni V, Pelkman CL, Fahey Jr GC, Swanson KS. Agave inulin supplementation affects the fecal microbiota of healthy adults participating in a randomized, double-blind, placebo-controlled, crossover trial. *J Nutr*. 2015;145(9):2025–2032.
653. Kolho KL, Korpela K, Jaakkola T, et al. Fecal microbiota in pediatric inflammatory bowel disease and its relation to inflammation. *Am J Gastroenterol*. 2015;110(6):921–930.
654. Korem T, Zeevi D, Zmora N, et al. Bread affects clinical parameters and induces gut microbiome-associated personal glycemic responses. *Cell Metab*. 2017;25(6):1243–1253.e1245.
655. Parks BW, Nam E, Org E, et al. Genetic control of obesity and gut microbiota composition in response to high-fat, high-sucrose diet in mice. *Cell Metab*. 2013;17(1):141–152.
656. Salonen A, Lahti L, Salojärvi J, et al. Impact of diet and individual variation on intestinal microbiota composition and fermentation products in obese men. *ISME J*. 2014;8(11):2218–2230.
657. Furlotte NA, Kang HM, Ye C, Eskin E. Mixed-model coexpression: calculating gene coexpression while accounting for expression heterogeneity. *Bioinformatics (Oxford, England)*. 2011;27(13):i288–i294.
658. Fan J, Lv J. A selective overview of variable selection in high dimensional feature space. *Stat Sin*. 2010;20(1):101–148.
659. Zhao N, Chen J, Carroll IM, et al. Testing in microbiome-profiling studies with MiRKAT, the microbiome regression-based kernel association test. *Am J Hum Genet*. 2015;96(5):797–807.

660. Zhao Q, Shi X, Huang J, Liu J, Li Y, Ma S. Integrative analysis of "-omics" data using penalty functions. *Wiley Interdiscip Rev Comput Stat.* 2015;7(1):99–108.
661. Xia Y, Sun J, Chen DG. Univariate community analysis. In: *Statistical Analysis of Microbiome Data with R*. Singapore: Springer; 2018:251–283.
662. Mankiewicz R. *The Story of Mathematics*. Paperback ed. Princeton, NJ: Princeton University Press; 2004.
663. Moreno I, Codoñer FM, Vilella F, et al. Evidence that the endometrial microbiota has an effect on implantation success or failure. *Am J Obstet Gynecol.* 2016;215(6):684–703.
664. Welch BL. The generalization of Student's problem when several different population variances are involved. *Biometrika.* 1947;34:28–35.
665. Ruxton GD. The unequal variance t-test is an underused alternative to Student's t-test and the Mann–Whitney U test. *Behav Ecol.* 2006;17(4):688–690.
666. Ciaccio CE, Barnes C, Kennedy K, Chan M, Portnoy J, Rosenwasser L. Home dust microbiota is disordered in homes of low-income asthmatic children. *J Asthma.* 2015;52(9):873–880.
667. Kononikhin AS, Brzhozovskiy AG, Ryabokon AM, et al. Proteome profiling of the exhaled breath condensate after long-term spaceflights. *Int J Mol Sci.* 2019;20(18):4518.
668. Kourosh A, Luna RA, Balderas M, et al. Fecal microbiome signatures are different in food-allergic children compared to siblings and healthy children. *Pediatr Allergy Immunol.* 2018;29(5):545–554.
669. Spencer MD, Hamp TJ, Reid RW, Fischer LM, Zeisel SH, Fodor AA. Association between composition of the human gastrointestinal microbiome and development of fatty liver with choline deficiency. *Gastroenterology.* 2011;140(3):976–986.
670. Kruskal WH. Historical notes on the Wilcoxon unpaired two-sample test. *J Am Stat Assoc.* 1957;52(279):356–360.
671. Falony G, Joossens M, Vieira-Silva S, Wang J, Darzi Y, Faust K. Population-level analysis of gut microbiome variation. *Science.* 2016;352(6285):560–564.
672. Kovatcheva-Datchary P, Nilsson A, Akrami R, et al. Dietary fiber-induced improvement in glucose metabolism is associated with increased abundance of prevotella. *Cell Metab.* 2015;22(6):971–982.
673. Kreznar JH, Keller MP, Traeger LL, et al. Host genotype and gut microbiome modulate insulin secretion and diet-induced metabolic phenotypes. *Cell Rep.* 2017;18(7):1739–1750.
674. Roager HM, Licht TR, Poulsen SK, Larsen TM, Bahl MI. Microbial enterotypes, inferred by the prevotella-to-bacteroides ratio, remained stable during a 6-month randomized controlled diet intervention with the new nordic diet. *Appl Environ Microbiol.* 2014;80(3):1142–1149.
675. Suez J, Korem T, Zeevi D, et al. Artificial sweeteners induce glucose intolerance by altering the gut microbiota. *Nature.* 2014;514(7521):181–186.
676. Zhao L, Zhang F, Ding X, et al. Gut bacteria selectively promoted by dietary fibers alleviate type 2 diabetes. *Science.* 2018;359(6380):1151–1156.
677. Bouhnik Y, Raskine L, Simoneau G, et al. The capacity of nondigestible carbohydrates to stimulate fecal bifidobacteria in healthy humans: a double-blind, randomized, placebo-controlled, parallel-group, dose-response relation study. *Am J Clin Nutr.* 2004;80(6):1658–1664.
678. Santacruz A, Marcos A, Wärnberg J, et al. Interplay between weight loss and gut microbiota composition in overweight adolescents. *Obesity.* 2009;17(10):1906–1915.
679. Fisher RA. The correlation between relatives on the supposition of mendelian inheritance. *Earth Environ Sci Trans R Soc Edinb.* 1918;52:399–433.
680. Allali I, Arnold JW, Roach J, et al. A comparison of sequencing platforms and bioinformatics pipelines for compositional analysis of the gut microbiome. *BMC Microbiol.* 2017;17(1):194.

681. Daniel WW. Kruskal–Wallis one-way analysis of variance by ranks. In: *Applied Nonparametric Statistics*. 2nd ed. Boston: PWS-Kent; 1990:226–234.
682. Dao MC, Everard A, Aron-Wisnewsky J, et al. Akkermansia muciniphila and improved metabolic health during a dietary intervention in obesity: relationship with gut microbiome richness and ecology. *Gut*. 2016;65(3):426–436.
683. Mobini R, Tremaroli V, Stahlman M, et al. Metabolic effects of Lactobacillus reuteri DSM 17938 in people with type 2 diabetes: a randomized controlled trial. *Diabetes Obes Metab*. 2017;19(4):579–589.
684. Possemiers S, Bolca S, Eeckhaut E, Depypere H, Verstraete W. Metabolism of isoflavones, lignans and prenylflavonoids by intestinal bacteria: producer phenotyping and relation with intestinal community. *FEMS Microbiol Ecol*. 2007;61(2):372–383.
685. Zmora N, Zilberman-Schapira G, Suez J, et al. Personalized gut mucosal colonization resistance to empiric probiotics is associated with unique host and microbiome features. *Cell*. 2018;174(6):1388–1405.e1321.
686. Liss MA, Leach RJ, Rourke E, et al. Microbiome diversity in carriers of fluoroquinolone resistant Escherichia coli. *Investig Clin Urol*. 2019;60(2):75–83.
687. McArdle B, Anderson M. Fitting multivariate models to community data: a comment on distance-based redundancy analysis. *Ecology*. 2001;82(1):290–297.
688. Bhattacharya R, Xu F, Dong G, Li S, Tian C, Ponugoti B. Effect of bacteria on the wound healing behavior of oral epithelial cells. *PLoS One*. 2014;9(2):e89475.
689. Koh H. An adaptive microbiome α-diversity-based association analysis method. *Sci Rep*. 2018;8(1):18026.
690. Wu C, Chen J, Kim J, Pan W. An adaptive association test for microbiome data. *Genome Med*. 2016;8(1):56.
691. Koh H, Blaser MJ, Li H. A powerful microbiome-based association test and a microbial taxa discovery framework for comprehensive association mapping. *Microbiome*. 2017;5(1):45.
692. Radhakrishna Rao C. Large sample tests of statistical hypotheses concerning several parameters with applications to problems of estimation. *Math Proc Camb Philos Soc*. 1948;44(1):50–57.
693. Shannon CE. A mathematical theory of communication. *Bell Syst Tech J*. 1948;27(3):379–423.
694. Simpson EH. Measurement of diversity. *Nature*. 1949;163(4148):688.
695. Faith DP. Conservation evaluation and phylogenetic diversity. *Biol Conserv*. 1992;61(1):1–10.
696. Allen B, Kon M, Bar-Yam Y. A new phylogenetic diversity measure generalizing the shannon index and its application to phyllostomid bats. *Am Nat*. 2009;174(2):236–243.
697. Rao CR. Diversity and dissimilarity coefficients: a unified approach. *Theor Popul Biol*. 1982;21(1):24–43.
698. Warwick RM, Clarke KR. New 'biodiversity' measures reveal a decrease in taxonomic distinctness with increasing stress. *Mar Ecol Prog Ser*. 1995;129(1/3):301–305.
699. Koh H, Livanos AE, Blaser MJ, Li H. A highly adaptive microbiome-based association test for survival traits. *BMC Genomics*. 2018;19(1):210.
700. Pan W, Kim J, Zhang Y, Shen X, Wei P. A powerful and adaptive association test for rare variants. *Genetics*. 2014;197(4):1081–1095.
701. Koh H, Li Y, Zhan X, Chen J, Zhao N. A distance-based kernel association test based on the generalized linear mixed model for correlated microbiome studies. *Front Genet*. 2019;10:458.
702. Laird NM, Ware JH. Random-effects models for longitudinal data. *Biometrics*. 1982;38(4):963–974.
703. Breslow NE, Clayton DG. Approximate inference in generalized linear mixed models. *J Am Stat Assoc*. 1993;88(421):9–25.

704. Hoque MN, Istiaq A, Clement RA, et al. Resistome diversity in bovine clinical mastitis microbiome, a signature concurrence. *bioRxiv*. 2019. https://doi.org/10.1101/829283.
705. Zhan X. Relationship between MiRKAT and coefficient of determination in similarity matrix regression. *Processes*. 2019;7(2):79.
706. Mantel N. The detection of disease clustering and a generalized regression approach. *Cancer Res*. 1967;27(2):209–220.
707. Mantel N, Valand RS. A technique of nonparametric multivariate analysis. *Biometrics*. 1970;26(3):547–558.
708. Lisboa FJG, Peres-Neto PR, Chaer GM, et al. Much beyond Mantel: bringing Procrustes association metric to the plant and soil ecologist's toolbox. *PLoS One*. 2014;9(6):e101238.
709. Li T, Long M, Li H, et al. Multi-omics analysis reveals a correlation between the host phylogeny, gut microbiota and metabolite profiles in cyprinid fishes. *Front Microbiol*. 2017;8:454.
710. Zhou J, Yao Y, Jiao K, et al. Relationship between gingival crevicular fluid microbiota and cytokine profile in periodontal host homeostasis. *Front Microbiol*. 2017;8:2144.
711. Zhu D, An XL, Chen QL, et al. Antibiotics disturb the microbiome and increase the incidence of resistance genes in the gut of a common soil collembolan. *Environ Sci Technol*. 2018;52(5):3081–3090.
712. Clarke KR. Non-parametric multivariate analyses of changes in community structure. *Aust J Ecol*. 1993;18(1):117–143.
713. Kakumanu ML, Reeves AM, Anderson TD, Rodrigues RR, Williams MA. Honey bee gut microbiome is altered by in-hive pesticide exposures. *Front Microbiol*. 2016;7:1255.
714. Li KJ, Chen ZL, Huang Y, et al. Dysbiosis of lower respiratory tract microbiome are associated with inflammation and microbial function variety. *Respir Res*. 2019;20(1):272.
715. Marsilio S, Pilla R, Sarawichitr B, et al. Characterization of the fecal microbiome in cats with inflammatory bowel disease or alimentary small cell lymphoma. *Sci Rep*. 2019;9(1):19208.
716. Anderson MJ. A new method for non-parametric multivariate analysis of variance. *Austral Ecol*. 2001;26(1):32–46.
717. Mielke PW. 34 Meteorological applications of permutation techniques based on distance functions. In: *Handbook of Statistics*. Elsevier; 1984:813–830. vol. 4.
718. Mielke PW. The application of multivariate permutation methods based on distance functions in the earth sciences. *Earth Sci Rev*. 1991;31(1):55–71.
719. Warton D, Wright S, Wang Y. Distance-based multivariate analyses confound location and dispersion effects. *Methods Ecol Evol*. 2012;3:89–101.
720. Mielke Jr PW, Berry KJ. *Permutation Methods: A Distance Fuction Approach*. New York, NY: Springer; 2007.
721. McCune B, Grace JB. *Analysis of Ecological Communities*. Gleneden Beach, OR: MJM Software Design; 2002.
722. Falk MW, Seshan H, Dosoretz C, Wuertz S. Partial bioaugmentation to remove 3-chloroaniline slows bacterial species turnover rate in bioreactors. *Water Res*. 2013;47(19):7109–7119.
723. Li F, Hullar MAJ, Schwarz Y, Lampe JW. Human gut bacterial communities are altered by addition of cruciferous vegetables to a controlled fruit- and vegetable-free diet. *J Nutr*. 2009;139(9):1685–1691.
724. Morissette B, Talbot G, Beaulieu C, Lessard M. Growth performance of piglets during the first two weeks of lactation affects the development of the intestinal microbiota. *J Anim Physiol Anim Nutr (Berl)*. 2018;102(2):525–532.

725. Reese AT, Dunn RR. Drivers of microbiome biodiversity: a review of general rules, feces, and ignorance. *mBio*. 2018;9(4):e01294–01218.
726. Bacon-Shone J. Discrete and continuous compositions. In: Daunis-i Estadella J, Fernández JE, eds. *Proceedings of CODAWORK'08, The 3rd Compositional Data Analysis Workshop. M-23*, Girona: University of Girona; 2008.
727. Anders S, McCarthy DJ, Chen Y, et al. Count-based differential expression analysis of RNA sequencing data using R and bioconductor. *Nat Protoc*. 2013;8(9):1765–1786.
728. Kuczynski J, Lauber CL, Walters WA, Parfrey LW, Clemente JC, Gevers D. Experimental and analytical tools for studying the human microbiome. *Nat Rev Genet*. 2012;13:47–58.
729. Xu L, Paterson AD, Turpin W, Xu W. Assessment and selection of competing models for zero-inflated microbiome data. *PLoS One*. 2015;10(7):e0129606.
730. Xia Y, Morrison-Beedy D, Ma J, Feng C, Cross W, Tu X. Modeling count outcomes from HIV risk reduction interventions: a comparison of competing statistical models for count responses. *AIDS Res Treat*. 2012;2012:593569.
731. Feng C, Wang H, Han Y, Xia Y, Lu N, Tu XM. Some theoretical comparisons of negative binomial and zero-inflated poisson distributions. *Commun Stat Theory Methods*. 2015;44(15):3266–3277.
732. Mosimann JE. On the compound multinomial distribution, the multivariate β-distribution, and correlations among proportions. *Biometrika*. 1962;49(1/2):65–82.
733. Mosimann JE. On the compound negative multinomial distribution and correlations among inversely sampled pollen counts. *Biometrika*. 1963;50(1/2):47–54.
734. Holmes I, Harris K, Quince C. Dirichlet multinomial mixtures: generative models for microbial metagenomics. *PLoS One*. 2012;7(2):e30126.
735. Chen J, Li H. Kernel methods for regression analysis of microbiome compositional data. In: Liu Y, Hu M, Lin J, eds. *Topics in Applied Statistics*. New York, NY: Springer; 2013:55. Springer Proceedings in Mathematics & Statistics.
736. Chen J, Li H. Variable selection for sparse Dirichlet-multinomial regression with an application to microbiome data analysis. *Ann Appl Stat*. 2013;7(1):418–442. https://doi.org/10.1214/1212-AOAS1592.
737. Wadsworth WD, Argiento R, Guindani M, Galloway-Pena J, Shelburne SA, Vannucci M. An integrative Bayesian Dirichlet-multinomial regression model for the analysis of taxonomic abundances in microbiome data. *BMC Bioinf*. 2017;18(1):94.
738. Wang T, Zhao H. Constructing predictive microbial signatures at multiple taxonomic levels. *J Am Stat Assoc*. 2017;112(519):1022–1031.
739. Wang T, Zhao H. A Dirichlet-tree multinomial regression model for associating dietary nutrients with gut microorganisms. *Biometrics*. 2017;73(3):792–801.
740. O'Brien JD, Record NR, Countway P. The power and pitfalls of Dirichlet-multinomial mixture models for ecological count data. *bioRxiv*. 2016. https://doi.org/10.1101/045468.
741. Sankaran K, Holmes SP. Latent variable modeling for the microbiome. *Biostatistics*. 2018;20(4):599–614.
742. Shi P, Li H. A model for paired-multinomial data and its application to analysis of data on a taxonomic tree. *Biometrics*. 2017;73(4):1266–1278.
743. Tang Z-Z, Chen G, Alekseyenko AV, Li H. A general framework for association analysis of microbial communities on a taxonomic tree. *Bioinformatics (Oxford, England)*. 2017;33(9):1278–1285.
744. Tang Z-Z, Chen G. Robust and powerful differential composition tests for clustered microbiome data. *Stat Biosci*. 2019. https://doi.org/10.1007/s12561-019-09251-5.
745. Tang Z-Z, Chen G. Zero-inflated generalized Dirichlet multinomial regression model for microbiome compositional data analysis. *Biostatistics*. 2018;20(4):698–713.

746. Xia F, Chen J, Fung WK, Li H. A logistic normal multinomial regression model for microbiome compositional data analysis. *Biometrics*. 2013;69(4):1053–1063.
747. Nowicka M, Robinson MD. DRIMSeq: a Dirichlet-multinomial framework for multivariate count outcomes in genomics. *F1000Res*. 2016;5:1356.
748. Harrison JG, Calder WJ, Shastry V, Buerkle CA. Dirichlet-multinomial modelling outperforms alternatives for analysis of microbiome and other ecological count data. *Mol Ecol Resour*. 2020;20(2):481–497.
749. Wang C, Hu J, Blaser MJ, Li H. Estimating and testing the microbial causal mediation effect with high-dimensional and compositional microbiome data. *Bioinformatics*. 2020;36(2):347–355.
750. Weiss S, Xu ZZ, Peddada S, et al. Normalization and microbial differential abundance strategies depend upon data characteristics. *Microbiome*. 2017;5(1):27.
751. Bouguila N. Count data modeling and classification using finite mixtures of distributions. *IEEE Trans Neural Netw*. 2011;22(2):186–198.
752. Sjolander K, Karplus K, Brown M, et al. Dirichlet mixtures: a method for improved detection of weak but significant protein sequence homology. *Comput Appl Biosci*. 1996;12(4):327–345.
753. Ye X, Yu Y-K, Altschul SF. Compositional adjustment of Dirichlet mixture priors. *J Comput Biol*. 2010;17(12):1607–1620.
754. Song Y, Zhao H, Wang T. An adaptive independence test for microbiome community data. *Biometrics*. 2019. https://doi.org/10.1111/biom.13154.
755. Chu DM, Ma J, Prince AL, Antony KM, Seferovic MD, Aagaard KM. Maturation of the infant microbiome community structure and function across multiple body sites and in relation to mode of delivery. *Nat Med*. 2017;23(3):314–326.
756. Vandeputte D, Falony G, Vieira-Silva S, Tito RY, Joossens M, Raes J. Stool consistency is strongly associated with gut microbiota richness and composition, enterotypes and bacterial growth rates. *Gut*. 2016;65(1):57–62.
757. Lin W, Shi P, Feng R, Li H. Variable selection in regression with compositional covariates. *Biometrika*. 2014;101(4):785–797.
758. Tang Y, Ma L, Nicolae DL. A phylogenetic scan test on a Dirichlet-tree multinomial model for microbiome data. *Ann Appl Stat*. 2018;12(1):1–26.
759. Dennis SY. On the hyper-dirichlet type 1 and hyper-liouville distributions. *Commun Stat Theory Methods*. 1991;20(12):4069–4081.
760. Bradley P, Nayfach S, Pollard K. Phylogeny-corrected identification of microbial gene families relevant to human gut colonization. *PLoS Comput Biol*. 2018;14: e1006242.
761. Li Z, Lee K, Karagas MR, et al. Conditional regression based on a multivariate zero-inflated logistic-normal model for microbiome relative abundance data. *Stat Biosci*. 2018;10(3):587–608.
762. Connor RJ, Mosimann JE. Concepts of independence for proportions with a generalization of the dirichlet distribution. *J Am Stat Assoc*. 1969;64(325):194–206.
763. Tang Z-Z, Chen G, Hong Q, et al. Multi-omic analysis of the microbiome and metabolome in healthy subjects reveals microbiome-dependent relationships between diet and metabolites. *Front Genet*. 2019;10:454.
764. Tang Y, Nicolae DL. *Mixed Effect Dirichlet-Tree Multinomial for Longitudinal Microbiome Data and Weight Prediction*. arXiv; 2017, 1706.06380v1 [stat.AP] 20 Jun 2017.
765. Mao J, Chen Y, Ma L. Bayesian graphical compositional regression for microbiome data. *J Am Stat Assoc*. 2019;1–15. https://doi.org/10.1080/01621459.2019.1647212.
766. Goedecke JH, Mendham AE, Clamp L, et al. An exercise intervention to unravel the mechanisms underlying insulin resistance in a cohort of black south African women: protocol for a randomized controlled trial and baseline characteristics of participants. *JMIR Res Protoc*. 2018;7(4):e75.

767. Yang Y, Chen N, Chen T. Inference of environmental factor-microbe and microbe-microbe associations from metagenomic data using a hierarchical Bayesian statistical model. *Cell Syst.* 2017;4:129–137.e125.
768. Kurtz ZD, Bonneau R, Müller CL. Disentangling microbial associations from hidden environmental and technical factors via latent graphical models. *bioRxiv.* 2019. https://doi.org/10.1101/2019.12.21.885889.
769. Tackmann J, Rodrigues JFM, von Mering C. Rapid inference of direct interactions in large-scale ecological networks from heterogeneous microbial sequencing data. *Cell Syst.* 2019;9(3):286–296.e288.
770. Yuan H, He S, Deng M. Compositional data network analysis via lasso penalized D-trace loss. *Bioinformatics.* 2019;35(18):3404–3411.
771. Liu S, Hua K, Chen S, Zhang X. Comprehensive simulation of metagenomic sequencing data with non-uniform sampling distribution. *Quant Biol.* 2018;6(2):175–185.
772. Wong SH, Yu J. Gut microbiota in colorectal cancer: mechanisms of action and clinical applications. *Nat Rev Gastroenterol Hepatol.* 2019;16(11):690–704.
773. Larson NB, Chen J, Schaid DJ. A review of kernel methods for genetic association studies. *Genet Epidemiol.* 2019;43(2):122–136.
774. Lee S, Emond MJ, Bamshad MJ, et al. Optimal unified approach for rare-variant association testing with application to small-sample case-control whole-exome sequencing studies. *Am J Hum Genet.* 2012;91(2):224–237.
775. Wu MC, Lee S, Cai T, Li Y, Boehnke M, Lin X. Rare-variant association testing for sequencing data with the sequence kernel association test. *Am J Hum Genet.* 2011;89(1): 82–93.
776. Li S, Cui Y. Gene-centric gene–gene interaction: a model-based kernel machine method. *Ann Appl Stat.* 2012;6(3):1134–1161.
777. Lin X, Lee S, Christiani DC, Lin X. Test for interactions between a genetic marker set and environment in generalized linear models. *Biostatistics (Oxford, England).* 2013;14(4):667–681.
778. Schaid DJ, McDonnell SK, Sinnwell JP, Thibodeau SN. Multiple genetic variant association testing by collapsing and kernel methods with pedigree or population structured data. *Genet Epidemiol.* 2013;37(5):409–418.
779. Choi S, Lee S, Cichon S, et al. FARVAT: a family-based rare variant association test. *Bioinformatics.* 2014;30(22):3197–3205.
780. Saad M, Wijsman EM. Combining family- and population-based imputation data for association analysis of rare and common variants in large pedigrees. *Genet Epidemiol.* 2014;38(7):579–590.
781. Wang K. Boosting the power of the sequence kernel association test by properly estimating its null distribution. *Am J Hum Genet.* 2016;99(1):104–114.
782. Wu B, Pankow JS. Sequence kernel association test of multiple continuous phenotypes. *Genet Epidemiol.* 2016;40(2):91–100.
783. Schweiger R, Weissbrod O, Rahmani E, et al. RL-SKAT: an exact and efficient score test for heritability and set tests. *Genetics.* 2017;207(4):1275–1283.
784. Chen J, Chen W, Zhao N, Wu MC, Schaid DJ. Small sample kernel association tests for human genetic and microbiome association studies. *Genet Epidemiol.* 2016;40(1):5–19.
785. Zhan X, Tong X, Zhao N, Maity A, Wu MC, Chen J. A small-sample multivariate kernel machine test for microbiome association studies. *Genet Epidemiol.* 2017;41(3): 210–220.
786. Zhan X, Xue L, Zheng H, et al. A small-sample kernel association test for correlated data with application to microbiome association studies. *Genet Epidemiol.* 2018;42(8): 772–782.
787. Lumley T, Brody J, Peloso G, Morrison A, Rice K. FastSKAT: sequence kernel association tests for very large sets of markers. *Genet Epidemiol.* 2018;42(6):516–527.

788. Yan Q, Fang Z, Chen W. KMgene: a unified R package for gene-based association analysis for complex traits. *Bioinformatics (Oxford, England)*. 2018;34(12):2144–2146.
789. Plantinga A, Zhan X, Zhao N, Chen J, Jenq RR, Wu MC. MiRKAT-S: a community-level test of association between the microbiota and survival times. *Microbiome*. 2017;5(1):17.
790. Tang Z-Z, Chen G, Alekseyenko AV. PERMANOVA-S: association test for microbial community composition that accommodates confounders and multiple distances. *Bioinformatics (Oxford, England)*. 2016;32(17):2618–2625.
791. Benjamini Y, Drai D, et al. Controlling the false discovery rate in behavior genetics research. *Behav Brain Res*. 2001;125(1):279–284.
792. Benjamini Y. Discovering the false discovery rate. *J R Stat Soc Series B Stat Methodol*. 2010;72(4):405–416.
793. Parks DH, Tyson GW, Hugenholtz P, Beiko RG. STAMP: statistical analysis of taxonomic and functional profiles. *Bioinformatics (Oxford, England)*. 2014;30(21): 3123–3124.
794. Benjamini Y, Yekutieli D. The control of the false discovery rate in multiple testing under dependency. *Ann Stat*. 2001;29(4):1165–1188.
795. Hu Y-J, Satten GA. Testing hypotheses about the microbiome using the linear decomposition model. *bioRxiv*. 2019; https://doi.org/10.1101/229831.
796. Wu C. Multi-trait genome-wide analyses of the brain imaging phenotypes in UK Biobank. *bioRxiv*. 2019; https://doi.org/10.1101/758326.
797. Sun R, Lin X. *Set-Based Tests for Genetic Association Using the Generalized Berk-Jones Statistic*. arXiv Preprint; 2017, arXiv:171002469.
798. Kwak I-Y, Pan W. Adaptive gene- and pathway-trait association testing with GWAS summary statistics. *Bioinformatics (Oxford, England)*. 2016;32(8):1178–1184.
799. Mika M, Maurer J, Korten I, et al. Influence of the pneumococcal conjugate vaccines on the temporal variation of pneumococcal carriage and the nasal microbiota in healthy infants: a longitudinal analysis of a case–control study. *Microbiome*. 2017;5(1):85.
800. Hu J, Koh H, He L, Liu M, Blaser MJ, Li H. A two-stage microbial association mapping framework with advanced FDR control. *Microbiome*. 2018;6(1):131.
801. Yekutieli D. Hierarchical false discovery rate–controlling methodology. *J Am Stat Assoc*. 2008;103(481):309–316.
802. Yekutieli D, Reiner-Benaim A, Benjamini Y, et al. Approaches to multiplicity issues in complex research in microarray analysis. *Stat Neerl*. 2006;60(4):414–437.
803. Benjamini Y, Yekutieli D. Quantitative trait Loci analysis using the false discovery rate. *Genetics*. 2005;171(2):783–790.
804. Zehetmayer S, Bauer P, Posch M. Two-stage designs for experiments with a large number of hypotheses. *Bioinformatics*. 2005;21(19):3771–3777.
805. Reiner-Benaim A, Yekutieli D, Letwin NE, et al. Associating quantitative behavioral traits with gene expression in the brain: searching for diamonds in the hay. *Bioinformatics*. 2007;23(17):2239–2246.
806. Srinivasan A, Xue L, Zhan X. Compositional knockoff filter for high-dimensional regression analysis of microbiome data. *bioRxiv*. 2019. https://doi.org/10.1101/851337.
807. Aitchison J. The statistical analysis of compositional data (with discussion). *J R Stat Soc Series B Stat Methodol*. 1982;44(2):139–177.
808. Billheimer D, Guttorp P, Fagan WF. Statistical interpretation of species composition. *J Am Stat Assoc*. 2001;96(456):1205–1214.
809. Grantham NS, Guan Y, Reich BJ, Borer ET, Gross K. MIMIX: a Bayesian mixed-effects model for microbiome data from designed experiments. *J Am Stat Assoc*. 2019;1–16.

810. Li Z, Tian L, O'Malley A, et al. *IFAA: Robust Association Identification and Inference for Absolute Abundance in Microbiome Analyses*. arXiv; 2019, 1909.10101v3 [stat.AP].
811. Xia Y, Sun J, Chen D-G. *Statistical Analysis of Microbiome Data with R*. Springer Singapore: Singapore; 2018.
812. Principal Coordinates Analysis, Encyclopedia of Biostatistics. Zero-inflated beta regression for differential abundance analysis with metagenomics data. *J Comput Biol*. 2016;23(2):102–110.
813. Ospina R, Ferrari SLP. A general class of zero-or-one inflated beta regression models. *Comput Stat Data Anal*. 2012;56(6):1609–1623.
814. Storey JD, Tibshirani R. Statistical significance for genomewide studies. *Proc Natl Acad Sci USA*. 2003;100(16):9440–9445.
815. Chen EZ, Li H. A two-part mixed-effects model for analyzing longitudinal microbiome compositional data. *Bioinformatics*. 2016;32(17):2611–2617.
816. Liu, Z. and S. Lin (2018). Sparse Treatment-Effect Model for Taxon Identification with High-Dimensional Metagenomic Data. Microbiome Analysis. R. G. Beiko;, W. Hsiao; and J. Parkinson. New York, NY, USA, Springer Nature.
817. Chai H, Jiang H, Lin L, Liu L. A marginalized two-part Beta regression model for microbiome compositional data. *PLoS Comput Biol*. 2018;14(7):e1006329.
818. Bourke CD, Gough EK, Pimundu G, et al. Cotrimoxazole reduces systemic inflammation in HIV infection by altering the gut microbiome and immune activation. *Sci Transl Med*. 2019;11(486):eaav0537.
819. Nolan-Kenney R, Wu F, Hu J, et al. The association between smoking and gut microbiome in Bangladesh. *Nicotin Tob Res*. 2019. https://doi.org/10.1093/ntr/ntz220.
820. Liang K-Y, Zeger SL. Longitudinal data analysis using generalized linear models. *Biometrika*. 1986;73(1):13–22.
821. Zhang C-H. Nearly unbiased variable selection under minimax concave penalty. *Ann Stat*. 2010;38(2):894–942.
822. Randolph TW, Zhao S, Copeland W, Hullar M, Shojaie A. Kernel-penalized regression for analysis of microbiome data. *Ann Appl Stat*. 2018;12(1):540–566.
823. Rong R, Jiang S, Xu L, et al. MB-GAN: microbiome simulation via generative adversarial network. *bioRxiv*. 2019; https://doi.org/10.1101/863977.
824. Coker M, Hoen A, Dade E, et al. Specific class of intrapartum antibiotics relates to maturation of the infant gut microbiota: a prospective cohort study. *BJOG*. 2020;127(2):217–227.
825. Hoen AG, Madan JC, Li Z, et al. Sex-specific associations of infants' gut microbiome with arsenic exposure in a US population. *Sci Rep*. 2018;8(1):12627.
826. Banerjee K, Zhao N, Srinivasan A, et al. An adaptive multivariate two-sample test with application to microbiome differential abundance analysis. *Front Genet*. 2019;10:350.
827. Hawinkel S, Mattiello F, Bijnens L, Thas O. A broken promise: microbiome differential abundance methods do not control the false discovery rate. *Brief Bioinform*. 2017;20(1):210–221.
828. Sohn MB, Du R, An L. A robust approach for identifying differentially abundant features in metagenomic samples. *Bioinformatics (Oxford, England)*. 2015;31(14):2269–2275.
829. Cao Y, Lin W, Li H. Two-sample tests of high-dimensional means for compositional data. *Biometrika*. 2017;105(1):115–132.
830. Gretton A, Borgwardt K, Rasch M, Schölkopf B, Smola A. A kernel method for the two-sample problem. In: *NIPS*. Cambridge,MA: MIT Press; 2007:513–520.
831. Gretton A, Borgwardt KM, Rasch MJ, Schölkopf B, Smola A. A kernel two-sample test. *J Mach Learn Res*. 2012;13:723–773, JMLR.org.
832. Mishra A, Müller C. *Robust Regression With Compositional Covariates*. 2019.

833. Aitchison J, Bacon-Shone J. Log contrast models for experiments with mixtures. *Biometrika.* 1984;71(2):323–330.
834. Combettes PL, Müller CL. *Regression Models for Compositional Data: General Log-Contrast Formulations, Proximal Optimization, and Microbiome Data Applications.* arXiv Preprint; 2019, arXiv:1903.01050.
835. Martins EP, Hansen TF. Phylogenies and the comparative method: a general approach to incorporating phylogenetic information into the analysis of interspecific data. *Am Nat.* 1997;149(4):646–667.
836. Crawford J, Greene CS. Incorporating biological structure into machine learning models in biomedicine. *Curr Opin Biotechnol.* 2020;63:126–134.
837. Liu J, Li Y, Feng Y, et al. Patterned progression of gut microbiota associated with necrotizing enterocolitis and late onset sepsis in preterm infants: a prospective study in a Chinese neonatal intensive care unit. *PeerJ.* 2019;7:e7310.
838. Liu L, Gu H, Van Limbergen J, Kenney T. *SuRF: A New Method for Sparse Variable Selection, With Application in Microbiome Data Analysis.* arXiv e-prints; 2019.
839. Kim KJ, Park J, Park S-C, Won S. Phylogenetic tree-based microbiome association test. *Bioinformatics.* 2020;36(4):1000–1006.
840. Law CW, Chen Y, Shi W, Smyth GK. voom: precision weights unlock linear model analysis tools for RNA-seq read counts. *Genome Biol.* 2014;15(2):R29.
841. Chen L, Liu H, Kocher J-PA, Li H, Chen J. glmgraph: an R package for variable selection and predictive modeling of structured genomic data. *Bioinformatics (Oxford, England).* 2015;31(24):3991–3993.
842. Ning J, Beiko RG. Phylogenetic approaches to microbial community classification. *Microbiome.* 2015;3:47.
843. Tanaseichuk O, Borneman J, Jiang T. Phylogeny-based classification of microbial communities. *Bioinformatics.* 2013;30(4):449–456.
844. Cox DR. Regression models and life-tables. *J R Stat Soc Ser B Methodol.* 1972;34(2):187–220.
845. Han MK, Zhou Y, Murray S, et al. Lung microbiome and disease progression in idiopathic pulmonary fibrosis: an analysis of the COMET study. *Lancet Respir Med.* 2014;2(7):548–556.
846. Peters BA, Hayes RB, Goparaju C, Reid C, Pass HI, Ahn J. The microbiome in lung cancer tissue and recurrence-free survival. *Cancer Epidemiol Biomark Prev.* 2019;28(4):731–740.
847. Peters BA, Wilson M, Moran U, et al. Relating the gut metagenome and metatranscriptome to immunotherapy responses in melanoma patients. *Genome Med.* 2019;11:61. https://doi.org/10.1186/s13073-019-0672-4.
848. Diggle PJ, Heagerty P, Liang K-Y, Zeger SL. *Analysis of Longitudinal Data.* Oxford University Press; 2002.
849. Fitzmaurice GM, Laird NM, Ware JH. *Applied Longitudinal Analysis.* Wiley; 2004.
850. Ernest B, Gooding JR, Campagna SR, Saxton AM, Voy BH. MetabR: an R script for linear model analysis of quantitative metabolomic data. *BMC Res Notes.* 2012;5:596.
851. Fabregat-Traver D, Sharapov SZ, Hayward C, et al. High-performance mixed models based genome-wide association analysis with omicABEL software. *F1000Res.* 2014;3:200.
852. Zhao X, Niu L, Clerici C, Russo R, Byrd M, Setchell KDR. Data analysis of MS-based clinical lipidomics studies with crossover design: a tutorial mini-review of statistical methods. *Clin Mass Spectrom.* 2019;13:5–17.
853. Zhang X, Yi N. Fast zero-inflated negative binomial mixed modeling approach for analyzing longitudinal metagenomics data. *Bioinformatics.* 2020. https://doi.org/10.1093/bioinformatics/btz973, pii: btz973.

854. Cho I, Yamanishi S, Cox L, et al. Antibiotics in early life alter the murine colonic microbiome and adiposity. *Nature*. 2012;488(7413):621–626.
855. Cox LM, Yamanishi S, Sohn J, et al. Altering the intestinal microbiota during a critical developmental window has lasting metabolic consequences. *Cell*. 2014;158(4):705–721.
856. Ruan Q, Dutta D, Schwalbach MS, Steele JA, Fuhrman JA, Sun F. Local similarity analysis reveals unique associations among marine bacterioplankton species and environmental factors. *Bioinformatics*. 2006;22(20):2532–2538.
857. Xia LC, Steele JA, Cram JA, et al. Extended local similarity analysis (eLSA) of microbial community and other time series data with replicates. *BMC Syst Biol*. 2011;5(2):S15.
858. Xia LC, Ai D, Cram J, Fuhrman JA, S. F. Efficient statistical significance approximation for local similarity analysis of high-throughput time series data. *Bioinformatics*. 2013;29:230–237.
859. Shaw GT-W, Pao Y-Y, Wang D. MetaMIS: a metagenomic microbial interaction simulator based on microbial community profiles. *BMC Bioinf*. 2016;17(1):488.
860. Bucci V, Tzen B, Li N, et al. MDSINE: Microbial Dynamical Systems INference Engine for microbiome time-series analyses. *Genome Biol*. 2016;17(1):121.
861. Baksi KD, Kuntal BK, Mande SS. 'TIME': a web application for obtaining insights into microbial ecology using longitudinal microbiome data. *Front Microbiol*. 2018;9:36.
862. Lugo-Martinez J, Ruiz-Perez D, Narasimhan G, Bar-Joseph Z. Dynamic interaction network inference from longitudinal microbiome data. *Microbiome*. 2019;7(1):54.
863. Shields-Cutler RR, Al-Ghalith GA, Yassour M, Knights D. SplinectomeR enables group comparisons in longitudinal microbiome studies. *Front Microbiol*. 2018;9:785.
864. Gerber GK. The dynamic microbiome. *FEBS Lett*. 2014;588(22):4131–4139.
865. Zhang Y, Davis R. Principal trend analysis for time-course data with applications in genomic medicine. *Ann Appl Stat*. 2013;7(4):2205–2228.
866. Holter NS, Maritan A, Cieplak M, Fedoroff NV, Banavar JR. Dynamic modeling of gene expression data. *Proc Natl Acad Sci USA*. 2001;98(4):1693–1698.
867. Kimeldorf GS, Wahba G. A correspondence between Bayesian estimation on stochastic processes and smoothing by splines. *Ann Math Stat*. 1970;41(2):495–502.
868. Ilan Y. Why targeting the microbiome is not so successful: can randomness overcome the adaptation that occurs following gut manipulation? *Clin Exp Gastroenterol*. 2019;12:209–217.
869. Zhang H, Chen J, Li Z, Liu L. Testing for mediation effect with application to human microbiome data. *Stat Biosci*. 2019;1–16. https://doi.org/10.1007/s12561-019-09253-3.
870. Fu J, Bonder MJ, Cenit MC. The gut microbiome contributes to a substantial proportion of the variation in blood lipids. *Circ Res*. 2015;117(9):817–824.
871. Liu F, Wang C, Wu Z, Zhang Q, Liu P. A zero-inflated Poisson model for insertion tolerance analysis of genes based on Tn-seq data. *Bioinformatics*. 2016;32(11): 1701–1708.
872. Zhang X, Pei Y-F, Zhang L, et al. Negative binomial mixed models for analyzing longitudinal microbiome data. *Front Microbiol*. 2018;9:1683.
873. Lee J, Sison-Mangus M. A Bayesian semiparametric regression model for joint analysis of microbiome data. *Front Microbiol*. 2018;9:522.
874. van der Merwe S. A method for bayesian regression modelling of composition data. *S Afr Stat J*. 2019;53(1):55–64.
875. Kaul A, Mandal S, Davidov O, Peddada SD. Analysis of microbiome data in the presence of excess zeros. *Front Microbiol*. 2017;8:2114.
876. Abe K, Hirayama M, Ohno K, Shimamura T. A latent allocation model for the analysis of microbial composition and disease. *BMC Bioinf*. 2018;19(19):519.
877. Wang C, Hu J, Blaser MJ, Li H. Microbial trend analysis for common dynamic trend, group comparison and classification in longitudinal microbiome study. *bioRxiv*. 2020. https://doi.org/10.1101/2020.01.30.926824.

878. Gregory KE, Samuel BS, Houghteling P, et al. Influence of maternal breast milk ingestion on acquisition of the intestinal microbiome in preterm infants. *Microbiome.* 2016;4(1):68.
879. Fang R, Wagner B, Harris J, Fillon S. Zero-inflated negative binomial mixed model: an application to two microbial organisms important in oesophagitis. *Epidemiol Infect.* 2016;144(11):2447–2455.
880. Zhang X, Mallick H, Yi N. Zero-inflated negative binomial regression for differential abundance testing in microbiome studies. *J Bioinf Genomics.* 2016;2:2.
881. Chen J, King E, Deek R, et al. An omnibus test for differential distribution analysis of microbiome sequencing data. *Bioinformatics.* 2017;34(4):643–651.
882. Zheng X, Qin G, Tu D. A generalized partially linear mean-covariance regression model for longitudinal proportional data, with applications to the analysis of quality of life data from cancer clinical trials. *Stat Med.* 2017;36(12):1884–1894.
883. D'Agata AL, Wu J, Welandawe MKV, Dutra SVO, Kane B, Groer MW. Effects of early life NICU stress on the developing gut microbiome. *Dev Psychobiol.* 2019;61(5):650–660.
884. Gorshein E, Wei C, Ambrosy S, et al. Lactobacillus rhamnosus GG probiotic enteric regimen does not appreciably alter the gut microbiome or provide protection against GVHD after allogeneic hematopoietic stem cell transplantation. *Clin Transplant.* 2017;31(5):e12947.
885. Sitarik AR, Havstad S, Levin AM, et al. Dog introduction alters the home dust microbiota. *Indoor Air.* 2018;28(4):539–547.
886. Zhai J, Knox K, Twigg HL, Zhou H, Zhou JJ. Exact tests of zero variance component in presence of multiple variance components with application to longitudinal microbiome study. *bioRxiv.* 2018. https://doi.org/10.1101/281246.
887. Zhai J, Knox K, Twigg III HL, Zhou H, Zhou JJ. Exact variance component tests for longitudinal microbiome studies. *Genet Epidemiol.* 2019;43(3):250–262.
888. Brooks M, Kristensen K, van Benthem K, et al. glmmTMB balances speed and flexibility among packages for zero-inflated generalized linear mixed modeling. *R Journal.* 2017;9:378–400.
889. Rizopoulos, D. (2019). "GLMMadaptive: Generalized Linear Mixed Models Using Adaptive Gaussian Quadrature." R Package Version 0.6-0. https://drizopoulos.github.io/GLMMadaptive/ (9 January 2020, date last accessed).
890. Zhang X, Mallick H, Tang Z, Zhang L, Cui X, Benson AK. Negative binomial mixed models for analyzing microbiome count data. *BMC Bioinf.* 2017;18, Article 4.
891. Layeghifard M, Hwang DM, Guttman DS. Constructing and analyzing microbiome networks in R. In: *Microbiome Analysis.* Springer; 2018:243–266.
892. Bokulich NA, Dillon MR, Zhang Y, et al. q2-longitudinal: longitudinal and paired-sample analyses of microbiome data. *mSystems.* 2018;3(6):e00219-00218.
893. Lindstrom MJ, Bates DM. Newton-Raphson and EM algorithms for linear mixed-effects models for repeated-measures data. *J Am Stat Assoc.* 1988;83(404):1014–1022.
894. Guijarro KH, Aparicio V, De Gerónimo E, et al. Soil microbial communities and glyphosate decay in soils with different herbicide application history. *Sci Total Environ.* 2018;634:974–982.
895. Mahnert A, Haratani M, Schmuck M, Berg G. Enriching beneficial microbial diversity of indoor plants and their surrounding built environment with biostimulants. *Front Microbiol.* 2018;9:2985.
896. Cristianini N, Shawe-Taylor J. *An Introduction to Support Vector Machines and Other Kernel-based Learning Methods.* Cambridge: Cambridge University Press; 2000.
897. Lin X. Variance component testing in generalised linear models with random effects. *Biometrika.* 1997;84(2):309–326.

898. Plantinga AM, Chen J, Jenq RR, Wu MC. pldist: ecological dissimilarities for paired and longitudinal microbiome association analysis. *Bioinformatics*. 2019;35(19): 3567–3575.
899. Gower JC. A general coefficient of similarity and some of its properties. *Biometrics*. 1971;27(4):857–871.
900. Bray JR, Curtis JT. An ordination of the upland forest communities of southern Wisconsin. *Ecol Monogr*. 1957;27(4):325–349.
901. Jaccard P. The distribution of the flora in the alpine zone.1. *New Phytol*. 1912;11(2):37–50.
902. Williams J, Bravo H, Tom J, Paulson J. microbiomeDASim: simulating longitudinal differential abundance for microbiome data [version 1; peer review: 1 approved, 1 approved with reservations]. *F1000Res*. 2019;8:1769.
903. Foster G, Collins MD, Lawson PA, Buxton D, Murray FJ, Sime A. Actinobacillus seminis as a cause of abortion in a UK sheep flock. *Vet Rec*. 1999;144:479–480.
904. Osaka T, Moriyama E, Arai S, Date Y, Yagi J, Kikuchi J. Meta-analysis of fecal microbiota and metabolites in experimental colitic mice during the inflammatory and healing phases. *Nutrients*. 2017;9(12):E1329.
905. Smith MF, Geisert RD, Parrish JJ. Reproduction in domestic ruminants during the past 50 yr: discovery to application. *J Anim Sci*. 2018;96(7):2952–2970.
906. Raes J, Bork P. Molecular eco-systems biology: towards an understanding of community function. *Nat Rev Microbiol*. 2008;6:693–699.

Index

Note: Page numbers followed by "*f*" indicate figures and "*t*" indicate tables.

A

D

E

F

P

R

S

T

FSC
www.fsc.org
MIX
Papier | Fördert
gute Waldnutzung
FSC® C083411

Kolibri 360 GmbH
Ferdinand-Jühlke-Straße 7
99095 Erfurt, Deutschland
produktsicherheit@kolibri360.de